Undergraduate Texts in Mathematics

For other titles published in this series, go to
www.springer.com/series/666

Ronald W. Shonkwiler
James Herod

# Mathematical Biology

## An Introduction with Maple and Matlab

Second Edition

Ronald W. Shonkwiler
School of Mathematics
Georgia Institute of Technology
Atlanta, GA 30332-0160
USA
shonkwiler@math.gatech.edu

James Herod
School of Mathematics
Georgia Institute of Technology
Atlanta, GA 30332-0160
USA
jherod@tds.net



ISBN 978-1-4899-8281-0 ISBN 978-0-387-70984-0 (eBook)
DOI 10.1007/978-0-387-70984-0
Springer Dordrecht Heidelberg London New York

Mathematics Subject Classification (2000): 92-01, 92BXX



Softcover re-print of the Hardcover 2nd edition 2009
Based in part on *An Introduction to the Mathematics of Biology: With Computer Algebra Methods*, Edward K. Yeargers, Ronald W. Shonkwiler, and James V. Herod, Birkhäuser Boston, 1996.



Printed on acid-free paper

Springer is part of Springer Science+Business Media (www.springer.com)

# Preface

Biology is a source of fascination for most scientists, whether their training is in the life sciences or not. In particular, there is a special satisfaction in discovering an understanding of biology in the context of another science like mathematics. Fortunately there are plenty of interesting problems (and fun) in biology, and virtually all scientific disciplines have become the richer for it. For example, two major journals, *Mathematical Biosciences* and *Journal of Mathematical Biology*, have tripled in size since their inceptions 20–25 years ago.

More recently, the advent of genomics has spawned whole new fields of study in the biosciences, fields such as proteomics, comparative genomics, genomic medicine, pharmacogenomics, and structural genomics among them. These new disciplines are as much mathematical as biological.

The various sciences have a great deal to give to one another, but there are still too many fences separating them. In writing this book we have adopted the philosophy that mathematical biology is not merely the intrusion of one science into another, but that it has a unity of its own, in which both biology and mathematics should be equal, complete, and flow smoothly into and out of one another. There is a timeliness in calculating a protocol for administering a drug. Likewise, the significance of bones being "sinks" for lead accumulation while bonemeal is being sold as a dietary calcium supplement adds new meaning to mathematics as a *life science*. The dynamics of a compartmentalized system are classical; applications to biology can be novel. Exponential and logistic population growths are standard studies; the delay in the increase of AIDS cases behind the increase in the HIV-positive population is provocative.

With these ideas in mind we decided that our book would have to possess several important features. For example, it would have to be *understandable to students of either biology or mathematics*, the latter referring to any science students who normally take more than two years of calculus, i.e., majors in mathematics, physics, chemistry, and engineering.

A prime objective of this text is to introduce students of mathematics to the interesting mathematical problems and future challenges in biology.

*No prior study of biology would be necessary.*

Mathematics students rarely take biology as part of their degree programs, but our experience has been that very rapid progress is possible once a foundation has been laid. Thus the *coverage of biology would be extensive*, considerably more than actually needed to put the mathematics of the book into context. This would permit mathematics students to have much greater latitude in subsequent studies, especially in the "what-if" applications of a computer algebra system. It would also help to satisfy the intense intellectual interest that mathematics students have in the life sciences, as has been manifested in our classes.

Genomics is proving that mathematics is as much a part of biology as it is of physics. We urge biology students to equip themselves with two years' study of mathematics that includes calculus with linear algebra, differential equations, and some discrete mathematics. For the student with one year's study of calculus with linear algebra, we can say that our exposition of mathematics beyond that level is complete and self-contained. Thus we offer a focused expansion of your mathematical knowledge. Our biology students have had no problems with this approach.

We have divided the book into three parts:

Part I: Cells, Signals, Growth, and Populations;
Part II: Systems and Diseases;
Part III: Genomics.

One reason for this is that the mathematics of genomics is more abstract and advanced. *Moreover, an objective of this text is to introduce both biology and mathematics students to the new field of algebraic statistics.*

To help ease the burden of coping with the mathematics of the book we offer a chapter, Mathtools, dedicated as a refresher or an introduction to the mathematics needed for Parts I and II. At the same time, this chapter serves as a tutorial for the two computer systems, MAPLE and MATLAB that we use to accompany the mathematical derivations. All the computer syntax for the remainder of the text is illustrated here. In order to make this more useful, we construct a "code index," preceding the usual term index, that shows on which page of the text various code techniques are used.

*Every chapter should have "mathematical laboratory biology experiments."*

This is another important goal of the text. It is the computer algebra system that makes the mathematics accessible to nonmathematicians. More than that, powerful mathematical software essentially allows for interactive experimentation with biological models. These systems incorporate hundreds of mathematical techniques and algorithms and perform all the laborious and time-consuming calculations in seconds.

Once a biological system has been modeled, one can then perform "biology" experiments on the model using the computer algebra system. Often these are experiments that could not be done easily on the real system, such as ascertaining the long-term effects of lead ingestion or the effect of reduced sodium conductance to an action potential. Of course one should always be aware that the real biology is the final arbiter of any such experiment. This hands-on approach provides a rich source of information through the use of "what-if" input and thus allows students to grasp important biological and mathematical concepts in a way that is not possible otherwise.

A note about these exercises/experiments. We start out at the beginning in Chapter 2 with very simple and basic computer algebra commands and progress to more elaborate ones later in the book. By means of line-by-line comments we encourage the student to learn how to master the software and use it as a powerful tool. In this spirit, we also provide a "code index" as mentioned above. Even so, this book is not about programming. At the end of each modeling section, we may ask that the student perform computational experiments on the mathematical models developed in the section. The computer syntax provided there can be downloaded and used for this purpose. The websites are

www.springer.com/978-0-387-70983-3,
www.math.gatech.edu/~herod,
www.math.gatech.edu/~shenk.

At other times, we ask the student to construct, investigate, and report on a model similar to or an extension of one in the section. Generally, we provide the necessary computer code. It might seem that it is only necessary to download the code from our webpage and press the return key a few times. But in fact, the science is in observing and interpreting the computed results, and in going beyond that by posing questions about the phenomenon that may or may not be answerable by the model.

*Most importantly, the biology and mathematics would be integrated.*

Each chapter deals with a major topic, such as lead poisoning, and we begin by presenting a thorough foundation of fundamental biology. This leads into a discussion of a related mathematical concept and its elucidation with the computer algebra system. Thus for each major topic, the biology and the mathematics are combined into an integrated whole.

To summarize, we hope that mathematics students will look at this book as a way to learn enough biology to make good models and that biology students will see it as an opportunity to understand the dynamics of a biological system. For both these students and their engineering classmates, perhaps this book can present a new perspective for a life's work.

In teaching the material ourselves, we usually spend a week (three class periods) per chapter. On the first day we discuss the biology of the chapter, and on the second we talk about the mathematics and derivations. The third day is "lab" day, in which the students attempt the assigned exercises/experiments. This may be performed in an actual computer laboratory where help with MAPLE or MATLAB is available. Also, students can break up into small groups and work together; preferably the group memberships should change every week. Some of the derivations in the text are involved. We include them for completeness and for the interested student. The one day we spend in class on the derivations is tailored to the level of mathematical depth we think is appropriate.

## Acknowledgments

We are deeply indebted to Professor Edward Yeargers for allowing us to use the biology material he wrote for our earlier book, *An Introduction to the Mathematics of Biology*. Now in retirement but still the consumate teacher, Professor Yeargers wished us well in this new endeavor and expressed satisfaction that his pedagogical influence might reach a new generation of students.

We also wish to thank Ann Kostant, who communicated to us the greatest enthusiasm and encouragement for a major expansion of our earlier work to incorporate topics in the new and burgeoning field of genomics. And we would like to thank John Spiegelman for his monumental effort and high-quality results at preparing the tables and figures and reformatting the computer codes. Of course, since we reviewed the syntax carefully before going to press, any errors therein are ours alone.

Finally, we give a special thanks to our students, who show an abiding interest in the material and ask insightful and thought-provoking questions. Some of them keep in touch after having entered and established themselves in a career in this or some related field. Ultimately, it is for them that our effort is sustained.

*Ronald W. Shonkwiler*
*James V. Herod*

## Contents

# 1

# Biology, Mathematics, and a Mathematical Biology Laboratory

## 1.1 The Natural Linkage Between Mathematics and Biology

*Mathematics and biology have a synergistic relationship. Biology produces interesting problems, mathematics provides models to understand them, and biology returns to test the mathematical models. Recent advances in computer algebra systems have facilitated the manipulation of complicated mathematical systems. This has made it possible for scientists to focus on understanding mathematical biology, rather than on the formalities of obtaining solutions to equations.*

*What is the function of mathematical biology?*

Our answer to this question, and the guiding philosophy of this book, is simple: The function of mathematical biology is to exploit the natural relationship between biology and mathematics. The linkage between the two sciences is embodied in these reciprocal contributions that they make to each other: Biology generates complex problems and mathematics can provide ways to understand them. In turn, mathematical models suggest new lines of inquiry that can only be tested on real biological systems.

We believe that an understanding of the relationship between two subjects must be preceded by a thorough understanding of the subjects themselves. Indeed, the excitement of mathematical biology begins with the discovery of an interesting and uniquely biological problem. The excitement grows when we realize that mathematical tools at our disposal can profitably be applied to the problem. The interplay between mathematical tools and biological problems constitutes mathematical biology.

*The time is right for integrating mathematics and biology.*

Biology is a rapidly expanding science; research advances in the life sciences leave virtually no aspects of our public and private lives untouched. Newspapers bombard us with information about *in vitro* fertilization, bioengineering, DNA testing, genetic manipulation, environmental degradation, AIDS, and forensics.

R.W. Shonkwiler and J. Herod, *Mathematical Biology: An Introduction with Maple and Matlab*, Undergraduate Texts in Mathematics, DOI: 10.1007/978-0-387-70984-0_1,

Quite separately from the news pouring onto us from the outside world, we have an innate interest in biology. We have a natural curiosity about ourselves. Every day we ask ourselves a nonstop series of questions: What happens to our bodies as we get older? Where does our food go? How do poisons work? Why do I look like my mother? What does it mean to "think"? Why are HIV infections spreading so rapidly in certain population groups?

Professional biologists have traditionally made their livings by trying to answer these kinds of questions. But scientists with other kinds of training have also seen ways that they could enter the fray. As a result, chemists, physicists, engineers, and mathematicians have all made important contributions to the life sciences. These contributions often have been of a sort that required specialized training or a novel insight that only specialized training could generate.

In this book we present some mathematical approaches to understanding biological systems. This approach has the hazard that an in-depth analysis could quickly lead to unmanageably complex numerical and symbolic calculations. However, technical advances in the computer hardware and software industries have put powerful computational tools into the hands of anyone who is interested. Computer algebra systems allow scientists to bypass some of the details of solving mathematical problems. This then allows them to spend more time on the interpretation of biological phenomena, as revealed by the mathematical analysis.[1]

## 1.2 The Use of Models in Biology

*Scientists must represent real systems by models. Real systems are too complicated, and besides, observation may change the real system. A good model should be simple and it should exhibit the behaviors of the real system that interest us. Further, it should suggest experimental tests of itself that are so revealing that we must eventually discard the model in favor of a better one. We therefore measure scientific progress by the production of better and better models, not by whether we find some absolute truth.*

*A model is a representation of a real system.*

The driving force behind the creation of models is this admission: Truth is elusive, but we can gradually approximate it by creating better and better representations.

There are at least two reasons why the truth is so elusive in real systems. The first reason is obvious: The universe is extremely complicated. People have tried unsuccessfully to understand it for millennia, running up countless blind alleys and only occasionally finding enlightenment. Claims of great success abound, usually followed by their demise. Physicists in the late nineteenth century advised their students that Maxwell's equations had summed up everything important about physics, and that further research was useless. Einstein then developed the theory of general

[1] References [1]–[4] at the end of this chapter are some articles that describe the importance of mathematical biology.

relativity, which contained Maxwell's equations as a mere subcategory. The unified field theory ("The Theory of Everything") will contain Einstein's theory as a subcategory. Where will it end?

The second reason for the elusivity of the truth is a bit more complicated: It is that we tend to change reality when we examine any system too closely. This concept, which originates in quantum mechanics, suggests that the disturbances that inevitably accompany all observations will change the thing being observed. Thus "truth" will be changed by the very act of looking for it.[2] At the energy scale of atoms and molecules the disturbances induced by the observer are especially severe. This has the effect of rendering it impossible to observe a single such particle without completely changing some of the particle's fundamental properties. There are macroscopic analogues to this effect. For example, what is the "true" color of the paper in this book? The answer depends on the color of the light used to illuminate the paper, white light being merely a convenience; most other colors would also do. Thus you could be said to have chosen the color of the paper by your choice of observation method.

Do these considerations make a search for ultimate explanations hopeless? The answer is, "No, because what is really important is the progress of the search, rather than some ultimate explanation that is probably unattainable anyway."

*Science is a rational, continuing search for better models.*

Once we accept the facts that a perfect understanding of very complex systems is out of reach and that the notion of "ultimate explanations" is merely a dream, we will have freed ourselves to make scientific progress. We are then able to take a reductionist approach, fragmenting big systems into small ones that are individually amenable to understanding. When enough small parts are understood, we can take a holistic approach, trying to understand the relationships among the parts, thus reassembling the entire system.

In this book we reduce complicated biological systems to relatively simple mathematical models, usually of one to several equations. We then solve the equations for variables of interest and ask whether the functional dependencies of those variables predict salient features of the real system.

There are several things we expect from a good model of a real system:

(a) It must exhibit properties that are similar to those of the real system, and those properties must be the ones in which we are interested.[3] A six-inch replica of a 747 airliner, after adjusting for Reynolds' number, may have the exact fluid-dynamical properties of the real plane, but would be useless in determining the comfort of the seats of a real 747.

[2] This situation is demonstrated by the following exchange: *Question*: How would you decide which of two gemstones is a real ruby and which is a cheap imitation? *Answer*: Tap each sharply with a hammer. The one that shatters used to be the real ruby.

[3] One characteristic of the real system that we definitely do *not* want is its response to the observation process, described earlier. In keeping with the concept of a model as an idealization, we want the model to represent the real system in a "native state," divorced from the observer.

(b) It must self-destruct. A good model must suggest tests of itself and predict their outcomes. Eventually a good model will suggest a very clever experiment whose outcome will not be what the model predicted. The model must then be discarded in favor of a new one.

The search for better and better models thus involves the continual testing and replacement of existing models. This search must have a rational foundation, being based on phenomena that can be directly observed. A model that cannot be tested by the direct collection of data, and which therefore must be accepted on the basis of faith, has no place in science.

*Many kinds of models are important in understanding biological phenomena.*

Models are especially useful in biology. The most immediate reason is that living systems are much too complicated to be truly understood as whole entities. Thus to design a useful model, we must strip away irrelevant, confounding behaviors, leaving only those that directly interest us. We must walk a fine line here: In our zeal to simplify, we may strip away important features of the living system, and at the other extreme, a too-complicated model is intractable and useless.

Models in biology span a wide spectrum of types. Here are some that are commonly used:

| **Model** | **What the model represents** |
|---|---|
| aa × Aa | Gene behavior in a genetic cross. |
| $\frac{dA}{dt} = -kA$ | Rate of elimination of a drug from the blood. |
| $\boxed{\text{R}} \rightarrow \boxed{\text{C}} \rightarrow \boxed{\text{E}}$ | Reflex arc involving a stimulus Receptor, the Central nervous system, and an Effector muscle. |
| a camera | The eye of a vertebrate or of an octopus. |

*Why is there so much biological information in this book?*

It is possible to write a mathematical biology book that contains only a page or two of biological information at the beginning of each chapter. We see that format as the source of two problems: First, it is intellectually limiting. A student cannot apply the powerful tools of mathematics to biological problems he or she does not understand. This limitation can be removed by a thorough discussion of the underlying biological systems, which can suggest further applications of mathematics. Thus a strong grounding in biology helps students to move further into mathematical biology.

Second, giving short shrift to biology reinforces the misconception that each of the various sciences sits in a vacuum. In fact, it has been our experience that many students of mathematics, physics, and engineering have a genuine interest in biology, but little opportunity to study it. Taking our biological discussions well beyond the barest facts can help these students to understand the richness of biology, and thereby encourage interdisciplinary thinking.

## 1.3 What Can Be Derived from a Model and How Is It Analyzed?

*A model is more than the sum of its parts. Its success lies in its ability to discover new results, results that transcend the individual facts built into it. One result of a model can be the observation that seemingly dissimilar processes are in fact related. In an abstract form, the mathematical equations of the process might be identical to those of other phenomena. In this case the two disciplines reinforce: A conclusion difficult to see in one might be an easy consequence in the other.*

To analyze the mathematical equations that arise, we draw on the fundamentals of matrix calculations, counting principles for permutations and combinations, the calculus, and fundamentals of differential equations. However, we will make extensive use of the power of numerical and symbolic computational software—a computer algebra system. The calculations and graphs in this text are done using such software.

Syntax for both MAPLE and MATLAB accompanies the mathematical derivations in the text. This code should be treated something like a displayed equation. Like an equation, code is precise and technical. On first reading, it is often best to work through a line of reasoning, with only a glance at any included code, to understand the points being made. Then a critical examination of an equation or piece of code will make more sense, having the benefit of context and intended goal. The computer algebra syntax is displayed and set off in a distinctive font in order for the reader to be able to quickly find its beginning and ending. Where possible, equivalent syntax for MAPLE and MATLAB are presented together in tandem. It should be noted that the basic MATLAB system is numerical and does not perform symbolic computations. Thus equivalent MATLAB code is omitted in this case. An accessory package is available for MATLAB that can perform symbolic manipulation. And conveniently, this package is created by the same people who created MAPLE.

*Deriving consequences: The other side of modeling.*

Once a model has been formulated and the mathematical problems defined, then they must be solved. In this symbolic form, the problem takes on a life of its own, no longer necessarily tied to its physical origins. In symbolic form, the system may even apply to other, totally unexpected, phenomena. What do the seven bridges at Königsberg have to do with discoveries about DNA? The mathematician Euler formed an abstract model of the bridges and their adjoining land masses and founded the principles of Eulerian graphs on this model. Today, Eulerian graphs are used, among other ways, to investigate the ancestry of living things by calculating the probability of matches of DNA base pair sequences (see Kandel [5]). We take up the subject of phylogeny in Chapter 15. The differential equations describing spring–mass systems and engineering vibrations are identical to those governing electrical circuits with capacitors, inductors, and resistors. And again these very same equations pertain to the interplay between glucose and insulin in humans. The abstract and symbolic treatment of these systems through mathematics allows the transfer of intuition between them. Through mathematics, discoveries in any one of these areas can lead to a breakthrough in the others. But mathematics and applications are mutually reinforcing: The abstraction can uncover truths about the application, suggesting questions to ask and experiments to try; the application can foster mathematical intuition and form the basis of the results from which mathematical theorems are distilled.

In symbolic form, a biological problem is amendable to powerful mathematical processing techniques, such as differentiation or integration, and is governed by mathematical assertions known as theorems. Theorems furnish the conclusions that may be drawn about a model so long as their hypotheses are fulfilled. Assumptions built into a model are there to allow its equations to be posed and its conclusions to be mathematically sound. The validity of a model is closely associated with its assumptions, but experimentation is the final arbiter of its worth. The assumption underlying the exponential growth model, namely, $\frac{dy}{dt} = ky$ (see Section 2.4 and Chapter 3), is unlikely to be precisely fulfilled in any case, yet exponential growth is widely observed for biological populations. However, exponential growth ultimately predicts unlimited population size, which never materializes precisely due to a breakdown in the modeling assumption. A model is robust if it is widely applicable. In every case, the assumptions of a model *must* be spelled out and thoroughly understood. The validity of a model's conclusions must be experimentally confirmed. Limits of applicability, robustness, and regions of failure need to be determined by carefully designed experiments.

Some biological systems involve only a small number of entities or are greatly influenced by a few of them, maybe even one. Consider the possible DNA sequences 100 base pairs long. Among the possibilities, one or two base pairs might be critical to life. (It is known that tRNA molecules can have as few as 73 nucleotide residues (Lehninger [6]).) Or consider the survival prospects of a clutch of Canadian geese blown off migratory course to the Hawaiian islands. Their survival analysis must keep track of detailed events for each goose and possibly even details of their individual genetic makeups, for the loss of a single goose or the birth of defective goslings could spell extinction for the small colony. (The nene, indiginous to Hawaii, is thought to be related to the Canadian geese.) This is the mathematics of discrete systems, i.e., the mathematics of a finite number of states. The main tools we will need here are knowledge of matrices and their arithmetic, counting principles for permutations and combinations, and some basics of probability calculations.

Other biological systems or processes involve thousands, even millions, of entities, and the fate of a few of them has little influence on the entire system. Examples are the diffusion process of oxygen molecules or the reproduction of a bacterial colony. In these systems, individual analysis gives way to group averages. An average survival rate of 25% among goslings of a large flock of Canadian geese still ensures exponential growth of the flock in the absence of other effects; but this survival probability sustained by exactly four offspring of an isolated clutch might not result in exponential growth at all but rather total loss instead. When there are large numbers involved, the mathematics of the continuum may be brought to bear, principally calculus and differential equations. This greatly simplifies the analysis. The techniques are powerful and mature, and a great many are known.

*Computer algebra systems make the mathematics accessible.*

It is a dilemma: Students in biology and allied fields such as immunology, epidemiology, or pharmacology need to know how to quantify concepts and to make models. Yet, these students typically have only one year of undergraduate study in mathematics. (Hopefully this will change is our postgenomics world.) This one year may be very general and not involve any examples from biology. When the need arises, they are likely to accept the models and results of others, perhaps without deep understanding.

On the other side of campus, students in mathematics read in the popular technical press of biological phenomena, and wish they could see how to use their flair for mathematics to get them into biology. The examples they typically see in mathematics classes have their roots in physics. Applications of mathematics to biology seem far away.

How can this dilemma be resolved? Should the biology students be asked to take a minor in mathematics in order to be ready to use the power of differential equations for modeling? And what of algebraic models, discrete models, probabilistic models, or statistics? Must the mathematics students take a course in botany, and then zoology, before they can make a model for the level to which the small vertebrate population must be immunized in a geographic region in order to reduce the size of the population of ticks carrying Lyme disease? Such a model is suggested by Kantor [7].

There is an alternative. Computer algebra systems create a new paradigm for designing, analyzing, and drawing conclusions from models in science and engineering. The technology in the computer algebra systems allows the concepts to be paramount while computations and details become less important. With such a computational engine it is possible to read about models that are being actively explored in the current literature and do a computer analysis of these new models.

The theorems from which our conclusions are derived often result from carefully tracking evolving system behavior over many iterations in discrete systems or infinite time in continuous ones. Where possible, the mathematical equations are solved and the solutions exhibited. Predictions of the model are made under a range of starting conditions and possibly unusual parameter regimes. These are the bases of "what if" experiments. For example, given a satisfactory model for a fishery, *what if* one imposes various levels of harvesting? To answer this and related questions, the computer algebra system can carry out the technical computations: calculate roots, differentiate symbolically or numerically, integrate and solve differential equations, perform matrix arithmetic, track system evolution, and graphically display results.

In this book we will use computational packages to do the "heavy lifting." MATLAB is a very powerful system for general numerical computation. In addition, accessory "toolboxes" are available providing the specialized computations used in several disciplines. MAPLE is a system for both numerical and symbolic calculations. MAPLE is quite complete in its mathematical coverage and especially strong in symbolic computations. In addition to these two, recently created software packages are available to perform the computations of the emerging field of algebraic statistics.

These packages are for the most part available free of charge. We will encounter *BLAST* and *SINGULAR* in the genomics sections of the book.

## References

[1] THE FUTURE OF MATHEMATICAL BIOLOGY:
*Mathematics and Biology: The Interface, Challenges, and Opportunities*, National Science Foundation workshop, Lawrence Berkeley Laboratory, Berkeley, CA, U.S. Department of Energy, Washington, DC, 1992.

[2] THE IMPORTANCE OF MATHEMATICS IN THE LIFE SCIENCES:
L. J. Gross, Quantitative training for life-science students, *Biosci.*, **44**-2 (1994), 59.

[3] MODELING IN BIOLOGY:
W. D. Hillis, Why physicists like models and why biologists should, *Curr. Biol.*, **3**-2 (1993), 79–81.

[4] APPLICATIONS OF MATHEMATICAL BIOLOGY:
F. Hoppensteadt, Getting started in mathematical biology, *Not. Amer. Math. Soc.*, **42**-9 (1995), 969.

[5] DNA SEQUENCES:
D. Kandel, Y. Matias, R. Unger, and P. Winkler, *Shuffling Biological Sequences*, preprint, AT&T Bell Laboratories, Murray Hill, NJ, 1995.

[6] BIOCHEMISTRY OF NUCLEIC ACIDS:
A. Lehninger, *Biochemistry*, Worth Publishers, New York, 1975, 935.

[7] CONQUERING LYME DISEASE:
F. S. Kantor, Disarming Lyme disease, *Sci. Amer.*, **271** (1994), 34–39.

[8] THE PROMISE OF GENOMICS:
G. Smith, *Genomics Age*, AMACOM, New York, 2005.

# 2

# Some Mathematical Tools

## Introduction

This book is about biological modeling—the construction of mathematical abstractions intended to characterize biological phenomena and the derivation of predictions from these abstractions under real or hypothesized conditions. A model must capture the essence of an event or process but at the same time not be so complicated as to be intractable or to otherwise dilute its most important features. In this regard, differential equations have been widely invoked across the broad spectrum of biological modeling. Future values of the variables that describe a process depend on their rates of growth or decay. These in turn depend on present, or past, values of these same variables through simple linear or power relationships. These are the ingredients of a differential equation. We discuss linear and power laws between variables and their derivatives in Section 2.1 and differential equations in Section 2.4.

Sometimes a differential equation model is inappropriate because the phenomenon being studied is quantified in discrete units such as population size. If such sizes are very large, differential equations may still give correct results. Otherwise, difference equations may be more appropriate. We take up the basic principles of difference equations in Section 2.5.

Once formulated, a model contains parameters that must be specialized to the particular instance of the process being modeled. This requires gathering and treating experimental data. It requires determining values of the parameters of a model so as to agree with, or fit, the data. The universal technique for this is the method of least squares, which is the subject of Sections 2.2 and 2.3. Even though experimental data is subject to small random variations, or *noise*, and imprecision, least squares is designed to deal with this problem.

Describing noisy data and other manifestations of variation is the province of statistics. Distributions of values can be graphically portrayed as histograms or distilled to a single number, the average or mean. The most widely occurring distribution in the natural world is the normal, or Gaussian, distribution. These topics are taken up in Section 2.7.

R.W. Shonkwiler and J. Herod, *Mathematical Biology: An Introduction with Maple and Matlab*, Undergraduate Texts in Mathematics, DOI: 10.1007/978-0-387-70984-0_2,

Finally, to a greater extent in biological phenomena than in other fields of science and engineering, random processes play a significant role in shaping the course of events. This is true at all scales from diffusion at the atomic level to random combinations of genes to the behavior of whole organisms. Being in the wrong place at the wrong time can mean being a victim (or finding a meal). In Section 2.8 we discuss the basics of probabilities.

Fortunately, while an understanding of these mathematical tools is required for this book, deep knowledge of mathematical techniques is not. This is a consequence of the fruition of mathematical software. We will use the power of this software to execute calculations, invoke special functions, simplify algebra, solve differential equations, and generally perform the technical work. Above all, the software can make pictures of what is happening within the phenomenon in detail. Thereby, the curious are free to let their imaginations roam and focus on perfecting and exercising the models themselves.

As noted in the preface, you will be executing a lot of mathematical software code. As an aid to entering code, all the code in this book is posted on our webpages. Springer maintains the webpage

www.springer.com/978-0-387-70983-3,

Professor Herod's webpage is

www.math.gatech.edu/~herod,

and Professor Shonkwiler's webpage is

www.math.gatech.edu/~shenk.

In addition, as an aid to creating your own code, we provide a "code index" at the back of the book referencing the place in the text for syntax performing various mathematical and computer housekeeping tasks.

## 2.1 Linear Dependence

*The simplest, nonconstant, relationship between two variables is a linear one. The simplest linear relationship is one of proportionality: if one of the variables doubles or triples or halves in value, the other does likewise. Proportionality between variables $x$ and $y$ is expressed as $y = kx$ for some constant $k$. Proportionality can apply to derivatives of variables as well as to variables themselves, since they are just rates of change. Historically, one of the major impacts of calculus is the improved ability to model by the use of derivatives in just this way.*

*Relationships among variables can be graphically visualized.*

In studying almost any phenomenon, among the first observations to be made about it are its changing attributes. A tropical storm gains in wind speed as it develops;

the intensity of sound decreases with distance from its source; living things increase in weight in their early period of life. The measurable quantities associated with a given phenomenon are referred to as *constants, variables*, or *parameters*. Constants are unchanging quantities such as the mathematical constant $\pi = 3.14159\ldots$ or the physical constant named after Boltzmann: $k = 1.38 \times 10^{-16}$ ergs per degree. Variables are quantitative attributes of a phenomenon that can change in value, such as the wind speed of a tropical storm or the intensity of sound or the weight of an organism.

Parameters are quantities that are constant for a particular instance of a phenomenon, but can be different in another instance. For example, the strength of hair fibers is greater for thicker fibers and the same holds for spider web filaments, but the latter has a much higher strength per unit cross-section.[1] Strength per unit cross-section is a property of material that tends to be constant for a given type of material but varies over different materials.

Often two variables of a phenomenon are *linearly related*, that is, a graphical representation of their relationship is a straight line. Temperature as measured on the Fahrenheit scale, $F$, and on the Celsius scale, $C$, are related in this way; see Figure 2.1.1. Knowing that the temperatures $C = 0$ and $C = 100$ correspond to $F = 32$ and $F = 212$, respectively, allows one to derive their linear relationship, namely,

$$F = \frac{9}{5}C + 32. \tag{2.1.1}$$

In this, both $C$ and $F$ have *power* or *degree* one, that is, their exponent is 1. (Being understood, the 1 is not explicitly written.) When two variables are algebraically related and all terms in the equation are of degree one (or constant), then the graph of the equation will be a straight line. The multiplier, or *coefficient*, $\frac{9}{5}$ of $C$ in (2.1.1) is the *slope* of the straight line, or the *constant of proportionality*,between the variables. The constant term 32 in the equation is the *intercept* of the straight line, or *translational term* of the equation. These parameters are shown graphically in Figure 2.1.1.

We can isolate the constant of proportionality by appropriate translation. Absolute zero on the Celsius scale is $-273.15C$, which is usually expressed in degrees *Kelvin* $K$. Translation from degrees $K$ to degrees $C$ involves subtracting the fixed amount 273.15:

$$C = K - 273.15. \tag{2.1.2}$$

From (2.1.1), we calculate absolute zero on the Fahrenheit scale as

$$F = \frac{9}{5}(-273.15) + 32 = -459.67,$$

or about $-460$ degrees *Rankine* $R$. That is,

$$F = R - 459.67. \tag{2.1.3}$$

Hence, substituting equations (2.1.2) and (2.1.3) into (2.1.1), we find that $R$ is related to $K$ by

[1] The strength of a material per unit cross-section is known as *Young's modulus*.

MAPLE

```
#number sign # introduces a comment
#statements must be ended by a semicolon or by a colon (suppresses printing) but can span multiple lines
> plot([C,9/5*C+32,C=0..100],-10..100,-30..220,tickmarks=[5,2]);
```

MATLAB

```
% percent sign introduces a comment in Matlab
% an end of line completes a command, or semicolon ends a command and suppresses printing results
> C=(0:1:100); % C=vector of values from 0 to 100 by ones
> F=(9/5)*C+32; % F=vector, this arithmetic to each C value
> plot(C,F); % plot the Fs vs. the Cs
> xlabel('Temperature degrees C'); %label horizontal axis
> ylabel('Temperature degrees F'); %label vertical axis
> axis([-10,110,-30,220]); % x scale from -10 to 110, y from -30 to 220
```

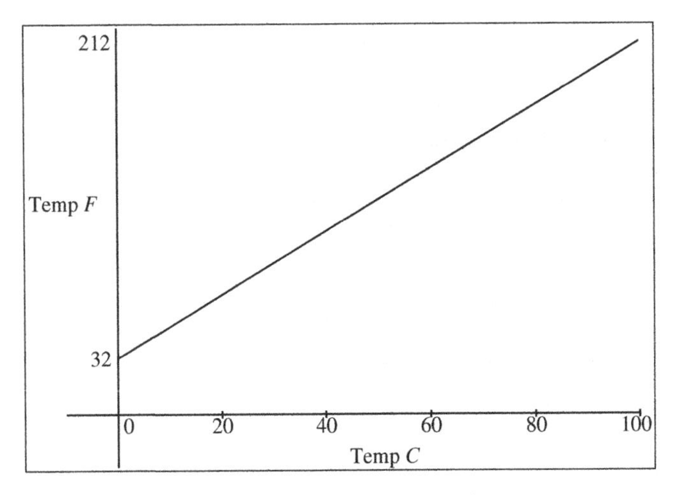


**Fig. 2.1.1.** Temperature conversion.

$$R = \frac{9}{5}K.$$

Thus $R$ is proportional to $K$ and both are zero at the same time, so there is no translational term.

One often observes that the relationship between two variables is one of proportionality but the constant is not yet known. Thus if variables $x$ and $y$ are linearly related (and both are zero at the same time), we write

$$y = kx$$

with the constant of proportionality $k$ to be subsequently determined (see Section 2.2 on least squares).

*Power laws can be converted to linear form.*

The area of a circle does not vary linearly with radius but rather quadratically, $A = \pi r^2$; the power, or degree, of $r$ is two. Heat radiates in proportion to the fourth power of absolute temperature, gravitational force varies in proportion to the inverse square power of distance, and diffusivity varies with the one-third power of density (see

Chapter 6). These are examples in which the relationship between variables is by a power law with the power different from one. There are many more.

In general, a power law is of the form

$$y = Ax^k \tag{2.1.4}$$

for some constants $A$ and $k$. Due to the particular ease of graphing linear relationships, it would be advantageous if this equation could be put into linear form. This can be done by taking the logarithm of both sides of the equation. Two popular bases for logarithms are 10 and $e = 2.718281828459\ldots$; the former is often denoted by log and the latter by ln. (MATLAB uses log for logarithm to the base $e$.) Either will work:

$$\log y = k \log x + \log A; \tag{2.1.5}$$

the relationship between $\log y$ and $\log x$ is linear. Plotting pairs of $(x, y)$ data values on special *log-log paper* will result in a straight line with slope $k$. Of course, on a log-log plot there is no point corresponding to $x = 0$ or $y = 0$. However, if $A = 1$ then $\log y$ is proportional to $\log x$ and the graph goes through the point $(1, 1)$. In general, $A$ appears on the graph of (2.1.4) as the value of $y$ when $x = 1$.

Another frequently encountered relationship between variables is an *exponential* one given by

$$y = Ca^x. \tag{2.1.6}$$

Note that the variable $x$ is now in the exponent. Exponential functions grow (or decay) much faster than polynomial functions; that is, if $a > 1$, then as an easy consequence of L'Hopital's rule, for any power $k$,

$$\lim_{x \to \infty} \frac{x^k}{a^x} = 0, \tag{2.1.7}$$

or in MAPLE,

```
MAPLE
> assume(a>1); assume(k>0);
> limit(x^k/a^x,x=infinity);
```

Figure 2.1.2 demonstrates this with $k = 3$ and $a = 2$. We have drawn graphs of $y = x^3$, $y = 2^x$, and $y = 100 \cdot \frac{x^3}{2^x}$. The graphs of the first two cross twice, the last time about $x \approx 10$:

```
MAPLE
> sol:=solve(x^3=2^x,x);
> evalf({sol[1],sol[2]});
```

```
MATLAB
% make a file named fig212.m with the following two lines (without the % signs);
% MATLAB requires functions be defined in external files and finds them via the MATLAB PATH
%  function y=fig212(x);
%  y=x.^3 - 2.^x;
% resume this calculation
> fzero('fig212',10) %no semicolon to print ans.
```

$$1.3734, \qquad 9.939.$$

Taking logarithms of (2.1.6) to base $e$ gives

MAPLE
```
> plot({[x,x^3,x=0..12],[x,2^x,x=0..12],[x,100*x^3/2^x,x=0..14]},x=0..14,y=0..4000);
```

MATLAB
```
> x=linspace(0,14); % 100 equally spaced values 0 to 14
> y=100*x.^3./2.^x; % .^means term by term power, ./ and .* mean term by term div. and mult.
> plot(x,y)
> hold on % keep axis, scale, etc., of the graph fixed
> x=linspace(0,12);
> plot(x,x.^3); % plot overlaid on the previous plot
> plot(x,2.^x); % ditto
```

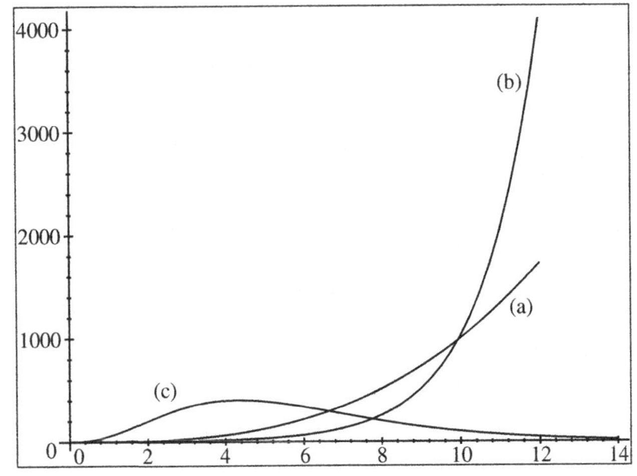


**Fig. 2.1.2.** Exponential vs. polynomial rate of growth graphs of (a) $x^3$, (b) $2^x$, and (c) $100\frac{x^3}{2^x}$.

$$\ln y = x \ln a + \ln C. \tag{2.1.8}$$

If the constant $a$ is $e$, then $\ln a = \ln e = 1$. Also note that any positive number can be written as some exponent of $e$, namely, $\ln a$. Thus $a = e^{\ln a} = e^r$ if we put $r = \ln a$. In the form of (2.1.8), it is $\ln y$ that is proportional to $x$. A *semilog plot* of exponentially related variables, as in (2.1.8), produces a straight line whose slope is $\ln a$.

By defining $r = \ln a$ and exponentiating both sides of (2.1.8), we get

$$y = Ce^{rx}, \quad \text{where } r = \ln a. \tag{2.1.9}$$

This is an alternative form of the relationship given in equation (2.1.6) and shows that an exponential relationship can be expressed in base $e$ if desired.

*Proportionality can pertain to derivatives, too.*

A natural and simplifying assumption about the growth of a population is that *the number of offspring born at any given time is proportional to the number of adults at that time* (see Chapter 3). This expresses a linear relationship between the number of offspring and the number of adults. Let $y(t)$ (or just $y$ in brief) denote the number of adults at time $t$. In any given small interval of time $\Delta t$, the number of offspring in that time represents the change in the population $\Delta y$. The ratio $\frac{\Delta y}{\Delta t}$ is the average rate of growth of the population over the time period $\Delta t$. The derivative $\frac{dy}{dt}$ is the instantaneous rate of growth at time $t$, or just the rate of growth at time $t$, instantaneous being understood. Making the questionable, but simplifying, assumption that new offspring are immediately adults leads to a mathematical expression of the italicized statement above:

$$\frac{dy}{dt} = ky$$

for some constant of proportionality $k$. That is, the derivative or rate of growth is proportional to the number present.

This particular differential equation is easily solved by integration,

$$\frac{dy}{y} = kdt \quad \text{or} \quad \ln y = kt + \ln A,$$

with constant of integration $\ln A$. Exponentiating both sides gives

$$y = Ae^{kt}.$$

This situation is typical, and we will encounter similar ones throughout the book.

## Exercises

1. Proportionality constants associated with changes in units are often used in making conversions after measurements have been made. Convert from the specified units to the indicated units.

   (a) Convert the following: $x$ inches to centimeters, $y$ pounds per gallon to kilograms per liter, $z$ miles per hour to kilometers per hour.

```
MAPLE
#Change of units is built-in
#type: ?convert.
> convert(x*inches,metric);
> convert(y*pounds/gallon,metric,US);
> convert(z*miles/hour,metric);

MATLAB
% some US to metric conversions
% Length: 1 inch = 2.54 cm (exactly), 39.3700 inch = 1 meter
% Mass: 1 lb = .45359237 kg (avoirdupois pound)
% Volume: 1 gallon = 3.785411784 liter (US gallon)
> x=0:10; y=2.54*x; plot(x,y) % plot cm vs. inch
% to plot kg/liter vs. pounds/gallon one finds the number of the former per 1 of the latter;
% use this 1 lb/gal = (1 lb/gal)*(1 gal/3.78 lit)*(.453 kg/lb)
% cancel units so that 1 lb/gal = .45359237/3.785411784 kg/lit.
```

   (b) Sketch three graphs similar to Figure 2.1.1 that show the changes in units indicated above. Syntax similar to that which generated Figure 2.1.1 can be used here.

2. In this exercise, we compare graphs of exponential and power law relations with standard graphs, log graphs, and log-log graphs. For this exercise, please type

the commands manually (rather than executing pretyped commands downloaded from the Web) and view the results of each command one by one. This will help internalize the commands and aid in connecting each with its action.

(a) Sketch the graphs of $\pi r^2$ and $\frac{4}{3}\pi r^3$ on the same graph. Then sketch both of these as log-log plots.

(b) Sketch the graphs of $3x^5$ and $5x^3$ on the same graph. Then sketch both these as log plots.

```
MAPLE
> plot({Pi*r^2,4/3*Pi*r^3},r=0..1);
> plots[loglogplot]({Pi*r^2,4/3*Pi*r^3},r=0.1..1);
> plot({3*x^5,5*x^3},x=0..1);
> plots[logplot]({3*x^5,5*x^3},x=0..1);

MATLAB
> r=0:.1:1; % create vector of r values
> plot(r,pi*r.^2)
  % plot pi r squared vs. r, use .^(dot hat, not ^)
  % to get term by term r squared, no need for .* (dot star) since pi is a constant
> hold on % to overlay this graph
> plot(r,pi*(4/3)*r.^3);
> hold off % begin new plot
> loglog(r,pi*r.^2) % MATLAB automatically avoided r=0
> hold on
> loglog(r,(4/3)*pi*r.^3)
> hold off
> x=linspace(0,1); % divide 0 to 1 into 100 subdivisions
> plot(x,3*x.^5); hold on
> plot(x,5*x.^3)
```

**3.** This exercise examines limits of quotients of polynomials and exponentials. Sketch the graphs of $3x^2 + 5x + 7$ and $2^x$ on the same axis. Also, sketch the graph of their quotients. Evaluate the limit of this quotient.

```
MAPLE
> plot({3*x^2+5*x+7,2^x},x=0..7);
> plot((3*x^2+5*x+7)/2^x,x=0..10,y=0..10);
> limit((3*x^2+5*x+7)/2^x,x=infinity);

MATLAB
> x=linspace(0,7); % vector of 100 x values
> plot(x,3*x.^2+5*x+7); hold on
> plot(x,2.^x)
  % or make a matrix whose first row=polynomial and second row=exponential
> M=[3*x.^2+5*x+7; 2.^x]; % note the semicolon in M
> hold off; plot(x,M) % and plot both at once
> plot(x,M(1,:)./M(2,:))
  % quotient of first row/second row term by term
  % observe the limit is 0 graphically
```

**4.** This exercise solves differential equations such as we encounter in Section 2.1. Give the solution and plot the graph of the solution for each of these differential equations:

$$\frac{dy}{dt} = 3y(t), \qquad y(0) = 2,$$
$$\frac{dy}{dt} = 2y(t), \qquad y(0) = 3,$$
$$\frac{dy}{dt} = 2y(t), \qquad y(0) = -3,$$

$$\frac{dy}{dt} = -2y(t), \qquad y(0) = 3.$$

Here is syntax that will do the first problem and will undo the definition of $y$ to prepare for the remaining problems.

```
MAPLE
> eq:=diff(y(t),t)=3*y(t);
> sol:=dsolve({eq,y(0)=2},y(t));
> y:=unapply(rhs(sol),t); plot(y(t),t=0..1);
> y:='y';

MATLAB
% for the 1st DE make an m-file, ex214a.m, say, containing
%  function yprime=ex214a(t,y); yprime=3*y;
> [t,y]=ode23('ex214a',[0 1],2);
> plot(t,y)
```

## 2.2 Linear Regression, the Method of Least Squares

In this section we introduce the method of least squares for fitting straight lines to experimental data. By transformation, the method can be made to work for data related by power laws and exponential laws as well as for linearly related data.

The method is illustrated with two examples.

*The method of least squares calculates a linear fit to experimental data.*

Imagine performing the following simple experiment: Record the temperature of a bath as shown on two different thermometers, one calibrated in Fahrenheit and the other in Celsius, as the bath is heated. We plot the temperature $F$ against the temperature $C$. Surprisingly, if there are three or more data points observed to high precision, they will not fall on a single straight line because the mathematical line established by two of the points will dictate infinitely many digits of precision for the others—no measuring device is capable of infinite precision. This is one source of error, and there are others. Thus experimental data, even data for linearly related variables, are not expected to fall perfectly on a straight line.

How then can we conclude experimentally that two variables are linearly related, and if they are, how can the slope and intercept of the correspondence be determined? The answer to the latter question is by the method of least squares fit and is the subject of this section; the answer to the first involves theoretical considerations and the collective judgment of scientists familiar with the phenomenon.

Assume that the variables $x$ and $y$ are suspected to be linearly related and we have three experimental points for them, for example $C$ and $F$ in the example above. For the three data points $(x_1, y_1)$, $(x_2, y_2)$, and $(x_3, y_3)$ shown in Figure 2.2.1, consider a possible straight line fit, $\ell(x)$. Let $e_1$, $e_2$, and $e_3$ be the errors

$$e_i = y_i - \ell(x_i), \quad i = 1, \ldots, 3,$$

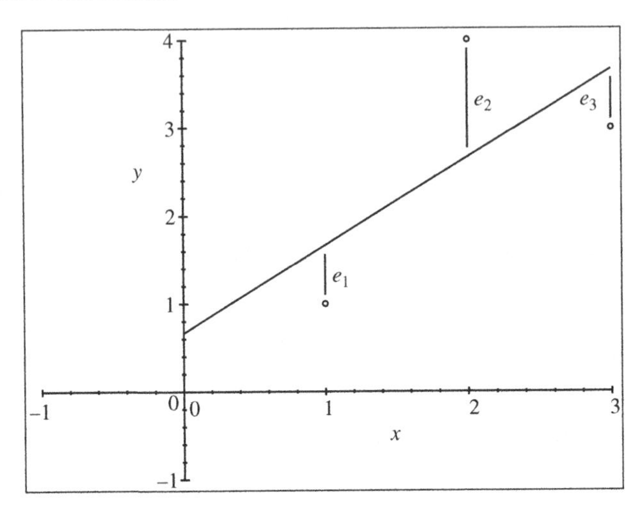


**Fig. 2.2.1.** The differences $e_i = y_i - \ell(x_i)$.

defined as the difference between the data value $y_i$ and the linear value $\ell(x_i)$ for each point. Note that we assume that all $x$-data values are exact and that the errors are in the $y$-values only. This is reasonable because $x$ is the independent variable; the $x$-values are the ones determined by the experimenter.

We want to choose a line $\ell$ that minimizes all of the errors at the same time; thus a first attempt might be to minimize the sum $e_1 + e_2 + e_3$. The difficulty with this idea is that these errors can cancel because they are signed values. Their sum could even be zero. But squaring each error eliminates this problem. And we choose the line $\ell$ so as to minimize

$$E = \sum_{i=1}^{3} e_i^2 = \sum_{i=1}^{3} [y_i - \ell(x_i)]^2,$$

that is, the least of the squared errors.

A line is determined by two parameters, slope $m$ and intercept $b$, $\ell(x) = mx + b$. Therefore the mathematical problem becomes, find $m$ and $b$ to minimize

$$E(m, b) = \sum_{i=1}^{n} [y_i - (mx_i + b)]^2 \tag{2.2.1}$$

for $n$ equal to the number of data points, three in this example. We emphasize that this error $E$ is a function of $m$ and $b$ (not $x$ and $y$; the $x_i$ and $y_i$ are specified numbers at the outset). Solving such a minimization problem is standard practice: Set the derivatives of $E$ with respect to its variables $m$ and $b$ equal to zero and solve for

$m$ and $b$,[2]

$$0 = \frac{\partial E}{\partial m} = -2\sum_{i=1}^{n}[y_i - (mx_i + b)]x_i,$$
$$0 = \frac{\partial E}{\partial b} = -2\sum_{i=1}^{n}[y_i - (mx_i + b)].$$

These equations simplify to

$$\begin{aligned} 0 &= \sum_{i=1}^{n} x_i y_i - m\sum_{i=1}^{n} x_i^2 - b\sum_{i=1}^{n} x_i, \\ 0 &= \sum_{i=1}^{n} y_i - m\sum_{i=1}^{n} x_i - nb, \end{aligned} \tag{2.2.2}$$

which may be easily solved.[3] The least squares solution is

$$\begin{aligned} m &= \frac{n\sum_{i=1}^{n} x_i y_i - \left(\sum_{i=1}^{n} x_i\right)\left(\sum_{i=1}^{n} y_i\right)}{n\sum_{i=1}^{n} x_i^2 - \left(\sum_{i=1}^{n} x_i\right)^2}, \\ b &= \frac{\left(\sum_{i=1}^{n} x_i^2\right)\left(\sum_{i=1}^{n} y_i\right) - \left(\sum_{i=1}^{n} x_i\right)\left(\sum_{i=1}^{n} x_i y_i\right)}{n\sum_{i=1}^{n} x_i^2 - \left(\sum_{i=1}^{n} x_i\right)^2}. \end{aligned} \tag{2.2.3}$$

The expression for $b$ simplifies to[4]

$$b = \bar{y} - m\bar{x}, \quad \text{where } \bar{y} = \frac{1}{n}\sum_{i=1}^{n} y_i \quad \text{and} \quad \bar{x} = \frac{1}{n}\sum_{i=1}^{n} x_i.$$

We will illustrate the least squares method with two examples.

**Example 2.2.1.** Juvenile height vs. age is only approximately linear.

In Table 2.2.1, we show age and average height data for children.

With $n = 7$, age and height interpreted as $x$ and $y$, respectively, in (2.2.1), and using the data of the table, parameters $m$ and $b$ can be evaluated from the equations in (2.2.3):

---

[2] Since $E$ is a function of two independent variables $m$ and $b$, it can vary with $m$ while $b$ is held constant or vice versa. To calculate its derivatives, we do just that: Pretend $b$ is a constant and differentiate with respect to $m$ as usual; this is called the *partial derivative* with respect to $m$ and is written $\frac{\partial E}{\partial m}$ in deference to the variables held fixed. Similarly, hold $m$ constant and differentiate with respect to $b$ to get $\frac{\partial E}{\partial b}$. At a minimum point of $E$, both derivatives must be zero, since $E$ will be momentarily stationary with respect to each variable.

[3] Verify this solution by substituting $m = \frac{nE-BF}{nA-BC}$ and $b = \frac{AF-cE}{nA-BC}$ into $mA + bB = E$ and $mC + nb = F$.

[4] Starting from $\bar{y} - m\bar{x}$ with $m$ from (2.2.3), make a common denominator and cancel the terms $-(\sum x_i)^2\bar{y} + \bar{x}\sum x_i \sum y_i$, and the expression for $b$ emerges.

**Table 2.2.1.** Average height vs. age for children. (Source: D. N. Holvey, ed., *The Merck Manual of Diagnosis and Therapy*, 15th ed., Merck, Sharp, and Dohme Research Laboratories, Rahway, NJ, 1987.)

| **Height (cm)** | 75 | 92 | 108 | 121 | 130 | 142 | 155 |
|---|---|---|---|---|---|---|---|
| **Age** | 1 | 3 | 5 | 7 | 9 | 11 | 13 |

```
MAPLE
> ht:=[75,92,108,121,130,142,155]; age:=[1,3,5,7,9,11,13];
> sumy:=sum(ht[n],n=1..7); sumx:=sum(age[n],n=1..7);
> sumx2:=sum(age[n]^2,n=1..7);
> sumxy:=sum(age[n]*ht[n],n=1..7);
> m:=evalf((7*sumxy-sumx*sumy)/(7*sumx2-sumx^2));
> b:=evalf((sumx2*sumy-sumx*sumxy)/(7*sumx2-sumx^2));

MATLAB
> ht=[75 92 108 121 130 142 155];
> age=[1 3 5 7 9 11 13];
> sumy=sum(ht);
> sumx=sum(age);
> age2=age.*age;
> sumx2=sum(age2);
> ageht=age.*ht;
> sumxy=sum(ageht);
> m=(7*sumxy-sumx*sumy)/(7*sumx2-sumx^2)
> b=(sumx2*sumy-sumx*sumxy)/(7*sumx2-sumx^2)
```

$$m = 6.46 \quad \text{and} \quad b = 72.3.$$

These data are plotted in Figure 2.2.2 along with the least squares fit for an assumed linear relationship $\text{ht} = m \cdot \text{age} + b$ between height and age.

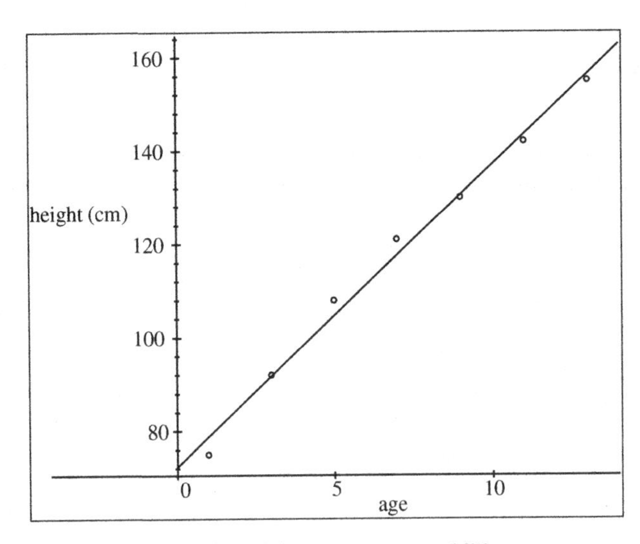


**Fig. 2.2.2.** Height vs. age among children.

Finding a least square fit is so important that it has its own routine in MAPLE called `fit[leastsquare]`. In MATLAB a least square fit is performed by a simple matrix statement. The mathematics of the matrix approach is the subject of the next section. Here then is the shortcut syntax for accomplishing what was done above.

```
MAPLE
> m:='m'; b:='b'; # clears m and b (single quotes/apostrophy)
  # next create an array of (age,ht) pairs;
> pts:=[seq([age[i],ht[i]],i=1..7)]:
> with(plots): with(stats):
> Data:=plot(pts,style=POINT,symbol=CIRCLE):
> fit[leastsquare[[x,y],y=m*x+b]]([age,ht]);
  # result in y=m*x+b form, m*x is the first operand on the right-hand side
> m:=op(1,op(1,rhs(%))); # strip off x too
> b:=op(2,rhs(%%)); # use %% to get second statement back
> Fit:=plot(m*x+b,x=0..14):
> display({Data,Fit});
```

```
MATLAB
% Now the matrix solution
% matrix of independent variable
% experimental values as columns
> MT=[1 3 5 7 9 11 13; 1 1 1 1 1 1 1]; % two rows
> M=MT'; % transpose to columns
% M = transpose of MT
% dependent variable data next, as col. vec.
> Y=[75; 92; 108; 121; 130; 142; 155];
> s=M\Y % MATLAB syntax for leastsquare
> m=s(1); b=s(2); % plot data and fit for comparison, Figure 2.2.2
> plot(age,ht,'o') % point plot ht vs. age with circles
> hold on
> fit=m*age+b;
> plot(age,fit); xlabel('age'); ylabel('Height (cm)');
```

This demonstrates the mechanics of the least squares method. But it must be kept in mind that the method is merely statistical; it can demonstrate that data are consistent or not with a linear assumption, but it cannot prove linearity. In this example, a linear fit to the data is reasonably good, but no rationale for a linear relationship has been provided.

**Example 2.2.2.** The number of AIDS cases increases cubically.

As we saw in the first part of this section, when the data are obviously not linear, we can try to fit a power law of the form $y = Ax^k$. Consider the following data as reported in the HIV/AIDS Surveillance Report published by the U.S. Department of Health and Human Services concerning the reported cases of AIDS by half-year shown in Table 2.2.2. The third column is the sum of all the cases reported to that time, i.e., the Cumulative AIDS Cases (CAC).

This cumulative AIDS cases data is shown later in Figure 2.2.4. The circle symbols of the figure give the CAC data vs. year; the solid curve is the least squares fit, which we discuss next. In this figure, CAC is measured in thousands and $t$ is decades from 1980, that is, $t = \frac{\text{year}-1980}{10}$.

We begin by first reading in the data:

```
MAPLE
> restart:
> AIDS:=([97, 206, 406, 700, 1289, 1654, 2576, 3392, 4922, 6343, 8359, 9968, 12990, 14397, 16604,
     17124, 19585, 19707, 21392, 20846, 23690, 24610, 26228, 22768, 4903]);
```

**Table 2.2.2.** Total and reported cases of AIDS in the U.S.

| Year | Reported cases of AIDS | Cumulative AIDS cases (thousands) |
|---|---|---|
| 1981 | 97 | 0.097 |
| 1981.5 | 206 | 0.303 |
| 1982 | 406 | 0.709 |
| 1982.5 | 700 | 1.409 |
| 1983 | 1289 | 2.698 |
| 1983.5 | 1654 | 4.352 |
| 1984 | 2576 | 6.928 |
| 1984.5 | 3392 | 10.320 |
| 1985 | 4922 | 15.242 |
| 1985.5 | 6343 | 21.585 |
| 1986 | 8359 | 29.944 |
| 1986.5 | 9968 | 39.912 |
| 1987 | 12990 | 52.902 |
| 1987.5 | 14397 | 67.299 |
| 1988 | 16604 | 83.903 |
| 1988.5 | 17124 | 101.027 |
| 1989 | 19585 | 12.0612 |
| 1989.5 | 19707 | 140.319 |
| 1990 | 21392 | 161.711 |
| 1990.5 | 20846 | 181.557 |
| 1991 | 23690 | 206.247 |
| 1991.5 | 24610 | 230.857 |
| 1992 | 26228 | 257.085 |
| 1992.5 | 22768 | 279.853 |

```
> CAC:=[seq(sum(AIDS[j]/1000.0, j=1..i),i=1..24)];
> Time:=[seq(1981+(i-1)/2,i=1..24)]:
```

MATLAB

```
% year by year cases; note that ellipses continue the line
> AIDS=[97 206 406 700 1289 1654 2576 3392 4922 6343 8359 9968 12990 14397 16604 17124 19585 ...
        19707 21392 20846 23690 24610 26228 22768];
> CAC=cumsum(AIDS)/1000; % cumulative sum (scaled down 1000)
  % housekeeping to get the sequence 0,0.5,1,1.5,...
> s=size(AIDS); % number of half-years
> count=[0:s(2)-1];
> time =1981+count/2;
```

To produce the fit we proceed as before using (2.2.1), but this time performing least squares on $y = \ln(\text{CAC})$ vs. $x = \ln t$:

$$\ln(\text{CAC}) = k * \ln t + \ln A. \tag{2.2.4}$$

Here we rescale time to be decades after 1980 and calculate the logarithm of the data:

MAPLE

```
> LnCAC:=map(ln,CAC);
> Lntime:=map(ln,[seq((i+1)/2/10,i=1..24)]);
```

MATLAB

```
% shifted and scaled time
```

```
> scaledTime=(time-1980)/10
  % log the data to do a log-log plot
> lnCAC=log(CAC)
> lnTime=log(scaledTime)
```

It remains to calculate the coefficients:

```
MAPLE
> with(stats):
> fit[leastsquare[[x,y],y=k*x+LnA]]([Lntime,LnCAC]);
> k:=op(1,op(1,rhs(%))); LnA:=(op(2,rhs(%%))); A:=exp(LnA);

MATLAB
% form the coefficient matrix for lnCAC = k*lnTime + b fit
> MT=[lnTime; ones(1,24)] % second row is ones
> M=MT';
> params=M\(lnCAC') % do the leastsquares
> k=params(1)
> A=exp(params(2))
```

$$k = 3.29, \quad \text{and} \quad \ln A = 5.04, \quad A = 155.$$

We draw the graph of Ln(CAC) vs. Ln(time) to emphasize that their relationship is nearly a straight line. The log-log plot of best fit is shown in Figure 2.2.3 and is drawn as follows:

```
MAPLE
> Lndata:=plot([seq([Lntime[i],LnCAC[i]],i=1..24)],style=POINT,symbol=CIRCLE):
> Lnfit:=plot(k*x+ln(A),x=-2.5..0.5):
> plots[display]({Lndata,Lnfit});

MATLAB
% now compare the fit to the data in log-log space
> plot(lnTime,lnCAC,'o')
> lnFit= params(1).*lnTime+params(2)
> plot(lnTime,lnFit)
```

The curve of best fit is, from (2.2.4),

$$\text{CAC} = 155t^{3.29}.$$

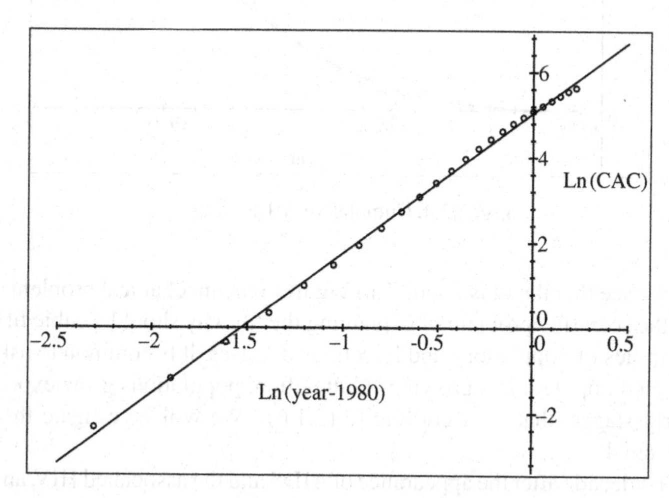


**Fig. 2.2.3.** Log-log plot of cumulative AIDS cases and its fit.

But we want an integer exponent; hence the exponent for the comparative graph to the data will be taken as

```
MAPLE
> n:=trunc(k);
```

$$n = 3,$$

$$\text{CAC} = 155t^3 = 155\left(\frac{\text{year} - 1980}{10}\right)^3.$$

Figure 2.2.4 is drawn as an overlay of the data and this fit.

```
MAPLE
> pts:=[seq([Time[i], CAC[i]], i=1..24)];
> Fit:=plot(A*((t-1980)/10)^n,t=1980..1993):
> Data:=plot(pts,style=POINT,symbol=CIRCLE):
> plots[display](Fit,Data);
```

```
MATLAB
% and compare in regular space
> hold off; plot(time,CAC)
> CACFit=exp(params(2)).*scaledTime.^params(1)
> plot(time,CACFit)
```

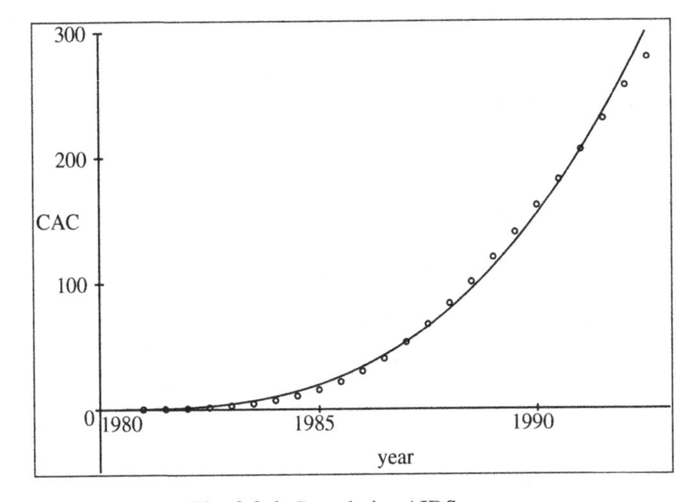


**Fig. 2.2.4.** Cumulative AIDS cases.

Again, we see that the fit is good. Turning from the mechanical problem of fitting the data to the scientific problem of explaining the fit, why should a cubic fit so well?

In the studies of populations and infectious diseases, it is common to ask at what rate an infected population is growing. Quite often, populations grow exponentially in their early stages, that is, according to (2.1.6). We will investigate this idea in Chapters 3 and 4.

In the first decade after the appearance of AIDS and the associated HIV, an analysis of the data for the total number of reported cases of AIDS led to the announcement that the population was growing cubically as a function of time. This was a relief of sorts because the growth was not exponential as expected, since exponential growth is much faster than polynomial growth; see (2.1.7).

Colgate et al. [2] constructed a model for HIV infection that led to the result that the growth rate should be cubic in the early stages. A central idea in the model is the recognition that the disease spreads at different rates in different "risk groups," and that there is a statistically predictable rate at which the disease crosses risk groups.

In the exercises, we attempt an exponential fit to these data.

## Exercises

**1.** Ideal weights for medium-build males are listed in Table 2.2.3 from [3].

**Table 2.2.3.** Ideal weights for medium-build males.

| Height (in) | Weight (lb) |
|---|---|
| 62 | 128 |
| 63 | 131 |
| 64 | 135 |
| 65 | 139 |
| 66 | 142 |
| 67 | 146 |
| 68 | 150 |
| 69 | 154 |
| 70 | 158 |
| 71 | 162 |
| 72 | 167 |
| 73 | 172 |

(a) Show that a linear fit for these data is

$$\text{wt} = 4.04 \cdot \text{ht} + 124.14.$$

(b) In many geometric solids, volume changes with the cube of the height. Give a cubic fit for these data.

(c) Using the techniques of Example 2.2.2, find $n$ and $A$ such that

$$\text{wt} = A \cdot (\text{ht} - 60)^n.$$

The following code can be used for Exercise 1(b). A modification of one line can be used for 1(a). For 1(c), modify the code for Example 2.2.2.

```
MAPLE
> ht:=[62,63,64,65,66,67,68,69,70,71,72,73,74];
> wt:=[128,131,135,139,142,146,150,154,158,162,167,172,177];
> with(stats): fit[leastsquare[[x,y], y=a*x^3+b*x^2+c*x+d]]([ht,wt]);
> y:=unapply(rhs(%),x);
```

```
> pts:=[seq([ht[i],wt[i]],i=1..13)];
> J:=plot(pts,style=POINT,symbol=CROSS):K:=plot(y(x),x=62..74):
> with(plots): display({J,K});
> errorLinear:=sum('(4.04*ht[i]-124.14- wt[i])^2','i'=1..13);
> errorcubic:=sum('(y(ht[i])-wt[i])^2','i'=1..13);
> evalf(%);
```

MATLAB

```
> ht=[62,63,64,65,66,67,68,69,70,71,72,73,74];
> wt=[128,131,135,139,142,146,150,154,158,162,167,172,177];
> MT=[ht.^3; ht.^2; ht; ones(1,13)];
> params=MT'\wt'; % MT prime, wt prime
> plot(ht,wt,'x'); hold on
> fit=params(1)*ht.^3+params(2)*ht.^2+params(3)*ht+params(4);
> plot(ht,fit)
> errorLinear=sum((4.04*ht-124.14-wt).^2)
> errorcubic=sum((fit-wt).^2)
```

**2.** Changes in the human life span are illustrated graphically on p. 110 of the October 1994 issue of *Scientific American*. These data appear in Table 2.2.4 in three rows: The first row indicates the age category. The next two rows indicate the percentage of people who survived to that age in the United States in the years 1900 and 1960. The last row is the percentage of people who survived to that age in ancient Rome. Get a least squares fit for these data sets. Syntax that provides such a fit is given for the 1960 data.

**Table 2.2.4.** Survival rates for recent U.S. and ancient Rome.

| **Age** | 0 | 10 | 20 | 30 | 40 | 50 | 60 | 80 | 100 |
|---|---|---|---|---|---|---|---|---|---|
| **1900** | 100 | 82 | 78 | 75 | 74 | 60 | 43 | 19 | 3 |
| **1960** | 100 | 98.5 | 98 | 96.5 | 95 | 92.5 | 79 | 34 | 4 |
| **Rome** | 90 | 73 | 50 | 40 | 30 | 22 | 15 | 5 | 0.5 |

MAPLE

```
> restart:
> age60:=[0,10,20,30,40,50,60,80,100]:
> percent60:=[100,98.5,98,96.5,95,92.5,79,34,4]:
> with(stats):
> fit[leastsquare[[x,y],y=a*x^4+b*x^3+c*x^2+d*x+e]]([age60,percent60]);
> yfit60:=unapply(rhs(%),x):
> pts60:=[seq([age60[i],percent60[i]],i=1..9)]:
> J6:=plot(pts60,style=POINT,symbol=CROSS):
> K6:=plot(yfit60(x),x=0..100):
> with(plots): display({J6,K6});
```

MATLAB

```
> age60=[0,10,20,30,40,50,60,80,100];
> percent60=[100,98.5,98,96.5,95,92.5,79,34,4];
> MT=[age60.^4; age60.^3;age60.^2; age60; ones(size(age60))];
> parms=MT'\percent60' % note the primes
> fit=parms(1)*age60.^4+parms(2)*age60.^3+parms(3)*age60.^2+parms(4)*age60+parms(5);
> plot(age60,percent60,age60,fit)
```

**3.** We have found a fit for the cumulative U.S. AIDS data as a cubic polynomial. We saw that, in a sense, a cubic polynomial is the appropriate choice. On first looking at the data as shown in Figure 2.2.4, one might guess that the growth is exponential. Find an exponential fit for those data. Such a fit would use (2.1.8).

Computer code to perform the calculations is only slightly different from that for the cubic fit:

```
MAPLE
> restart:
> AIDS:=([97, 206, 406, 700, 1289, 1654, 2576, 3392, 4922, 6343, 8359, 9968, 12990, 14397, 16604,
17124, 19585, 19707, 21392, 20846, 23690, 24610, 26228, 22768, 4903]);
> CAC:=[seq(sum(AIDS[j]/1000.0,j=1..i),i=1..24)];
> Time:=[seq(1981+(i-1)/2,i=1..24)]:
> pts:=[seq([Time[i],CAC[i]],i=1..24)]:
> LnCAC:=map(ln,CAC);
> Times:=[seq((i+1)/2/10,i=1..24)];
> with(stats):
> fit[leastsquare[[x,y],y=m*x+b]]([Times,LnCAC]);
> k:=op(1,op(1,rhs(%)));A:=op(2,rhs(%%));
> y:=t->exp(A)*exp(k*t);
> J:=plot(y((t-1980)/10),t=1980..1992):
> K:=plot(pts,style=POINT,symbol=CIRCLE):
> plots[display]({J,K});
```

```
MATLAB
> AIDS=[97, 206, 406, 700, 1289, 1654, 2576, 3392, 4922, 6343, 8359, 9968, 12990, 14397, 16604,...
        17124, 19585, 19707, 21392, 20846, 23690, 24610, 26228, 22768];
> CAC=cumsum(AIDS)/1000;
> s=size(AIDS); % number of half-years
> count=[0:s(2)-1];
> Time =1981+count/2;
> pts=[Time' CAC'];
> plot(pts(:,1),pts(:,2)); hold on
> Times=(Time-1980)/10; LnCAC=log(CAC);
> MT=[Times; ones(1,s(2))]; % note the space
> params=MT'\LnCAC'
> k=params(1); A=params(2);
> y=exp(A)*exp(k.*Times);
> plot(10*Times+1980,y)
```

**4.** Table 2.2.5 presents unpublished data that was gathered by Dr. Melinda Millard-Stafford at the Exercise Science Laboratory in the Department of Health and Performance Sciences at Georgia Tech. It relates the circumference of the forearm with grip strength. The first two columns are for a group of college women, and the following two columns are for college men. Find *regression lines* (that is, least square fits) for both sets of data:

```
MAPLE
> CW:=[24.2,22.9,27.,21.5,23.5,22.4, 23.8, 25.5, 24.5,25.5,22.,24.5];
> GSW:=[38.5,26.,34.,25.5,37.,30.,34.,43.5,30.5, 36.,29.,32];
> with(stats):
> fit[leastsquare[[x,y],y=m*x+b]]([CW,GSW]);
> pts:=[seq([CW[i],GSW[i]],i=1..12)];
> J:=plot(pts,style=POINT,symbol=CROSS):
> K:=plot(2.107*x-17.447,x=21..28):
> CM:=[28.5,24.5,26.5,28.25,28.2,29.5,24.5,26.9,28.2,25.6,28.1,27.8,29.5,29.5,29];
> GSM:=[45.8,47.5,50.8,51.5,55.0,51.,47.5,45.,56.0,49.5,57.5,51.,59.5, 58.,68.25];
> fit[leastsquare[[x,y],y=m*x+b]]([CM,GSM]);
> pts:=[seq([CM[i],GSM[i]],i=1..15)];
> L:=plot(pts,style=POINT,symbol=CIRCLE):
> M:=plot(2.153*x-6.567,x=24..30):
> with(plots): display({J,K,L,M});
```

```
MATLAB
> CW=[24.2,22.9,27.,21.5,23.5,22.4,23.8,25.5,24.5,25.5,22.,24.5];
> GSW=[38.5,26.,34.,25.5,37.,30.,34.,43.5,30.5,36.,29.,32];
> MT=[CW; ones(size(CW))];
> parmsW=MT'\GSW';
> plot(CW,GSW,'x'); hold on
```

**Table 2.2.5.** Forearm and grip strength, males/females.

| Females | | Males | |
|---|---|---|---|
| **Circumference (cm)** | **Grip (kg)** | **Circumference (cm)** | **Grip (kg)** |
| 24.2 | 38.5 | 28.5 | 45.8 |
| 22.9 | 26.0 | 24.5 | 47.5 |
| 27.0 | 34.0 | 26.5 | 50.8 |
| 21.5 | 25.5 | 28.25 | 51.5 |
| 23.5 | 37.0 | 28.2 | 55.0 |
| 22.4 | 30.0 | 29.5 | 51.0 |
| 23.8 | 34.0 | 24.5 | 47.5 |
| 25.5 | 43.5 | 26.9 | 45.0 |
| 24.5 | 30.5 | 28.2 | 56.0 |
| 25.5 | 36.0 | 25.6 | 49.5 |
| 22.0 | 29.0 | 28.1 | 57.5 |
| 24.5 | 32.0 | 27.8 | 51.0 |
| | | 29.5 | 59.5 |
| | | 29.5 | 58.0 |
| | | 29.0 | 68.25 |

```
> x=21:28; plot(x,parmsW(1)*x+parmsW(2))
  %%%
> CM=[28.5,24.5,26.5,28.25,28.2,29.5,24.5,26.9,28.2,25.6,28.1,27.8,29.5,29.5,29];
> GSM=[45.8,47.5,50.8,51.5,55.0,51.,47.5,45.,56.0,49.5,57.5,51.,59.5,58.,68.25];
> MT=[CM; ones(size(CM))];
> parmsM=MT'\GSM'
> plot(CM,GSM,'o')
> x=24:30;
> plot(x,parmsM(1)*x+parmsM(2))
```

## 2.3 Multiple Regression

*The least squares method extends to experimental models with arbitrarily many parameters. However, the model must be linear in the parameters. The mathematical problem of their calculation can be cast in matrix form, and as such, the parameters emerge as the solution of a linear system. The method is again illustrated with two examples.*

*Least squares can be extended to more than two parameters*

In the previous section, we learned how to perform linear regression, or least squares, on two parameters, to get the slope $m$ and intercept $b$ of a straight-line fit to data. We also saw that the method applies to other models for the data than just the linear model. By a *model* here we mean a mathematical formula of a given form involving unknown parameters. Thus the *exponential model* for $(x, y)$ data is

$$y = Ae^{rx}.$$

And to apply linear regression, we transform it to the form

$$\ln y = rx + \ln A,$$

by taking the logarithm of both sides (cf. (2.1.8)). Here the transformed data is $Y = \ln y$ and $X = x$, while the transformed parameters are $M = r$ and $B = \ln A$. The key requirement of a regression model is that it be linear in the parameters.

**Regression principle.** *The method of least squares can be adapted to calculate the parameters of a model if there is some transformation of the model that is linear in the transformed parameters.*

Consider the Michaelis–Menten equation for the initial reaction rate $v_0$ of the enzyme-catalyzed reaction of a substrate having a concentration denoted by $[S]$ (see Section 8.6),

$$v_0 = \frac{v_{\max}[S]}{K_m + [S]};$$

the parameters are $v_{\max}$ and $K_m$. By taking the reciprocal of both sides of this equation, we get the *Lineweaver–Burk equation*:

$$\frac{1}{v_0} = \frac{K_m}{v_{\max}}\frac{1}{[S]} + \frac{1}{v_{\max}}. \qquad (2.3.1)$$

Now the transformed model is linear in its parameters $M = \frac{K_m}{v_{\max}}$ and $B = \frac{1}{v_{\max}}$, and the transformed data are $Y = \frac{1}{v_0}$ and $X = \frac{1}{[S]}$. After determining the slope $M$ and intercept $B$ of a *double reciprocal plot* of $\frac{1}{v_0}$ vs. $\frac{1}{[S]}$ by least squares, then calculate $v_{\max} = \frac{1}{B}$ and $K_m = \frac{M}{B}$.

So far we have looked only at two-parameter models; but the principles apply to models of any number of parameters. For example, the *Merck Manual* (R. Berkow, ed., *The Merck Manual of Diagnosis and Therapy*, 14th ed., Merck, Sharp, and Dohme Research Laboratories, Rahway, NJ, 1982) gives a relationship between the outer surface area of a person as a function of height and weight as follows:

$$\text{surface area} = c \cdot \text{wt}^a \cdot \text{ht}^b,$$

with parameters $a$, $b$, and $c$ ($a$ and $b$ have been determined to be 0.425 and 0.725, respectively). A transformed model, linear in parameters, for this is

$$\ln(\text{surface area}) = a\ln(\text{wt}) + b\ln(\text{ht}) + \ln c.$$

The transformed data are triples of values $(X_1, X_2, Y)$, where $X_1 = \ln(\text{wt})$, $X_2 = \ln(\text{ht})$, and $Y = \ln(\text{surface area})$.

We now extend the method of least squares to linear models of $r$ generalized independent variables $X_1, \ldots, X_r$ and one generalized dependent or response variable $Y$,

$$Y = a_1X_1 + a_2X_2 + \cdots + a_rX_r.$$

Note that we can recover the two variable case of Section 2.2 by taking $r = 2$ and $X_2 = 1$. Assume that there are $n$ data points $(X_{1,i}, \ldots, X_{r,i}, Y_i)$, $i = 1, \ldots, n$. As before, let $e_i$ denote the error between the experimental value $Y_i$ and the predicted value,

$$e_i = Y_i - (a_1 X_{1,i} + \cdots + a_r X_{r,i}), \quad i = 1, \ldots, n.$$

And as before, we choose parameter values $a_1, \ldots, a_r$ to minimize the squared error,

$$E(a_1, \ldots, a_r) = \sum_{i=1}^{n} e_i^2 = \sum_{i=1}^{n} [Y_i - (a_1 X_{1,i} + \cdots + a_r X_{r,i})]^2.$$

To minimize $E$, differentiate it with respect to each parameter $a_j$ and set the derivative to zero,

$$0 = \frac{\partial E}{\partial a_j} = -2 \sum_{i=1}^{n} X_{j,i} [Y_i - (a_1 X_{1,i} + \cdots + a_r X_{r,i})], \quad j = 1, \ldots, r.$$

The resulting linear system for the unknowns $a_1, \ldots, a_r$ can be rearranged to the following form (compare with equations (2.2.2)):

$$\begin{aligned} a_1 \sum_i^n X_{1,i} X_{1,i} + \cdots + a_r \sum_i^n X_{1,i} X_{r,i} &= \sum_i^n X_{1,i} Y_i, \\ a_1 \sum_i^n X_{r,i} X_{1,i} + \cdots + a_r \sum_i^n X_{r,i} X_{r,i} &= \sum_i^n X_{r,i} Y_i. \end{aligned} \tag{2.3.2}$$

It is possible to write this system in a very compact way using matrix notation. Let $M^T$ be the matrix of data values of the independent variables,

$$M^T = \begin{bmatrix} X_{1,1} & X_{1,2} & \ldots & X_{1,n} \\ X_{2,1} & X_{2,2} & \ldots & X_{2,n} \\ \vdots & \vdots & \ldots & \vdots \\ X_{r,1} & X_{r,2} & \ldots & X_{r,n} \end{bmatrix}.$$

The $i$th row of the matrix is the vector of data values of $X_i$. Represent the data values of the dependent variable $Y$ as a column vector and denote the whole column by $\mathbf{Y}$,

$$\mathbf{Y} = \begin{bmatrix} Y_1 \\ Y_2 \\ \vdots \\ Y_n \end{bmatrix}.$$

Denoting by $M$ the transpose of $M^T$, the system of equations (2.3.2) can be written in matrix form as

$$M^T M \mathbf{a} = M^T \mathbf{Y}, \tag{2.3.3}$$

where $\mathbf{a}$ is the column vector of regression parameters.

**Example 2.3.1.** Can body mass and skin fold predict body fat?

Sparling et al. [4] investigate the possibility of predicting body fat from height, weight, and skin fold measurements for women. Percentage body fat can be estimated by two methods: hydrostatic weighing and bioelectric impedance analysis. As in standard practice, height and weight enter the prediction as the fixed combination of weight divided by height squared to form a factor called *body-mass index*,

$$\text{body-mass index} = \frac{\text{weight}}{\text{height}^2}.$$

The assumed relationship is taken as

$$\text{percent body fat} = a * \text{body-mass index} + b * \text{skin fold} + c$$

for some constants $a$, $b$, and $c$.

Table 2.3.1 gives a subset of data of Sparling [4] that we will use for this example to find these constants. The weight and height measurements were made in pounds and inches respectively; body-mass index is to be in kilograms per square meter, so the conversions 0.0254 meter = 1 inch and 2.2046 pounds = 1 kilogram have been done to calculate the body-mass index column of the table.

**Table 2.3.1.** Height, weight, skin fold, and % body fat for women.

| Height (in) | Weight (lbs) | Body mass (kg/m$^2$) | Skin fold | % Body fat |
|---|---|---|---|---|
| 63.0 | 109.3 | 19.36 | 86.0 | 19.3 |
| 65.0 | 115.6 | 19.24 | 94.5 | 22.2 |
| 61.7 | 112.4 | 20.76 | 105.3 | 24.3 |
| 65.2 | 129.6 | 21.43 | 91.5 | 17.1 |
| 66.2 | 116.7 | 18.72 | 75.2 | 19.6 |
| 65.2 | 114.0 | 18.85 | 93.2 | 23.9 |
| 70.0 | 152.2 | 21.84 | 156.0 | 29.5 |
| 63.9 | 115.6 | 19.90 | 75.1 | 24.1 |
| 63.2 | 121.3 | 21.35 | 119.8 | 26.2 |
| 68.7 | 167.7 | 24.98 | 169.3 | 33.7 |
| 68.0 | 160.9 | 24.46 | 170.0 | 36.2 |
| 66.0 | 149.9 | 24.19 | 148.2 | 31.0 |

We compute the third column of Table 2.3.1 from the first two:

MAPLE
```
> ht:=[63,65,61.7,65.2,66.2,65.2,70.0,63.9,63.2,68.7,68,66];
  wt:=[109.3,115.6,112.4,129.6,116.7,114.0,152.2,115.6,121.3,167.7,160.9,149.9];
> convert([seq(wt[i]*lbs/(ht[i]/12*feet)^2,i=1..12)],metric);
```

MATLAB
```
% (1 kg/2.2046 lb)/(0.0254 m/1 in)^2 = 703.076 kg-in^2/lb-m^2
ht=[63,65,61.7,65.2,66.2,65.2,70.0,63.9,63.2,68.7,68,66];
wt=[109.3,115.6,112.4,129.6,116.7,114.0,152.2,115.6,121.3,167.7,160.9,149.9];
bodymass=(wt./(ht.*ht))*703.076;
% this is the M1 in the next step
```

To apply (2.3.3), we take $X_1$ to be body-mass index, $X_2$ to be skin fold, and $X_3 = 1$ identically. From the table, $M^T$ is

$$M^T = \begin{bmatrix} 19.36 & 19.24 & 20.76 & 21.43 & 18.72 & \ldots & 24.19 \\ 86.0 & 94.5 & 105.3 & 91.5 & 75.2 & \ldots & 148.2 \\ 1 & 1 & 1 & 1 & 1 & \ldots & 1 \end{bmatrix},$$

and the response vector is

$$\mathbf{Y}^T = \begin{bmatrix} 19.3 \; 22.2 \; 24.3 \; 17.1 \; 19.6 \ldots \; 31.0 \end{bmatrix}.$$

Solving the system of equations (2.3.3) gives the values of the parameters. We continue the present example:

```
MAPLE
> BMI:=[19.36,19.24, 20.76, 21.43, 18.72, 18.85, 21.84, 19.90, 21.35, 24.98, 24.46, 24.19];
> SF:=[86.0, 94.5,105.3, 91.5, 75.2, 93.2, 156.0, 75.1, 119.8, 69.3, 170.0, 148.2];
> PBF:=[19.3, 22.2, 24.3, 17.1, 19.6, 23.9, 29.5, 24.1, 26.2, 33.7, 36.2, 31.0];
> with(stats):
> fit[leastsquare[[bdymass,sfld,c]]]([BMI,SF,PBF]);
> bdft:=unapply(rhs(%),(bdymass,sfld));
```

```
MATLAB
% matrix of X values (metric)
> M1=[19.36 19.24 20.76 21.43 18.72 18.85 21.84 19.9 21.35 24.98 24.46 24.19];
> M2=[86.0 94.5 105.3 91.5 75.2 93.2 156.0 75.1 119.8 169.3 170 148.2];
> MT=[M1; M2; ones(1,12)];
% now vector of corresponding Y values
> Y=[19.3; 22.2; 24.3; 17.1; 19.6; 23.9; 29.5; 24.1; 26.2; 33.7; 36.2; 31.0];
% do min. norm inversion (i.e., least squares)
> params=MT'\Y
```

$$a = .00656, \qquad b = .1507, \qquad c = 8.074.$$

Thus we find that

$$\begin{aligned} &\text{percent body fat} \\ &\qquad \approx .00656 \times \text{body-mass index} + .1507 \times \text{skin fold} + 8.074. \end{aligned} \tag{2.3.4}$$

To test the calculations, here is a data sample not used in the calculation. The subject is 64.5 inches tall, weighs 135 pounds, and has skin fold that measures 159.9 millimeters. Her body-fat percentage is 30.8 as compared to the predicted value of 32.3:

```
MAPLE
> convert(135*lbs/((64.5/12*ft)^2),metric);
> bdft(22.815,159.9);
```

```
MATLAB
% predict percent body fat for subject 64.5 inches tall, weight of 135 lbs, and skin fold of 159.9 mm
% 2.2046 lbs per kilogram and 39.37 inches per meter
> bmi= (135/2.2046)/(64.5/39.37)^2
% so percent body fat is predicted as
> pbf=params(1)*bmi+params(2)*159.9+params(3)
```

$$\text{bdft} = 32.3.$$

**Example 2.3.2.** Can thigh circumference and leg strength predict vertical jumping ability?

Unpublished data gathered by Dr. Millard-Stafford in the Exercise Science Laboratory at Georgia Tech relates men's ability to jump vertically to the circumference of the thigh and leg strength as measured by leg press. The correlation was to find $a$, $b$, and $c$ such that

$$\text{jump height} = a * (\text{thigh circumference}) + b * (\text{bench press}) + c.$$

Hence the generalized variable $X_1$ is thigh circumference, $X_2$ is bench press, and $X_3 = 1$.

Data from a sample of college-age men is shown in Table 2.3.2. From the table,

$$M^T = \begin{bmatrix} 58.5 & 50 & 59.5 & 58 & \ldots & 56.25 \\ 220 & 150 & 165 & 270 & \ldots & 200 \\ 1 & 1 & 1 & 1 & \ldots & 1 \end{bmatrix}$$

and

$$\mathbf{Y}^T = \begin{bmatrix} 19.5 & 18 & 22 & 19 & \ldots & 29 \end{bmatrix}.$$

Solutions for (2.3.3) for these data are approximately found:

```
MAPLE
> thigh:=[58.5, 50, 59.5, 58, 60.5, 57.5, 49.3, 53.6, 58.3, 51, 54.2, 54, 59.5, 57.5, 56.25];
> press:=[220,150,165,270,200,250,210,130,220,165,190,165,280,190,200];
> jump:=[19.5,18,22,19,21,22,29.5,18,20,20,25,17,26.5,23,29];
```

**Table 2.3.2.** Leg size, strength, and jumping ability for men.

| Thigh average circumference (cm) | Leg press (lbs) | Vertical jump (in) |
|---|---|---|
| 58.5 | 220 | 19.5 |
| 50.0 | 150 | 18.0 |
| 59.5 | 165 | 22.0 |
| 58.0 | 270 | 19.0 |
| 60.5 | 200 | 21.0 |
| 57.5 | 250 | 22.0 |
| 49.3 | 210 | 29.5 |
| 53.6 | 130 | 18.0 |
| 58.3 | 220 | 20.0 |
| 51.0 | 165 | 20.0 |
| 54.2 | 190 | 25.0 |
| 54.0 | 165 | 17.0 |
| 59.5 | 280 | 26.5 |
| 57.5 | 190 | 23.0 |
| 56.25 | 200 | 29.0 |

```
> with(stats):
> fit[leastsquare[[x,y,z], z=a*x+b*y+c, {a,b,c}]]([thigh,press,jump]);
```

```
MATLAB
> M1=[58.5 50.0 59.5 58.0 60.5 57.5 49.3 53.6 58.3 51.0 54.2 54.0 59.5 57.5 56.25];
> M2=[220 150 165 270 200 250 210 130 220 165 190 165 280 190 200];
> MT=[M1; M2; ones(1,15)];
  % now vector of corresponding Y values
> YT=[19.5 18.0 22.0 19.0 21.0 22.0 29.5 18.0 20.0 20.0 25.0 17.0 26.5 23.0 29.0];
  % min norm inversion
> params=MT'\(YT')
```

$$a = -.29, \qquad b = .044, \qquad c = 29.5.$$

Hence multilinear regression predicts that the height a male can jump is given by the formula

$$\begin{aligned} &\text{jump height} \\ &\qquad \approx -.029 \times (\text{thigh circumference}) + 0.044 \times (\text{bench press}) + 29.5. \end{aligned} \tag{2.3.5}$$

Surprisingly, the coefficient of the thigh circumference term is negative, which suggests that thick thighs hinder vertical jumping ability.

## Exercises

**1.** This exercise will review some of the arithmetic for matrices and vectors:

```
MAPLE
> with(LinearAlgebra);
> A:=Matrix([[a,b],[c,d],[e,f]]); C:=Vector([c1,c2]);
```

```
MATLAB
> a=1; b=2; c=3; d=4; e=5; f=6; c1=7; c2=8;
> A=[a,b; c,d; e,f]
> C=[c1; c2]
```

Multiplication of the matrix $A$ and the vector $c$ produces a vector:

```
MAPLE
> A.C;
```

```
MATLAB
> A*C
```

An interchange of rows and columns of $A$ produces the *transpose* of $A$. A matrix can be multiplied by its transpose:

```
MAPLE
> Transpose(A).A;
```

```
MATLAB
> A'*A
```

**2.** Compute the solution for Example 2.3.1 using the matrix structure. The following syntax will accomplish this:

```
MAPLE
> with(LinearAlgebra):
> M:=Matrix([[19.36, 86, 1], [19.24, 94.5, 1], [20.76, 105.3, 1], [21.43, 91.5, 1], [18.72, 75.2, 1],
        [18.85, 93.2, 1], [21.84, 156.0, 1], [19.9, 75.1, 1], [21.35, 119.8, 1], [24.98, 169.3, 1],
        [24.46, 170., 1], [24.19, 148.2, 1]]);
> evalm(transpose(M)); # or Transpose(M)
```

```
> A:=evalm(transpose(M).M);
> z:=vector([19.3, 22.2, 24.3, 17.1, 19.6, 23.9, 29.5, 24.1, 26.2, 33.7, 36.2, 31.0]);
> y:=evalm(transpose(M).z);
> evalm(A^(-1).y);
```

MATLAB

```
> M1=[19.36 19.24 20.76 21.43 18.72 18.85 21.84 19.9 21.35 24.98 24.46 24.19];
> M2=[86.0 94.5 105.3 91.5 75.2 93.2 156.0 75.1 119.8 169.3 170 148.2];
> MT=[M1; M2; ones(1,12)];
  % each row = multiplier of a parameter
> M=MT' % transpose of MT
> A= MT*M % square 3x3 matrix
> z=[19.3; 22.2; 24.3; 17.1; 19.6; 23.9; 29.5; 24.1; 26.2; 33.7; 36.2; 31.0]; % 12x1 vector
> y=MT*z % 3x1 vector
> params=inv(A)*y
> MT'\z % same thing
```

**3.** (a) In this exercise, we get a linear regression fit for some hypothetical data relating age, percentage body fat, and maximum heart rate. (See Table 2.3.3.) Maximum heart rate is determined by having an individual exercise until near complete exhaustion.

**Table 2.3.3.** Data for age, % body fat, and maximum heart rate.

| Age (years) | % Body fat | Maximum heart rate |
|---|---|---|
| 30 | 21.3 | 186 |
| 38 | 24.1 | 183 |
| 41 | 26.7 | 172 |
| 38 | 25.3 | 177 |
| 29 | 18.5 | 191 |
| 39 | 25.2 | 175 |
| 46 | 25.6 | 175 |
| 41 | 20.4 | 176 |
| 42 | 27.3 | 171 |
| 24 | 15.8 | 201 |

The syntax that follows will get a linear regression fit for these data. This syntax will also produce a plot of the regression plane. Observe that it shows a steep decline in maximum heart rate as a function of age and a lesser decline with increased percentage body fat.

(b) As an example of the use of this regression formula, compare the predicted maximum heart rate for two persons at age 40 where one has maintained 15% body fat and the other has gained weight to 25% body fat. Also, compare two people with 20% body fat where one is age 40 and the other is age 50:

MAPLE

```
> age:=[30,38,41,38,29,39,46,41,42,24];
> BF:= [21.3,24.1,26.7,25.3,18.5,25.2,25.6,20.4,27.3,15.8];
> hr:=[186,183,172,177,191,175,175,176,171,201];
> with(stats):
> fit[leastsquare[[a,b,c]]]([age,BF,hr]);
> h:=unapply(rhs(%),(a,b));
> plot3d(h(a,b),a=30..60,b=10..20,axes=NORMAL);
> h(40,15); h(40,25); h(40,20); h(50,20);
```

```
MATLAB
> age=[30,38,41,38,29,39,46,41,42,24];
> BF=[21.2,24.1,36.7,25.3,18.5,25.2,25.6,20.4,27.3,15.8];
> hrt=[186,183,172,177,191,175,175,176,171,201];
> MT=[age;BF; ones(size(age))];
> parms=MT'\hrt'
> [Xage YBF]=meshgrid(age,BF);
> R=parms(1)*Xage+parms(2)*YBF+parms(3);
> C=ones(size(R)); % for a uniform color
> surf(age,BF,R,C) % surface graph
> h=[40 15 1]*parms
> h=[40 25 1]*parms
> h=[40 20 1]*parms
> h=[50 20 1]*parms
```

**4.** Table 2.3.4 contains further data to relate leg size, strength, and the ability to jump. These data were gathered for college women.

**Table 2.3.4.** Leg size, strength, and jumping ability for women.

| Thigh circumference (cm) | Leg press (lbs) | Vertical jump (in) |
|---|---|---|
| 52.0 | 140 | 13.0 |
| 54.2 | 110 | 8.5 |
| 64.5 | 150 | 13.0 |
| 52.3 | 120 | 13.0 |
| 54.5 | 130 | 13.0 |
| 58.0 | 120 | 13.0 |
| 48.0 | 95 | 8.5 |
| 58.4 | 180 | 19.0 |
| 58.5 | 125 | 14.0 |
| 60.0 | 125 | 18.5 |
| 49.2 | 95 | 16.5 |
| 55.5 | 115 | 10.5 |

Find a least squares data fit for these data, which are from unpublished work by Dr. Millard-Stafford in the Health and Performance Science Department at Georgia Tech.

## 2.4 Modeling with Differential Equations

*Understanding a natural process quantitatively often leads to a differential equation model. Consequently, a great deal of effort has gone into the study of differential equations. The theory of linear differential equations, in particular, is well known, and not without reason, since this type occurs widely.*

*Besides their exact solution in terms of functions, numerical and asymptotic solutions are also possible when exact solutions are not available.*

*In differential equations, as with organisms, there is need of a nomenclature.*

In Section 2.1, we proposed a simple differential equation for mimicking the growth of a biological population, namely,

$$\frac{dy}{dt} = ky. \tag{2.4.1}$$

A *differential equation* refers to any equation involving derivatives. Other examples are

$$\frac{d^2y}{dt^2} - 4\frac{dy}{dt} + 4y = e^{-t} \tag{2.4.2}$$

and

$$\frac{dy}{dt} = y - \frac{y^2}{2 + \sin t} \tag{2.4.3}$$

and many others. If only first-order derivatives appear in a differential equation, then it is called a *first-order* equation. Both equations (2.4.1) and (2.4.3) are of first order, but (2.4.2) is a second-order equation. Every first-order differential equation can be written in the form

$$\frac{dy}{dt} = f(t, y) \tag{2.4.4}$$

for some function $f$ of two variables. Thus $f(t, y) = ky$ in the first equation above and $f(t, y) = y - \frac{y^2}{2+\sin t}$ in the third.

A *solution* of a differential equation means a function $y = y(t)$ that satisfies the equation for all values of $t$ (over some specified range of $t$ values). Thus $y = Ae^{kt}$ is a solution of (2.4.1) because then $\frac{dy}{dt} = kAe^{kt}$, and substitution into (2.4.1) gives

$$kAe^{kt} = k(Ae^{kt}),$$

true for all $t$. Note that $A$ is a parameter of the solution and can be any value, so it is called an *arbitrary constant*. Recalling Section 2.1, $A$ arose as the constant of integration in the solution of (2.4.1). In general, the solution of a first-order differential equation will incorporate such a parameter. This is because a first-order differential equation is making a statement about the slope of its solution rather than the solution itself.

To fix the value of the inevitable arbitrary constant arising in the solution of a differential equation, a point in the plane through which the solution must pass is also specified, for example at $t = 0$. A differential equation along with such a side condition is called an *initial value problem*,

$$\frac{dy}{dt} = f(t, y) \quad \text{and} \quad y(0) = y_0. \tag{2.4.5}$$

It is not required to specify the point for which $t = 0$. It could be any other value of $t$ for which $y(t)$ is known. The *domain of definition*, or simply domain, of the differential equation is the set of points $(t, y)$ for which the right-hand side of (2.4.4) is defined. Often this is the entire $(t, y)$-plane.

*Initial value problems can be solved analytically.*

Exact solutions are known for many differential equations; cf. Kamke [5]. For the most part, solutions derive from a handful of principles. Although we will not study solution techniques here to any extent, we make two exceptions and discuss methods for linear systems below and the method of separation of variables next.

Actually we have already seen variables separable at work in Section 2.1: The idea is to algebraically modify the differential equation in such a way that all instances of the independent variable are on one side of the equation and all those of the dependent variable are on the other. Then the solution results as the integral of the two sides. For example, consider

$$\frac{dy}{dt} = ay - by^2.$$

Dividing by the terms on the right-hand side and multiplying by $dt$ separates the variables, leaving only the integration to be done:

$$\int \frac{dy}{y(a - by)} = \int dt.$$

Instead of delving into solution methods further, our focus in this text is deciding what solutions mean and which equations should constitute a model in the first place. Happily, some of the solution techniques, such as separation of variables, are sufficiently mechanical that computers can handle the job, relieving us for higher-level tasks. Here then are (symbolic) solutions to equations (2.4.2) and (2.4.3):

```
MAPLE
> restart:
> dsolve(diff(y(t),t,t)-4*diff(y(t),t)+4*y(t)=exp(-t),y(t));
```

$$y(t) = \frac{1}{9} + C_1 e^{2t} + C_2 t e^{2t}$$

and

```
MAPLE
> dsolve(diff(y(t),t)=y(t)-y(t)^2/(2+sin(t)), y(t));
```

$$\frac{1}{y(t)} = e^{-t} \int \frac{e^t}{2 + \sin(t)} dt + e^{-t} C_1.$$

*Initial value problems can be solved numerically.*

As mentioned above, (2.4.4) specifies the slope of the solution required by the differential equation at every point $(t, y)$ in the domain. This may be visualized by plotting a short line segment having that slope at each point. This has been done in Figure 2.4.1 for (2.4.3). Such a plot is called a *direction field*. Solutions to the equation must follow the field and cannot cross slopes. With such a direction field it is possible to sketch solutions manually. Just start at the initial point $(0, y(0))$ and follow the direction field. Keep in mind that a figure such as Figure 2.4.1 is only a

```
MAPLE
> with(DEtools):
> dfieldplot(diff(y(t),t)=y(t)-y(t)^2/(2+sin(t)),y(t), t=0..5,y=-1..5);
```

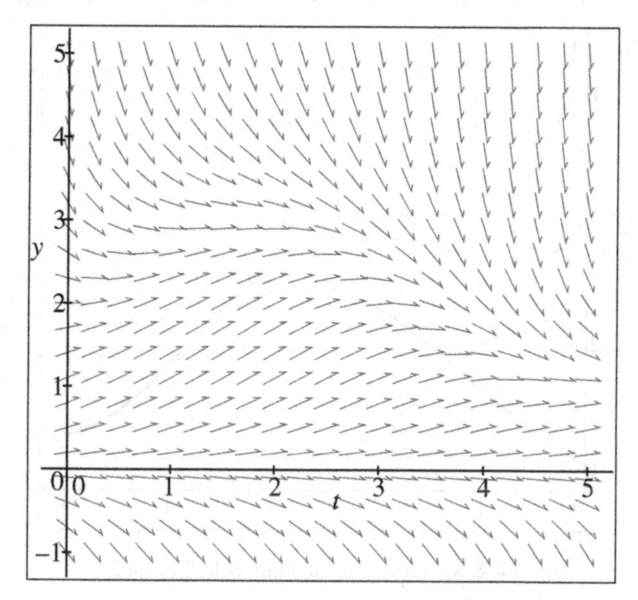


**Fig. 2.4.1.** Direction field for (2.4.3).

representation of the true direction field, that is to say, it shows only a small subset of the slope segments.

The mathematician Euler realized three centuries ago that the direction field could be used to numerically approximate solutions of an initial value problem in a precise way. Since Euler's time, techniques have improved—*Runge–Kutta methods* are used today—but the spirit of *Euler's method* is common to most of them; namely, the solution takes a small step $\Delta t$ to the right and $\Delta y$ up, where

$$\Delta y = f(t_i, y_i) \cdot \Delta t.$$

The idea is that $\frac{\Delta y}{\Delta t}$ approximates $\frac{dy}{dt}$. These increments are stepped off one after another,

$$y_{i+1} = y_i + \Delta y, \qquad t_{i+1} = t_i + \Delta t, \quad i = 0, 1, 2, \ldots,$$

with starting values $y_0 = y(0)$ and $t_0 = 0$. Figure 2.4.2 shows some numerical solutions of (2.4.3).

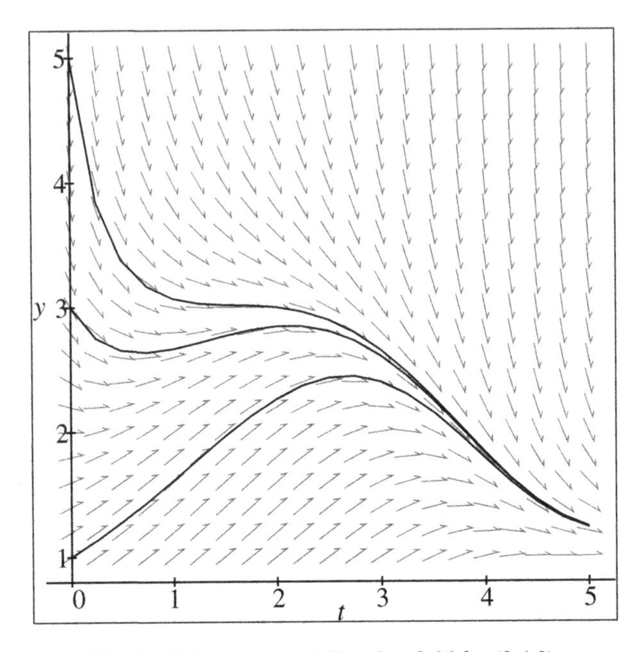


**Fig. 2.4.2.** Solutions and direction field for (2.4.3).

## Code 2.4.1.

```
MAPLE
> with(DEtools):
> DEplot(diff(y(t),t)=y(t)-y(t)^2/(2+sin(t)),y(t), t=0..5,{[0,1],[0,3],[0,5]}, linecolor=BLACK);

MATLAB
% make up an m-file, ode243.m, as follows
% function yprim=ode243(t,y)
% yprim = y - (y.^2./(2+sin(t)));
% now for the solution with initial value=1
> tspan=[0 5];
> [t1,y1]=ode23('ode243',tspan,1);
% and for initial value=3
> [t3,y3]=ode23('ode243',tspan,3);
% and for initial value=5
> [t5,y5]=ode23('ode243',tspan,5);
% plot them all
> plot(t1,y1,t3,y3,t5,y5);
```

*Linear differential equations are among the simplest kind.*

A differential equation that can be put into the form

$$a_n(t)\frac{d^n y}{dt^n} + \cdots + a_2(t)\frac{d^2 y}{dt^2} + a_1(t)\frac{dy}{dt} + a_0(t)y = r(t) \tag{2.4.6}$$

is *linear*. The coefficients $a_i(t)$, $i = 0, \ldots, n$, can be functions of $t$, as can the *right-hand side* $r(t)$. Equations (2.4.1) and (2.4.2) are linear but (2.4.3) is not. When there are multiplications among the derivatives or the dependent variable $y$, such as $y^2$, the differential equation will not be linear. If $y_1(t)$ and $y_2(t)$ are both solutions to a linear differential equation with right-hand side 0, then so is $Ay_1(t) + By_2(t)$ for any constants $A$ and $B$. Consider the first-order linear differential equation

$$\frac{dy}{dt} = my + R(t), \tag{2.4.7}$$

where we have taken $m = -\frac{a_0}{a_1}$ and $R(t) = \frac{r(t)}{a_1}$ in (2.4.6). Its solution is

$$y = Ae^{g(t)} + \Phi(t), \quad \text{where } g(t) = \int m dt. \tag{2.4.8}$$

In this, $A$ is the arbitrary constant and $\Phi$ is given below. To see this, first assume that $R$ is 0, and write the differential equation as

$$\frac{dy}{y} = m dt.$$

Now integrate both sides, letting $g(t) = \int m dt$ and $C$ be the constant of integration,

$$\ln y = g(t) + C, \quad \text{or} \quad y = Ae^{g(t)},$$

where $A = e^C$. By direct substitution, it can be seen that

$$\Phi = e^{g(t)} \int e^{-g(t)} R(t) dt \tag{2.4.9}$$

is a solution.[5] But it has no arbitrary constant, so add the two solutions, linearity allows this, to get (2.4.8). If $m$ is a constant, then $\int m dt = mt$.

To see that finding this solution is mechanical enough that a computer can handle the job, try these commands:

```
MAPLE
> dsolve(diff(y(t),t)=m(t)*y(t)+R(t),y(t));
> dsolve(diff(y(t),t)=m*y(t)+R(t),y(t));
```

*Systems of differential equations generalize their scalar counterparts.*

Quite often, modeling projects involve many more variables than two. Consequently it may require several differential equations to adequately describe the phenomenon. Consider the following model for small deviations about steady-state levels of a glucose/insulin system; $g$ denotes the concentration of glucose and $i$ the same for insulin,

$$\begin{aligned} \frac{dg}{dt} &= -\alpha g - \beta i + p(t), \\ \frac{di}{dt} &= \gamma g - \delta i. \end{aligned} \tag{2.4.10}$$

[5] A clever idea is to try a solution of the form $y = v(t)e^{g(t)}$ with $v(t)$ unknown and substitute this into (2.4.7) to get $v'e^{g(t)} = R(t)$, since the term $vg'e^{g(t)} = vme^{g(t)}$ drops out. Now solve for $v$.

As discussed in Section 2.1, the second equation expresses a proportionality relationship, namely, the rate of secretion of insulin increases in proportion to the concentration of glucose but decreases in proportion to the concentration of insulin. (Modeling coefficients are assumed to be positive unless stated otherwise.) The first equation makes a similar statement about the rate of removal of glucose, except that there is an additional term, $p(t)$, which is meant to account for ingestion of glucose. Because glucose and insulin levels are interrelated, each equation involves both variables. The equations define a system; the differential equations have to be solved simultaneously.

A system of differential equations can be written in vector form by defining a vector, say $\mathbf{Y}$, whose components are the dependent variables of the system. In vector notation, (2.4.10) becomes

$$\frac{d\mathbf{Y}}{dt} = M\mathbf{Y} + \mathbf{P}, \tag{2.4.11}$$

where the matrix $M$ and vector $\mathbf{P}$ are

$$M = \begin{bmatrix} -\alpha & -\beta \\ \gamma & -\delta \end{bmatrix}, \qquad \mathbf{P} = \begin{bmatrix} p(t) \\ 0 \end{bmatrix}.$$

Since the system (2.4.10) is linear, its vector expression takes on the simple matrix form of (2.4.11). Furthermore, this matrix system can be solved in the same way as the scalar differential equation (2.4.7). We have

$$\mathbf{Y} = e^{Mt}\mathbf{Y}_0 + e^{Mt}\int_0^t e^{-Ms}\mathbf{P}(\mathbf{s})ds. \tag{2.4.12}$$

Just as the exponential of the scalar product $mt$ is

$$e^{mt} = 1 + mt + \frac{m^2t^2}{2!} + \frac{m^3t^3}{3!} + \cdots, \tag{2.4.13}$$

so the exponential of the matrix product $Mt$ is

$$e^{Mt} = I + Mt + \frac{M^2t^2}{2!} + \frac{M^3t^3}{3!} + \cdots. \tag{2.4.14}$$

Since many properties of the exponential function stem from its power series expansion equation (2.4.13), the matrix exponential enjoys the same properties, in particular, the property that makes for the same form of solution,

$$\frac{d}{dt}e^{Mt}\mathbf{V(t)} = e^{Mt}\frac{d}{dt}\mathbf{V(t)} + e^{Mt}M\mathbf{V(t)}.$$

As in the case of a scalar differential equation, the system solutions can be plotted against $t$ to help us understand how the variables behave. For example, we could plot $g(t)$ and $i(t)$ using (2.4.12) (see Figure 2.4.3). But for a system there is an alternative; we can suppress $t$ and plot $i(t)$ against $g(t)$. This is done, conceptually, by making a table of values of $t$ and calculating the corresponding values of $g$ and $i$. But we only plot $(i, g)$ pairs. The coordinate plane of $i$ and $g$ is called the *phase plane* and the graph is called a *phase portrait* of the solution (see Figure 2.4.4).

MAPLE
```
> Gldeq:= diff(g(t),t)=-g(t)-i(t), diff(i(t),t)=-i(t)+g(t);
> sol:=dsolve({Gldeq, g(0)=1, i(0)=0},{g(t),i(t)}):
> g:= unapply(subs(sol,g(t)),t); i:= unapply(subs(sol,i(t)),t);
> plot({g(t),i(t)},t=0..5);
```

MATLAB
```
% Make up an m-file, fig243.m, as follows
% function Yprime=fig243(t,x)
% Yprime = [-x(1) - x(2); x(1) - x(2)];
% for the solution with initial value g=1 and i=0
> [t,Y]=ode23('fig243',[0 5],[1;0]); % semicolon for column vector
> plot(t,Y) % plot both columns of Y vs. t
```

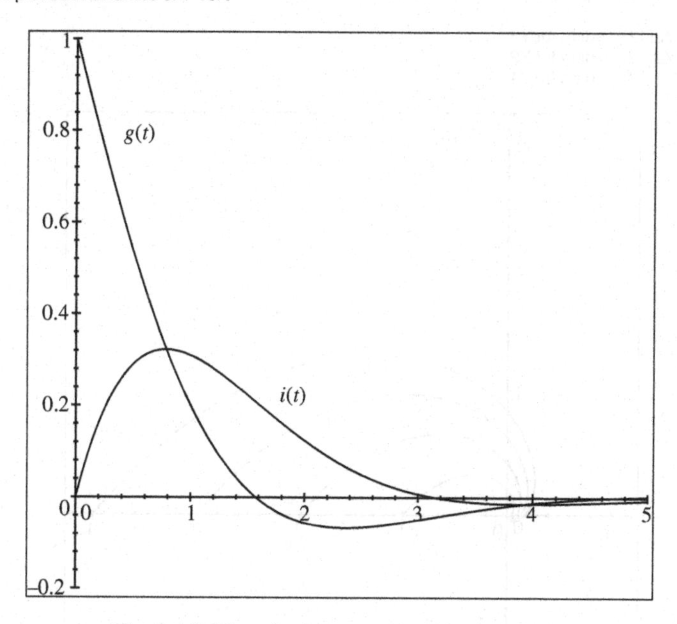


**Fig. 2.4.3.** Plot of solutions $g(t)$, $i(t)$ of (2.4.10).

*Asymptotics predict the ultimate course of the model.*

Often in science and engineering, we are interested in forecasting the future behavior of an observed process, $y(t)$. As $t$ becomes large there are several possibilities; among them are the following: $y$ can tend to a finite limit $y_\infty$, known as an *asymptotic limit*,

$$\lim_{t\to\infty} y(t) = y_\infty;$$

$y$ can tend to plus or minus infinity,

$$\lim_{t\to\infty} y(t) = \pm\infty;$$

$y$ can oscillate periodically; $y$ can oscillate unboundedly,

$$\lim_{t\to\infty} |y(t)| = \infty;$$

MAPLE
```
> restart:
> with(DEtools):
> GIdeq:= diff(g(t),t)=-g(t)-i(t), diff(i(t),t)=-i(t)+g(t);
> inits:={[0,1,0],[0,2,0],[0,3,0],[0,4,0]};
> phaseportrait([GIdeq],[g,i],t=0..4,inits, stepsize=.1,g=-1..4,i=-1..1.3);
```

MATLAB
```
> [t,Y4]=ode23('fig243',[0 5],[4;0]);
> [t,Y3]=ode23('fig243',[0 5],[3;0]);
> [t,Y2]=ode23('fig243',[0 5],[2;0]);
> [t,Y1]=ode23('fig243',[0 5],[1;0]);
> plot(Y4(:,1),Y4(:,2)) % plot the first component of Y4 against the second
> hold on
> plot(Y3(:,1),Y3(:,2)) %ditto for Y3
> plot(Y2(:,1),Y2(:,2)) %ditto for Y2
> plot(Y1(:,1),Y1(:,2)) %ditto for Y1
```

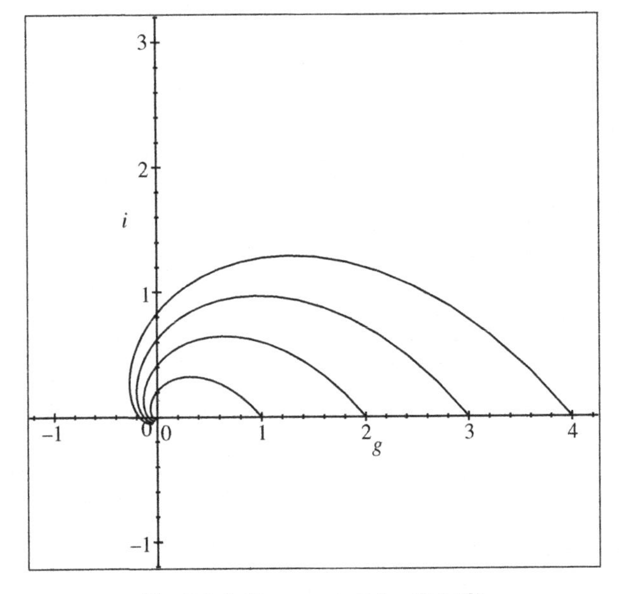


**Fig. 2.4.4.** Phase portrait for (2.4.10).

or $y$ can oscillate chaotically. If $y$ is part of a system, its fate will be linked to that of the other variables; in this case, we inquire about the vector solution $\mathbf{Y}$.

In the simplest case, $\mathbf{Y}$ has asymptotic limits. If the system is *autonomous*, meaning $t$ appears nowhere in the system (except, of course, in the form $\frac{d}{dt}$), then to find the asymptotic limits, set all the derivatives of the system to zero. Solutions of the resulting algebraic system are called *critical points* or *stationary points*.[6] In the glucose/insulin example, suppose the glucose ingestion term, $p(t)$, were constant at $p$; then setting the derivatives to zero leads to the algebraic system

$$\begin{aligned} 0 &= -\alpha g - \beta i + p, \\ 0 &= \gamma g - \delta i. \end{aligned} \tag{2.4.15}$$

[6] These are also called *equilibrium points* by some authors.

MAPLE

```
> solve({-alpha*g-beta*i+p=0,gamma*g-delta*i=0},{g,i});
```

Its one critical point is $g = -\frac{\delta p}{\gamma\beta+\alpha\delta}$, $i = \frac{\gamma p}{\gamma\beta+\alpha\delta}$. If this point is taken as the initial point of the system, then for all time, $g$ will be $\frac{\delta p}{\gamma\beta+\alpha\delta}$ and $i$ will be $\frac{\gamma p}{\gamma\beta+\alpha\delta}$.

It is not necessarily the case that a stationary point is also an asymptotic limit. Exponential growth, $\frac{dy}{dt} = y$, is an example, since $y = 0$ is a stationary point, but if $y(0) \neq 0$, then $y \to \infty$ as $t \to \infty$. On the other hand, when it can be shown that the solution of a system tends to an asymptotic limit, a giant step has been taken in understanding the system. For example, exponential decay, $\frac{dy}{dt} = -y$, has asymptotic limit 0 for any starting point $y(0)$, for if $y > 0$, then $\frac{dy}{dt}$ is negative, so $y$ will decrease. Similarly, if $y < 0$, then $\frac{dy}{dt} > 0$, so $y$ will increase. Either way, 0 is the asymptotic limit.

A complication here is that the existence or the value of the asymptotic limit can often depend on the starting point $\mathbf{Y}(0)$. Given that there is an asymptotic limit, $\mathbf{Y}_\infty$, the set of all starting points for which the solution tends to $\mathbf{Y}_\infty$ is called its *basin of attraction*, $\mathbf{B}_{\mathbf{Y}_\infty}$,

$$\mathbf{B}_{\mathbf{Y}_\infty} = \left\{\mathbf{Y}_0 : \lim_{t\to\infty} \mathbf{Y}(t) = \mathbf{Y}_\infty \text{ when } \mathbf{Y}(0) = \mathbf{Y}_0\right\}.$$

If the basin of attraction of a system is essentially the entire domain of definition, the asymptotic limit is said to be *global*. By way of example, the differential equation $\frac{dy}{dt} = -y(1-y)$ has asymptotic limit $y = 0$ for solutions starting from $-\infty < y_0 < 1$; but when the starting point is beyond 1, solutions tend to infinity.

Periodicity is a more complicated asymptotic behavior. Further, just as in the asymptotic limit case, the solution can start out periodic, or can asymptotically tend to periodicity. An example of the former is $\frac{dy}{dt} = \cos t$, while the latter behavior is demonstrated by $\frac{dy}{dt} = -y + \cos t$. This second differential equation is solved by (2.4.8), $y = Ae^{-t} + \frac{1}{2}(\cos t + \sin t)$; $A$ depends on the initial condition, but the whole term tends to zero. A well-known periodic system is the one due to Lotka and Volterra modeling predator–prey interaction. We study this in Section 4.4.

## Exercises

**1.** Here are four differential equations with the same initial conditions:

$$\frac{d^2y}{dt^2} + 6y(t) = 0, \quad y(0) = 1, \quad y'(0) = 0;$$

$$\frac{d^2y}{dt^2} - 6y(t) = 0, \quad y(0) = 1, \quad y'(0) = 0;$$

$$\frac{d^2y}{dt^2} + 2\frac{dy}{dt} + 6y(t) = 0, \quad y(0) = 1, \quad y'(0) = 0;$$

$$\frac{d^2y}{dt^2} - 2\frac{dy}{dt} + 6y(t) = 0, \quad y(0) = 1, \quad y'(0) = 0.$$

While these differential equations have a similar appearance, they have radically different behaviors. Sketch the graphs of all four equations with the same initial values. Here is syntax that will draw the graphs:

```
MAPLE
> dsolve({diff(y(t),t,t)+6*y(t)=0, y(0)=1, D(y)(0)=0},y(t));
> y1:=unapply(rhs(%),t);
> dsolve({diff(y(t),t,t)-6*y(t)=0, y(0)=1, D(y)(0)=0},y(t));
> y2:=unapply(rhs(%),t);
> dsolve({diff(y(t),t,t)+2*diff(y(t),t)+6*y(t)=0, y(0)=1,D(y)(0)=0},y(t));
> y3:=unapply(rhs(%),t);
> dsolve({diff(y(t),t,t)-2*diff(y(t),t)+6*y(t)=0, y(0) = 1,D(y)(0)= 0},y(t));
> y4:=unapply(rhs(%),t);
> plot([y1(t),y2(t),y3(t),y4(t)],t=0..4,y=-5..5, color=[black,blue,green,red]);
```

```
MATLAB
% to deal with a second-order differential equation, it has to be made into a vector-valued first-order
% DE as follows: the first component Y1 is y and the second Y2 is dy/dt. Then d^2y/dt^2+6y=0
% becomes the vector system dY1/dt=Y2; dY2/dt = -6Y1;
% so make an m-file, exer241a.m, as follows:
%    function Yprime=exer241a(t,Y); Yprime = [Y(2); -6*Y(1)];
> [t,Y]=ode23('exer241a',[0 4],[1; 0]);
> plot(t,Y(:,1))
%%%
% DE (b) converts to first-order vector system dY1/dt=Y2; dY2/dt=6*Y1;
% DE (c) converts to first-order vector system dY1/dt=Y2; dY2/dt=-6*Y1-2*Y2;
% DE (d) converts to first-order vector system dY1/dt=Y2; dY2/dt=-6*Y1+2*Y2;
% We leave it to the reader to obtain the numerical solutions and plot as above.
```

**2.** We illustrate four ways to visualize solutions to a single second-order differential equation in order to emphasize that different perspectives provide different insights. We use the same equation in all four visualizations:

$$\frac{d^2y}{dt^2} + \frac{y(t)}{5} = \cos(t).$$

(a) Find and graph an analytic solution that starts at $y(0) = 0$.

```
MAPLE
> dsolve({diff(y(t),t,t)+y(t)/5=cos(t), y(0)=0,D(y)(0)=0},y(t));
> y:=unapply(rhs(%),t);
> plot(y(t),t=0..4*Pi);
```

```
MATLAB
% make an m-file, exer242.m, with
%   function Yprime=exer242(t,Y); Yprime=[Y(2); -Y(1)/5+cos(t)];
% then solve and plot with
> [t,Y]=ode23('exer242',[0 4*pi],[0;0]);
> plot(t,Y(:,1))
```

(b) Give a direction field for the equation.

```
MAPLE
> restart: with(DEtools):
> dfieldplot(diff(y(t),t)+y(t)/5=cos(t),y(t),t=0..4*Pi,y=-1..5);
```

```
MATLAB
% No built-in direction field in Matlab; see DFIELD from http://math.rice.edu/~dfield.
```

(c) Give several trajectories overlaid in the direction field.

```
MAPLE (direction field)
> restart:
> with(DEtools):
> DEplot(diff(y(t),t)+y(t)/5=cos(t),y(t),t=0..4*Pi,{[0,1],[0,3],[0,5]});
```

(d) Give an animation to show the effect of the coefficient of $y(t)$ changing.

```
MAPLE (animation)
> restart: with(plots):
> for n from 1 to 8 do
  a:=n/10:
  dsolve({diff(y(t),t,t)+a*y(t)/5=cos(t),y(0)=0},y(t)):
  y:=unapply(rhs(%),t):
  P[n]:=plot([t,y(t),t=0..10*Pi],t=0..10*Pi):
  y:='y':
  od:
> display([seq(P[n],n=1..8)],insequence=true);
```

**3.** Find the critical points for each of the following equations. Plot a few trajectories to confirm where the basins of attractions are.

(a) $\frac{dy}{dt} = -y(t)(1 - y(t))$.

```
MAPLE
> solve(y*(1-y)=0,y);
> with(DEtools):
> de:=diff(y(t),t)=-y(t)*(1-y(t));
> DEplot(de,y(t),t=0..5,{[0,-1],[0,-1/2],[0,1/2]},y=-1..2);

MATLAB
% make an m-file, exer243a.m, with
%   function yprime=exer243a(t,y); yprime=-y.*(1-y);
> p=[1 -1 0]; % coefficients of p(y)=-y(1-y)
> roots(p)
> [t,y]=ode23('exer243a',[0 5],-1);
> plot(t,y); hold on
> [t,y]=ode23('exer243a',[0 5],-1/2);
> plot(t,y)
> [t,y]=ode23('exer243a',[0 5],1/2);
> plot(t,y)
```

(b) $x' = 4x(t) - x\hat{\ }2(t) - x(t)y(t)$; $y' = 5y(t) - 2y\hat{\ }2(t) - x(t)y(t)$.

```
MAPLE
> solve({4*x-x^2-x*y=0, 5*y-2*y^2-y*x=0}, {x,y});
> with(DEtools):
> deq1:=diff(x(t),t)=4*x(t)-x(t)^2-x(t)*y(t);
> deq2:=diff(y(t),t)= 5*y(t)-2*y(t)^2-y(t)*x(t);
> inits:={[0,1,1],[0,1,4],[0,4,1],[0,4,4]};
> DEplot([deq1,deq2],[x,y],t=0..4,inits,x=-1..5,y=-1..5,stepsize=.05);

MATLAB
% contents of m-file, exer243b.m:
% function Yprime=exer243b(t,Y);
% Yprime=[4*Y(1)-Y(1).*Y(1)-Y(1).*Y(2); 5*Y(2)-2*Y(2).^2-Y(1).*Y(2)];
> [t,Y]=ode23('exer243b',[0 4],[1;1]);
> hold off; plot3(t,Y(:,1),Y(:,2))
> grid
> xlabel('x axis'); ylabel('y axis');
> zlabel('z axis'); hold on
> [t,Y]=ode23('exer243b',[0 4],[1;4]);
> plot3(t,Y(:,1),Y(:,2))
> [t,Y]=ode23('exer243b',[0 4],[4;1]);
> plot3(t,Y(:,1),Y(:,2))
> [t,Y]=ode23('exer243b',[0 4],[4;4]);
> plot3(t,Y(:,1),Y(:,2))
> view(30,30) % 30 deg CCW from negative y-axis, 30 deg elevation
> view(-100,30) % 100 deg CW from negative y-axis, 30 deg elevation
```

**4.** The solution for $Z' = AZ(t)$, $Z(0) = C$, with $A$ a constant square matrix and $C$ a vector is $\exp(At)C$. Compute this exponential in the case

$$A = \begin{pmatrix} -1 & -1 \\ 1 & -1 \end{pmatrix}.$$

Evaluate $\exp(At)C$, where $C$ is the vector

$$C = \begin{pmatrix} 1 \\ 0 \end{pmatrix}.$$

```
MAPLE
> with(LinearAlgebra):
> A:=Matrix([[-1,-1],[1,-1]]);
> MatrixExponential(A,t);
> evalm(%.[1,0]);

MATLAB
> A=[-1, -1; 1, -1]
> t=2; At=A*t; expm(At)
> t=5; At=A*t; expm(At)
> expm(At)*[1;0] % exp(At)*C, where C is a 2x1 column vector
```

## 2.5 Modeling with Difference Equations

Biological systems are not always continuous. Considering population growth, individuals come in discrete units, so a differential equation model for population growth is only an approximation. When population size is large, the approximation is sufficiently accurate to describe the model's behavior and asymptotics. But there are many biological phenomena whose analysis requires a treatment in terms of discrete units.

Difference equations are similar to differential equations except that the independent variable, time or space, is taken in discrete, indivisible units. Although difference equation analysis is often more difficult than its continuous counterpart, there is a striking analogy between the two theories.

Difference equations are one example of what is more generally known as *recurrence relations*. This refers to some quantity that is defined in terms of itself.

Just as numerical and asymptotic analyses are available for differential equations, the same holds for difference equations as well.

*A differential equation has a natural difference equation counterpart.*

In Section 2.1 we mentioned a differential equation model for population growth,

$$\frac{dy}{dt} = ky. \tag{2.5.1}$$

This model postulates that infinitesimal units of population, $dy$, are added to the general population over infinitesimal units of time, $dt$. Of course this can only be an approximation. And indeed it is an adequate one in many cases, for example, for describing a bacterial colony.

However, for a more accurate description, an approach respecting that biological units are discrete and reproductive intervals are also discrete is called for. We are led to the discrete version of (2.5.1),

$$y_{t+1} - y_t = r y_t.$$

Here the variable $t$ proceeds in discrete units $t = 0, 1, 2, \ldots$. As in the differential equation, a starting value $y_0$ is required to complete the description.

To solve the difference equation we recast it as a recurrence relation together with a starting value (denote this by $y_0$),

$$y_{t+1} = (1 + r) y_t, \quad y_0 = \text{starting value}.$$

The solution is easy to obtain by stepping through the generations recurrently,

$$y_1 = (1 + r) y_0,$$
$$y_2 = (1 + r) y_1 = (1 + r)^2 y_0,$$
$$y_3 = (1 + r) y_2 = (1 + r)^3 y_0,$$

and so on. It is easy to see that there is a closed (nonrecurrent) form for the $y_t$, namely,

$$y_t = (1 + r)^t y_0, \quad t = 0, 1, 2, \ldots.$$

Comparing this with the solution of the differential equation,

$$y = e^{kt} y_0 = (e^k)^t y_0,$$

shows that $e^k$ corresponds to $1 + r$. The relationship between the per period growth rate $r$ and the instantaneous growth rate $k$ is

$$r = e^k - 1 \quad \text{, or} \quad k = \log(1 + r). \tag{2.5.2}$$

A second-order differential equation such as

$$\frac{d^2 y}{dt^2} - 4\frac{dy}{dt} + 4y = 0 \tag{2.5.3}$$

can be written as a difference equation by noting how the second derivative converts. Since $\frac{d^2y}{dt^2} = \frac{d}{dt}(\frac{dy}{dt})$, we may write

$$\frac{d^2 y}{dt^2} \to \left.\frac{dy}{dt}\right|_{t+1} - \left.\frac{dy}{dt}\right|_t$$
$$\to (y_{t+2} - y_{t+1}) - (y_{t+1} - y_t) = y_{t+2} - 2y_{t+1} + y_t.$$

Then (2.5.3) becomes

$$y_{t+2} - 2y_{t+1} + y_t - 4(y_{t+1} - y_t) + 4y_t = 0.$$

This may be written as the linear recurrence relation

$$y_{t+2} - 6y_{t+1} + 9y_t = 0.$$

Just as a second-order differential equation requires two initial values for a complete solution, so also a second-order recurrence relation requires two initial values for a complete solution.

The general second-order (homogeneous) recurrence relation is

$$c_2 y_{t+2} + c_1 y_{t+1} + c_0 y_t = 0 \tag{2.5.4}$$

for some constants $c_2$, $c_1$, and $c_0$. On the strength of what we saw above, we expect a solution of the form $y_t = Ar^t$ for some $r$. Substitute this into (2.5.4):

$$c_2 A r^{t+2} + c_1 A r^{t+1} + c_0 A r^t = 0.$$

Factoring out $Ar^t$ gives

$$Ar^t(c_2 r^2 + c_1 r + c_0) = 0.$$

This is satisfied trivially if $A = 0$ or if $r$ solves the quadratic equation

$$c_2 r^2 + c_1 r + c_0 = 0. \tag{2.5.5}$$

This is called the *auxiliary equation.*

Suppose (2.5.5) has two distinct real roots, $r = r_1$ and $r = r_2$, then the homogeneous equation has the solution

$$y_t = A r_1^t + B r_2^t \tag{2.5.6}$$

for some constants $A$ and $B$. These will be determined by the initial conditions.

Consider the equation due to Fibonacci for the growth of a rabbit population. He stated that the size of the population in terms of reproducing pairs at generation $t$ is the sum of the sizes of the last two generations, that is,

$$y_t = y_{t-1} + y_{t-2}, \quad t = 3, 4, \ldots, \tag{2.5.7}$$

or equivalently,

$$y_{t+2} = y_{t+1} + y_t, \qquad t = 1, 2, \ldots.$$

Starting with one (juvenile) pair, after one breeding period these become adults, so there is still one pair. But in the next breeding period they produce one new juvenile pair, so now there are two pairs of rabbits. In general, the population sequence according to (2.5.7) is

$$1, 1, 2, 3, 5, 8, 13, 21, 34, \ldots.$$

To find a closed-form solution, we use the method above. Transpose the terms on the right side of the equal sign to the left. That leads us to solve the quadratic equation

$$r^2 - r - 1 = 0.$$

From the quadratic formula, the roots are

$$r = \frac{1}{2}(1 \pm \sqrt{1 - (-4)}),$$

and so the solution is

$$y_t = A\left(\frac{1+\sqrt{5}}{2}\right)^t + B\left(\frac{1-\sqrt{5}}{2}\right)^t. \tag{2.5.8}$$

Using the starting values $y_1 = y_2 = 1$ as above, substitute into (2.5.8), first with $t = 1$ and then with $t = 2$:

$$1 = A\left(\frac{1+\sqrt{5}}{2}\right) + B\left(\frac{1-\sqrt{5}}{2}\right),$$

$$1 = B\left(\frac{1+\sqrt{5}}{2}\right)^2 + B\left(\frac{1-\sqrt{5}}{2}\right)^2.$$

Finally, solve this system of two equations in two unknowns (using Code 2.5.1, for example) to get $A = \frac{1}{\sqrt{5}}$ and $B = -\frac{1}{\sqrt{5}}$.

**Code 2.5.1.**

MAPLE

```
> eq1:=1=A*((1+sqrt(5))/2)+B*((1-sqrt(5))/2);
> eq2:=1=A*((1+sqrt(5))/2)^2+B*((1-sqrt(5))/2)^2;
> solve({eq1,eq2},{A,B});
```

MATLAB

```
> M=[(1+sqrt(5))/2 (1-sqrt(5))/2; ((1+sqrt(5))/2)^2 ((1-sqrt(5))/2)^2]
> b=[1;1]
> sol=M\b
```

Hence

$$y_t = \frac{1}{\sqrt{5}}\left(\left(\frac{1+\sqrt{5}}{2}\right)^t - \left(\frac{1-\sqrt{5}}{2}\right)^t\right). \tag{2.5.9}$$

What happens to $y_t$ as $t \to \infty$? Since $\frac{1-\sqrt{5}}{2} = -0.618\ldots$ is less than 1 in absolute value, this quantity raised to the $t$th power tends to 0 as $t \to \infty$. Therefore,

$$y_t \approx \frac{1}{\sqrt{5}}\left(\frac{1+\sqrt{5}}{2}\right)^t$$

for large $t$. In fact, rounding this approximation to the nearest integer is exact for all $t$.

If the roots of the auxiliary equation are repeated, say $r = r_1$ with multiplicity 2, then one must use a solution of the form

$$y_t = Ar_1^t + Btr_1^t$$

instead of (2.5.6). As before, use the starting values to find the constants $A$ and $B$.

*Systems of equations lead to a higher-order single-variable equation.*

Consider the following system of two recurrence relations:

$$x_{t+1} = c_{11}x_t + c_{12}y_t, \tag{2.5.10a}$$

$$y_{t+1} = c_{21}x_t + c_{22}y_t. \tag{2.5.10b}$$

The first may be written as

$$x_{t+2} = c_{11}x_{t+1} + c_{12}y_{t+1}.$$

Now the second may be substituted into this to give

$$x_{t+2} = c_{11}x_{t+1} + c_{12}(c_{21}x_t + c_{22}y_t).$$

Finally, use (2.5.10a) to eliminate $y_t$ from this equation:

$$\begin{aligned} x_{t+2} &= c_{11}x_{t+1} + c_{12}c_{21}x_t + c_{22}(x_{t+1} - c_{11}x_t) \\ &= (c_{11} + c_{22})x_{t+1} - (c_{11}c_{22} - c_{12}c_{21})x_t. \end{aligned}$$

*Chaos*

Consider the *logistic recurrence relation*

$$y_{t+1} = \lambda y_t(1 - y_t), \tag{2.5.11}$$

where $\lambda$ is a constant. This equation arises in the study of population growth. For values of $\lambda$ less than 3, this equation converges to a unique asymptotic value. But if $\lambda$ is greater than 3, strange behavior is exhibited. For example, if $\lambda$ is 4 or greater, the $y_t$s are seemingly random values. More precisely, this is called *chaos* rather than random because the values are correlated; truly random values must be uncorrelated. For $3 \le \lambda < 1 + \sqrt{6}$, the $y_t$s asymptotically oscillate between two values, called a *2-cycle*. For values of $\lambda$ between $1 + \sqrt{6}$ and 4, cycles of various periods are encountered. The following code produces fully chaotic behavior:

```
MAPLE
> lam:=4:
> chaos:=proc() global y;
> y:= lam*y*(1-y);
> RETURN(y);
> end:
> y:=.05:
> for i from 1 to 24 do chaos();
> od;
```

```
MATLAB
> lam=4; y=0.05; for i=1:24 y=lam*y*(1-y)
> end
```

## 2.6 Matrix Analysis

*The easiest kind of matrix to understand and with which to calculate is a diagonal matrix $J$, that is, one whose $ik$th term is zero, $j_{ik} = 0$, unless $i = k$. The product of* *two diagonal matrices is again diagonal. The diagonal terms of the product are just the products of the diagonal terms of the factors. This pattern extends to all powers, $J^r$, as well. As a consequence, the exponential of a diagonal matrix is just the matrix of exponentials of the diagonal terms.*

*It might seem that diagonal matrices are rare, but the truth is quite to the contrary. For most problems involving a matrix, say A, there is a* change of basis matrix *P such that $PAP^{-1}$ is diagonal. We exploit this simplification to make predictions about the asymptotic behavior of solutions of differential equations.*

*Eigenvalues predict the asymptotic behavior of matrix models.*

Every $n \times n$ matrix $A$ has associated with it a unique set of $n$ complex numbers, $\lambda_1, \lambda_2, \ldots, \lambda_n$, called *eigenvalues*. Repetitions are possible, so the eigenvalues for $A$ might not be distinct, but even with repetitions, there are always exactly $n$ in number. In turn, each eigenvalue $\lambda$ has associated with it a nonunique vector $\mathbf{e}$ called an *eigenvector*. An eigenvalue–eigenvector pair $\lambda, \mathbf{e}$ is defined by the matrix equation

$$A\mathbf{e} = \lambda\mathbf{e}. \tag{2.6.1}$$

An eigenvector for $\lambda$ such as $\mathbf{e}$ is not unique, because for every number $a$, the vector $\mathbf{e}' = a\mathbf{e}$ is also an eigenvector, as is easily seen from (2.6.1).

**Example 2.6.1.** The matrix

$$A = \begin{bmatrix} 1 & 3 \\ 0 & -2 \end{bmatrix}$$

has eigenvalues $\lambda_1 = 1$ and $\lambda_2 = -2$ with corresponding eigenvectors $\mathbf{e}_1 = \binom{1}{0}$ and $\mathbf{e}_2 = \binom{1}{-1}$. Before invoking the computer on this one (see Exercise 1 in this section), work through it by hand.

*Eigenvalues and eigenvectors play a central role in every mathematical model embracing matrices.*

This statement cannot be overemphasized. The reason is largely a consequence of the following theorem.

**Theorem 1.** *Let the $n \times n$ matrix $A$ have $n$ distinct eigenvalues; then there exists a nonsingular matrix $P$ such that the matrix*

$$J = PAP^{-1} \tag{2.6.2}$$

*is the diagonal matrix of the eigenvalues of $A$,*

$$J = \begin{bmatrix} \lambda_1 & 0 & \ldots & 0 \\ 0 & \lambda_2 & \ldots & 0 \\ \vdots & \vdots & \ldots & \vdots \\ 0 & 0 & \ldots & \lambda_n \end{bmatrix}.$$

*The columns of $P$ are the eigenvectors of $A$ taken in the same order as the list of eigenvalues.*

If the eigenvalues are not distinct, then we are not guaranteed that there will be a completely diagonal form; it can happen that there is not one. But even if not, there is an almost diagonal form, called the *Jordan canonical form* (or just Jordan form), which has a pattern of 1s above the main diagonal. By calculating the Jordan form of a matrix, we get the diagonal form if the matrix has one. We will not need to discuss Jordan form here, except to say that the computer algebra system can compute it.

The matrix product of this theorem, $PAP^{-1}$, is a change of basis modification of $A$; in other words, by using the eigenvectors as the reference system, the matrix $A$ becomes the diagonal matrix $J$. Note that if $J = PAP^{-1}$, then the $k$th power of $J$ and $A$ are related as the $k$-fold product of $PAP^{-1}$,

$$J^k = (PAP^{-1})(PAP^{-1})\cdots(PAP^{-1}) = PA^kP^{-1}, \tag{2.6.3}$$

since the interior multiplications cancel.

Diagonal matrices are especially easy to work with; for example, to raise $J$ to a power $J^k$ becomes raising the diagonal entries to that power:

$$J^k = \begin{bmatrix} \lambda_1^k & 0 & \dots & 0 \\ 0 & \lambda_2^k & \dots & 0 \\ \vdots & \vdots & \dots & \vdots \\ 0 & 0 & \dots & \lambda_n^k \end{bmatrix}.$$

As a result, the exponential of $J$ is just the exponential of the diagonal entries. From (2.4.14),

$$\begin{aligned} e^{Jt} &= I + Jt + \frac{J^2t^2}{2!} + \frac{J^3t^3}{3!} + \cdots \\ &= \begin{bmatrix} \left(1 + \lambda_1 t + \frac{\lambda_1^2t^2}{2!} + \cdots\right) & 0 & \dots & 0 \\ 0 & \left(1 + \lambda_2 t + \frac{\lambda_2^2t^2}{2!} + \cdots\right) & \dots & 0 \\ \vdots & \vdots & \dots & \vdots \\ 0 & 0 & \dots & \left(1 + \lambda_n t + \frac{\lambda_n^2t^2}{2!} + \cdots\right) \end{bmatrix} \\ &= \begin{bmatrix} e^{\lambda_1 t} & 0 & \dots & 0 \\ 0 & e^{\lambda_2 t} & \dots & 0 \\ \vdots & \vdots & \dots & \vdots \\ 0 & 0 & \dots & e^{\lambda_n t} \end{bmatrix}. \end{aligned} \tag{2.6.4}$$

We illustrate the way in which these results are used.

The age structure of a population can be modeled so that it evolves as dictated by a matrix $L$, such as the following (see Chapter 5):

$$L = \begin{bmatrix} 0 & 0 & 0 & 0 & 0.08 & 0.28 & 0.42 \\ .657 & 0 & 0 & 0 & 0 & 0 & 0 \\ 0 & .930 & 0 & 0 & 0 & 0 & 0 \\ 0 & 0 & .930 & 0 & 0 & 0 & 0 \\ 0 & 0 & 0 & .930 & 0 & 0 & 0 \\ 0 & 0 & 0 & 0 & .935 & 0 & 0 \\ 0 & 0 & 0 & 0 & 0 & .935 & 0 \end{bmatrix}.$$

After $k$ generations, the pertinent matrix is the $k$th power of $L$. From the theorem, there exists a matrix $P$ such that $J = PLP^{-1}$, and according to (2.6.3),

$$L^k = P^{-1} J^k P.$$

Letting $\lambda_1$ be the largest eigenvalue of $L$ in absolute value, it is easy to see that

$$\frac{1}{\lambda_1^k} J^k = \begin{bmatrix} 1 & 0 & \dots & 0 \\ 0 & \left(\frac{\lambda_2}{\lambda_1}\right)^k & \dots & 0 \\ \vdots & \vdots & \dots & \vdots \\ 0 & 0 & \dots & \left(\frac{\lambda_n}{\lambda_1}\right)^k \end{bmatrix}$$

$$\longrightarrow \begin{bmatrix} 1\,0 \dots 0 \\ 0\,0 \dots 0 \\ \vdots\,\vdots \dots \vdots \\ 0\,0 \dots 0 \end{bmatrix} \quad \text{as } k \to \infty.$$

In other words, for large $k$, $L^k$ is approximately $\lambda_1^k$ times a fairly simple fixed matrix related to its eigenvectors; thus it grows or decays like $\lambda_1^k$.

In another example, consider the matrix form of the linear differential equation (2.4.11) of Section 2.4. From above, the matrix exponential $e^{Mt}$ can be written as

$$e^{Mt} = P^{-1} e^{Jt} P,$$

where $e^{Jt}$ consists of exponential functions of the eigenvalues. If all those eigenvalues are negative, then no matter what $P$ is, every solution will tend to 0 as $t \to \infty$. But if one or more eigenvalues are positive, then at least one component of a solution will tend to infinity.

In Chapter 9, we will consider compartment models. A *compartment* matrix $C$ is defined as one whose terms $c_{ij}$ satisfy the following conditions:

1. All diagonal terms $c_{ii}$ are negative or zero.
2. All other terms are positive or zero.
3. All column sums $\sum_i c_{ij}$ are negative or zero.

Under these conditions, it can be shown that the eigenvalues of $C$ have negative or zero real parts and so the asymptotic result above applies.

The fact that the eigenvalues have negative real parts under the conditions of a compartment matrix derives from *Gershgorin's circle theorem.*

**Theorem 2.** *If $A$ is a matrix and $S$ is the following union of circles in the complex plane,*

$$S = \bigcup_m \left\{ complex\ z : |a_{mm} - z| \leq \sum_{j \neq m} |a_{jm}| \right\},$$

*then every eigenvalue of $A$ lies in $S$.*

Notice that the $m$th circle above has center $a_{mm}$ and radius equal to the sum of the absolute values of the other terms of the $m$th column.

## Exercises

1. For both the following matrices $A$, find the eigenvalues and eigenvectors. Then find the Jordan form. Plot solutions $[Z_1, Z_2, Z_3]$ for $Z' = AZ$. Note that the Jordan structure for the two is different:

$$A_1 = \begin{pmatrix} 0 & 0 & -2 \\ 1 & 2 & 1 \\ 1 & 0 & 3 \end{pmatrix} \quad \text{and} \quad A_2 = \begin{pmatrix} 3 & 1 & -1 \\ -1 & 2 & 1 \\ 2 & 1 & 0 \end{pmatrix}.$$

Here is the syntax for $A_1$. Define the following matrix:

```
MAPLE
> restart;
> with(LinearAlgebra):
> A:=Matrix([[0,0,-2],[1,2,1],[1,0,3]]);
```

```
MATLAB
> A=[0 0 -2; 1 2 1; 1 0 3]
```

(a) Find the eigenvalues and eigenvectors of $A$. (Note that both $x_1 = (-1, 0, 1)^t$ and $x_2 = (0, 1, 0)^t$ are eigenvectors for the eigenvalue 2; therefore, so is every linear combination $ax_1 + bx_2$.)

```
MAPLE
> ev:=Eigenvectors(A);
  # first column = eigenvalues, second “column” = matrix whose columns are eigenvectors
> evals:=ev[1]; evects:=ev[2];
  # evects[1] is a row, we want the column; transpose
> whattype(Transpose(evects)[1]); # a row vector, needs to be a colmn vector
> x1:=convert(Transpose(evects)[1],Vector[column]);
> x2:=convert(Transpose(evects)[2],Vector[column]);
> x3:=convert(Transpose(evects)[3],Vector[column]);
```

```
MATLAB
> [evect, eval]=eig(A)
  % evect is P inverse and eval is J
```

(b) Find the Jordan form and verify that the associated matrix $P$ has the property that

$$PAP^{-1} = J.$$

```
MAPLE (symbolic derivative)
> J:=JordanForm(A);
> Q:=JordanForm(A, output='Q');
> Q^(-1).A.Q;
```

MATLAB
```
> P=inv(evect)
> J=P*A*inv(P)
```

In order to get (2.6.2), take $P$ to be $Q^{-1}$.

MAPLE
```
> P:=Q^(-1); P.A.P^(-1);
```

**2.** In a compartment matrix, one or more of the column sums may be zero. In this case, one eigenvalue can be zero and solutions for the differential equations

$$Z' = CZ(t)$$

may have a limit different from zero.
If all the column sums are negative in a compartment matrix, the eigenvalues will have negative real part. All solutions for the differential equations

$$Z' = CZ(t)$$

will have limit zero in this case.
The following matrices contrast these two cases:

$$C_1 = \begin{pmatrix} -1 & 1 & 0 \\ 1 & -1 & 0 \\ 0 & 0 & -1 \end{pmatrix} \quad \text{and} \quad \begin{pmatrix} -1 & 0 & \frac{1}{2} \\ \frac{1}{2} & -1 & 0 \\ 0 & \frac{1}{2} & -1 \end{pmatrix}.$$

Let $C$ be the matrix defined below:

MAPLE
```
> with(LinearAlgebra):
  C:=Matrix([[-1,1,0],[1,-1,0],[0,0,-1]]);
```

MATLAB
```
> C=[-1 1 0; 1 -1 0; 0 0 -1]
```

(a) Find the eigenvalues and eigenvectors for $C$.

MAPLE
```
> Eigenvectors(C);
```

MATLAB
```
> [evects, evals] = eig(C)
```

(b) Graph each component of $z$ with $z(0) = [1, 1, 1]$.

MAPLE
```
> exptC:=MatrixExponential(C,t);
> U:=evalm( exptC.[1,0,1]);
> u:=unapply(U[1],t); v:=unapply(U[2],t); w:=unapply(U[3],t);
> plot({u(t),v(t),w(t)},t=0..2, color=[black,blue,green]);
```

MATLAB
```
% contents of the m-file exer252.m
% function Zprime=exer252(t,Z);
% Zprime=[-1*Z(1)+1*Z(2); 1*Z(1)-1*Z(2); -1*Z(3)];
> [t,Z]=ode23('exer252',[0 10],[1; 0; 1]);
> plot(t,Z)
```

## 2.7 Statistical Data

*Variation impacts almost everything. Variation can be quantified by describing its distribution. A* distribution *is the set of the fractions of observations having particular values with respect to the number of the possible values. For example, the distribution of word lengths of the previous sentence is* 3 *of length* 1, 4 *of length* 2, 2 *of length* 3, *and so on (all divided by* 18, *the number of words in the sentence). The graph of a distribution with the observations grouped or made discrete to some resolution is a histogram. Distributions are approximately described by their mean, or average, value and the degree to which the observations deviate from the mean, their standard deviation. A widely occurring distribution is the normal, or Gaussian. This bell-shaped distribution is completely determined by its mean and standard deviation.*

*Histograms portray statistical data.*

Given that the natural world is rife with variables, it is not surprising to find that variation is widespread. Trees have different heights, ocean temperatures change from place to place and from top to bottom, the individuals of a population have different ages, and so on. Natural selection thrives on variation. Variation is often due to chance events; thus the height of a tree depends on its genetic makeup, the soil in which it grows, rainfall, and sunlight among other things. Describing variation is a science all to its own.

Since pictures are worth many words, we start with histograms. Corresponding to the phenomenon under study, any variation observed occurs within a specific range of possibilities, a *sample space*. This range of possibilities is then partitioned or divided up into a number of subranges, or classes. A *histogram* is a graph of the fraction of observations falling within the various subranges plotted against those subranges.

Consider the recent age distribution data for the U.S. population, shown in Table 2.7.1. The possible range of ages, 0 to infinity, is partitioned into subranges or intervals of every five years from birth to age 84; a last interval, 85+, could be added if necessary for completeness. The table lists the percentage of the total population falling within the given interval; each percentage is also refined by sex. The cumulative percentage is also given, that is, the sum of the percentages up to and including the given interval. A histogram is a graph of these data; on each partition interval is placed a rectangle, or bar, whose width is that of the interval and whose height is the corresponding percentage (see Figure 2.7.1).

The resolution of a histogram is determined by the choice of subranges: Smaller and more numerous intervals mean better resolution and more accurate determination of the distribution; larger and fewer intervals entail less data storage and processing.

The cumulative values are plotted in Figure 2.7.2. Since the percentage values have a resolution of five years, a decision has to be made about where the increments should appear in the cumulative plot. For example, 7.2% of the population is in the first age interval counting those who have not yet reached their fifth birthday. Should this increment be placed at age 0, at age 5, or maybe at age 2.5 in the cumulative graph?

Table 2.7.1. Age distribution for the U.S. population.

| Age | % Female | % Male | % Population | Cumulative |
|---|---|---|---|---|
| 0–4 | 3.6 | 3.6 | 7.2 | 7.2 |
| 5–9 | 3.9 | 3.7 | 7.6 | 14.8 |
| 10–14 | 4.1 | 3.9 | 8.0 | 22.8 |
| 15–19 | 4.7 | 4.3 | 9.0 | 31.8 |
| 20–24 | 5.0 | 4.2 | 9.2 | 41.0 |
| 25–29 | 4.3 | 4.0 | 8.3 | 49.3 |
| 30–34 | 4.0 | 3.5 | 7.5 | 56.8 |
| 35–39 | 3.6 | 2.9 | 6.5 | 63.3 |
| 40–44 | 2.7 | 2.2 | 4.9 | 68.2 |
| 45–49 | 2.8 | 2.0 | 4.8 | 73.0 |
| 50–54 | 3.0 | 2.2 | 5.2 | 78.2 |
| 55–59 | 3.1 | 2.1 | 5.2 | 83.4 |
| 60–64 | 2.8 | 1.9 | 4.7 | 88.1 |
| 65–69 | 2.3 | 1.8 | 4.1 | 92.2 |
| 70–74 | 2.0 | 1.4 | 3.4 | 95.6 |
| 75–79 | 1.7 | 0.8 | 2.5 | 98.1 |
| 80–84 | 1.6 | 0.3 | 1.9 | 100 |

We have chosen to do something different, namely, to indicate this information as a line segment that is 0 at age 0 and is 7.2 at age 5. In like fashion, we indicate in the cumulative graph the second bar of the histogram of height 7.6% as a line segment joining the points 7.2 at age 5 with 14.8 ($= 7.2 + 7.6$) at age 10. Continuing this idea for the balance of the data produces the figure. Our rationale here is the assumption that the people within any age group are approximately evenly distributed by age in this group. A graph that consists of joined line segments is called a *polygonal graph* or a *linear spline*.

This graph of accumulated percentages is called the *cumulative distribution function*, or *cdf* for short. No matter what decision is made about placing the cumulative percentages, the cdf satisfies these properties:

1. it starts at 0,
2. it never decreases, and
3. it eventually reaches 1 (or, as a percentage, 100%).

*The mean and median approximately locate the center of the distribution.*

Sometimes it is convenient to summarize the information in a histogram. Of course, no single number or pair of numbers can convey all the information; such a summary is therefore a compromise, but nevertheless a useful one. First, some information about where the data lie is given by the *mean*, or *average*; it is frequently denoted by $\mu$. Given the $n$ values $x_1, x_2, \ldots, x_n$, their mean is

MAPLE

```
> mcent:=[3.6, 3.7, 3.9, 4.3, 4.2, 4.0, 3.5, 2.9, 2.2,2.0, 2.2, 2.1, 1.9, 1.8,1.4, 0.8, 0.3]:
  fcent:=[3.6, 3.9, 4.1, 4.7, 5.0, 4.3, 4.0, 3.6, 2.7, 2.8, 3.0, 3.1, 2.8, 2.3, 2.0, 1.7, 1.6]:
  tot:=[seq(mcent[i]+fcent[i],i=1..17)]:
> ranges:=[0..5, 5..10, 10..15, 15..20, 20..25, 25..30, 30..35, 35..40, 40..45, 45..50, 50..55, 55..60, 60..65,
           65..70, 70..75, 75..80, 80..85]:
> with(stats): with(plots):
> mpop:=[seq(Weight(ranges[i], 5*mcent[i]),i=1..17)]:
> fpop:=[seq(Weight(ranges[i], 5*fcent[i]),i=1..17)]:
> pop:=[seq(Weight(ranges[i], 5*tot[i]),i=1..17)]:
> statplots[histogram](pop);
```

MATLAB

```
> mcent=[3.6 3.7 3.9 4.3 4.2 4.0 3.5 2.9 2.2 2.0 2.2 2.1 1.9 1.8 1.4 0.8 0.3];
> fcent=[3.6 3.9 4.1 4.7 5.0 4.3 4.0 3.6 2.7 2.8 3.0 3.1 2.8 2.3 2.0 1.7 1.6];
> total=mcent+fcent;
> x=[5:5:85]; % 5, 10, 15, ..., 85
> bar(x,total) % bars centered on the x values
> xlabel('Age(years)')
> ylabel('Percent in age bracket');
```

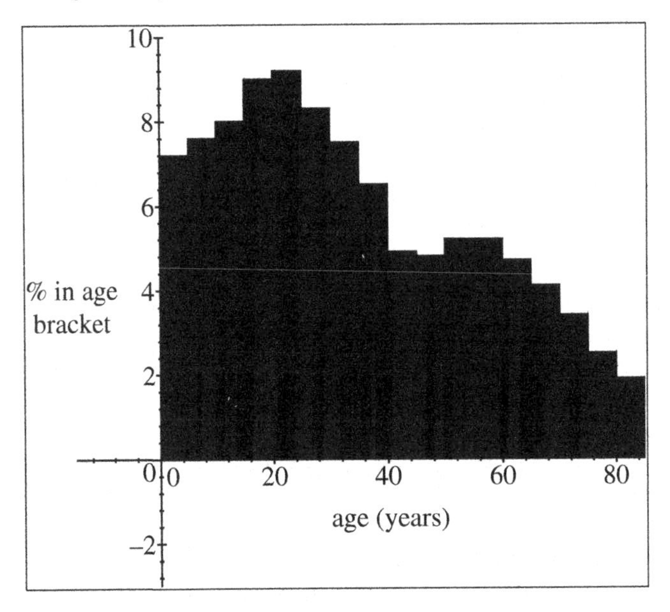


**Fig. 2.7.1.** Histogram for the U.S. population distributed by age.

$$\mu = \frac{x_1 + x_2 + \cdots + x_n}{n} = \frac{1}{n}\sum_{i=1}^{n} x_i. \tag{2.7.1}$$

Another popular notation for this quotient is $\bar{x}$. It is necessarily true that some values $x_i$ are smaller than the mean and some are larger. (Either that or all $x$s are equal.) In fact, one understands the mean to be in the center of the $x$ values in a sense made precise by (2.7.1). Given $\bar{x}$ and $n$, the sum of the $x$s is easily computed:

$$\sum_{i=1}^{n} x_i = n\bar{x}.$$

MAPLE
```
> age:=[2.5, 7.5, 12.5, 17.5, 22.5, 27.5, 32.5, 37.5, 42.5, 47.5, 52.5, 57.5, 62.5, 67.5, 72.5, 77.5, 82.5];
> cummale:=[seq(sum('mcent[i]','i'=1..n),n=1..17)]:
> cumfale:=[seq(sum('fcent[i]','i'=1..n),n=1..17)]:
> cumtot:=[seq(sum('tot[i]','i'=1..n),n=1..17)]:
> ptsm:=[seq([age[i],cummale[i]],i=1..17)];
> ptsf:=[seq([age[i],cumfale[i]],i=1..17)]:
> ptsT:=[seq([age[i],cumtot[i]],i=1..17)]:
> plot({ptsm,ptsf,ptsT},color=BLACK);
```

MATLAB
```
> cumM=cumsum(mcent);
> cumF=cumsum(fcent);
> cumTot=cumsum(total)
> plot(x,cumTot,x,cumM,x,cumF)
```

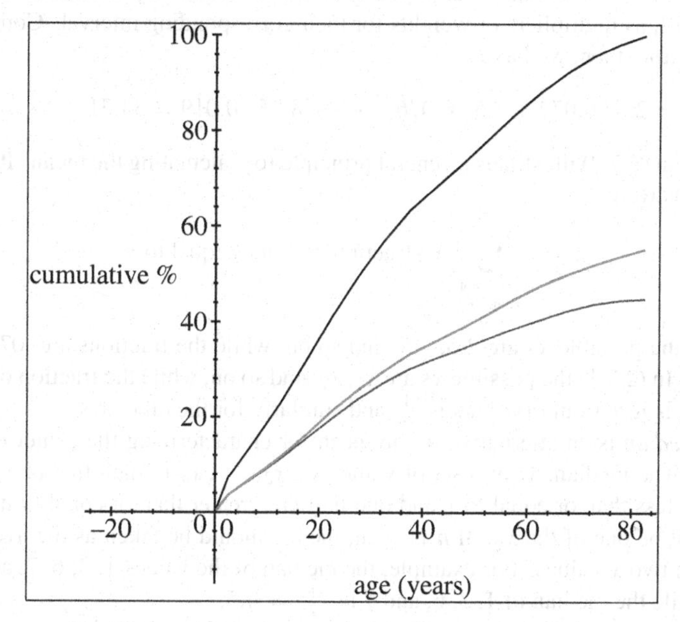


**Fig. 2.7.2.** Cumulative populations (% of the total vs. age).

Computing the mean of a histogram goes somewhat differently. Suppose the total number of people referred to by Table 2.7.1 to be 100 million. (It no doubt corresponds to many more than that, but it will be more convenient to calculate percentages using 100 million, and we will see that in the end, this choice is irrelevant.) Then the 7.2% in the first group translates into 7.2 million people. We do not know their individual ages, but as above, if they were evenly distributed over ages 0 to 4.999..., then counting all 7.2 million as 2.5 gives the same result. Hence in (2.7.1) these people contribute a value of 2.5 for 7.2 million such people, or

$$\text{contribution of "0 to 5" group} = 2.5 \cdot 7.2 = \frac{0+5}{2} \cdot 7.2$$

in millions. Similarly the second group contributes

$$\text{contribution of "5 to 10" group} = 7.5 \cdot 7.8 = \frac{5+10}{2} \cdot 7.8.$$

Continuing in this way we get, where we are counting in millions,

$$\sum_{i=1}^{n} x_i = 2.5 \cdot 7.2 + 7.5 \cdot 7.6 + 12.5 \cdot 8.0 + \cdots + 82.5 \cdot 1.9 = 3431.0 \text{ (million)}.$$

Divide the result by 100 (million) to obtain the mean. But dividing by 100 million means a quotient such as $\frac{7.2 \text{ million}}{100 \text{ million}}$ is just the fraction .072 (or 7.2%). In other words, we do not need to know the total population size; instead, we just use the fractions, such as .072, as multipliers or weights for their corresponding interval. Completing the calculation, then, we have

$$\bar{x} = 2.5 \cdot 0.072 + 7.5 \cdot 0.076 + \cdots + 82.5 \cdot 0.019 = 34.31. \qquad (2.7.2)$$

Equation (2.7.2) illustrates a general principle for calculating the mean. It applies to (2.7.1) as well:

$$\mu = \sum_{\substack{\text{over possible} \\ \text{values } x}} x \cdot \text{fraction of values equal to } x. \qquad (2.7.3)$$

In (2.7.2) the possible $x$s are 2.5, 7.5, and so on, while the fractions are .072, .076, and so on. In (2.7.1) the possible $x$s are $x_1, x_2$, and so on, while the fraction of values that are $x_1$ is just 1 out of $n$, that is, $\frac{1}{n}$, and similarly for the other $x_i$s.

The *median* is an alternative to the mean for characterizing the center of a distribution. The median, $\hat{x}$, of a set of values $x_1, x_2, \ldots, x_n$ is such that one-half the values are less than or equal to $\hat{x}$ and one-half are greater than or equal to it. If $n$ is odd, $\hat{x}$ will be one of the $x$s. If $n$ is even, then $\hat{x}$ should be taken as the average of the middle two $x$ values. For example, the median of the values 1, 3, 6, 7, and 15 is $\hat{x} = 6$, while the median of 1, 3, 6, and 7 is $\frac{3+6}{2} = 4.5$.

The median is sometimes preferable to the mean because it is a more typical value. For example, for the values 3, 3, 3, 3, and 1000, the mean is 506, while the median is 3.

In the population data, the median age for men and women is between 29 and 30. This can be seen from an examination of the last column of Table (2.7.1). Contrast this median age with the average age; thus for men,

$$\begin{aligned}\text{average age for men} &= \frac{\sum_{n=1}^{17} [\text{percentage men at age } n] \cdot [\text{age } [n]]}{\text{total percentage of men}} \\ &= 32.17.\end{aligned}$$

In a similar manner, the average age for women in this data set is about 35.5, and the average age for the total population is about 33.8. The averages for these three sets of data—male population age distribution, female population age distribution, and total population age distribution—can be found with simple computer algebra commands and agree with our paper-and-pen calculations.

```
MAPLE
> Sum('age[j]'*'tot[j]',j=1..17)=sum(age[j]*tot[j],j=1..17);
> Sum('mcent[n]*age[n]','n'=1..17)/Sum('mcent[n]','n'=1..17)
  =sum('mcent[n]*age[n]','n'=1..17)/sum('mcent[n]','n'=1..17);
> with(describe): mean(pop); median(pop);
```

```
MATLAB
> xmid=[2.5:5:82.5];
> pop=xmid.*total; % term by term mult. = percentage weighted ranges
> muTotal=sum(pop)/100 % divide by 100 as data is in percent
> muM=sum(xmid.*mcent)/sum(mcent)
> muF=sum(xmid.*fcent)/sum(fcent)
```

*Variance and standard deviation measure dispersion.*

As mentioned above, a single number will not be able to capture all the information in a histogram. The data set 60, 60, 60, 60 has a mean of 60, as does the data set 30, 0, 120, 90. If these data referred to possible speeds in miles per hour for a trip across Nevada by bus for two different bus companies, then we might prefer our chances with the first company. The *variance* of a data set measures how widely the data is dispersed from the mean; for $n$ values $x_1, x_2, \ldots, x_n$, their variance $v$, or sometimes $\sigma^2$, is defined as

$$v = \frac{1}{n}\sum_{i=1}^{n}(x_i - \bar{x})^2, \tag{2.7.4}$$

where $\bar{x}$ is the mean as before.[7] Thus the speed variance for bus company 1 is 0 and that for bus company 2 is

$$\frac{1}{4}[(30-60)^2 + (0-60)^2 + (120-60)^2 + (90-60)^2] = 2{,}250.$$

As before, a more general equation for variance, one suitable for histograms, for example, is the following:

$$v = \sum_{\substack{\text{over possible}\\ \text{values } x}} (x - \bar{x})^2 \cdot \text{fraction of values equal to } x. \tag{2.7.5}$$

A problem with variance is that it corresponds to squared data values, making it hard to interpret its meaning in terms of the original data. If the data has units, like miles per hour, then variance is in the square of those units. Closely related to variance is *standard deviation*, denoted by $\sigma$. Standard deviation is defined as the square root of variance,

$$\text{standard deviation} = \sqrt{\text{variance}}.$$

[7] For data representing a sample drawn from some distribution, $\bar{x}$ is only an estimate of the distribution's mean, and for that reason, this definition of variance is a biased estimator of the distribution's variance. Divide by $n-1$ in place of $n$ for an unbiased estimator. Our definition is, however, the maximum likelihood estimator of the variance for normal distributions. Furthermore, this definition is consistent with the definition of variance for probability distributions (see Section 2.8), and for that reason we prefer it.

Standard deviation is a measure of the dispersion of data on the same scale as the data itself. The standard deviation of bus speeds for company 2 is 47.4 miles per hour. This is not saying that the average (unsigned) deviation of the data from the mean is 47.4 (for that would be $\frac{1}{n}\sum_1^n |x_i - \bar{x}| = 45$), but this is, in spirit, what the standard deviation measures. For the bus companies, we make these calculations:

```
MAPLE
> bus1:=[60,60,60,60]; bus2:=[30,0,120,90];
> range(bus1), range(bus2);
> median(bus1), median(bus2);
> mean(bus1), mean(bus2);
> variance(bus1), variance(bus2);
> standarddeviation(bus1), standarddeviation(bus2);
```

```
MATLAB
> bus1=[60 60 60 60]; bus2=[30 0 120 90];
> max(bus1), min(bus1)
> max(bus2), min(bus2)
> median(bus1), median(bus2)
> mean(bus1), mean(bus2)
> cov(bus1), cov(bus2)
> std(bus1), std(bus2)
```

We can perform similar calculations for the U.S. census data of Table 2.7.1. The results are given in Table 2.7.2.

```
MAPLE
> range(mpop), range(fpop), range(pop);
> median(mpop), median(fpop), median(pop);
> mean(mpop), mean(fpop), mean(pop);
> variance(mpop), variance(fpop), variance(pop);
> standarddeviation(mpop), standarddeviation(fpop),
> standarddeviation(pop);
```

```
MATLAB
> v=(xmid-muTotal).^2 % unweighted vector of deviations squared
> var=sum(v.*total)/100 % variance of total population
> sqrt(var) % std dev of the total population
```

**Table 2.7.2.** Summary for the U.S. age distribution

| | Range | Median | Mean | Standard deviation |
|---|---|---|---|---|
| **Male** | 0–84 | 29 | 31.7 | 21.16 |
| **Female** | 0–84 | 30 | 35.6 | 22.68 |
| **Total** | 0–84 | 29 | 33.8 | 22.10 |

*The normal distribution is everywhere.*

It is well known that histograms are often bell-shaped. This is especially true in the biological sciences. The mathematician Carl Friedrich Gauss discovered the explanation for this, and it is now known as the *central limit theorem* (see Hogg and Craig [6]).

**Central limit theorem.** *The accumulated result of many independent random outcomes, in the limit, tends to a* Gaussian, *or* normal, *distribution given by*

$$G(x) = \frac{1}{\sqrt{2\pi}\sigma} e^{-\frac{1}{2}(\frac{x-\mu}{\sigma})^2}, \quad -\infty < x < \infty,$$

*where* $\mu$ *and* $\sigma$ *are the* mean *and* standard deviation *of the distribution.*

The normal distribution is a continuous distribution, meaning that its resolution is infinitely fine; its histogram, given by $G(x)$, is smooth (see Figure 2.7.3). The two parameters mean $\mu$ and standard deviation $\sigma$ completely determine the normal distribution. Likewise, even though a given histogram is not Gaussian, nevertheless its description is often given in terms of just its mean and variance or standard deviation.

In Figure 2.7.3(a), we show three curves with the same mean but different standard deviations. In Figure 2.7.3(b), the three curves have the same standard deviation but different means.

```
MAPLE
> y:=(sigma,mu,x)->exp(-(x-mu)^2/(2*sigma^2))/(sqrt(2*Pi)*sigma);
> plot({y(1,0,x),y(2,0,x),y(3,0,x)},x=-10..10);
> plot({y(1,-4,x),y(1,0,x),y(1,4,x)},x=-10..10);
```

```
MATLAB
% make up an m-file, gaussian.m:
% function y=gaussian(x,m,s);
  %% m=mean, s=stddev
  %% note 1/sqrt(2*pi)=.3989422803
  % y=(.3989422803/s)*exp(-0.5*((x-m)./s).^2);
> x=[-10:.1:10];
> y=gaussian(x,0,1); plot(x,y);hold on;
> y=gaussian(x,0,2); plot(x,y);
> y=gaussian(x,0,4); plot(x,y);
> hold off
> y=gaussian(x,0,1); plot(x,y);hold on
> y=gaussian(x,-5,1); plot(x,y);
> y=gaussian(x,5,1); plot(x,y);
```

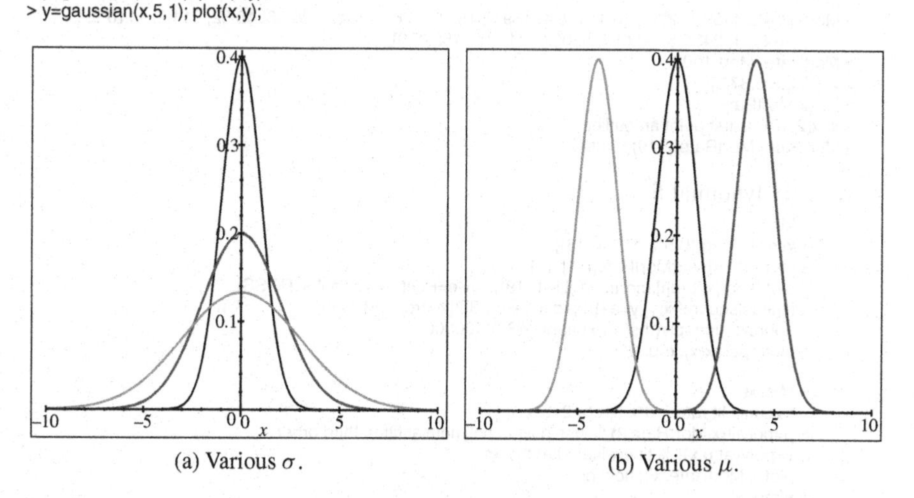


(a) Various $\sigma$. (b) Various $\mu$.

**Fig. 2.7.3.**

## Exercises

1. In the February 1994 Epidemiology Report published by the Alabama Department of Public Health, the data in Table 2.7.3 were provided as Age-Specific Mortality. Make a histogram for these data. While the data are given over age ranges, get a fit for the data so that one could predict the death rate for intermediate years. Find the median, mean, and standard deviation for the data.

**Table 2.7.3.**

| | | | |
|---|---|---|---|
| 0–1 | 1122.4 | 40–45 | 287.8 |
| 1–5 | 55.1 | 45–50 | 487.2 |
| 5–10 | 27.5 | 50–55 | 711.2 |
| 10–15 | 33.4 | 55–60 | 1116.9 |
| 15–20 | 118.4 | 60–65 | 1685.1 |
| 20–25 | 139.6 | 65–70 | 2435.5 |
| 25–30 | 158.0 | 70–75 | 3632.4 |
| 30–35 | 196.4 | 75–80 | 5300.0 |
| 35–40 | 231.0 | 80–85 | 8142.0 |
| | | 85+ | 15279.0 |

MAPLE
```
> with(stats): with (plots): with(describe):
> Mort:=[1122.4, 55.1, 27.5, 33.4, 118.4, 139.6, 158.0, 196.4, 231.0, 287.8, 487.2, 711.2, 1116.9,
        1685.1, 2435.5, 3632.4, 5300.0, 8142.0, 15278.0]:
> MortRate:=[seq(Mort[i]/100000,i=1..19)];
> ranges:=[seq(5*i..5*(i+1),i=1..17)];
> mortdata:=[Weight(0..1,MortRate[1]), Weight(1..5,4*MortRate[2]),
        seq(Weight(ranges[i],5*MortRate[2+i]), i=1..17)]:
> statplots[histogram](mortdata);
```

MATLAB
```
> Mort=[1122.4, 55.1, 27.5, 33.4, 118.4, 139.6, 158.0, 196.4, 231.0, 287.8, 487.2, 711.2, 1116.9, ...
        1685.1, 2435.5, 3632.4, 5300.0, 8142.0, 15278.0];
> MortRate=Mort/1000;
> x=[.5,2.5:5:87.5];
> bar(x,MortRate)
> x=x(2:19) % first point an outlier
> MortRate=MortRate(2:19) % ditto
```

(a) A polynomial fit:

MAPLE
```
> xcord:=[seq(3+5*(i-1),i=1..18)];
> mortrate:=[seq(MortRate[i+1], i=1..18)];
> plot([seq([xcord[i],mortrate[i]],i=1..18)], style=POINT, symbol=CROSS);
> fit[leastsquare[[x,y],y=a+b*x+c*x^2+d*x^3]]([xcord,mortrate]);
> approx:=unapply(rhs(%),x);approx(30)*100000;
> plot(approx(x),x=0..90);
```

MATLAB
```
% cubic fit rate = d*x^3+c*x^2+b*x+a
> p=polyfit(x,MortRate,3) % use built-in polynomial fitter, third order
> y=polyval(p,x); % fit evaluated at the xs
> plot(x,MortRate,'x'); hold on
> plot(x,y)
% or use the general leastsquares model
```

```
> MT=[x.^3; x.^2; x; ones(size(x))];
> cubic=MT'\MortRate'
> y=polyval(cubic,x); plot(x,y)
```

(b) An exponential fit:

```
MAPLE
> Lnmortrate:=map(ln,mortrate);
> fit[leastsquare[[x,y],y=m*x+b]]([xcord,Lnmortrate]);
> k:=op(1,op(1,rhs(%))); A:=op(2,rhs(%%));
> expfit:=t->exp(A)*exp(k*t); expfit(30)*100000;
> J:=plot(expfit(t),t=0..85):
  K:=plot([seq([xcord[i],MortRate[i+1]],i=1..18)],style=POINT,symbol=CROSS):
> display({J,K});

MATLAB
% exponential fit log(MortRate)=a+b*x or MortRate=exp(a)*exp(bx)
> Lnmortrate=log(MortRate);
> MT=[ones(size(x)); x];
> expon=MT'\Lnmortrate'
> hold off
> plot(x,MortRate,'x'); hold on
> plot(x,exp(expon(1))*exp(expon(2)*x))
```

(c) A linear spline for the data (see the discussion in this section):

```
MAPLE
> readlib(spline):
> linefit:=spline(xcord,mortrate,x,linear):
> y:=unapply(linefit.x): y(30)*100000;
> J:=plot(y(t), t=0..85):
> display({J,K});

MATLAB
% linear spline fit = straight line between points, usual MATLAB method
> hold off
> plot(x,MortRate,x,MortRate,'x')
```

Give the range, median, mean, and standard deviation of the mortality rates. Note that the first entry is applicable to humans in an age group of width one year and the second is in a group of width four years. Each of the others applies to spans of five years. Thus we set up a weighted sum:

```
MAPLE
> summary:=[Weight(Mort[1],1),Weight(Mort[2],4),seq(Weight(Mort[i],5),i=3..19)];
> range(summary); median(summary); mean(summary);
> standarddeviation(summary);

MATLAB
% to interpolate any desired value, use interp1, e.g., rate=interp1(x,MortRate,70)
% interpolated value at x=70
% mean, median, and standard deviation (of Mortality weighted by age)
> size(Mort)
> wt=[1,4,5*ones(1,17)]
> wtSum = Mort*wt' % dot product
> mu=wtSum/sum(wt)
> median(Mort) % picks out the middle value, no duplicates here
> v=(Mort-mu).^2; % vector of squared differences
> var=sum(v.*wt)/sum(wt);
> std=sqrt(var)
```

**2.** What follows in Table 2.7.4 are data for the heights of a group of males. Determine a histogram for these data. Find the range, median, mean, and standard deviation for the data. Give a normal distribution with the same mean and standard deviation as the data. Plot the data and the distribution on the same graph.

**Table 2.7.4.**

| **Number of students** | 2 | 1 | 2 | 7 | 10 | 14 | 7 | 5 | 2 | 1 |
|---|---|---|---|---|---|---|---|---|---|---|
| **Height (in)** | 66 | 67 | 68 | 69 | 70 | 71 | 72 | 73 | 74 | 75 |

```
MAPLE
> with(stats): with(plots): with(describe):
> htinches:=[seq(60+i,i=1..15)];
> numMales:=[0,0,0,0,0,2,1,2,7,10,14,7,5,2,1];
> ranges:=[seq(htinches[i]..htinches[i]+1, i=1..15)];
> maledata:=[seq(Weight(ranges[i],numMales[i]), i=1..15)];
> statplots[histogram](maledata);
> range(maledata); median(maledata); mean(maledata); standarddeviation(maledata);
  # note the use of back quotes in the next for formatted printing
> 'The average height is',floor(%%/12), 'feet and',floor(frac(%%%/12)*12), 'inches';
> 'The standard deviation is',floor(frac(%%%/12)*12), 'inches';

MATLAB
> htinches=61:75;
> numMales=[0,0,0,0,0,2,1,2,7,10,14,7,5,2,1];
> bar(htinches,numMales)
> min(htinches)
> max(htinches) % range = from min to max
> unrolled=[]; % dup. each height by its #cases
> s=size(htinches);
> for k=1:s(2)
>   j=numMales(k);
>   while j>0
>     unrolled=[unrolled, htinches(k)];
>     j=j-1;
>   end
> end
> median(unrolled)
> mu=mean(unrolled+.5) % e.g., height 66 counts as 66.5
  % alternatively
> mu=dot((htinches+.5),numMales)/sum(numMales)
> v=(htinches+.5-mu).^2;
> var=sum(v.*numMales)/sum(numMales)
> std=sqrt(var)
```

In what follows, we give a normal distribution that has the same mean and standard deviation as the height data:

```
MAPLE
> mu:=mean(maledata);
> sigma:=standarddeviation(maledata);
> ND:=x->exp(-(x-mu)^2/(2*sigma^2))/(sigma*sqrt(2*Pi));
> J:=plot(mu*ND(x),x=60..76):
> K:=statplots[histogram](maledata):
> plots[display]({J,K});

MATLAB
> x=60:.1:76;
> y=exp(-((x-mu)/std).^2/2)/(std*sqrt(2*pi));
> bar(htinches,numMales/sum(numMales))
> hold on; plot(x,y)
```

To the extent that the graph $K$ is an approximation for the graph $J$, the heights are normally distributed about the mean.

3. Table 2.7.5 contains population data estimates for the United States (in thousands) as published by the U.S. Bureau of the Census, Population Division, release PPL-21 (1995).

**Table 2.7.5.**

| Five-year age groups | 1990 | 1995 | Five-year age groups | 1990 | 1995 |
|---|---|---|---|---|---|
| 0–5 | 18,849 | 19,662 | 50–55 | 11,368 | 13,525 |
| 5–10 | 18,062 | 19,081 | 55–60 | 10,473 | 11,020 |
| 10–15 | 17,189 | 18,863 | 60–65 | 10,619 | 10,065 |
| 15–20 | 17,749 | 17,883 | 65–70 | 10,077 | 9,929 |
| 20–25 | 19,133 | 18,043 | 70–75 | 8,022 | 8,816 |
| 25–30 | 21,232 | 18,990 | 75–80 | 6,145 | 6,637 |
| 30–35 | 21,907 | 22,012 | 80–85 | 3,934 | 4,424 |
| 35–40 | 19,975 | 22,166 | 85–90 | 2,049 | 2,300 |
| 40–45 | 17,790 | 20,072 | 90–95 | 764 | 982 |
| 45–50 | 13,820 | 17,190 | 95–100 | 207 | 257 |
| | | | 100+ | 37 | 52 |

Find the median and mean ages. Estimate the number of people at ages 21, 22, 23, 24, and 25 in 1990 and in 1995. Make a histogram for the percentages of the population in each age category for both population estimates.

4. In (2.7.3), we stated that the mean $\mu$ is defined as

$$\mu = \sum_{\text{all possible } xs} x \cdot f(x),$$

where $f(x)$ is the fraction of all values that are equal to $x$. If these values are spread continuously over all numbers, $\mu$ can be conceived as an integral. In this sense, this integral of the normal distribution given by (2.7.3) yields

$$\mu = \int_{-\infty}^{\infty} x \frac{1}{\sigma\sqrt{2\pi}} \exp\left(-\frac{1}{2}\left(\frac{x-\mu}{\sigma}\right)^2\right) dx.$$

In a similar manner,

$$\sigma^2 = \int_{-\infty}^{\infty} (x-\mu)^2 \frac{1}{\sigma\sqrt{2\pi}} \exp\left(-\frac{1}{2}\left(\frac{x-\mu}{\sigma}\right)^2\right) dx.$$

Here is a way to evaluate the integrals:

MAPLE
```
> sigma:='sigma': mu='mu':
> f:=x->exp(-(x-mu)^2/(2*sigma^2))/(sigma*sqrt(2*Pi));
> assume(sigma > 0);
> int(x*f(x),x=-infinity..infinity);
```

```
> int((x-mu)^2*f(x),x=-infinity..infinity);

MATLAB
% to integrate with MATLAB one can use trapz(x,y) on x and y vectors or use Simpson's rule, quad(), ...
% but this requires an m-file.
% Here we will use the trapzoidal rule.
> x=linspace(-3,3); % simulates -infinity to +infinity here
> y=exp(-x.^2/2)/sqrt(2*pi);
> trapz(x,y) % approximately 1
```

## 2.8 Probability

*The biosphere is a complicated place. One complication is its unpredictable events, such as when a tree will fall or exactly what the genome of an offspring will be. Probability theory deals with unpredictable events by making predictions in the form of relative frequency of outcomes. Histograms portray the distribution of these relative frequencies and serve to characterize the underlying phenomenon.*

*Statistics deals with the construction and subsequent analysis of histograms retroactively, that is, from observed data. Probability deals with the prediction of histograms by calculation. In this regard, important properties to look for in calculating probabilities are independence, disjointness, and equal likelihood.*

*Probabilities and their distributions.*

Probability theory applies mathematical principles to random phenomena in order to make precise statements and accurate predictions about seemingly unpredictable events. The probability of an event $E$, written $\Pr(E)$, is the fraction of times $E$ occurs in an infinitely long sequence of trials. (Defining probability is difficult to do without being circular and without requiring experimentation. A definition requiring the outcome of infinitely many trials is obviously undesirable. The situation is similar to that in geometry, where the term "point" is necessarily left undefined; despite this, geometry has enjoyed great success.) For example, let an "experiment" consist in rolling a single die for which each of the six faces has equal chance of landing facing up. Take event $E$ to mean a 3 or a 5 lands facing up. Evidently, the probability of $E$ is then $\frac{1}{3}$, $\Pr(E) = \frac{1}{3}$, that is, rolling a 3 or 5 will happen approximately one-third of the time in a large number of rolls.

More generally, by an *event* $E$ in a probabilistic experiment, we mean some designated set of outcomes of the experiment. The number of outcomes, or *cardinality*, of $E$ is denoted by $|E|$. The set of all possible outcomes of an experiment is its *universe*, and is denoted by $U$. Here are some fundamental laws.

**Principle of universality.** One of the possible outcomes of an experiment will occur with certainty:

$$\Pr(U) = 1. \tag{2.8.1}$$

**Principle of disjoint events.** If events $E$ and $F$ are *disjoint*, $E \cap F = \emptyset$, that is, they have no outcomes in common, then the probability that $E$ or $F$ will occur (sometimes written $E \cup F$) is the sum

$$\Pr(E \text{ or } F) = \Pr(E) + \Pr(F). \tag{2.8.2}$$

**Principle of equal likelihood.** Suppose each outcome in $U$ has the same chance of occurring, i.e., is *equally likely*. Then the probability of an event $E$ is the ratio of the number of outcomes making up $E$ to the total number of outcomes,

$$\Pr(E) = \frac{|E|}{|U|}. \tag{2.8.3}$$

To illustrate, consider the experiment of rolling a pair of dice, one red and one green. Any one of six numbers can come up on each die equally likely, so the total number of possibilities is 36; the first possibility could be 1 on red and 1 on green, the second: 1 on red and 2 on green and so on. In this scheme, the last would be 6 on red and 6 on green. So $|U| = 36$. There are two ways to roll an 11, a 5 on red and 6 on green or the other way around. So letting $E$ be the event that an 11 is rolled, we have $\Pr(E) = \frac{2}{36} = \frac{1}{18}$. Let $S$ be the event that a 7 is rolled; this can happen in six different ways, so $\Pr(S) = \frac{6}{36} = \frac{1}{6}$. Now the probability that a 7 or 11 is rolled is their sum

$$\Pr(S \cup E) = \Pr(S) + \Pr(E) = \frac{2+6}{36} = \frac{2}{9}.$$

Since probabilities are frequencies of occurrence, they share properties with statistical distributions. Probability distributions can be visualized by histograms and their mean and variance calculated. For example, let the variable $X$ denote the outcome of the roll of a pair of dice. Table 2.8.1 gives the possible outcomes of $X$ along with their probabilities. Figure 2.8.1 graphically portrays the table as a histogram. Just as in the previous section, the rectangle on $x$ represents the fraction of times a dice roll will be $x$.

**Table 2.8.1.** Probabilities for a dice roll.

| **Roll** | 2 | 3 | 4 | 5 | 6 | 7 | 8 | 9 | 10 | 11 | 12 |
|---|---|---|---|---|---|---|---|---|---|---|---|
| **Probability** | $\frac{1}{36}$ | $\frac{2}{36}$ | $\frac{3}{36}$ | $\frac{4}{36}$ | $\frac{5}{36}$ | $\frac{6}{36}$ | $\frac{5}{36}$ | $\frac{4}{36}$ | $\frac{3}{36}$ | $\frac{2}{36}$ | $\frac{1}{36}$ |

MAPLE

```
> with(stats): with(plots): with(describe):
> roll:=[seq(n,n=2..12)];
> prob:=[1/36,2/36,3/36,4/36,5/36,6/36,5/36,4/36,3/36,2/36,1/36];
> wtroll:=[seq(Weight(roll[i]-1/2..roll[i]+1/2, prob[i]),i=1..11)];
> statplots[histogram](wtroll);
```

MATLAB

```
> roll=ones(1,11);
> roll=cumsum(roll);
> roll=roll+1;
> prob=[1 2 3 4 5 6 5 4 3 2 1]/36;
> bar(roll,prob)
```

Equation (2.7.3) can be used to calculate the mean value $\bar{X}$ of the random variable $X$, also known as its *expected* value, $E(X)$,

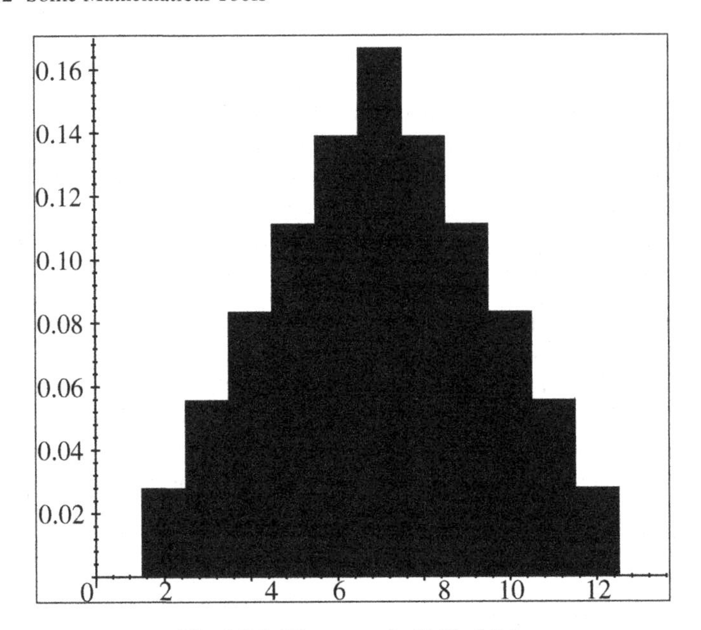


**Fig. 2.8.1.** Histogram for Table 2.8.1.

$$\bar{X} = \sum_{\substack{\text{over all possible}\\ \text{values } x \text{ of } X}} x \cdot \Pr(X = x). \tag{2.8.4}$$

From Table 2.8.1,

$$E(X) = 2 \cdot \frac{1}{36} + 3 \cdot \frac{2}{36} + 4 \cdot \frac{3}{36} + 5 \cdot \frac{4}{36} + 6 \cdot \frac{5}{36} + 7 \cdot \frac{6}{36} \\ + 8 \cdot \frac{5}{36} + 9 \cdot \frac{4}{36} + 10 \cdot \frac{3}{36} + 11 \cdot \frac{2}{36} + 12 \cdot \frac{1}{36} = 7.$$

MAPLE
```
> Sum('roll[i]*prob[i]','i'=1..11)=sum('roll[i]*prob[i]',i=1..11);
```

MATLAB
```
> weightedRoll=prob.*roll;
```

Similarly, the variance is defined as

$$V(X) = E(X - \bar{X})^2 = \sum_{\substack{\text{over all possible}\\ \text{values } x \text{ of } X}} (x - \bar{X})^2 \cdot \Pr(X = x). \tag{2.8.5}$$

For a dice roll,

$$V(X) = (2-7)^2 \frac{1}{36} + (3-7)^2 \frac{2}{36} + (4-7)^2 \frac{3}{36} + (5-7)^2 \frac{4}{36} \\ + (6-7)^2 \frac{5}{36} + (7-7)^2 \frac{6}{36} + (8-7)^2 \frac{5}{36} + (9-7)^2 \frac{4}{36}$$

$$+ (10-7)^2\frac{3}{36} + (11-7)^2\frac{2}{36} + (12-7)^2\frac{1}{36} = \frac{35}{6}.$$

MAPLE
```
> Sum('(roll[i]-7)^2*prob[i]','i'=1..11)=sum('(roll[i]-7)^2*prob[i]','i'=1..11);
> mean(wtroll);variance(wtroll);
```

MATLAB
```
> m=sum(weightedRoll)
> v=(roll-m).^2;
  % sum of squared deviations
> var=sum(v.*prob)
```

*Probability calculations can be simplified by decomposition and independence.*

Consider the experiment of tossing a fair coin in the air four times and observing the side landing up. Suppose we want to calculate the probability that heads will come up three of the four times. This grand experiment consists of four subexperiments, namely, the four individual coin tosses. Decomposing a probability experiment into subexperiments can often simplify making probability calculations. This is especially true if the subexperiments, and therefore their events, are *independent*. Two events $E$ and $F$ are independent when the fact that one of them has or has not occurred has no bearing on the other.

**Principle of independence.** If two events $E$ and $F$ are independent, then the probability that both will occur is the product of their individual probabilities,

$$\Pr(E \text{ and } F) = \Pr(E) \cdot \Pr(F).$$

One way three heads in four tosses can occur is by getting a head on the first three tosses and a tail on the last one; we will denote this by $HHHT$. Since the four tosses are independent, to calculate the probability of this outcome, we just multiply the individual probabilities of an $H$ the first time, an $H$ the second and also the third, and on the fourth, a $T$; each of these has probability $\frac{1}{2}$; hence

$$\Pr(HHHT) = \left(\frac{1}{2}\right)^4 = \frac{1}{16}.$$

There are three other ways that three of the four tosses will be $H$; they are $HHTH$, $HTHH$, and $THHH$. Each of these is also $\frac{1}{16}$ probable; therefore, by the principle of disjoint events,

$$\Pr(\text{three heads out of four tosses}) = 4 \cdot \frac{1}{16} = \frac{1}{4}.$$

*Permutations and combinations are at the core of probability calculations.*

The previous example raises a question: By direct enumeration, we found that there are four ways to get three heads (or, equivalently, one tail) in four tosses of a coin, but how can we conveniently calculate, for example, the number of ways to get eight heads in 14 coin tosses or, in general, $k$ heads in $n$ coin tosses? This is the problem of counting *combinations*.

To answer, consider the following experiment: Place balls labeled 1, 2, and so on to $n$ in a hat and select $k$ of them at random to decide where to place the $H$s. For instance, if $n = 4$ and $k = 3$, the selected balls might be 3, then 4, then 1, signifying the sequence $HTHH$.

As a subquestion, in how many ways can balls 1, 3, and 4 be selected—this is the problem of counting *permutations*, the various ways to order a set of objects. Actually, there are six permutations here; they are $(1, 3, 4)$, $(1, 4, 3)$, $(3, 1, 4)$, $(3, 4, 1)$, $(4, 1, 3)$, and $(4, 3, 1)$. The reasoning goes like this: There are three choices for the first ball from the possibilities 1, 3, 4. This choice having been made, there are two remaining choices for the second, and finally, only one possibility for the last. Hence the number of permutations of three objects $= 3 \cdot 2 \cdot 1 = 6$.

```
MAPLE
> with(combinat):
> permute([1,3,4]);
> numbperm(3);
```

More generally, the number of permutations of $n$ objects is

$$\text{number of permutations of } n \text{ objects} = n \cdot (n-1) \cdot (n-2) \cdots 2 \cdot 1 = n!.$$

As indicated, this product is written $n!$ and called *n factorial*.

So, in similar fashion, the number of ways to select $k$ balls from a hat holding $n$ balls is

$$n \cdot (n-1) \cdot (n-2) \cdot \cdots \cdot (n-k+1).$$

As we said above, the labels on the selected balls signify when the heads occur in the $n$ tosses. But each such choice has $k!$ permutations, all of which also give $k$ heads. Therefore, the number of ways of getting $k$ heads in $n$ tosses is

$$\frac{n(n-1)(n-2)\cdots(n-k+1)}{k(k-1)\cdots 2 \cdot 1}. \tag{2.8.6}$$

```
MAPLE
> with(combinat):
> numbcomb(6,3);
> binomial(6,3);
```

The value calculated by (2.8.6) is known as the number of combinations of $n$ objects taken $k$ at a time. This ratio occurs so frequently that there is a shorthand notation for it, $\binom{n}{k}$, or sometimes $C(n, k)$, called *n choose k*. An alternative form of $\binom{n}{k}$ is

$$\binom{n}{k} = \frac{n(n-1)\cdots(n-k+1)}{k(k-1)\cdots 2 \cdot 1} = \frac{n!}{k!(n-k)!}, \tag{2.8.7}$$

where the third member follows from the second by multiplying numerator and denominator by $(n-k)!$.

Some elementary facts about $n$ choose $k$ follow. For consistency in these formulas, zero factorial is defined to be 1,

$$0! = 1.$$

The first three combination numbers are

$$\binom{n}{0} = 1, \quad \binom{n}{1} = n, \quad \binom{n}{2} = \frac{n(n-1)}{2}.$$

There is a symmetry:

$$\binom{n}{k} = \binom{n}{n-k} \quad \text{for all } k = 0, 1, \ldots, n.$$

These numbers $n$ choose $k$ occur in the binomial theorem, which states that for any $p$ and $q$,

$$\sum_{k=0}^{n} \binom{n}{k} p^k q^{n-k} = (p+q)^n. \tag{2.8.8}$$

Finally, the probability of realizing $k$ heads in $n$ tosses of a fair coin is, denoting it by $H_n(k)$,

$$H_n(k) = \binom{n}{k} \left(\frac{1}{2}\right)^n, \quad k = 0, 1, \ldots, n. \tag{2.8.9}$$

The distribution $H_n(k)$ is shown in Figure 2.8.2 for $n = 60$. If the coin is not fair, say the probability of a heads is $p$ and that of a tails is $q = 1 - p$, then $H_n(k)$ becomes

$$H_n(k) = \binom{n}{k} p^k q^{n-k}, \quad k = 0, 1, \ldots, n. \tag{2.8.10}$$

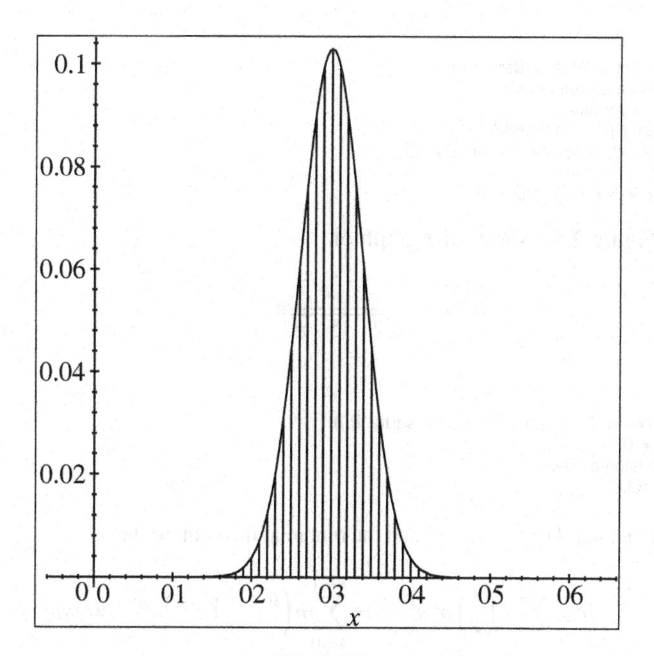


**Fig. 2.8.2.**

*Continuous variations require continuous distributions.*

In Figure 2.8.2, we show the heads distribution histogram $H_{60}(k)$ for 60 coin tosses. Notice that the distribution takes on the characteristic bell shape of the Gaussian distribution, as predicted by the central limit theorem, discussed in the previous section:

$$G(x) = \frac{1}{\sqrt{2\pi}\sigma} e^{-\frac{1}{2}(\frac{x-\mu}{\sigma})^2}, \quad -\infty < x < \infty, \tag{2.8.11}$$

where $\mu$ and $\sigma$ are the mean and standard deviation. In the figure, we have superimposed the Gaussian distribution on top of the histogram. In order to get the approximation right, we must match the means and variances of the two distributions. The mean of $H_n(k)$ for a biased coin, (2.8.10), is given by[8]

$$\mu = np. \tag{2.8.12}$$

And the variance of $H_n(k)$ is (see [8])

$$v = npq. \tag{2.8.13}$$

With $p = q = \frac{1}{2}$ and $n = 60$, we get $\mu = 30$ and $\sigma^2 = 15$.

```
MAPLE
> n:=60;
> flip:=[seq(binomial(n,i)*(1/2)^i*(1-1/2)^(n-i),i=0..n)]:
> wtflip:=[seq(Weight(i-1,flip[i]),i=1..n+1)]:
> with(stats); with(describe):
> mu:=evalf(mean(wtflip)); sigma:=standarddeviation(wtflip);
> sigma^2;
```

```
MATLAB
% use the previous m-file, gaussian.m:
% function y=gaussian(x,m,s);
% m=mean, s=stddev
% note 1/sqrt(2*pi)=.3989422803
% y=(.3989422803/s)*exp(-0.5*((x-m)./s).^2);
> x=[-10:.1:10];
> y=gaussian(x,30,sqrt(15)); plot(x,y)
```

Hence Figure 2.8.2 shows the graph of

$$G(x) = \frac{1}{\sqrt{2 \cdot 15 \cdot \pi}} e^{-\frac{1}{2}\frac{(x-30)^2}{15}}.$$

```
MAPLE
> G:=x->exp(-(x-mu)^2/(2*sigma^2))/(sigma*sqrt(2*Pi));
> J:=plot(G(x),x=0..n):
> K:=statplots[histogram](wtflip):
> plots[display]({J,K});
```

---

[8] Using the fact that $k\binom{n}{k} = n\binom{n-1}{k-1}$ and the binomial theorem, we have

$$\mu = \sum_{k=0}^{n} k\binom{n}{k} p^k q^{n-k} = \sum_{r=0}^{n-1} n\binom{n-1}{r} p^{r+1} q^{n-r} = np.$$

The normal or Gaussian distribution is an example of a continuous distribution. Any nonnegative function $f(x) \geq 0$ with total integral 1,

$$\int_{-\infty}^{\infty} f(x)dx = 1,$$

can define a probability distribution. The condition that the total integral be 1 is dictated by the universality principle, equation (2.8.1). In this role, such a function $f$ is called a *probability density function.* Probabilities are given as integrals of $f$. For example, let $X$ denote the outcome of the probabilistic experiment governed by $f$; then the probability that $X$ lies between 3 and 5, say, is exactly

$$\Pr(3 \leq X \leq 5) = \int_{3}^{5} f(x)dx.$$

Similarly, the probability that an outcome will lie in a *very* small interval of width $dx$ at the point $x$ is[9]

$$\Pr(X \text{ falls in an interval of width } dx \text{ at } x) = f(x)dx. \tag{2.8.14}$$

This shows that outcomes are more likely to occur where $f$ is large and less likely to occur where $f$ is small.

The simplest continuous distribution is the *uniform* distribution,

$$u(x) = \text{constant}.$$

Evidently, for an experiment governed by the uniform distribution, an outcome is just as likely to be at one place as another. For example, butterflies fly in a kind of random flight path that confounds their predators. As a first approximation, we might hypothesize that a butterfly makes its new direction somewhere within 45 degrees of its present heading uniformly. Let $\Theta$ denote the butterfly's directional change; $\Theta$ is governed by the uniform probability law

$$u(\Theta) = \begin{cases} \text{constant} & \text{if } -45 \leq \Theta \leq 45, \\ 0 & \text{otherwise.} \end{cases} \tag{2.8.15}$$

By the universality principle,

$$\int_{-45}^{45} u(\Theta)d\Theta = 1;$$

therefore the constant must be $\frac{1}{90}$ in (2.8.15).

[9] This equation is interpreted in the same spirit as the concept "velocity at a point" in dynamics, which is the ratio of infinitesimals $\frac{ds}{dt}$.

## Exercises

**1.** An undergraduate student in mathematics wants to apply to three of six graduate programs in mathematical biology. She will make a list of three programs in the order of her preferences. Since the order is important, this is a problem of permutations. How many such choices can she make?

MAPLE
```
> restart;
> with(combinat):
  #list the permutations and count
> permute([a,b,c,d,e,f],3);nops(%);
  #calculate directly
> numbperm(6,3);
  #use the formula
> 6!/3!;
```

MATLAB
```
  % No built-in combinatorics in MATLAB but it is easy to do factorials and hence permutations and
  %  combination calculations
  %  permutations of six things taken three at a time
> n6=1:6; n3=1:3;
> perm6t3=prod(n6)/prod(n3)
```

The student must send a list of three references to any school to which she applies. There are six professors who know her abilities well, of whom she must choose three. Since the order is not important, this is a problem of combinations. How many such lists can she make?

MAPLE
```
> with(combinat):
> choose([a,b,c,d,e,f],3);nops(%);
> numbcomb(6,3);
> 6!/(3!*(6-3)!);
```

MATLAB
```
  % combinations of six things taken three at a time
> comb6t3=perm6t3/prod(n3)
```

**2.** Five patients need heart transplants and three hearts for transplant surgery are available. How many ways are there to make a list of recipients? How many ways are there to make a list of the two of the five who must wait for further donors? (The answer to the previous two questions should be the same.) How many lists can be made for the possible recipients in the order in which the surgery will be performed?

MAPLE
```
> with(combinat):
> numbcomb(5,3); numbcomb(5,2);
> numbperm(5,3);
```

MATLAB
```
  % combinations of five things taken two at a time
> comb5t2=prod(1:5)/(prod(1:2)*prod(1:3))
> comb5t3=prod(1:5)/(prod(1:3)*prod(1:2))
> perm5t3=prod(1:5)/prod(1:2)
```

**3.** Choose an integer in the interval $[1, 6]$. If a single die is thrown 300 times, one would expect to get the number chosen about 50 times. Do this experiment and record how often each face of the die appears.

```
MAPLE
> with(stats): with(describe):
> die:=rand(1..6);
> for i from 1 to 6 do
  count[i]:=0
  od:
> for i from 1 to 300 do
  n:=die():
  count[n]:=count[n]+1:
  od:
> for i from 1 to 6 do
  print(count[i]);
  od;
> i:='i':
```

```
MATLAB
  % rand(1,300) is a random vector with components between 0 and up to but not including 1; then 6
  %   times this gives numbers from 0 up to 6; add 1 and get numbers 1 up to 7; finally, fix() truncates
  %   the fractional part
> die=fix(6*rand(1,300)+1);
  % now count the number of 3s
> count3s=1./(die-3); % gives infinity at every 3
> count3s=isinf(count3s); % 1 for infinity, 0 otherwise
> number3s=sum(count3s)
```

**4.** Simulate throwing a pair of dice for 360 times using a random number generator and complete Table 2.8.2 using the sums of the top faces.

**Table 2.8.2.**

| Sums | Predicted | Simulated |
|---|---|---|
| 2 | 10 | ________ |
| 3 | 20 | ________ |
| 4 | 30 | ________ |
| 5 | 40 | ________ |
| 6 | 50 | ________ |
| 7 | 60 | ________ |
| 8 | 50 | ________ |
| 9 | 40 | ________ |
| 10 | 30 | ________ |
| 11 | 20 | ________ |
| 12 | 10 | ________ |

Calculate the mean and standard deviation for your sample using the appropriate equations of Section 2.7 and compare this with the outcome probabilities. Draw a histogram for the simulated throws on the same graph as the normal distribution defined by (2.8.11); use the mean and the standard deviation you just calculated. The following syntax may help:

```
MAPLE
> with(stats): with(describe):
> red:=rand(1..6):
  blue:=rand(1..6):
> for i from 2 to 12 do
  count[i]:=0:
  od:
```

```
> for i from 1 to 360 do
  n:=red()+blue():
  count[n]:=count[n]+1;
  od:
> for i from 2 to 12 do
  print(count[i]);
  od;
> inter:=seq(n-1/2..n+1/2,n=2..12);
> throws:=[seq(Weight(inter[i-1],count[i]),i=2..12)];
> mean(throws)=evalf(mean(throws));
> standarddeviation(throws)=evalf(standarddeviation(throws));
> theory:=[Weight(inter[1],10), Weight(inter[2],20), Weight(inter[3],30), Weight(inter[4],40),
          Weight(inter[5],50), Weight(inter[6],60), Weight(inter[7],50), Weight(inter[8],40),
          Weight(inter[9],30), Weight(inter[10],20), Weight(inter[11],10)];
> mu:=mean(theory);
> sigma:=standarddeviation(theory);evalf(sigma);
> y:=x->360*exp(-(x-mu)^2/(2*sigma^2))/(sigma*sqrt(2*Pi));
> J:=statplots[histogram](throws):
> K:=plot([x,y(x),x=0..14]):
> plots[display]({J,K});
```

MATLAB

```
> red=fix(6*rand(1,360)+1);
> blue=fix(6*rand(1,360)+1);
> pairDice=red+blue;
> x=2:12;
> hist(pairDice,x)
> hold on
> h=hist(pairDice,x)
> mu=dot(x,h)/sum(h)
  % weight each int by its fraction of outcomes, add
> v=(x-mu).^2; % vector of diffs squared
> var = dot(v,h)/sum(h); % variance
> sigma=sqrt(var)
> t=linspace(2,12);
> y=360*exp(-(t-mu).^2/(2*sigma^2))/(sigma*sqrt(2*pi));
> plot(t,y)
  % the theoretical probability for seeing 2 is 1/36, same for 12, for seeing 3 is 2/36, same for 11, etc.,
  % for seeing 7 is 6/36.
  % compare with h above
> theory=[10 20 30 40 50 60 50 40 30 20 10];
> mu=dot(x,theory)/sum(theory)
> v=(x-mu).^2;var=dot(v,theory)/sum(theory);
> sigma=sqrt(var)
> y=360*exp(-(t-mu).^2/(2*sigma^2))/(sigma*sqrt(2*pi));
> hold off
> plot(t,y); hold on
> hist(pairDice,x)
```

**5.** This exercise is a study of independent events. Suppose a couple's genetic makeup makes the probability that a child they conceive will have brown eyes equal to $\frac{3}{4}$. Assume that the eye color for two children is a pair of independent events.

(a) What is the probability that the couple will have two blue-eyed children? One blue-eyed and one brown-eyed? Two brown-eyed children? What is the sum of these probabilities?

MAPLE

```
> binomial(2,0)*1/4*1/4;
> binomial(2,1)*3/4*1/4;
> binomial(2,2)*3/4*3/4;
> sum(binomial(2,j)*(3/4)^j*(1/4)^(2-j),j=0..2);
```

MATLAB

```
% #ways for two blue eyed is C(2,2)
```

```
  % (2 choose 2)=2!/(2!*0!) so the probability is that times (1/4)^2, etc.
> blublu=prod(1:2)/(prod(1:2)*1)*(1/4)^2 % blu/blu children
> Bwnblu=prod(1:2)/(prod(1:1)*prod(1:1))*(3/4)*(1/4) % Bwn/Blu
> BwnBwn=prod(1:2)/(1*prod(1:2))*(3/4)^2 % Bwn/Bwn children
> blublu+Bwnblu+BwnBwn
```

(b) Suppose that the couple have five children. What is the probability that among the five, exactly two will have brown eyes?

```
  MAPLE
> binomial(5,2)*(3/4)^2*(1/4)^3;

  MATLAB
  % exactly two are brown eyed is (5 choose 2)*(3/4)^2*(1/4)^3
> exact2=prod(1:5)/(prod(1:2)*prod(1:3))*(3/4)^2*(1/4)^3
```

(c) What is the probability that among the five children, there are at least two with brown eyes?

```
  MAPLE
> sum(binomial(5,j)*(3/4)^j*(1/4)^(5-j),j=2..5);

  MATLAB
> exact3=prod(1:5)/(prod(1:3)*prod(1:2))*(3/4)^3*(1/4)^2
> exact4=prod(1:5)/(prod(1:4)*prod(1:1))*(3/4)^4*(1/4)^1
> exact5=prod(1:5)/(prod(1:5)*1)*(3/4)^5
> atleast2=exact2+exact3+exact4+exact5
```

## References and Suggested Further Reading


[1] AIDS CASES IN THE U.S.:
*HIV/AIDS Surveillance Report*, Division of HIV/AIDS, Centers for Disease Control, U.S. Department of Health and Human Services, Atlanta, GA, July, 1993.

[2] CUBIC GROWTH OF AIDS:
S. A. Colgate, E. A. Stanley, J. M. Hyman, S. P. Layne, and C. Qualls, Risk-behavior model of the cubic growth of acquired immunodeficiency syndrome in the United States, *Proc. Nat. Acad. Sci. USA*, **86** (1989), 4793–4797.

[3] IDEAL HEIGHT AND WEIGHT:
S. R. Williams, *Nutrition and Diet Therapy*, 2nd ed., Mosby, St. Louis, 1973, 655.

[4] GEORGIA TECH EXERCISE LABORATORY:
P. B. Sparling, M. Millard-Stafford, L. B. Rosskopf, L. Dicarlo, and B. T. Hinson, Body composition by bioelectric impedance and densitometry in black women, *Amer. J. Human Biol.*, **5** (1993), 111–117.

[5] CLASSICAL DIFFERENTIAL EQUATIONS:
E. Kamke, *Differentialgleichungen Lösungsmethoden und Lösungen*, Chelsea, New York, 1948.

[6] THE CENTRAL LIMIT THEOREM:
R. Hogg and A. Craig, *Introduction to Mathematical Statistics*, Macmillan, New York, 1965.

[7] MORTALITY TABLES FOR ALABAMA:
*Epidemiology Report* IX (*Number* 2), Alabama Department of Public Health, Montgomery, AL, February, 1994.

[8] BASIC COMBINATORICS:
R. P. Grimaldi, *Discrete and Combinatorial Mathematics*, Addison–Wesley, New York, 1998.

# Part I

## Cells, Signals, Growth, and Populations

# 3

# Reproduction and the Drive for Survival

## Introduction

This chapter is an introduction to cell structure and biological reproduction and the effects that these have on the survival of species according to the Darwinian model of evolution. The Darwinian model of evolution postulates that all living systems must compete for resources that are too limited to sustain all the organisms that are born. Those organisms possessing properties that are best suited to the environment can survive and may pass the favored properties to their offspring.

A system is said to be alive if it has certain properties. These life properties, e.g., metabolism, reproduction, and response to stimuli, interact with each other, and indeed, the interactions themselves must be part of the list of life properties.

Cells contain organelles, which are subcellular inclusions dedicated to performing specific tasks such as photosynthesis and protein synthesis. Membranes are organelles that are components of other organelles and are functional in their own right—they regulate material transport into and out of cells. Prokaryotic organisms (bacteria and blue-green algae) lack most organelles. Eukaryotic organisms (protozoa, fungi, plants, and animals) have cells with a wide range of organelles.

A cell's genetic information is contained along the length of certain organelles called chromosomes. In asexual reproduction, genetic material of one cell is exactly replicated and the identical copies are partitioned among two daughter cells. Thus the daughter cells end up with genetic information identical to that of the parent cell, a decided advantage if the environment is one in which the parent cell thrived. In multicellular organisms, certain genes may be "turned off" in mitosis; the result will be cells with different behaviors, which leads to the various tissues found in multicellular organisms. Genetic information is not lost in this way; it is merely inactivated, often reversibly. Mitosis also decreases the surface-to-volume ratio of cells, which allows the cell to take up food and release waste more easily.

Sexual reproduction, the combining of genetic information from two parents into one or more offspring, leads to variations among the offspring. This is achieved by the production of novel combinations of genetic information and by complex interactions between genetic materials affecting the same property. The result is the possibility

R.W. Shonkwiler and J. Herod, *Mathematical Biology: An Introduction with Maple and Matlab*, Undergraduate Texts in Mathematics, DOI: 10.1007/978-0-387-70984-0_3,

for immense variation, which is one of the empirical observations at the heart of the Darwinian model.

Left unchecked, populations would grow exponentially, but factors in the environment always control the sizes of populations.

## 3.1 The Darwinian Model of Evolution

*We introduce the Darwinian model of evolution, a model that ties all biology together. Finite resources of all kinds place limits on the reproduction and growth of organisms. All must compete for these resources and most will not get enough. Those that survive may pass their favorable properties to their offspring.*

*The diversity of organisms is represented by taxonomic categories.*

A group of organisms is said to represent a *species* if there is real or potential exchange of genetic material among its members and they are reproductively isolated from all other such groups. Thus members of a single species are capable of interbreeding and producing fertile offspring. By inference, if individuals are very similar but reproductively isolated from one another, they are in different species. The definition above makes good sense in most cases: Horses and cows, different species, live in the same area but never mate; horses and donkeys, different species, may live in the same area and interbreed, but their offspring are sterile mules; lions and tigers, also different species, do not live in the same area, but have interbred in zoos to give sterile offspring. The definition also produces some odd results: St. Bernard dogs and chihuahuas would be in different species by the reproductive-isolation criterion, although both might be in the same species as, say, a fox terrier. English sparrows in the United States and in England would have to be put into different species, even though they are essentially identical. There are other, somewhat different definitions of species. For an in-depth discussion, see [5].

A group of species is a *genus* and a group of genera is a *family*. Higher levels are *orders*, *classes*, *phyla* (called *divisions* in plants), and *kingdoms*. To identify an organism, its generic and specific names are usually given in the following format: *Homo sapiens* (humans) or *Acer rubrum* (red maple trees).

*Living systems operate under a set of powerful constraints.*

1. *Available space is finite.* Some organisms can survive a few kilometers into the air or under water and others live a few meters under the ground, but that does not change the basic premise: Our planet is a sphere of fixed surface area and everything alive must share that area for all its needs, including nutrient procurement and waste disposal.
2. *The temperature range for life is very restricted.* Most living systems cannot function if their internal temperature is outside a range of about 0° to 50°C, the lower limitation being imposed by the destructive effect of ice crystals on

cell membranes, and the upper limit being imposed by heat inactivation of large molecules. Some organisms can extend this range a bit with special mechanisms, e.g., antifreeze-like substances in their bodies, but this temperature limitation is generally not too flexible.

3. *Energetic resources are limited.* The only energy sources originating on earth are geothermal, radioactive, and that which is available in some inorganic compounds. Some organisms, said to be *chemoautotrophic*, can use the latter compounds, but these organisms are exceptional. By far, the majority of the energy available for life comes from the sun. While the sun's energy is virtually inexhaustible, it tends not to accumulate in any long-term biological form on earth. This limitation lies in an empirical observation—the second law of thermodynamics—that energy becomes less useful as it undergoes transformation from one form to another. The transformations that solar energy undergoes are described by a food chain: the sun's energy is captured and used by photosynthetic plants, which are eaten by herbivores, which are eaten by carnivores, which die and are broken down by decomposing organisms. At each step, much of the useful energy is lost irreversibly to the immediate creation of disorder and/or to heat, which radiates away and creates disorder elsewhere. Thus the sun's radiant energy does not accumulate in living systems for longer than a single organism's lifetime, and must be constantly replenished. (See Yeargers [1] for further discussion.)

4. *Physical resources are finite.* Obviously, there is more mass to the inorganic world than to the organic one. The problem is that most of the earth's nonorganic mass is not available to the organisms that inhabit the earth's surface. For example, only tiny fractions of our planet's inventory of such critical materials as carbon, oxygen, and nitrogen are actually available to life. The rest is either underground or tied up in the form of compounds not chemically accessible to life.

*The Darwinian model of evolution correlates biological diversity and the survival of species.*

The four constraints listed above would not be so serious if living organisms were different from what they are. We might picture a world in which every organism was nonreproducing, had a constant size, and was immortal. Perhaps the organisms would be photosynthetic and would have unlimited supplies of oxygen, carbon dioxide, nitrogen, and other important inorganic substances. They would have infinite sinks for waste materials or would produce little waste in the first place.

The biological world just described is, of course, just the opposite of the real one, where there is rapid reproduction and a resultant competition for space and resources. Charles Darwin formulated a model to describe the nature and effect of this competition on living systems. This model may be presented as two empirical observations and two conclusions.

**Observation 1.** *More organisms are born than can survive to reproductive maturity.*

The high death toll among the young, from primitive plants to humans, is plain to see. There simply are not enough resources or space to go around, and the young are among the first to be affected.

**Observation 2.** *All organisms exhibit innate variability.*

While we are easily able to spot differences between humans or even other mammals, it is not easy for us to identify differences between members of a group of daffodils or coral snakes. The differences are there nonetheless, and if we observe the plants and snakes carefully, we will see that, because of the differences, some will thrive and others will not.

**Conclusion 1.** *The only organisms that will survive and reproduce are those whose individual innate variations make them well suited to the environment.*

Note the importance of context here: an organism suited to one environment may be totally unsuited to another. Note also the importance of reproduction; it is not enough to live—one must pass one's genes to subsequent generations. The ability to produce fertile offspring is called *fitness*. This combines the ability to attract a mate with the fertility of offspring. If Tarzan were sterile, he would have zero fitness in spite of mate attraction.

**Conclusion 2.** *Properties favored by selection can be passed on to offspring.*

Selection winnows out the unfit, i.e., those individuals whose innate properties make them less competitive in a given environmental context. The survivors can pass on favored characteristics to their progeny.

*Reproductive isolation can generate new species.*

Suppose that a population, or large, freely interbreeding group, of a species becomes divided in half, such that members of one half can no longer breed with the other half. Genetic mutations and selection in one half may be independent of that in the other half, leading to a divergence of properties between the two halves. After enough time passes, the two groups may accumulate enough differences to become different species, as defined in the previous section. This is the usual method for species creation (see also Section 15.1). An example is found at the Grand Canyon; the squirrels at the north and south rims of the canyon have evolved into different species by virtue of their geographical separation.

The idea of reproductive isolation may suggest geographical separation, but many other forms of separation will work as well. For example, one part of the population may mate at night and the other during the day, even if they occupy the same geographical area. As a second example, we return to dogs: St. Bernards and chihuahuas are reproductively isolated from each other.

## 3.2 Cells

*A cell is not just a bag of sap. It is a mass of convoluted membranes that separate the inside of a cell from the outside world. These membranes also form internal structures that perform specialized tasks in support of the entire cell. Certain primitive cells, e.g., bacteria and some algae, have not developed most of these internal structures.*

*Organelles are cellular inclusions that perform particular tasks.*

A cell is not a bag of homogeneous material. High-resolution electron microscopy shows that the interiors of cells contain numerous simple and complex structures, each functionally dedicated to one or more of the tasks that a cell needs carried out. The cell is thus analogous to a society, each different organelle contributing to the welfare of the whole. The sizes of organelles can range from about one-thousandth of a cell diameter to half a cell diameter, and the number of each kind can range from one to many thousands. The kinds of organelles that cells contain provide the basis for one of the most fundamental taxonomic dichotomies in biology: prokaryotes vs. eukaryotes.

*Eukaryotes have many well-defined organelles and an extensive membrane system.*

The group called the *eukaryotes*[1] include virtually all the kinds of organisms in our everyday world. Mammals, fish, worms, sponges, amoebas, trees, fungi, and most algae are in this group. As the name implies, they have obvious, membrane-limited nuclei. Among their many other organelles, all formed from membranes, one finds an *endoplasmic reticulum* for partitioning off internal compartments of the cell, *chloroplasts* for photosynthesis, *mitochondria* to get energy from food, *ribosomes* for protein synthesis, and an external membrane to regulate the movement of materials into and out of the cell.

*Prokaryotic cells have a very limited set of organelles.*

The organisms called the *prokaryotes*[2] include only two groups, the bacteria and the blue-green algae. They lack a matrix of internal membranes and most other organelles found in eukaryotes. They have genetic material in a more-or-less localized region, but it is not bounded by a membrane; thus prokaryotes lack true nuclei. Prokaryotes have *ribosomes* for protein synthesis, but they are much simpler than those of eukaryotes. The function of prokaryotic mitochondria—getting energy from foods—is performed in specialized regions of the plasma membrane, and the chlorophyll of photosynthetic prokaryotes is not confined to chloroplasts.

---

[1] The word means "with true nuclei."

[2] Prokaryotes lack true nuclei.

## 3.3 Replication of Living Systems

*Living systems can be understood only in terms of the integration of elemental processes into a unified whole. It is the organic whole that defines life, not the components.*

*Asexual reproduction can replace those members of a species that die. The new organisms will be genetically identical to the parent organism. To the extent that the environment does not change, the newly generated organisms should be well suited to that environment.*

*Sexual reproduction results in offspring containing genetic material from two parents. It not only replaces organisms that die, but provides the new members with properties different from those of their parents. Thus Darwinian selection will maximize the chance that some of the new organisms will fit better into their environment than did their parents.*

*What do we mean by a "living system"?*

To deal with this question we need to back up conceptually and ask how we know whether something is alive in the first place. This question causes at least mild embarrassment to every thinking biologist. All scientists know that the solution of any problem must begin with clear definitions of fundamental terms and yet a definition of "life" is as elusive as quicksilver.

If we start with the notion that a definition of a "living system" must come before anything else in biology, then that definition should use only nonbiological terms. However, one virtually always sees living systems defined by taking a group of things everyone has already agreed to be living things, and then listing properties they have in common. Examples of these life properties are organization, response to stimuli, metabolism, growth, evolution, and, of course, reproduction. A system is said to be alive if it has these properties (and/or others) because other systems that have these properties are, by consensus, alive. Thus living systems end up being defined in terms of living systems. This definition is a recursive one: The first case is simply given, and all subsequent cases are defined in terms of one or more preceding ones.

The list of life properties against which a putative living system would be compared is an interesting one because no one property is sufficient. For example, a building is organized, dynamite responds to stimuli, many metabolic reactions can be carried out in a test tube, salt crystals grow, mountain ranges evolve, and many chemical reactions are autocatalytic, spawning like reactions. Of course, we could always insist that the putative system should exhibit two or more of the properties, but clever people will find a nonliving exception.

In spite of these objections, definition by precedent, applied to living systems, has an appealing practicality and simplicity—most six-year-olds are quite expert at creating such definitions. At a more intellectual level, however, recursion always leaves us with the bothersome matter of the first case, which must be accepted as axiomatic—an idea foreign to biology—or accepted as a matter of faith, an idea that makes most scientists cringe.

One way out of this dilemma is to drop the pretense of objectivity. After all, almost everyone, scientist or lay person, will agree with each other that something is or isn't alive. One wag has said, "It's like my wife—I can't define her, but I always know her when I see her." There is, however, a more satisfying way to handle this problem, and that is to note the unity of the life properties list: The listed properties are related to each other. For instance, only a highly organized system could contain enough information to metabolize and therefore to respond to stimuli. A group of organisms evolves and/or grows when some of its members respond to stimuli in certain ways, leading some to thrive and some not. Reproduction, which requires metabolism and growth, can produce variation upon which selection acts. Selection, in turn, requires reproduction to replace those organisms that were weeded out by selection.

We see then that living systems perform numerous elemental processes, none of which is unique to living systems. *What is unique to living systems is the integration of all these processes into a unified, smoothly functioning whole.* Any attempt to limit our focus to one process in isolation will miss the point; for example, we must view reproduction as one part of a highly interacting system of processes. This does not preclude discussion of the individual processes—but it is their *mutual interactions* that characterize life. In Chapter 8, we will further discuss the importance of organization to biological systems by considering biomolecular structure.

*Why do living systems reproduce?*

To try to answer this question we must first lay some groundwork by stating something that is obvious: Every organism is capable of dying. If an organism were incapable of any kind of reproduction, it would surely die at some point and would not be here for us to observe.[3] Reproduction is therefore required as part of any lifestyle that includes the possibility of death, i.e., it includes all living things.

The cause of an organism's death may be built-in, i.e., its life span may be genetically preprogrammed. Alternatively, the organism may wear out, a notion called the "wear-and-tear" theory, suggesting that we collect chemical and physical injuries until something critical in us stops working. Finally, some other organism, ranging from a virus to a grizzly bear, may kill the organism in the course of disease or predation.

A number of reproductive modes have evolved since life began, but they may be collected into two broad categories—asexual and sexual. Asexual reproduction itself is associated with three phenomena: First, there is the matter of a cell's surface-to-volume ratio, which affects the cell's ability to take up food and to produce and release waste. Second, asexual reproduction allows the formation of daughter cells identical to the parent cell, thus providing for metabolic continuity under nonvarying environmental conditions. Third, asexual reproduction allows multicellular organisms to develop physiologically different tissues by allowing genetic information to be switched on and off. This provides for organ formation.

Sexual reproduction, on the other hand, rearranges genetic information by combining genetic contributions from two parents in novel ways; this provides a range

[3] This reasoning is analogous to the "anthropic principle" of cosmology, in response to the question "Why does our universe exist?" The principle says that if any *other* kind of universe existed, we would not be here to observe it. (We do not wish to get too metaphysical here.)

of variations in offspring upon which selection can act. In Chapter 13, we will describe the details of asexual and sexual reproduction in cells. Here we will restrict our discussion to general principles.

*Simple cell division changes the surface-to-volume ratio $\frac{S}{V}$ of a cell.*

An interesting model connects asexual cell division to waste management. Consider a metabolizing spherical cell of radius $R$: The amount of waste the cell produces ought to be roughly proportional to the mass, and therefore to the volume, of the cell. The volume $V$ of a sphere is proportional to $R^3$. On the other hand, the ability of the cell to get rid of waste ought to be proportional to the surface area of the cell, because waste remains in the cell until it crosses the outer cell membrane on the way out. The surface area $S$ is proportional to $R^2$. As a result, the ratio $\frac{S}{V}$, a measure of the cell's ability to get rid of its waste to the cell's production of waste, is proportional to $R^{-1}$. For each kind of cell there must be some minimum value permitted for the ratio $\frac{S}{V} = \frac{1}{R}$, a value at which waste collects faster than the cell can get rid of it. This requires that the cell divide, thus decreasing $R$ and increasing $\frac{S}{V}$. A similar model, describing the ability of a cell to take up and utilize food, should be obvious.

*Asexual reproduction maintains the genetic material of a single parent in its offspring.*

In general, asexual reproduction leads to offspring that are genetically identical to the parent cell. This will be especially useful if the environment is relatively constant; the offspring will thrive in the same environment in which the parent thrived.

Most eukaryotic cells replicate asexually by a process called *mitosis*.[4] In mitosis, a cell's genetic material is copied and each of two daughter cells gets one of the identical copies. At the same time, the cytoplasm and its organelles are divided equally among the daughter cells. Single-celled organisms, such as amoebas, divide asexually by mitosis, as do the individual cells of multicellular organisms like daisies and humans. The details of mitosis are spelled out in Chapter 13, where we also describe how the various cells of a multicellular organism get to be different, in spite of their generation by mitosis.

Entire multicellular organisms can reproduce asexually. A cut-up starfish can yield a complete starfish from each piece. Colonies of trees are generated by the spreading root system of a single tree. These and similar processes create offspring that are genetically identical to the parent.

*The various tissues of multicellular organisms are created by turning genes on and off.*

A human has dozens of physiologically and anatomically different kinds of cell types. Virtually all of them result from mitosis in a fertilized egg. Thus we might expect them all to be identical because they have the same genes.[5]

The differences between the cells is attributable to different active gene sets. The active genes in a liver cell are not the same ones active in a skin cell. Nevertheless,

[4] Bacteria reproduce asexually by a somewhat different process, called binary fission. We will not go into it.

the liver cell and the skin cell contain the same genes, but each cell type has turned off those not appropriate to that cell's function.

*Sexual reproduction provides for variation in offspring.*

Sexual reproduction is characterized by offspring whose genetic material is contributed by two different parents. The interesting thing about the two contributions is that they do not simply add to one another. Rather, they combine in unexpected ways to yield offspring that are often quite different from either parent. Further, each offspring will generally be different from the other offspring. We have only to compare ourselves to our parents and siblings to verify this.

The variations induced by sexual reproduction maximize the chance that at least a few progeny will find a given environment to be hospitable. Of course, this also means that many will die, but in nature that is no problem because those that die will serve as food for some other organism. Note the lack of mercy here—many variants are tried by sexual reproduction and most die. The few survivors perpetuate the species.

Sexual reproduction is found in a very wide variety of organisms, ranging from humans to single-celled organisms such as amoebas and bacteria. In fact, organisms whose life cycles exclude sexual reproduction are so unusual that they are generally put into special taxonomic categories based solely on that fact. In simple organisms, sexual reproduction may not result in increased numbers, but the offspring will be different from the parent cells. Chapter 13 and references [2] and [3] contain detailed discussions of sexual reproduction and genetics.

## 3.4 Population Growth and Its Limitations

*The size of a population, and its trend, has vital significance for that population, for interacting populations, and for the environment. It is believed that the Polynesian population of Easter Island grew too large to be supported by the island's resources, with disastrous consequences for most of the flora and fauna of the island. A large seagull population living near a puffin rookery spells high chick losses for the puffins. And in another example, at the height of a disease, the pathogen load on the victim can reach $10^9$ organisms per milliliter.*

*We now combine the topics of the two previous sections of this chapter, namely, the increase in an organism's numbers and the struggle among them for survival. The result is that in a real situation, population growth is limited.*

*Unchecked growth of a population is exponential.*

One of the observations of the Darwinian model of evolution is that more organisms

[5] As always, there are notable exceptions. Mammalian red blood cells have nuclei when they are first formed, but lose them and spend most of their lives anucleate, therefore without genes.

are born than can possibly survive. (We use the word "born" here in a very broad sense to include all instances of sexual and asexual reproduction.) Let us suppose for a moment that some organism is capable of unchecked reproduction, doubling its numbers at each reproductive cycle. One would become two, then two would become four, four would become eight, etc. After $N$ reproductive cycles, there would be $2^N$ organisms. If the organism's numbers increased $M$-fold at each reproductive cycle, there would be $M^N$ organisms after $N$ reproductive cycles. This kind of growth is exponential, and it can rapidly lead to huge numbers. Table 3.4.1 shows the numbers generated by an organism that doubles at each cycle.

**Table 3.4.1.**

| $N$: | 0 | 1 | 2 | 10 | 25 | 40 | 72 |
|---|---|---|---|---|---|---|---|
| **Number of organisms:** | 1 | 2 | 4 | 1024 | $3.4 \times 10^7$ | $1.1 \times 10^{12}$ | $4.7 \times 10^{21}$ |

Many bacteria can double their numbers every 20 minutes. Each cell could therefore potentially generate $4.7 \times 10^{21}$ cells per day. To put this number into perspective, a typical bacterium has a mass on the order of $10^{-12}$ grams, and a day of reproduction could then produce a mass of $4.7 \times 10^9$ grams of bacteria from each original cell. Assuming that the cells have the density of water, 1 gm/cm$^3$, $10^9$ grams is the mass of a solid block of bacteria about 1.6 meters on a side. Obviously, no such thing actually happens.

*Real life: Population growth meets environmental resistance.*

Every population has births (in the broad sense described above) and it has deaths. The net growth in numbers is (births − deaths). The *per capita growth rate*, $r$, is defined by[6]

$$r = \frac{\text{birth rate} - \text{death rate}}{\text{population size}}.$$

The maximum value that $r$ can have for an organism is $r_{\max}$, called the *biotic potential*. Estimates of $r_{\max}$ have been made by Brewer [4]. They range from about 0.03 per year for large mammals to about 10 per year for insects and about 10,000 per year for bacteria. These numbers are all positive, and we therefore expect organisms growing at their biotic potential to increase in numbers over time, not so dramatically as described by Table 3.4.1, but constantly increasing nevertheless.

We must remember that $r_{\max}$ is the rate of natural increase under optimal conditions, which seldom exist. Under suboptimal conditions, the birth rate will be low and the death rate high, and these conditions may even lead the value of $r$ to be negative. In any case, the value of $r$ will drop as inimical environmental factors make

[6] The units of birth rate are (numbers of births per time) and those of death rate are (number of deaths per time). The units of population are (numbers of individuals) and $r$ is in units of (time)$^{-1}$.

themselves felt. These factors are collectively called *environmental resistance*, and they are responsible for the fact that we are not waist-deep in bacteria or, for that matter, dogs or crabgrass.

From our discussion of evolution, we now understand that some organisms have a higher tolerance for environmental resistance than do others. Those with the highest tolerance will prosper at the expense of those with low tolerance. Our experience, however, is that every species is ultimately controlled at *some* level by environmental resistance.

## 3.5 The Exponential Model for Growth and Decay

Despite its simplicity, most populations do in fact increase exponentially at some time over their existence. There are two parameters governing the process, initial population size and the per capita growth rate. Both or either may be easily determined from experimental data by least squares.

If the growth rate parameter is negative, then the process is exponential decay. Although populations sometimes collapse catastrophically, they can also decline exponentially. Moreover, exponential decay pertains to other phenomena as well, such as radioactive decay. In conjunction with decay processes, it is customary to recast the growth rate as a half-life.

*A constant per capita growth rate leads to exponential growth.*

By a population we mean an interbreeding subpopulation of a species. Often this implies geographical localization, for example, the Easter Island community, or a bacterial colony within a petri dish. The first published model for predicting population size was by Thomas Malthus in 1798, who assumed that the growth rate of a population is proportional to their numbers $y$, that is,

$$\frac{dy}{dt} = ry, \tag{3.5.1}$$

where $r$ is the constant of proportionality. By dividing (3.5.1) by $y$, we see that $r$ is the per capita grow rate,

$$\frac{1}{y}\frac{dy}{dt} = r$$

with units of per time, e.g., per second. Hence Malthus's law assumes the per capita growth rate to be constant. For a no-growth, replacement-only colony, $r$ will be zero.

Malthus's model is a vast oversimplification of survival and reproduction. Although population size can only be integer-valued in reality, by incorporating the derivative $\frac{dy}{dt}$, $y$ is necessarily a continuous variable in this model; it can take on any nonnegative value. Further, the parameter $r$ must be taken as an average value over all population members. Therefore, (3.5.1) is a continuum model and does not apply to extremely small populations. Nevertheless, it captures a germ of truth about population dynamics and is mathematically tractable. It is a significant first step from which better models emerge and to which other models are compared.

If these approximations are a concern, one can opt for a discrete model. The differential equation is replaced by a difference equation instead:

$$y_{t+1} - y_t = \rho y_t, \qquad t = 0, 1, 2, \ldots. \tag{3.5.2}$$

This model provides for increments to the population at the discrete times indicated in (3.5.2). The size of the population at time $t$ is $y_t$. The relationship between the instantaneous growth rate $r$ of the continuous model and that, $\rho$, of this one was worked out in Section 2.5; we repeat it here:

$$e^r = 1 + \rho. \tag{3.5.3}$$

In Table 3.5.1, we see that the two models agree closely for growth rates up to 6% or so.

**Table 3.5.1.** Discrete vs. continuous growth rates.

| $r$ | $1+r$ | $e^r$ |
|---|---|---|
| 0 | 1 | 1 |
| 0.01 | 1.01 | 1.010 |
| 0.02 | 1.02 | 1.020 |
| 0.03 | 1.03 | 1.030 |
| 0.04 | 1.04 | 1.041 |
| 0.05 | 1.05 | 1.051 |
| 0.06 | 1.06 | 1.062 |

From Section 2.5, the solution of the discrete model is given by

$$y_t = (1 + \rho)^t y_0,$$

where $y_0$ is the initial population size. For the solution of the continuous model, we must solve its differential equation.

Equation (3.5.1) can be solved by separating the $y$ and $t$ variables and integrating,

$$\frac{dy}{y} = r\,dt \quad \text{or} \quad \int \frac{dy}{y} = \int r\,dt.$$

These integrals evaluate to

$$\ln y = rt + c,$$

where $c$ is the constant of integration. Exponentiate both sides to obtain

$$y = e^{rt+c}, \quad \text{or} \quad y = y_0 e^{rt}, \tag{3.5.4}$$

where $y_0 = e^c$. In this, the parameter $y_0$ is the value of $y$ when $t = 0$; cf. Section 2.4.

Under Malthus's law (3.5.4), a population increases *exponentially*, ash shown by the solid curve in Figure 3.5.5. While exponential growth cannot continue indefinitely, it is observed for populations when resources are abundant and population density is low; compare the data designated by the circles in Figure 3.5.5. Under these conditions, populations approximate their biotic potential, $r_{\max}$; cf. Section 3.4.

The per capita growth rate parameter $r$ is often given in terms of *doubling time*. Denote by $T_2$ the time when the population size reaches twice its initial value; from (3.5.4), we get

$$2y_0 = y_0 e^{rT_2}.$$

Divide out $y_0$ and solve for $T_2$ by first taking the logarithm of both sides,

$$\ln 2 = rT_2,$$

and then dividing by $r$; the doubling time works out to be

$$T_2 = \frac{\ln 2}{r} \approx \frac{0.7}{r}. \tag{3.5.5}$$

Thus the per capita growth rate and doubling time are inversely related, a higher per capita growth rate makes for a shorter doubling time and conversely. Rearranging (3.5.5) gives the per capita growth rate in terms of doubling time,

$$r = \frac{\ln 2}{T_2}. \tag{3.5.6}$$

Growth rate parameters are not always positive. In the presence of serious adversity, a population can die off exponentially. To make it explicit that the parameter is negative, the sign is usually written out,

$$y = y_0 e^{-\mu t}, \tag{3.5.7}$$

where $\mu > 0$. Such an exponential decay process is characterized by its *half-life* $T_{1/2}$, given by

$$\frac{1}{2} y_0 = y_0 e^{-\mu T_{1/2}},$$

or

$$T_{1/2} = \frac{\ln 2}{\mu} \approx \frac{0.7}{\mu}. \tag{3.5.8}$$

Exponential growth and decay apply to other natural phenomena as well as biological processes. One of these is radioactive decay, where emissions occur in proportion to the amount of radioactive material remaining. The activity of a material in this regard is measured in terms of half-life; for example, the half-life of $^{14}C$ is about 5700 years. Radioactive decay is the scientific basis behind various artifact dating methods using different isotopes. Figure 3.5.1 is a chart for carbon-14.

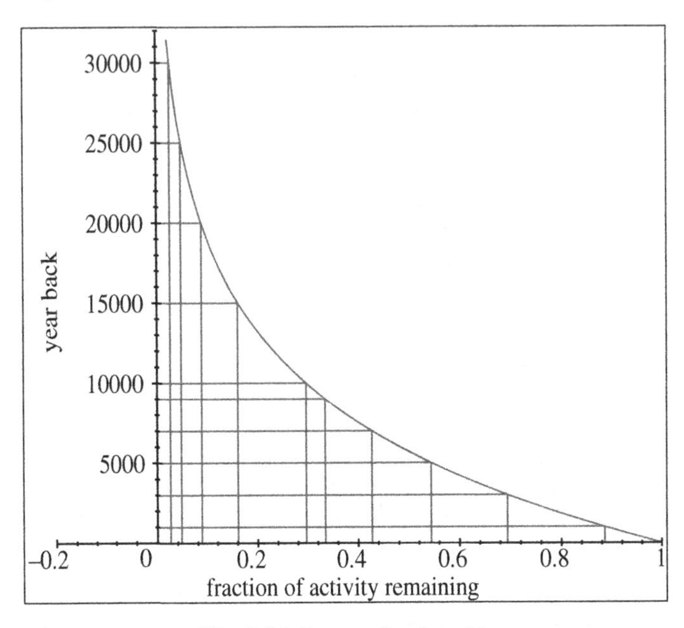


**Fig. 3.5.1.** Decay of carbon-14.

As observed above, Malthus's assumption of immediate reproduction embodied in (3.5.1) hardly seems accurate. Mammals, for instance, undergo a lengthy maturation period. Further, since no real population grows unboundedly, the assumption of constant per capita growth breaks down eventually for all organisms. Nevertheless, there is often a phase in the growth of populations, even populations of organisms with structured life cycles, during which exponential grow is in fact observed. This is referred to as the *exponential growth phase* of the population.

It is possible to mathematically account for a maturation period and hence more accurately model population growth. This is done by the incorporation of a *delay*, $\tau$, between the time offspring are born and the time they reproduce. In differential equation form, we have

$$\left.\frac{dy}{dt}\right|_t = r \cdot y(t-\tau); \tag{3.5.9}$$

in words, the growth rate at the present time is proportional to the population size $\tau$ time units ago (births within the last $\tau$ period of time do not contribute offspring). Equation (3.5.9) is an example of a *delay differential equation*. An initial condition for the equation must prescribe $y(t)$ for $-\tau \le t \le 0$. As an illustration, let $r = 1$, $\tau = 1$, and $y(t) = e^{t/10}$ for $-1 \le t \le 0$ as an initial condition. Begin by setting $f_0(t)$ to this initial function and solving

$$\frac{dy}{dt} = f_0(t-1)$$

for $y$ on the interval $[0, 1]$, that is, for $0 \le t \le 1$. In this, solving means integrating, since the right-hand side is a given function of $t$. Define this solution to be $f_1(t)$ and repeat the procedure with it to get a solution on $[1, 2]$. Continue in this way, moving ahead by steps of length 1.

Code 3.5.1 is a computational procedure that produces the graph of a solution for (3.5.9).

**Code 3.5.1.**

```
MAPLE
> f0:=t->exp(t/10);dsolve({diff(y(t),t)=f0(t-1),y(0)=f0(0)},y(t));
> f1:=unapply(rhs(%),t);dsolve({diff(y(t),t)=f1(t-1),y(1)=f1(1)},y(t));
> f2:=unapply(rhs(%),t);dsolve({diff(y(t),t)=f2(t-1),y(2)=f2(2)},y(t));
> f3:=unapply(rhs(%),t);
> plot({[t,f0(t),t=-1..0],[t,f1(t),t=0..1],[t,f2(t),t=1..2],[t,f3(t),t=2..3]},t=-1..3,y=-1..6,color=black);

MATLAB
% make an m-file, delayFcn0.m:
% function y=delayFcn0(t)
% y = exp(t/10);
> N=10; % # steps per unit interval
> delT=1/N; % so delta t=0.1
% t is now linked to index i by t=-1+(i-1)*delT
% set initial values via delay fcn f0
> for i=1:N+1
> t=-1+(i-1)*delT; f(i)=delayFcn0(t);
> end
% work from t=0 in steps of delT
% ending time tfinal = 2, ending index is n
% solve tfinal=-1+(n-1)*delT for n
> n=(2+1)*N+1;
> for i=N+1:n-1
> t=-1+(i-1)*delT;
> delY=f(i-N)*delT; % N back = delay of 1
> f(i+1)=f(i)+delY; % Euler's method
> end
> t=-1:delT:2; plot(t,f);
```

Delay can also be incorporated into the discrete model in the same way. Equation (3.5.2) is modified to

$$y_{t+1} - y_t = \rho y_{t-\tau}, \quad t = 0, 1, 2, \ldots.$$

Just as for the continuous model, the values $y_{-\tau}, y_{-\tau+1}, \ldots, y_0$ must be prescribed.

A comparison of delay vs. no delay for both the continuous and discrete models is presented in Figure 3.5.2. Although the increase in population size is reduced under grow with delay, the population still follows an exponential-like growth law. The extent to which this is "exponential" is examined in the exercises.

*Growth parameters can be determined from experimental data.*

Exponential growth entails two parameters, initial population size $y_0$ and growth rate $r$. Given $n$ experimental data values, $(t_1, y_1), (t_2, y_2), \ldots, (t_n, y_n)$, we would like to find the specific parameter values for the experiment. As discussed in Section 2.2, this is done by the method of least squares. We first put the equation into a form linear with respect to the parameters; take the logarithm of both sides of (3.5.4):

$$\ln y = \ln y_0 + rt. \tag{3.5.10}$$

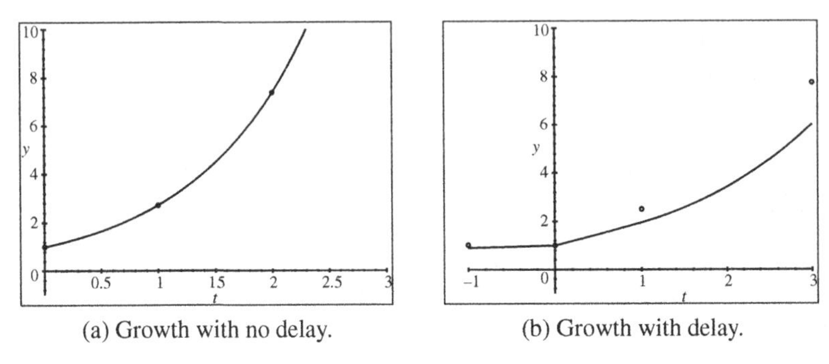


(a) Growth with no delay. (b) Growth with delay.

**Fig. 3.5.2.** The solid curves show the continuous model; circles show growth for the discrete model.

Now use equations (2.2.3) with $\ln y$ playing the role of $y$ and $t$ playing the role of $x$; in turn, the parameters are $\ln y_0$ playing the role of $b$ and $r$ standing in for $m$. We get

$$\ln y_0 = \frac{\sum_{i=1}^n t_i^2 \sum_{i=1}^n \ln y_i - \sum_{i=1}^n t_i \sum_{i=1}^n t_i \ln y_i}{n \sum_{i=1}^n t_i^2 - \left(\sum_{i=1}^n t_i\right)^2}$$

and

$$r = \frac{n \sum_{i=1}^n t_i \ln y_i - \sum_{i=1}^n t_i \sum_{i=1}^n \ln y_i}{n \sum_{i=1}^n t_i^2 - \left(\sum_{i=1}^n t_i\right)^2}.$$

A slightly different problem presents itself when we are sure of the initial population size, $y_0$, and only want to determine $r$ by fit. If there were no experimental error, only one data value $(t_1, y_1)$, besides the starting one, would be needed for this; thus

$$y_1 = y_0 e^{rt_1},$$

so

$$r = \frac{\ln y_1 - \ln y_0}{t_1}.$$

Unfortunately, however, experimental error invariably affects data, and performing this calculation using two data values will likely result in two different (but close) values of $r$. Given $n$ data values (beyond the starting one), $(t_1, y_1), (t_2, y_2), \ldots, (t_n, y_n)$, there will be $n$ corresponding calculations of $r$. Which is the right one?

To solve this, we use a specialization of the least squares method. As above, we use the logarithm form of Malthus's equation, (3.5.10); squared error is then given by

$$E = \sum_{i=1}^n [\ln y_i - (\ln y_0 + rt_i)]^2. \tag{3.5.11}$$

As before, differentiate $E$ with respect to $r$ and set the derivative to zero:

$$2\sum_{i=1}^{n}[\ln y_i - (\ln y_0 + rt_i)](-t_i) = 0.$$

Now solve this for $r$ and get

$$r = \frac{\sum_{i=1}^{n} t_i(\ln y_i - \ln y_0)}{\sum_{i=1}^{n} t_i^2}. \quad (3.5.12)$$

Alternatively, we can let the computer algebra system derive (3.5.12) as follows: Suppose the starting value is known, $y(0) = A$, and we have data given symbolically as

$$\{[a[1], b[1]], [a[2], b[2]], [a[3], b[3]]\}.$$

We find the value of $r$ given in (3.5.12) for this general problem in the following manner:

```
MAPLE
> xval:=[seq(a[i],i=1..3)];
> yval:=[seq(b[i],i=1..3)];
> lny:=map(ln,yval);
> with(stats);
> fit[leastsquare[[x,y],y=r*x+ln(A),{r}]]([xval,lny]);
> coeff(rhs(%),x);
> combine(simplify(%));
```

```
MATLAB
% least squares for r in y=A*exp(rt)
> A=2;
> t=[1 2 3]; % time data
> y=[3 5 9]; % corresponding y data
> lny=log(y)-log(A); % map to log and subtract bias
> M=t'; % set up independent variables M matrix
> r=M\(lny') % and solve
```

**Example 3.5.1 (the U.S. census data).** To illustrate these ideas, we determine the per capita growth rate for the U.S. over the years 1790–1990. In Table 3.5.2, we give the U.S. census for every 10 years, the period required by the U.S. Constitution.

**Table 3.5.2.** U.S. population census. (Source: *Statistical Abstracts of the United States*, 113th ed., Bureau of the Census, U.S. Department of Commerce, Washington, DC, 1993.)

| | | | | | |
|---|---|---|---|---|---|
| 1790 | 3929214 | 1860 | 31433321 | 1930 | 122775046 |
| 1800 | 5308483 | 1870 | 39818449 | 1940 | 131669275 |
| 1810 | 7239881 | 1880 | 50155783 | 1950 | 151325798 |
| 1820 | 9638453 | 1890 | 62947714 | 1960 | 179323175 |
| 1830 | 12866020 | 1900 | 75994575 | 1970 | 203302031 |
| 1840 | 17069453 | 1910 | 91972266 | 1980 | 226545805 |
| 1850 | 23191876 | 1920 | 105710620 | 1990 | 248709873 |

First, we plot the data to note that it does seem to grow exponentially. We read in population data as millions of people and plot the data in order to see that the population apparently is growing exponentially.

```
MAPLE
> restart;
> tt:=[seq(1790+i*10,i=0..20)];
> pop:=[3.929214, 5.308483, 7.239881, 9.638453, 12.866020, 17.069453, 23.191876, 31.433321,
      39.818449, 50.155783, 62.947714, 75.994575, 91.972266, 105.710620, 122.775046,
      131.669275, 151.325798,179.323175, 203.302031, 226.545805, 248.709873];
> data:=[seq([tt[i],pop[i]],i=1..21)];
> plot(data,style=POINT,symbol=CROSS,tickmarks=[4,5]);
```

```
MATLAB
> tt=[1790:10:1990];
> pop=[3.929214 5.308483 7.239881 9.638453 12.866020 17.069453 23.191876 31.433321 39.818449 ...
      50.155783 62.947714 75.994575 91.972266 105.710620 122.775046 131.669275 151.325798 ...
      179.323175 203.302031 226.545805 248.709873];
> plot(tt,pop,'o');
```

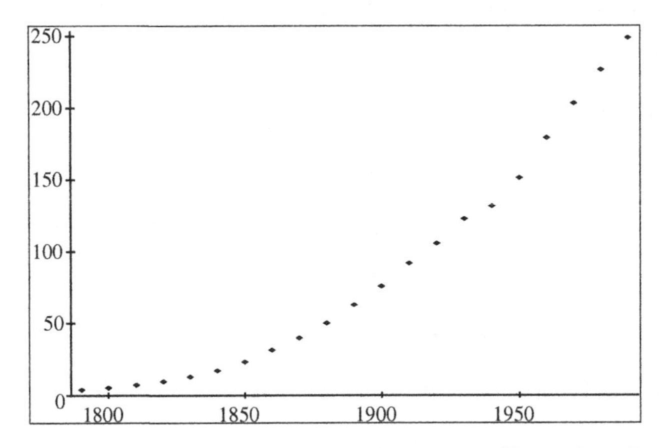


**Fig. 3.5.3.** Population data for the U.S.: years vs. millions of people.

In order to make the data manageable, we rescale the time data by taking 1790 as year zero. A plot of the rescaled data should look exactly the same:

```
MAPLE
> tzeroed=seq[((i-1)*10,i=1..21)];
```

```
MATLAB
> stt=0:10:200; % translated time
```

It appears that the growth of the U.S. population is exponential until about 1940. We will try to get an exponential fit between 1790 and 1930. We take the logarithm of the data. The plot of the logarithm of the data should be approximately a straight line (see Figure 3.5.4):

```
MAPLE
> lnpop:=[seq(ln(pop[i]),i=1..21)];
> plot([seq([tzeroed[i],lnpop[i]],i=1..21)],style=POINT,symbol=CIRCLE);
```

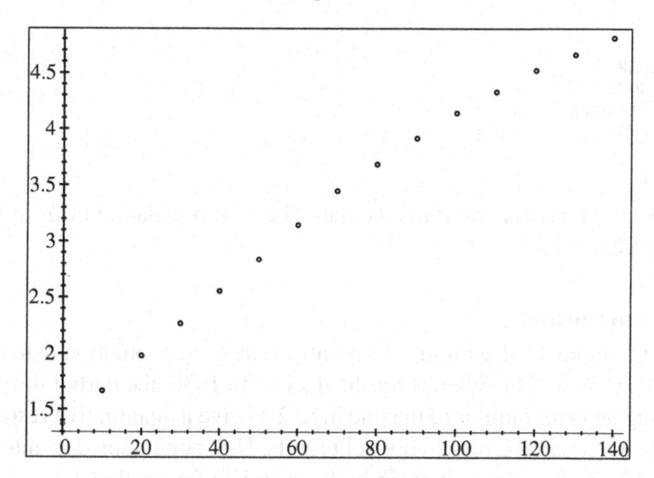


**Fig. 3.5.4.** Logarithm of population data.

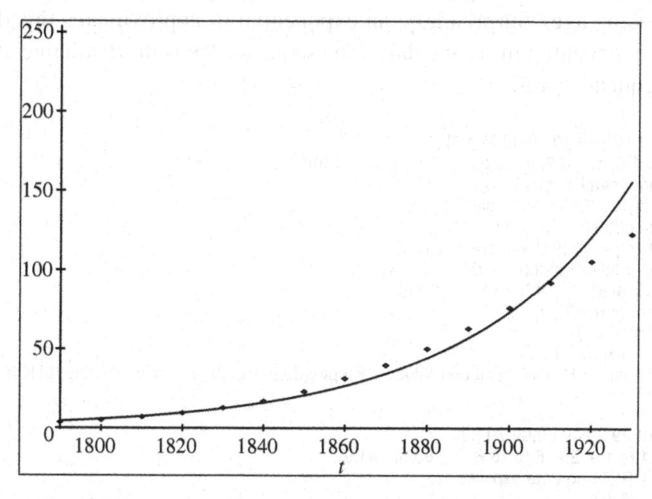


**Fig. 3.5.5.** Exponential growth data fit between 1790 and 1930.

```
MATLAB
> lnPop=log(pop);
> plot(stt,lnPop,'o')
```

The exponential of this linear fit will approximate the data. (See Figure 3.5.5.) Recall that these techniques were used in Chapter 2.

```
MAPLE
> with(stats):
> fit[leastsquare[[t,y],y=m*t+b]]([tzeroed,lnpop]);
> y:=unapply(rhs(%),t);
> J:=plot(exp(y(t-1790)),t=1790..1930,tickmarks=[4,5]):
> K:=plot(data,style=POINT,symbol=CROSS,tickmarks=[4,5]):
> plots[display]({J,K});
```

```
MATLAB
> MT=[stt; ones(1,21)];
> params=MT'\(lnPop')
> lnFit=params(2)+params(1)*stt
> plot(stt,exp(lnFit))
> hold on
> plot(stt,pop,'o')
```

To the degree that this graph fits the data, the U.S. population prior to 1930 was growing exponentially.

### Exercises/Experiments

1. Repeat Example 3.5.1 with just 15 points of the U.S. population data instead of all of them. Which fit is better for the data up to 1930, the partial fit or the total fit? Using an *error* similar to the one in (2.2.1) give a quantitative response (i.e., compute the squared errors). (If this fit for the U.S. population data interests you, note that we will return to it again in the exercises for Section 4.3.)

2. We present below the expected number of deaths per 1000 people as a function of increasing age. Surprisingly, an exponential fit approximates this data well. Find an exponential fit for the data. The sequence `DR` is the death rate at the ages in the sequence `yrs`.

```
MAPLE
> yrs:=([9,19,29,39,49,59,69,79,89]);
> DR:=([.3,1.5, 1.9, 2.9, 6.5, 16.5, 37.0, 83.5, 181.9]);
> pts:=[seq([yrs[i],DR[i]],i=1..9)];
> plot(pts,style=POINT, symbol=CROSS);
> lnpts:=[seq([yrs[i],ln(DR[i]) ], i=1..9)];
> plot(lnpts,style=POINT,symbol=CIRCLE);
> with(stats): lnDR:=map(ln,DR);
> fit[leastsquare[[t,y],y=a*t+b]]([yrs,lnDR]);
> a:=op(1,op(1,rhs(%)));
> b:=op(2,rhs(%%));
> death:=t->exp(a*t+b);
> J:=plot(pts,style=POINT, symbol=CROSS): K:=plot(death(t),t=0..90): plots[display]({J,K});
```

```
MATLAB
> yrs=[9,19,29,39,49,59,69,79,89];
> DR=[.3,1.5, 1.9, 2.9, 6.5, 16.5, 37, 83.5, 181.9];
> plot(yrs,DR); % so data exp. like
> lnDR= log(DR);
> MT=[yrs;ones(size(yrs))];
  % matrix of independent variable data
> params=MT'\lnDR';
> a=params(1); b=params(2);
> fit=exp(b)*exp(a*yrs);
> plot(yrs,DR,yrs,fit);
```

3. Using the least squares methods of this section, and by sampling nine data points on the interval [0, 3], determine whether the growth of the solution for the delay (3.5.12) depicted in Figure 3.5.2(b) is exponential.

4. In Section 2.2 we gave a cubic polynomial fit for the cumulative number of AIDS cases in the U.S. Find an exponential fit for those data. Determine which fit has the smaller error—the cubic polynomial fit or the exponential fit.

5. What would the U.S. population be today if the growth rate from 1790 were (a) 2% higher? (b) 5% higher? (c) 2% lower? (d) 5% lower?

**Questions for Thought and Discussion**

1. What is the surface-to-volume ($\frac{S}{V}$) ratio of a spherical cell with a radius of 2? What is the radius of a spherical cell with $\frac{S}{V} = 4$? A spherical cell with $\frac{S}{V} = 3$ divides exactly in two. What is the $\frac{S}{V}$ ratio of each of the daughter cells?
2. Name some factors that might prevent a population from reaching its biotic potential.
3. Variations induced by sexual reproduction generally lead to the early deaths of many, if not most, of the organisms. What could be advantageous about such a ruthless system?

## References and Suggested Further Reading

[1] BIOPHYSICS OF LIVING SYSTEMS:
E. K. Yeargers, *Basic Biophysics for Biology*, CRC Press, Boca Raton, FL, 1992.

[2] CELL DIVISION AND REPRODUCTION:
W. T. Keeton and J. L. Gould, *Biological Science*, 5th ed., Norton, New York, 1993.

[3] CELL DIVISION AND REPRODUCTION:
W. S. Beck, K. F. Liem, and G. G. Simpson, *Life: An Introduction to Biology*, 3rd ed., Harper–Collins, New York, 1991.

[4] POPULATION BIOLOGY:
R. Brewer, *The Science of Ecology*, 2nd ed., Saunders College Publishing, Fort Worth, TX, 1988.

[5] DEFINITION OF SPECIES:
C. Zimmer, What is a species?, *Sci. Amer.*, **298**-6 (2008), 72–79.

# 4

# Interactions Between Organisms and Their Environment

## Introduction

This chapter is a discussion of the factors that control the growth of populations of organisms.

Evolutionary fitness is measured by the ability to have fertile offspring. Selection pressure is due to both biotic and abiotic factors and is usually very subtle, expressing itself over long time periods. In the absence of constraints, the growth of populations would be exponential, rapidly leading to very large population numbers. The collection of environmental factors that keep populations in check is called *environmental resistance*, which consists of density-independent and density-dependent factors. Some organisms, called $r$-strategists, have short reproductive cycles marked by small prenatal and postnatal investments in their young and by the ability to capitalize on transient environmental opportunities. Their numbers usually increase very rapidly at first, but then decrease very rapidly when the environmental opportunity disappears. Their deaths are due to climatic factors that act independently of population numbers.

A different lifestyle is exhibited by $K$-strategists, who spend a lot of energy caring for their relatively infrequent young, under relatively stable environmental conditions. As the population grows, density-dependent factors such as disease, predation, and competition act to maintain the population at a stable level. A moderate degree of crowding is often beneficial, however, allowing mates and prey to be located. From a practical standpoint, most organisms exhibit a combination of $r$- and $K$-strategic properties.

The composition of plant and animal communities often changes over periods of many years, as the members make the area unsuitable for themselves. This process of succession continues until a stable community, called a *climax community*, appears.

## 4.1 How Population Growth Is Controlled

*In Chapter* 3, *we saw that uncontrolled growth of a biological population is exponential. In natural populations, however, external factors control growth. We can distinguish two extremes of population growth kinetics, depending on the nature of*

R.W. Shonkwiler and J. Herod, *Mathematical Biology: An Introduction with Maple and Matlab*, Undergraduate Texts in Mathematics, DOI: 10.1007/978-0-387-70984-0_4,

*these external factors, although most organisms are a blend of the two. First, r-*strategists *exploit unstable environments and make a small investment in the raising of their young. They produce many offspring, which often are killed off in large numbers by climatic factors. Second, K-*strategists *have few offspring, and invest heavily in raising them. Their numbers are held at some equilibrium value by factors that are dependent on the density of the population.*

*An organism's environment includes biotic and abiotic factors.*

An *ecosystem* is a group of interacting living and nonliving elements. Every real organism sits in such a mixture of living and nonliving elements, interacting with them all at once. A famous biologist, Barry Commoner, has summed this up with the observation that "Everything is connected to everything else." Living components of an organism's environment include those organisms that it eats, those that eat it, those that exchange diseases and parasites with it, and those that try to occupy its space. The nonliving elements include the many compounds and structures that provide the organism with shelter, that fall on it, that it breathes, and that poison it. (See [1, 2, 3, 4] for discussions of environmental resistance, ecology, and population biology.)

*Density-independent factors regulate r-strategists' populations.*

Figure 4.1.1 shows two kinds of population growth curves, in which an initial increase in numbers is followed by either a precipitous drop (curve (a)) or a period of zero growth (curve (b)). The two kinds of growth curves are generated by different kinds of environmental resistance.[1]

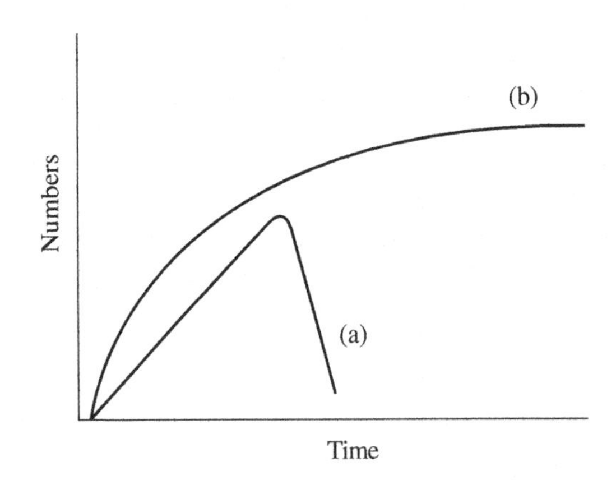


**Fig. 4.1.1.** A graph of the number of individuals in a population vs. time for (a) an idealized $r$-strategist and (b) an idealized $K$-strategist. $r$-strategists suffer rapid losses when density-independent factors like the weather change. $K$-strategists' numbers tend to reach a stable value over time because density-dependent environmental resistance balances birth rate.

[1] Note that the vertical axis in Figure 4.1.1 is the total number of individuals in a population; thus it allows for births, deaths, and migration.

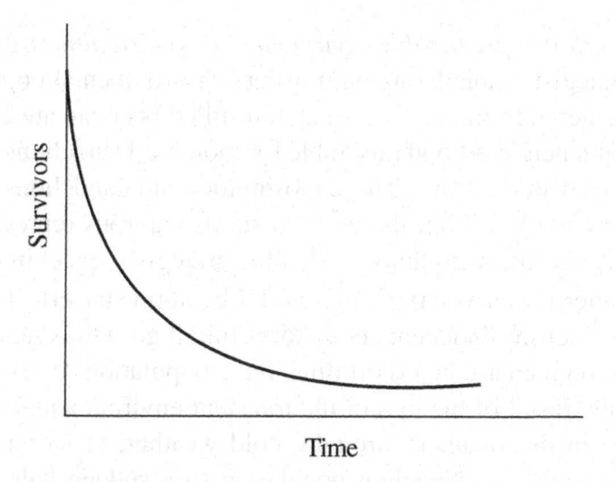


**Fig. 4.1.2.** An idealized survivorship curve for a group of $r$-strategists. The graph shows the number of individuals surviving as a function of time, beginning with a fixed number at time $t = 0$. Lack of parental investment and an opportunistic lifestyle lead to a high mortality rate among the young.

Organisms whose growth kinetics resemble curve (a) of Figure 4.1.1 are called *r-strategists*, and the environmental resistance that controls their numbers is said to be *density-independent*.[2] This means that the organism's numbers are limited by factors that do not depend upon the organism's population density. Climatic factors, such as storms or bitter winters, and earthquakes and volcanoes are density-independent factors in that they exert their effects on dense and sparse populations alike.

Two characteristics are helpful in identifying $r$-strategists:

1. *Small parental investment in their young.* The concept of "parental investment" combines the energy and time dedicated by the parent to the young in both the prenatal and the postnatal periods. Abbreviation of the prenatal period leads to the birth of physiologically vulnerable young, while abbreviation of postnatal care leaves the young unprotected. As a result, an $r$-strategist must generate large numbers of offspring, most of whom will not survive long enough to reproduce themselves. Enough, however, will survive to continue the population. Figure 4.1.2 is a *survivorship curve* for an $r$-strategist; it shows the number of survivors from a group as a function of time.[3] Note the high death rate during early life.

Because of high mortality among its young, an $r$-strategist must produce many offspring, which makes death by disease and predation numerically unimportant, inasmuch as the dead ones are quickly replaced. On the other hand, the organism's short life span ensures that the availability of food and water do not become limiting factors either. Thus density-dependent factors such as predation and resource availability do not affect the population growth rates of $r$-strategists.

---

[2] The symbol $r$ indicates the importance of the rate of growth, which is also symbolized by $r$.

[3] Note that the vertical axes in Figures 4.1.2 and 4.1.4 are the numbers of individuals surviving from an initial, fixed group; thus they allow only for deaths.

2. *The ability to exploit unpredictable environmental opportunities rapidly.* It is common to find $r$-strategists capitalizing on transient environmental opportunities. The mosquitoes that emerge from one discarded, rain-filled beer can are capable of making human lives in a neighborhood miserable for months. Dandelions can quickly fill up a small patch of disturbed soil. These mosquitoes and dandelions have exploited situations that may not last long; therefore, a short, vigorous reproductive effort is required. Both organisms, in common with all $r$-strategists, excel in that regard.

We can now interpret curve (a) of Figure 4.1.1 by noting the effect of *environmental resistance*, i.e., density-independent factors. Initial growth is rapid and it results in a large population increase in a short time, but a population "crash" follows. This crash is usually the result of the loss of the transient environmental opportunity because of changes in the weather: drought, cold weather, or storms can bring the growth of the mosquito or dandelion population to a sudden halt. By this time, however, enough offspring have reached maturity to propagate the population.

*Density-dependent factors regulate the populations of $K$-strategists.*

Organisms whose growth curve resembles that of curve (b) of Figure 4.1.1 are called *$K$-strategists*, and their population growth rate is regulated by population *density-dependent* factors. As with $r$-strategists, the initial growth rate is rapid, but as the density of the population increases, certain resources such as food and space become scarce, predation and disease increase, and waste begins to accumulate. These negative conditions generate a feedback effect: Increasing population density produces conditions that slow down population growth. An equilibrium situation results in which the population growth curve levels out; this long-term, steady-state population is the *carrying capacity* of the environment.

The carrying capacity of a particular environment is symbolized by $K$; hence the name "$K$-strategist" refers to an organism that lives in the equilibrium situation described in the previous paragraph. The growth curve of a $K$-strategist, shown as (b) in Figure 4.1.1, is called a *logistic curve*. Figure 4.1.3 is a logistic curve for a more realistic situation.

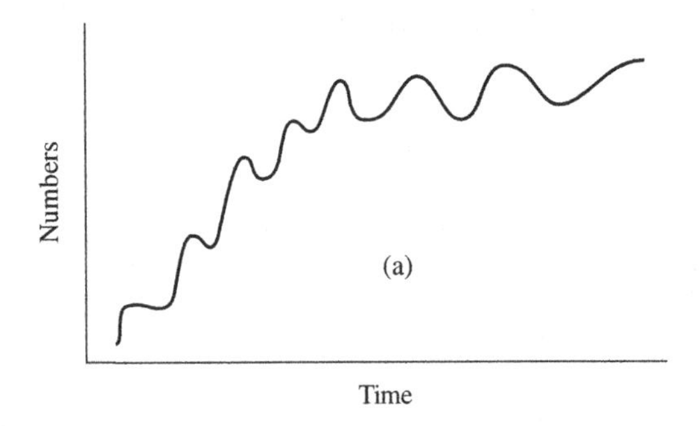


**Fig. 4.1.3.** A more realistic growth curve of a population of $K$-strategists. The numbers fluctuate around an idealized curve, as shown. Compare this with Figure 4.1.1(b).

Two characteristics are helpful in identifying $K$-strategists:

1. *Large parental investment in their young.* $K$-strategists reproduce slowly, with long gestation periods, to increase physiological and anatomical development of the young, who therefore must be born in small broods. After birth, the young are tended until they can reasonably be expected to fend for themselves. One could say that $K$-strategists put all their eggs in one basket and then watch that basket very carefully.

Figure 4.1.4 is an idealized survivorship curve for a $K$-strategist. Note that infant mortality is low (compared to $r$-strategists—see Figure 4.1.2).

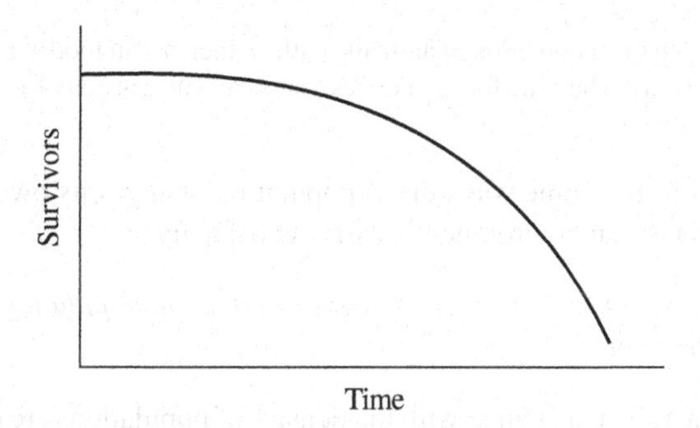


**Fig. 4.1.4.** An idealized survivorship curve for a group of $K$-strategists. The graph shows the number of individuals surviving as a function of time, beginning with a fixed number at time $t = 0$. High parental investment leads to a low infant mortality rate.

2. *The ability to exploit stable environmental situations.* Once the population of a $K$-strategist has reached the carrying capacity of its environment, the population size stays relatively constant. This is nicely demonstrated by the work of H. N. Southern, who studied mating pairs of tawny owls in England [5]. The owl pairs had adjacent territories, with each individual pair occupying a territory that was its own and which was the right size to provide it with nesting space and food (mainly rodents). Every year some adults in the area died, leaving one or more territories that could be occupied by new mating pairs. Southern found that while the remaining adults could have more than replaced those who died, only enough owlets survived in each season to keep the overall numbers of adults constant. The population control measures at work were failure to breed, reduced clutch size, death of eggs and chicks, and emigration. These measures ensured that the total number of adult owls was about the same at the start of each new breeding season.

As long as environmental resistance remains the same, so will population numbers. But if the environmental resistance changes, the carrying capacity of the environment will, too. For example, if the amount of food is the limiting factor, a new value of $K$ is attained when the amount of food increases. This is shown in Figure 4.1.5.

The density-dependent factors that, in conjunction with the organism's reproductive drive, maintain a stabilized population are discussed in the next section. In a later

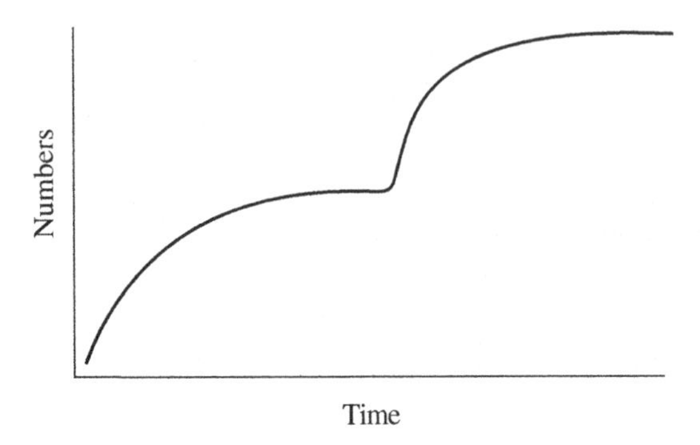


**Fig. 4.1.5.** The growth of a population of animals, with an increase in food availability midway along the horizontal axis. The extra food generates a new carrying capacity for the environment.

section, we will discuss some ways that a population changes its own environment, and thereby changes that environment's carrying capacity.

*Some density-dependent factors exert a negative effect on populations and can thus help control K-strategists.*

Environmental factors that change with the density of populations are of many kinds. This section is a discussion of several of them.

*Predation.* The density of predators, free-living organisms that feed on the bodies of other organisms, would be expected to increase or decrease with the density of prey populations. Figure 4.1.6 shows some famous data, the number of hare and lynx pelts brought to the Hudson Bay Company in Canada over a period of approximately 90 years. Over most of this period, changes in the number of hare pelts led to changes in the number of lynx pelts, as anticipated. After all, if the density of hares increased we would expect the lynx density to follow suit. A detailed study of the data, however, reveals that things were not quite that simple, because in the cycles beginning in 1880 and 1900 the lynxes led the hares. Analysis of this observation can provide us with some enlightening information.

Most importantly, prey population density may depend more strongly on its own food supplies than on predator numbers. Plant matter, the food of many prey species, varies in availability over periods of a year or more. For example, Figure 4.1.7 shows how a tree might partition its reproductive effort (represented by nut production) and its vegetative effort (represented by the size of its annual tree rings). Note the cycles of abundant nut production (called *mast years*) alternating with periods of vigorous vegetative growth; these alternations are common among plants. We should expect that the densities of populations of prey, which frequently are herbivores, would increase during mast years and decrease in other years, independently of predator density (see [2]).

There are some other reasons why we should be cautious about the Hudson Bay data: First, in the absence of hares, lynxes might be easier to catch because, being hungry, they would be more willing to approach baited traps. Second, the naive

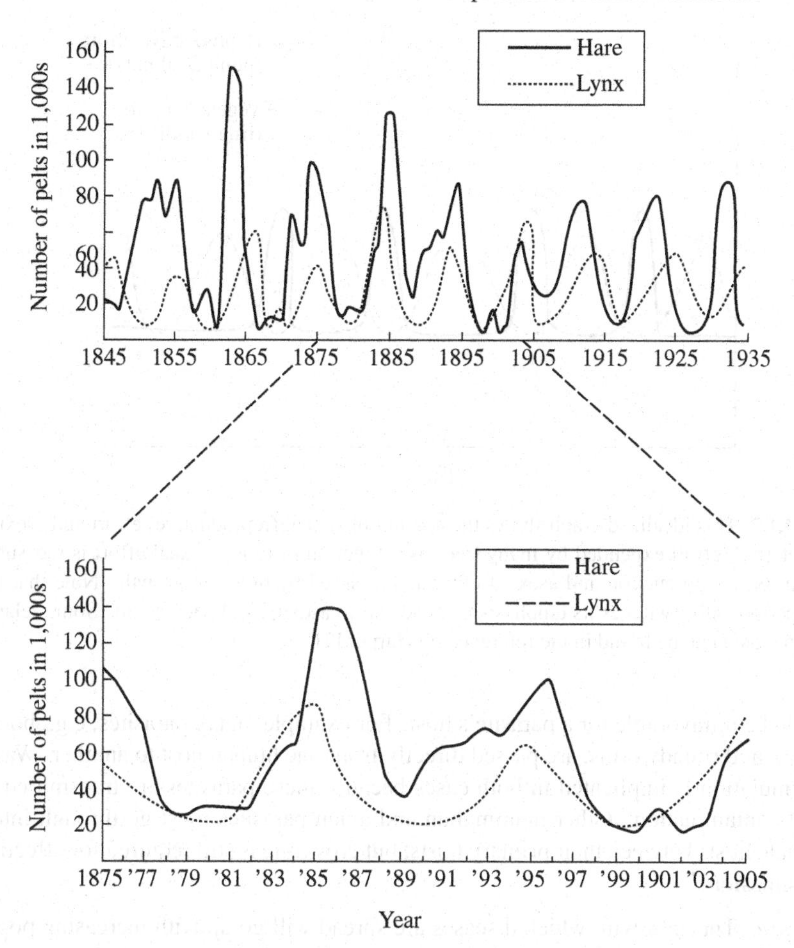


**Fig. 4.1.6.** The Hudson's Bay Company data. The curve shows the number of predator (lynx) and prey (hare) pelts brought to the company by trappers over a 90-year period. Note that from 1875 to 1905, changes in the lynxes sometimes precede changes in the hares. (Redrawn from D. A. McLulich, Sunspots and abundance of animals, *J. Roy. Astronom. Soc. Canada*, **30** (1936), 233. Used with permission.)

interpretation of Figure 4.1.6 assumes equal trapping efficiencies of prey and predator. Third, for the data to be interpreted accurately, the hares whose pelts are enumerated in Figure 4.1.6 should consist solely of a subset of all the hares that could be killed by lynxes, and the lynxes whose pelts are enumerated in the figure should consist solely of a subset of all the lynxes that could kill hares. The problem here is that very young and very old lynxes, many of whom would have contributed pelts to the study, may not kill hares at all (e.g., because of infirmity they may subsist on carrion).

*Parasitism. Parasitism* is a form of interaction in which one of two organisms benefits and the other is harmed but not generally killed. A high population density

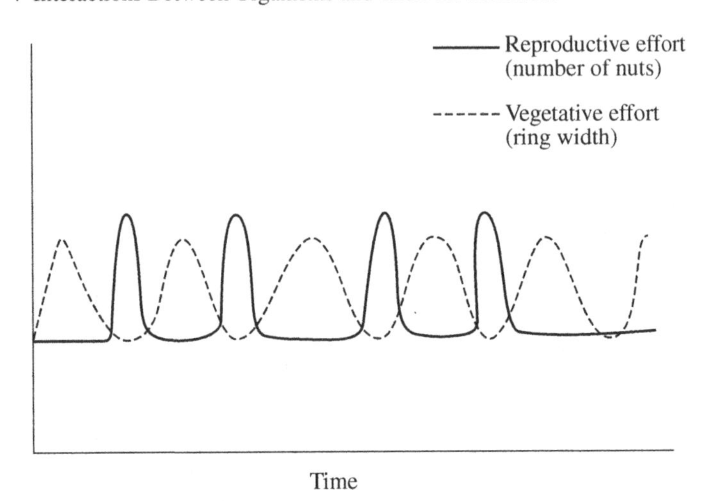


**Fig. 4.1.7.** This idealized graph shows the amount of sexual (reproductive) effort and asexual (vegetative) effort expended by many trees as a function of time. Sexual effort is measured by nut (seed) production and asexual effort is measured by tree ring growth. Note that the tree periodically switches its emphasis from sexual to asexual and back again. Some related original data can be found in the reference by Harper [2].

would be unfavorable for a parasite's host. For example, many parasites, e.g., hookworms and roundworms, are passed directly from one human host to another. Waste accumulation is implicated in both cases because these parasites are transmitted in fecal contamination. Other mammalian and avian parasites must go through intermediate hosts between their primary hosts, but crowding is still required for effective transmission.

*Disease.* The ease with which diseases are spread will go up with increasing population density. The spread of colds through school populations is a good example.

An important aggravating factor in the spread of disease is the accumulation of waste. For example, typhoid fever and cholera are easily carried between victims by fecal contamination of drinking water.

*Interspecific competition.* Every kind of organism occupies an *ecological niche*, which is the functional role that organism plays in its community. An organism's niche includes a consideration of all of its behaviors, their effects on the other members of the community, and the effects of the behaviors of other members of the community on the organism in question.

An empirical rule in biology, *Gause's law*, states that no two species can long occupy the same ecological niche. What will happen is that differences in fitness, even very subtle ones, will eventually cause one of the two species to fill the niche, eliminating the other species. This concept is demonstrated by Figure 4.1.8. When two organisms compete in a uniform habitat, one of the two species always becomes extinct. The "winner" is usually the species having a numerical advantage at the outset of the experiment. (Note the role of luck here—a common and decisive variable in

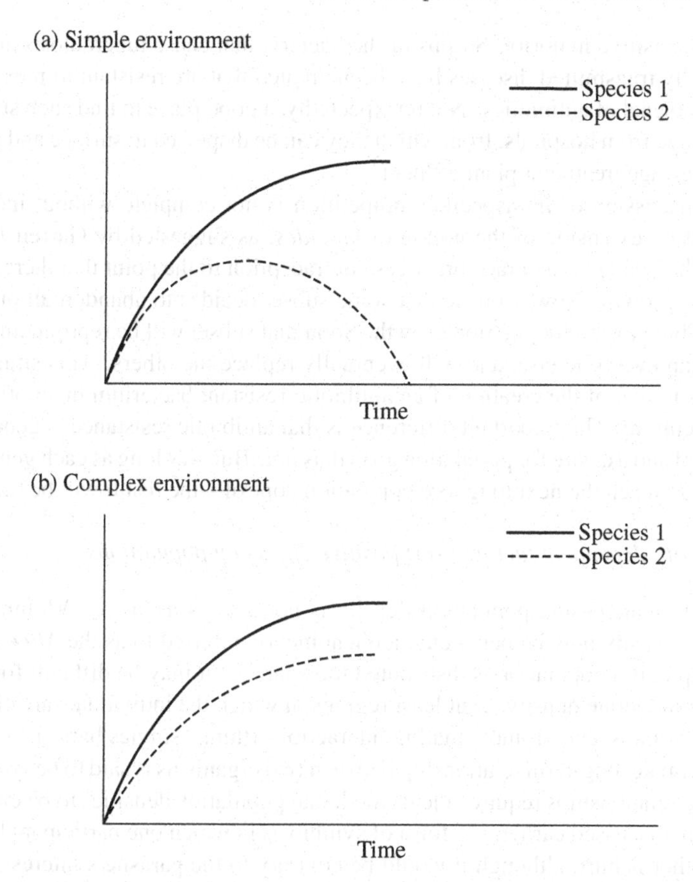


**Fig. 4.1.8.** Graphs showing the effect of environmental complexity on interspecies relationships. The data for (a) are obtained by counting the individuals of two species in a pure growth medium. The data for (b) are obtained by counting the individuals of the two species in a mechanically complex medium where, for example, pieces of broken glass tubing provide habitats for species 2. The more complex environment supports both species, while the simpler environment supports only one species.

Darwinian evolution.) On the other hand, when the environment is more complex, both organisms can thrive because each can fit into its own special niche.

*Intraspecific competition.* As individuals die, they are replaced by new individuals who are presumably better suited to the environment than their predecessors. The general fitness of the population thus improves because it becomes composed of fitter individuals.

The use of antibiotics to control bacterial diseases has contributed immeasurably to the welfare of the human species. Once in a while, however, a mutation occurs in a bacterium that confers on it resistance to that antibiotic. The surviving bacterium can then exploit its greater fitness to the antibiotic environment by reproducing rapidly, making use of the space and nutritional resources provided by the deaths of the antibiotic-sensitive majority. Strains of the bacteria that cause tuberculosis and several sexually transmitted diseases have been created that are resistant to most of the available arsenal of antibiotics. Not unexpectedly, a good place to find such strains is in the sewage from hospitals, from which they can be dispersed to surface and ground water in sewage treatment plant effluent.

This discussion of intraspecific competition is not complete without including an interesting extension of the notion of *biocides*, as suggested by Garrett Hardin. Suppose the whole human race practices contraception to the point that there is zero population growth. Now suppose that some subset decides to abandon all practices that contribute to zero population growth. Soon that subset will be reproducing more rapidly than everyone else, and will eventually replace the others. This situation is analogous to that of the creation of an antibiotic-resistant bacterium in an otherwise sensitive culture. The important difference is that antibiotic resistance is genetically transmitted and a desire for population growth is not. But—as long as each generation continues to teach the next to ignore population control—the result will be the same.

*Some density-dependent factors exert positive effects on populations.*

The effect of increasing population density is not always negative. Within limits, increasing density may be beneficial, a phenomenon referred to as the *Allee effect*.[4] For example, if a population is distributed too sparsely, it may be difficult for mates to meet; a moderate density, or at least regions in which the individuals are clumped into small groups, can promote mating interactions (think "singles bars").

An intimate long-term relationship between two organisms is said to be *symbiotic*. Symbiotic relationships require at least moderate population densities to be effective. Parasitism, discussed earlier, is a form of symbiosis in which one participant benefits and the other is hurt, although it would be contrary to the parasite's interests to kill the host. The closeness of the association between parasite and host is reflected in the high degree of parasite–host specificity. For instance, the feline tapeworm does not often infect dogs, nor does the canine tapeworm often infect cats.

Another form of symbiosis is *commensalism*, in which one participant benefits and the other is unaffected. An example is the nesting of birds in trees: The birds profit from the association, but the trees are not affected.

The third form of symbiosis recognized by biologists is *mutualism*, in which both participants benefit. An example is that of termites and certain microorganisms that inhabit their digestive systems. Very few organisms can digest the cellulose that makes up wood; the symbionts in termite digestive systems are rare exceptions. The termites provide access to wood and the microorganisms provide digestion. Both can use the digestive products for food, so both organisms profit from the symbiotic association.

*It would be unexpected to find a pure $K$-strategist or a pure $r$-strategist.*

The discussions above, in conjunction with Figure 4.1.1, apply to idealized $K$- or

[4] Named for a prominent population biologist.

$r$-strategists. Virtually all organisms are somewhere in between the two, being controlled by a mixture of density-independent and density-dependent factors. For example, a prolonged drought is nondiscriminatory, reducing the numbers of both mosquitoes and rabbits. The density of mosquitoes might be reduced more than that of rabbits, but both will be reduced to some degree. On the other hand, both mosquitoes and rabbits serve as prey for other animals. There are more mosquitoes in a mosquito population than rabbits in a rabbit population, and the mosquitoes reproduce faster, so predation will affect the rabbits more. Still, both animals suffer from predation to some extent.

Density-independent factors may control a population in one context and density-dependent factors may control it in another context. A bitter winter could reduce rodent numbers for a while and then, as the weather warms up, predators, arriving by migration or arousing from hibernation, might assume control of the numbers of rodents. Even the growth of human populations can have variable outcomes, depending on the assumption of the model (see [6]).

*The highest sustainable yield of an organism is obtained during the period of most rapid growth.*

Industries like lumbering or fishing have, or should have, a vested interest in sustainable maintenance of their product sources. The key word here is "sustainable." It is possible to obtain a very high initial yield of lumber by clear-cutting a mature forest or by seining out all the fish in a lake. Of course, this is a one-time event and is therefore self-defeating. A far better strategy is to keep the forest or fish population at its point of maximal growth, i.e., the steepest part of the growth curve (b) in Figure 4.1.1. The population, growing rapidly, is then able to replace the harvested individuals. Any particular harvest may be small, but the forest or lake will continue to yield products for a long time, giving a high long-term yield. The imposition of bag limits on duck hunters, for instance, has resulted in the stable availability of wild ducks, season after season. Well-managed hunting can be viewed as a density-dependent population-limiting factor that replaces predation, disease, and competition, all of which would kill many ducks anyway.

## 4.2 Community Ecology

*There is a natural progression of plant and animal communities over time in a particular region. This progression occurs because each community makes the area less hospitable to itself and more hospitable to the succeeding community. This succession of communities will eventually stabilize into a climax community that is predictable for the geography and climate of that area.*

*Continued occupation of an area by a population may make that region less hospitable to them and more hospitable to others.*

Suppose that there is a *community* (several interacting populations) of plants in and around a small lake in north Georgia. Starting from the center of the lake and moving outward, we might find algae and other aquatic plants in the water, marsh plants and low shrubs along the bank, pine trees farther inland, and finally, hardwoods well removed from the lake. If one could observe this community for a hundred or so years, the pattern of populations would be seen to change in a predictable way.

As the algae and other aquatic plants died, their mass would fill up the lake, making it hostile to those very plants whose litter filled it. Marsh plants would start growing in the center of the lake, which would now be boggy. The area that once rimmed the lake would start to dry out as the lake disappeared, and small shrubs and pine trees would take up residence on its margins. Hardwoods would move into the area formerly occupied by the pine trees. This progressive change, called *succession*, would continue until the entire area was covered by hardwoods, after which no further change would be seen. The final, stable, population of hardwoods is called the *climax community* for that area. Climax communities differ from one part of the world to another, e.g., they may be rain forests in parts of Brazil and tundra in Alaska, but they are predictable.

If the hardwood forest described above is destroyed by lumbering or fire, a process called *secondary succession* ensues: Grasses take over, followed by shrubs, then pines, and then hardwoods again. Thus both primary and secondary succession lead to the same climax community.

Succession applies to both plant and animal populations, and as the above example demonstrates, it is due to changes made in the environment by its inhabitants. The drying of the lake is only one possible cause of succession; for instance, the leaf litter deposited by trees could change the pH of the soil beneath the trees, thus reducing mineral uptake by the very trees that deposited the litter. A new population of trees might then find the soil more hospitable, and move in. Alternatively, insects might drive away certain of their prey, making the area less desirable for the insects and more desirable for other animals.

## 4.3 Environmentally Limited Population Growth

*Real populations do not realize constant per capita growth rates. By engineering the growth rate as a function of the population size, finely structured population models can be constructed. Thus if the growth rate is taken to decrease to zero with increasing population size, then a finite limit, the carrying capacity, is imposed on the population. On the other hand, if the growth rate is assigned to be negative at small population sizes, then small populations are driven to extinction.*

*Along with the power to tailor the population model in this way comes the problem of its solution and the problem of estimating parameters. However, for one-variable models, simple sign considerations predict the asymptotic behavior and numerical methods can easily display solutions.*

*Logistic growth stabilizes a population at the environmental carrying capacity.*

As discussed in Sections 3.1 and 4.1, when a biological population becomes too large, the per capita growth rate diminishes. This is because the individuals interfere with each other and are forced to compete for limited resources. Consider the model, due to Pierre Verhulst in 1845, wherein the per capita growth rate decreases linearly with population size $y$:

$$\frac{1}{y}\frac{dy}{dt} = r\left(1 - \frac{y}{K}\right). \tag{4.3.1}$$

The profile of the right-hand side is depicted in Figure 4.3.1.

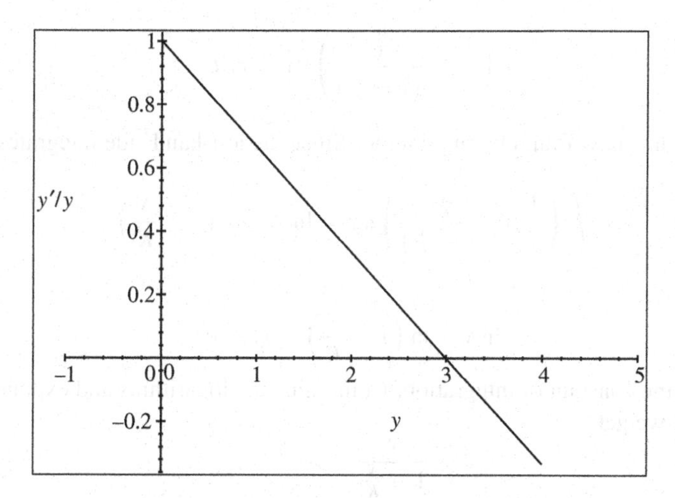


**Fig. 4.3.1.** Linearly decreasing per capita growth rate ($r = 1$, $K = 3$).

This differential equation is known as the *logistic (differential) equation*; two of its solutions are graphed later in Figure 4.3.2. Multiplying (4.3.1) by $y$ yields the alternative form

$$\frac{dy}{dt} = ry\left(1 - \frac{y}{K}\right). \tag{4.3.2}$$

From this equation we see that the derivative $\frac{dy}{dt}$ is zero when $y = 0$ or $y = K$. These are the *stationary points* of the equation (see Section 2.4). The stationary point $y = K$, at which the per capita growth rate becomes zero, is called the *carrying capacity* (of the environment).

When the population size $y$ is small, the term $\frac{y}{K}$ is nearly zero and the per capita growth rate is approximately $r$ as in exponential growth. Thus for small population size (but not so small that the continuum model breaks down), the population increases exponentially. Hence solutions are repelled from the stationary point $y = 0$. But as the population size approaches the carrying capacity $K$, the per capita growth rate decreases to zero and the population ceases to change in size. Further, if the population size ever exceeds the carrying capacity for some reason, then the per capita growth rate will be negative and the population size will decrease to $K$. Hence solutions are globally attracted to the stationary point $y = K$.

From the form (4.3.2) of the logistic equation we see that it is nonlinear, with a quadratic nonlinearity in $y$. Nevertheless, it can be solved by separation of variables (see Section 2.4). Rewrite (4.3.1) as

$$\frac{dy}{y\left(1-\frac{y}{K}\right)} = r dt.$$

The fraction on the left-hand side can be expanded by *partial fraction decomposition* and written as the sum of two simpler fractions (check this by reversing the step)

$$\left(\frac{1}{y} + \frac{\frac{1}{K}}{\left(1-\frac{y}{K}\right)}\right) dy = r dt.$$

The solution is now found by integration. Since the left-hand side integrates to

$$\int \left(\frac{1}{y} + \frac{\frac{1}{K}}{\left(1-\frac{y}{K}\right)}\right) dy = \ln y - \ln\left(1-\frac{y}{K}\right),$$

we get

$$\ln y - \ln\left(1-\frac{y}{K}\right) = rt + c, \tag{4.3.3}$$

where $c$ is the constant of integration. Combining the logarithms and exponentiating both sides, we get

$$\frac{y}{1-\frac{y}{K}} = Ae^{rt}, \tag{4.3.4}$$

where $A = e^c$, and $A$ is not the $t = 0$ value of $y$. Finally, we solve (4.3.4) for $y$. First, divide numerator and denominator of the left-hand side by $y$ and reciprocate both sides; this gives

$$\frac{1}{y} - \frac{1}{K} = \frac{1}{Ae^{rt}},$$

or, isolating $y$,

$$\frac{1}{y} = \frac{1}{Ae^{rt}} + \frac{1}{K}. \tag{4.3.5}$$

Now reciprocate both sides of this and get

$$y = \frac{1}{\frac{1}{Ae^{rt}} + \frac{1}{K}},$$

or equivalently,

$$y = \frac{Ae^{rt}}{1 + \frac{A}{K}e^{rt}}. \tag{4.3.6}$$

Equation (4.3.6) is the solution of the logistic equation (4.3.1). To emphasize that it is the concept of "logistic growth" that is important here, not these solution techniques, we show how a solution for (4.3.1) can be found (symbolically) by the computer algebra system. The initial value is taken as $y(0) = y_0$ in the following:

MAPLE

```
> dsolve({diff(y(t),t) = r*y(t)*(1-y(t)/k),y(0)=y0},y(t));
```

The output of this computation is

$$y(t) = \frac{k}{1 + \frac{e^{-rt}(k-y_0)}{y_0}}.$$

Clearing the compound denominator easily reduces this computer solution to

$$y(t) = \frac{ke^{rt}y_0}{y_0(e^{rt}-1)+k}.$$

Three members of the family of solutions (4.3.6) are shown in Figure 4.3.2 for different starting values $y_0$. We take $r = 1$ and $K = 3$ and find solutions for (4.3.2) with $y_0 = 1$, or 2, or 4.

*Logistic parameters can sometimes be estimated by least squares.*

Unfortunately, the logistic solution (4.3.6) is not linear in its parameters $A$, $r$, and $K$. Therefore, there is no straightforward way to implement least squares. However, if the data values are separated by fixed time periods, $\tau$, then it is possible to remap the equations so least squares will work.

Suppose the data points are $(t_1, y_1), (t_2, y_2), \ldots, (t_n, y_n)$ with $t_i = t_{i-1} + \tau$, $i = 2, \ldots, n$. Then $t_i = t_1 + (i-1)\tau$ and the predicted value of $\frac{1}{y_i}$, from (4.3.5), is given by

$$\frac{1}{y_i} = \frac{1}{Ae^{rt_1}e^{(i-1)r\tau}} + \frac{1}{K} = \frac{1}{e^{r\tau}}\left[\frac{1}{Ae^{rt_1}e^{(i-2)r\tau}} + \frac{e^{r\tau}}{K}\right]. \tag{4.3.7}$$

But by rewriting the term involving $K$ as

$$\frac{1}{K} + \frac{e^{r\tau}-1}{K},$$

and using (4.3.5) again, (4.3.7) becomes

$$\frac{1}{y_i} = \frac{1}{e^{r\tau}}\left[\frac{1}{y_{i-1}} + \frac{e^{r\tau}-1}{K}\right].$$

Now put $z = \frac{1}{y}$, and we have

$$z_i = e^{-r\tau}z_{i-1} + \frac{1-e^{-r\tau}}{K}, \quad \text{where } y = \frac{1}{z}. \tag{4.3.8}$$

A least squares calculation is performed on the points $(z_1, z_2), (z_2, z_3), \ldots, (z_{n-1}, z_n)$ to determine $r$ and $K$. With $r$ and $K$ known, least squares can be performed on, say (4.3.5), to determine $A$.

In the exercises we will illustrate this method and suggest another for U.S. population data.

```
MAPLE
> r:=1;k:=3;
> dsolve({diff(y(t),t)=r*y(t)*(1-y(t)/k),y(0)=1},y(t));
  y1:=unapply(rhs(%),t);
> dsolve({diff(y(t),t)=r*y(t)*(1-y(t)/k),y(0)=2},y(t));
  y2:=unapply(rhs(%),t);
> dsolve({diff(y(t),t)=r*y(t)*(1-y(t)/k),y(0)=4},y(t));
  y4:=unapply(rhs(%),t);
> plot({y1(t),y2(t),y4(t)},t=0..5,y=0..5);
```

```
MATLAB
% solve the logistic eqn for starting values y0=1, y0=2, y0=4
% Make up an m-file, fig432.m, as follows:
%   function yprime = fig432(t,y) % with r=1 and K=3
%   r=1; K=3; yprime=y.*r.*(1-y./K);
> tspan=[0 5];
> [t1,y1]=ode23('fig432',tspan,1);
> [t2,y2]=ode23('fig432',tspan,2);
> [t4,y4]=ode23('fig432',tspan,4);
> plot(t1,y1,t2,y2,t4,y4)
```

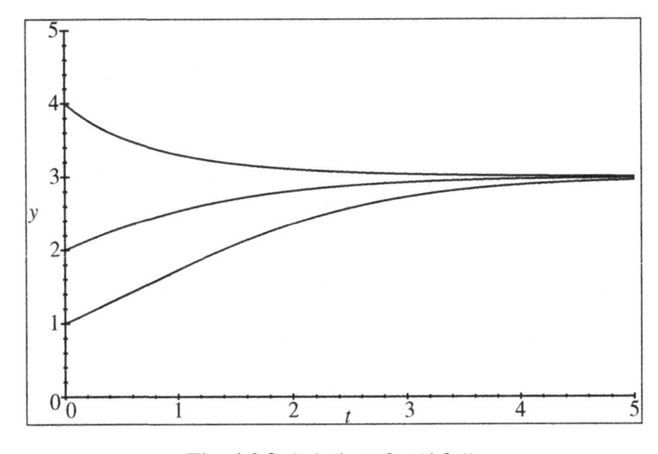


**Fig. 4.3.2.** Solutions for (4.3.1).

*The logistic equation has a discrete analogue.*

The corresponding discrete population model to (4.3.2) is

$$y_{t+1} - y_t = \rho y_t \left(1 - \frac{y_t}{K}\right). \tag{4.3.9}$$

By transposing $y_t$, we get an equivalent form,

$$y_{t+1} = y_t \left(1 + \rho - \frac{\rho y_t}{K}\right) = (1+\rho) y_t \left(1 - \frac{\rho}{1+\rho} \frac{y_t}{K}\right). \tag{4.3.10}$$

Recall that we encountered a similar recurrence relation, (2.5.11), in Section 2.5. From that discussion, we suspect that some values of $\rho$ may lead to chaos. In fact, with $K = 1$ and $\rho = 3$, we get the population behavior shown in Figure 4.3.3.

MAPLE
```
> K:=1;
> rho:=3.0;
> c:= (rho/K)/(1+rho):
> y[0]:=1/48.0;
> for i from 1 to 60 do y[i]:=(1+rho)*y[i-1]*(1-c*y[i-1]);
> od;
> pts:=[seq([i,y[i]],i=0..60)];
> plot(pts);
```
MATLAB
```
> K=1; rho=3; c=(rho/K)/(1+rho); y(1)=0.05; t=[1:60];
> for i = 2:60 y(i)=(1+rho)*y(i-1)*(1-c*y(i-1));
  end
> plot(t,y)
```

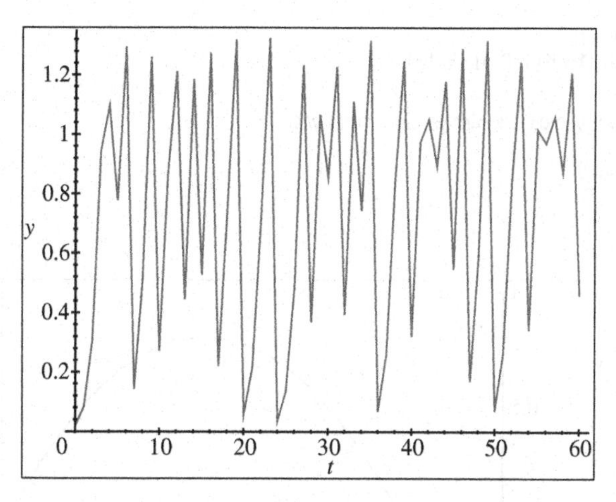


**Fig. 4.3.3.** Logistic growth, discrete model, $\rho$ large.

Does the chaos phenomenon extend to the continuous model too? Not strictly, according to the Verhulst equation (4.3.2). This is because as $y$ increases continuously, $\frac{y}{K}$ will increase to 1 without overshooting. Then continued population growth will stop, since $\frac{dy}{dt}$ will then be 0. However, if population increases are based not on the present population size but on the population size in the previous generation, say, then instability and chaos is possible.

In fact, real populations are sometimes chaotic. An unwelcome example is in the sardine population off the coast of California. In this case, the cause appears to be the practice of harvesting too many big fish. For details, see [12].

*Nonlinear per capita growth rates allow more complicated population behavior.*

Real populations are in danger of extinction if their size falls to a low level. Predation might eliminate the last few members completely, finding mates becomes more difficult, and lack of genetic diversity renders the population susceptible to epidemics. By constructing a per capita growth rate that is actually negative below some critical value $\theta$, there results a population model that tends to extinction if population size falls too low. Such a per capita growth rate is given as the right-hand side of the following modification of the logistic equation:

$$\frac{1}{y}\frac{dy}{dt} = r\left(\frac{y}{\theta} - 1\right)\left(1 - \frac{y}{K}\right), \tag{4.3.11}$$

where $0 < \theta < K$. This form of the per capita growth rate is pictured in Figure 4.3.4 using the specific parameters $r = 1$, $\theta = \frac{1}{5}$, and $K = 1$. It is sometimes referred to as the *predator pit*.

We draw the graph in Figure 4.3.4 with these parameters:

```
MAPLE
> restart
> r:=1; theta:=1/5; K:=1;
> plot([y,r*(y/theta-1)*(1-y/K),y=0..1],-.2..1,-1..1);
```

```
MATLAB
> r=1; theta=0.2; K=1; y=0:0.05:1; f=r.*(y/theta - 1).*(1-y/K);
> plot(y,f); hold on
> xaxis = zeros(size(y));
> plot(y,xaxis)
```

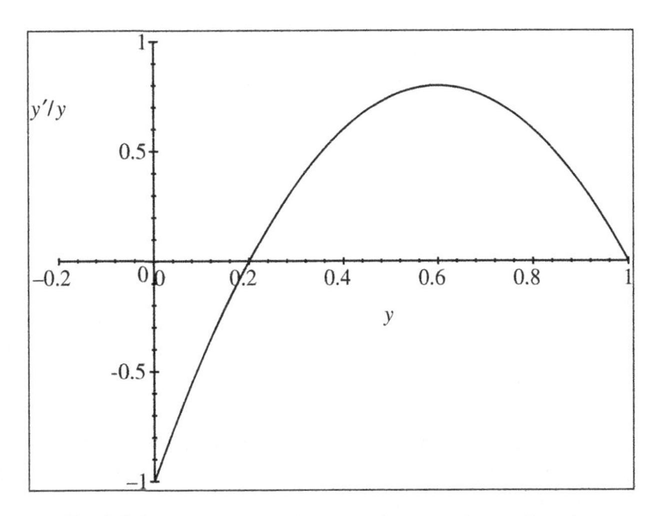


**Fig. 4.3.4.** The predator pit per capita growth rate function.

The stationary points of (4.3.11) are $y = 0$, $y = \theta$, and $y = K$. Unlike before, now $y = 0$ is asymptotically stable; that is, if the starting value $y_0$ of a solution is near enough to 0, then the solution will tend to 0 as $t$ increases. This follows because the sign of the right-hand side of (4.3.11) is negative for $0 < y < \theta$, causing $\frac{dy}{dt} < 0$. Hence $y$ will decrease. On the other hand, a solution starting with $y_0 > \theta$ tends to $K$ as $t$ increases. This follows because when $\theta < y < K$, the right-hand side of (4.3.11) is positive, so $\frac{dy}{dt} > 0$ also and hence $y$ will increase even more. As before, solutions starting above $K$ decrease asymptotically to $K$.

Some solutions to (4.3.11) are shown in Figure 4.3.5 with the following syntax:

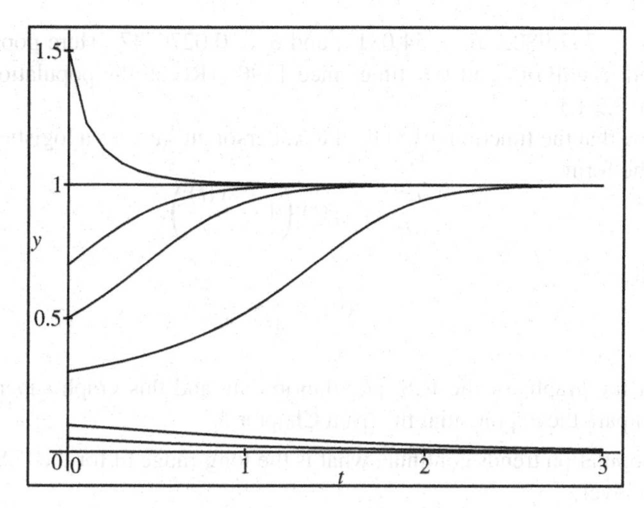


**Fig. 4.3.5.** Some solutions to the predator pit equation.

```
MAPLE
> r:=1; theta:=1/5; K:=1;
> inits:={[0,.05],[0,.1],[0,0.3],[0,.5],[0,1],[0,0.7],[0,1.5]};
> with(DEtools): DEplot(diff(y(t),t)=r*y(t)*(y(t)/theta-1)*(1-y(t)/K),y(t),t=0..3,inits, arrows=NONE,stepsize=0.1);
```

```
MATLAB
% Make up an m-file, fig434.m:
%  function yprime = fig434(t,y)
%  with r=1, theta=.2, and K=1.
%  r=1; theta=0.2; K=1;
%  yprime = y.*r.*(1-y./K).*(y/theta-1);
> tspan=[0 3];
> [t05,y05]=ode23('fig434',tspan,.05);
> [t1,y1]=ode23('fig434',tspan,.1);
> [t3,y3]=ode23('fig434',tspan,.3);
> [t5,y5]=ode23('fig434',tspan,.5);
> [t7,y7]=ode23('fig434',tspan,.7);
> [t15,y15]=ode23('fig434',tspan,1.5);
> plot(t05,y05,t1,y1,t3,y3,t5,y5,t7,y7,t15,y15)
```

As our last illustration, we construct a population model that engenders little population growth for small populations, rapid growth for intermediate sized ones, and low growth again for large populations. This is achieved by the quadratic per capita growth rate and given as the right-hand side of the differential equation

$$\frac{1}{y}\frac{dy}{dt} = ry\left(1 - \frac{y}{K}\right). \tag{4.3.12}$$

## Exercises/Experiments

**1.** At the meeting of the Southeastern Section of the Mathematics Association of America, Terry Anderson presented a MAPLE program that determined a logistic fit for the U.S. population data. His fit is given by

$$\text{U.S. population} \approx \frac{\alpha}{1 + \beta e^{-\delta t}},$$

where $\alpha = 387.9802$, $\beta = 54.0812$, and $\delta = 0.0270347$. Here population is measured in millions and $t$ = time since 1790. (Recall the population data of Example 3.5.1.)

(a) Show that the function given by the Anderson fit satisfies a logistic equation of the form

$$\frac{dy}{dt} = \delta y(t)\left(1 - \frac{y(t)}{\alpha}\right),$$

with

$$y(0) = \frac{\alpha}{1+\beta}.$$

(b) Plot the graphs of the U.S. population data and this graph superimposed. Compare the exponential fits from Chapter 3.

(c) If population trends continue, what is the long-range fit for the U.S. population level?

```
MAPLE
> Anderfit:=t->alpha/(1+beta*exp(-delta*t));
> dsolve({diff(y(t),t)-delta*y(t)*(1-y(t)/alpha)=0, y(0)=alpha/(1+beta)},y(t));
> alpha:=387.980205; beta:=54.0812024; delta:=0.02270347337;
> J:=plot(Anderfit(t),t=0..200):
> tt:=[seq(i*10,i=0..20)];
> pop:=[3.929214, 5.308483, 7.239881, 9.638453, 12.866020, 17.069453, 23.191876,
        31.433321, 39.818449, 50.155783, 62.947714, 75.994575, 91.972266, 105.710620,
        122.775046, 131.669275, 151.325798, 179.323175, 203.302031, 226.545805,
        248.709873];
> data:= [seq([tt[i],pop[i]],i=1..21)];
> K:=plot(data,style=POINT):
> plots[display]({J,K});
> expfit:= t->exp(0.02075384393*t+1.766257672);
> L:=plot(expfit(t),t=0..200):
> plots[display]({J,K,L});
> plot(Anderfit(t-1790),t=1790..2150);
```

```
MATLAB
> tt=0:10:200;
> pop=[3.929214, 5.308483, 7.239881, 9.638453, 12.866020, 17.069453, 23.191876,...
        31.433321, 39.818449, 50.155783, 62.947714, 75.994575, 91.972266, 105.710620,...
        122.775046, 131.669275, 151.325798, 179.323175, 203.302031, 226.545805,...
        248.709873];
> plot(tt,pop,'x'); hold on;
> alpha=387.980205; beta=54.0812024; delta=0.02270347337;
> Anderfit=alpha./(1+beta*exp(-delta*tt));
> plot(tt,Anderfit)
```

**2.** Using the method of (4.3.8), get a logistic fit for the U.S. population. Use the data in Example3.5.1.

**3.** Suppose that the spruce budworm, in the absence of predation by birds, will grow according to a simple logistic equation of the form

$$\frac{dB}{dt} = rB\left(1 - \frac{B}{K}\right).$$

Budworms feed on the foliage of trees. The size of the carrying capacity, $K$, will therefore depend on the amount of foliage on the trees; we take it to be constant for this model.

(a) Draw graphs for how the population might grow if $r$ were 0.48 and $K$ were 15. Use several initial values.

(b) Introduce predation by birds into this model in the following manner: Suppose that for small levels of worm population there is almost no predation, but for larger levels, birds are attracted to this food source. Allow for a limit to the number of worms that each bird can eat. A model for predation by birds might have the form

$$P(B) = a\frac{B^2}{b^2 + B^2},$$

where $a$ and $b$ are positive (see [7]). Sketch the graph for level of predation of the budworms as a function of the size of the population. Take $a$ and $b$ to be 2.

(c) A model for the budworm population size in the presence of predation could be modeled as

$$\frac{dB}{dt} = rB\left(1 - \frac{B}{K}\right) - a\frac{B^2}{b^2 + B^2}.$$

To understand the delicacy of this model and the implications for the care that needs to be taken in modeling, investigate graphs of solutions for this model with parameters $r = 0.48$, $a = b = 2$, and $K = 15$ or $K = 17$.

(d) Verify that in one case, there is a positive steady-state solution and in the other, the limit of the budworm population is zero.
The significance of the graph with $K = 17$ is that the worm population can rise to a high level. With $K = 15$, only a low level for the size of the budworms is possible. The birds will eat enough of the budworms to save the trees!
Here is the syntax for making this study with $K = 15$:

```
MAPLE
> K:= 15;
> h:=(t,B)->.48*B*(1-B/K)-2*B^2/(4+B^2);
> plot(h(0,B),B=0..20);
> inits:={[0,1], [0,2], [0,4], [0,5], [0,6], [0,8], [0,10], [0,12], [0,14], [0,16]};
> with(DEtools);
> DEplot(diff(y(t),t)=h(t,y(t)),y(t),t=0..30,inits,arrows=NONE,stepsize=0.1);
```

```
MATLAB
% make an m-file, exer43.m
% function Bprime=exer43(t,B); r=.48; K=15; a=2; b=2; Bprime=r*B.*(1-B/K)-a*B.^2./(b^2+B.^2);
> K=15; a=2; b=2; r=.48;
> B=0:.1:20; Bprime=exer43(0,B); plot(B,Bprime)
> [t,y1]=ode23('exer43',[0 30],1);plot(t,y1)
> hold on
> [t,y2]=ode23('exer43',[0 30],2);plot(t,y2)
> [t,y4]=ode23('exer43',[0 30],4);plot(t,y4)
> [t,y5]=ode23('exer43',[0 30],5);plot(t,y5)
> [t,y6]=ode23('exer43',[0 30],6);plot(t,y6)
> [t,y8]=ode23('exer43',[0 30],8);plot(t,y8)
> [t,y10]=ode23('exer43',[0 30],10);plot(t,y10)
> [t,y12]=ode23('exer43',[0 30],12);plot(t,y12)
> [t,y14]=ode23('exer43',[0 30],14);plot(t,y14)
> [t,y16]=ode23('exer43',[0 30],16);plot(t,y16)
```

**4.** The following is a logistic adaptation of Code 3.5.1. Experiment with the parameter $r$ and observe the behavior of the population size. Is the size chaotic for some values of $r$? It might be necessary to decrease delT (by increasing $N$ to, say, 20) to get valid results.

```
MAPLE
> N:=10: delT:=1/N:
  #t is now linked to index i by t=-1+(i-1)*delT
> for i from 1 to N+1 do f[i]:=1:
  od:
> #work from t=0 in steps of delT
> tfinal:=10: #end time tfinal, end index is n
> n:=(tfinal+1)*N+1:
> K:=3; r:=1.2;
> for i from N+1 to n-1 do t:=-1+(i-1)*delT:
    delY:=r*f[i-N]*(1-f[i-N]/K)*delT: #N back=delay of 1
    f[i+1]:=f[i]+delY: #Eulers method
> od;
> pts:=[seq([i,f[i]],i=0..n)];
> plot(pts);
```

```
MATLAB
  % make an m-file, delayFcn0.m:
  % function y=delayFcn0(t)
  % y = 1;
> N=10; % steps per unit interval
> delT=1/N; % so delta t=0.1
  % t is now linked to index i by t=-1+(i-1)*delT
  % set initial values via delay fcn f0
> for i=1:N+1
>   t=-1+(i-1)*delT; f(i)=delayFcn0(t);
> end
  % work from t=0 in steps of delT
  tfinal=10; % end time tfinal, end index is n
  % solve tfinal=-1+(n-1)*delT for n
> n=(tfinal+1)*N+1;
> K=3; r=1.2;
> for i=N+1:n-1
>   t=-1+(i-1)*delT;
>   delY=r*f(i-N)*(1-f(i-N)/K)*delT; % N back=delay of 1
>   f(i+1)=f(i)+delY; % Eulers method
> end
> t=-1:delT:tfinal; plot(t,f);
```

## 4.4 A Brief Look at Multiple Species Systems

*Without exception, biological populations interact with populations of other species. Indeed, the web of interactions is so pervasive that the entire field of Ecology is devoted to it. Mathematically, the subject began about 70 years ago with a simple two-species, predator–prey differential equation model. The central premise of this Lotka–Volterra model is a mass action–interaction term. While community differential equation models are difficult to solve exactly, they can nonetheless be analyzed by qualitative methods. One tool for this is to linearize the system of equations about their stationary solution points and to determine the eigenvalues of the resulting interaction, or community, matrix. The eigenvalues in turn predict the stability of the web. The Lotka–Volterra system has neutral stability at its nontrivial stationary point, which, like Malthus's unbounded population growth, is a shortcoming that indicates the need for a better model.*

*Interacting population models utilize a mass action–interaction term.*

Alfred Lotka (1925) and, independently, Vito Volterra (1926) proposed a simple model for the population dynamics of two interacting species (see [8]). The central assumption of the model is that the degree of interaction is proportional to the numbers, $x$ and $y$, of each species and hence to their product, that is,

$$\text{degree of interaction} = (\text{constant})xy.$$

The Lotka–Volterra system is less than satisfactory as a serious model because it entails neutral stability (see below). However, it does illustrate the basic principles of multispecies models and the techniques for their analysis. Further, like the Malthusian model, it serves as a point of departure for better models. The central assumption stated above is also used as the interaction term between reactants in the description of chemical reactions. In that context it is called the *mass action principle*. The principle implies that encounters occur more frequently in direct proportion to their concentrations.

The original Lotka–Volterra equations are

$$\begin{aligned} \frac{dx}{dt} &= rx - axy, \\ \frac{dy}{dt} &= -my + bxy, \end{aligned} \tag{4.4.1}$$

where the positive constants $r$, $m$, $a$, and $b$ are parameters. The model was meant to treat predator–prey interactions. In this, $x$ denotes the population size of the prey, and $y$ the same for the predators. In the absence of predators, the equation for the prey reduces to $\frac{dx}{dt} = rx$. Hence the prey population increases exponentially with rate $r$ in this case; see Section 3.5. Similarly, in the absence of prey, the predator equation becomes $\frac{dy}{dt} = -my$, dictating an exponential decline with rate $m$.

The sign of the interaction term for the prey, $-a$, is negative, indicating that interaction is detrimental to them. The parameter $a$ measures the average degree of the effect of one predator in depressing the per capita growth rate of the prey. Thus $a$ is likely to be large in a model for butterflies and birds but much smaller in a model for caribou and wolves. In contrast, the sign of the interaction term for the predators, $+b$, is positive, indicating that they are benefited by the interaction. As above, the magnitude of $b$ is indicative of the average effect of one prey on the per capita predator growth rate.

Besides describing predator–prey dynamics, the Lotka–Volterra system describes to a host–parasite interaction as well. Furthermore, by changing the signs of the interaction terms, or allowing them to be zero, the same basic system applies to other kinds of biological interactions as discussed in Section 4.1, such as mutualism, competition, commensalism, and amensalism.

Mathematically, the Lotka–Volterra system is not easily solved. Nevertheless, solutions may be numerically approximated and qualitatively described. Since the system has two dependent variables, a solution consists of a pair of functions $x(t)$ and $y(t)$ whose derivatives satisfy (4.4.1). Figure 4.4.1 is the plot of the solution to these equations with $r = a = m = b = 1$ and initial values $x(0) = 1.5$ and $y(0) = 0.5$. The figure is drawn with the following syntax:

```
MAPLE
> predprey:=diff(x(t),t)=r*x(t)-a*x(t)*y(t), diff(y(t),t)=-m*y(t)+b*x(t)*y(t);
> r:=1; a:=1; m:=1; b:=1;
> sol:=dsolve({predprey,x(0)=3/2,y(0)=1/2}, {x(t),y(t)},type=numeric, output=listprocedure);
> xsol:=subs(sol,x(t)); ysol:=subs(sol,y(t));
> plot([xsol,ysol],0..10,-1..3);
```

```
MATLAB
% make an m-file named predPrey44.m with:
% function Yprime=predPrey44(t,x)
% r=1; a=1; m=1; b=1;
% Yprime=[r*x(1)-a*x(1).*x(2); -m*x(2)+b*x(1).*x(2)];
> [t,Y]=ode23('predPrey44',[0 10],[1.5;0.5]); % ; for column vector
> plot(t,Y) % both curves as the columns of Y vs. t
```

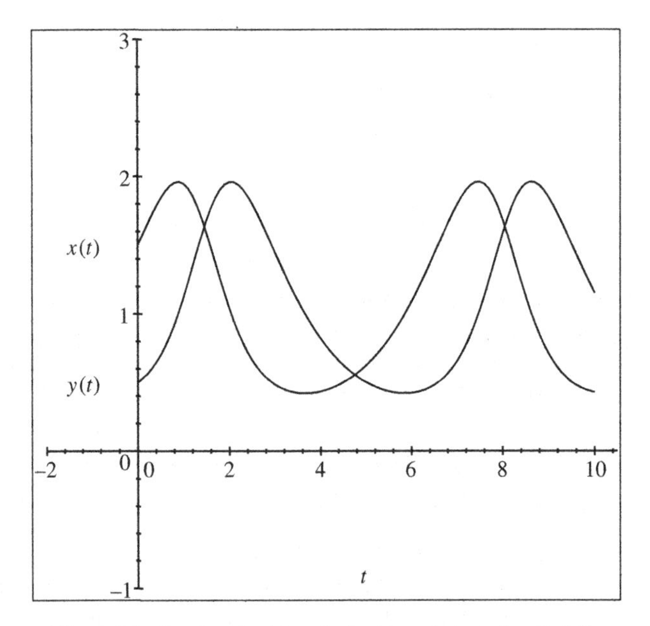


**Fig. 4.4.1.** Graphs of $x(t)$ and of $y(t)$, solutions for (4.4.1).

Notice that the prey curve leads the predator curve.[5] We discuss this next.

Although there are three variables in a Lotka–Volterra system, $t$ is easily eliminated by dividing $\frac{dy}{dt}$ by $\frac{dx}{dt}$; thus

[5] In Section 4.1, we have discussed a number of biological reasons why in a real situation, this model is inadequate.

MAPLE
```
> with(plots): with(DEtools):
> inits:={[0,3/2,1/2],[0,4/5,3/2]};
> phaseportrait({predprey},[x,y],t=0..10,inits,stepsize=.1);
```

MATLAB
```
> x=Y(:,1); y=Y(:,2); plot(x,y)
```

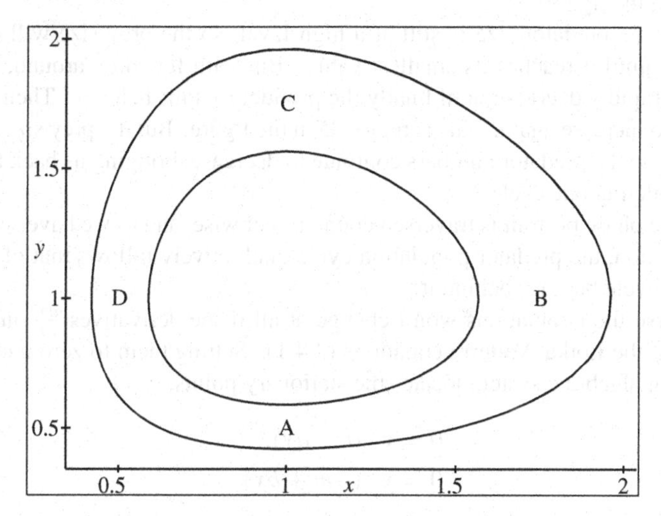


**Fig. 4.4.2.** A plot of two solutions of (4.4.1) in the $(x, y)$-plane.

$$\frac{dy}{dx} = \frac{-my + bxy}{rx - axy}.$$

This equation does not contain $t$ and can be solved exactly as an implicit relation between $x$ and $y$:[6]

MAPLE
```
> dsolve(diff(y(x),x)=(-y(x)+x*y(x))/(x-x*y(x)),y(x),implicit);
```

$$-\ln(y(x)) + y(x) - \ln(x) + x = C.$$

This solution gives rise to a system of closed curves in the $(x, y)$-plane called the *phase plane* of the system. These same curves, or *phase portraits*, can be generated from a solution pair $x(t)$ and $y(t)$ as above by treating $t$ as a parameter. In Figure 4.4.2, we show the phase portrait of the solution pictured in Figure 4.4.1.

Let us now trace this phase portrait. Start at the bottom of the curve, region A, with only a small number of prey and predators. With few predators, the population size of the prey grows almost exponentially. But as the prey size becomes large, the interaction term for the predators, $bxy$, becomes large and their numbers $y$ begin to grow. Eventually, the product $ay$ first equals and then exceeds $r$, in the first equation of (4.4.1), at which time the population size of the prey must decrease. This takes us to region B in the figure.

[6] *Implicit* means that neither variable $x$ nor $y$ is solved for in terms of the other.

However, the number of prey is still large, so predator size $y$ continues to grow, forcing prey size $x$ to continue declining. This is the upward and leftward section of the portrait. Eventually, the product $bx$ first equals and then falls below $m$ in the second equation of (4.4.1), whereupon the predator size now begins to decrease. This is point C in the figure.

At first, the predator size is still at a high level, so the prey size will continue to decrease until it reaches its smallest value. But with few prey around, predator numbers $y$ rapidly decrease until finally the product $ay$ falls below $r$. Then the prey size starts to increase again. This is region D in the figure. But the prey size is still at a low level, so the predator numbers continue to decrease, bringing us back to region A and completing one cycle.

Thus the phase portrait is traversed counterclockwise, and as we have seen in the above narration, the predator population cycle qualitatively follows that of the prey population cycle but lags behind it.

Of course the populations won't change at all if the derivatives $\frac{dx}{dt}$ and $\frac{dy}{dt}$ are both zero in the Lotka–Volterra equations (4.4.1). Setting them to zero and solving the resulting algebraic system locates the stationary points,

$$\begin{aligned} 0 &= x \cdot (r - ay), \\ 0 &= y \cdot (-m + bx). \end{aligned}$$

Thus if $x = \frac{m}{b}$ and $y = \frac{r}{a}$, the populations remain fixed. Of course, $x = y = 0$ is also a stationary point.

*Stability determinations are made from an eigenanalysis of the community matrix.*

Consider the stationary point $(0, 0)$. What if the system starts close to this point, that is, $y_0$ and $x_0$ are both very nearly 0? We assume that these values are so small that the quadratic terms in (4.4.1) are negligible, and we discard them. This is called *linearizing the system* about the stationary point. Then the equations become

$$\begin{aligned} \frac{dx}{dt} &= rx, \\ \frac{dy}{dt} &= -my. \end{aligned} \tag{4.4.2}$$

Hence $x$ will increase and $y$ will further decrease (but not to zero) and a phase portrait will be initiated as discussed above. The system will *not*, however, return to $(0, 0)$. Therefore, this stationary point is unstable.

We can come to the same conclusion by rewriting the system (4.4.2) in matrix form and examining the eigenvalues of the matrix on the right-hand side. This matrix is

$$\begin{bmatrix} r & 0 \\ 0 & -m \end{bmatrix}, \tag{4.4.3}$$

and its eigenvalues are $\lambda_1 = r$ and $\lambda_2 = -m$. Since one of these is real and positive, the conclusion is that the stationary point $(0, 0)$ is unstable.

Now consider the stationary point $x = \frac{m}{b}$ and $y = \frac{r}{a}$ and linearize about it as follows. Let $\xi = x - \frac{m}{b}$ and $\eta = y - \frac{r}{a}$. In these new variables, the first equation of the system (4.4.1) becomes

$$\frac{d\xi}{dt} = r\left(\xi + \frac{m}{b}\right) - a\left(\xi + \frac{m}{b}\right)\left(\eta + \frac{r}{a}\right) = -\frac{am}{b}\eta - a\xi\eta.$$

Again discard the quadratic term; this yields

$$\frac{d\xi}{dt} = -\frac{am}{b}\eta.$$

The second equation of the system becomes

$$\frac{d\eta}{t} = -m\left(\eta + \frac{r}{a}\right) + b\left(\xi + \frac{m}{b}\right)\left(\eta + \frac{r}{a}\right),$$
$$\frac{d\eta}{dt} = \frac{br}{a}\xi + b\xi\eta.$$

And discarding the quadratic term gives

$$\frac{d\eta}{dt} = \frac{br}{a}\xi.$$

Thus the equations in (4.4.1) become

$$\begin{aligned} \frac{d\xi}{dt} &= -\frac{am}{b}\eta, \\ \frac{d\eta}{dt} &= \frac{br}{a}\xi. \end{aligned} \tag{4.4.4}$$

The right-hand side of (4.4.4) can be written in matrix form:

$$\begin{bmatrix} 0 & -\frac{am}{b} \\ \frac{br}{a} & 0 \end{bmatrix} \begin{bmatrix} \xi \\ \eta \end{bmatrix}. \tag{4.4.5}$$

This time the eigenvalues of the matrix are imaginary, $\lambda = \pm i\sqrt{mr}$. This implies that the stationary point is *neutrally stable*.

Determining the stability at stationary points is an important problem. Linearizing about these points is a common tool for studying this stability, and has been formalized into a computational procedure. In the exercises, we give more applications that utilize the above analysis and that use a computer algebra system. Also, we give an example in which the procedure incorrectly predicts the behavior at a stationary point. The text by Steven H. Strogatz [9] explains conditions under which the procedure is guaranteed to work.

To illustrate a computational procedure for this predator–prey model, first create the vector function $V$:

```
MAPLE
> restart:
> with(LinearAlgebra):with(VectorCalculus):
> V:=Vector([r*x-a*x*y,-m*y+b*x*y]);
```

```
MATLAB
% We must compute derivates numerically
% make an m-file predPrey44.m with:
%   function Yprime=predPrey44(t,x);
%   r=1; a=1; m=1; b=1;
%   Yprime=[r*x(1)-a*x(1).*x(2); -m*x(2)+b*x(1).*x(2)];
```

Find the critical points of (4.4.1) by asking where this vector-valued function is zero (symbolically):

```
MAPLE
> solve({V[1]=0,V[2]=0}, {x,y});
```

This investigation provides the solutions $\{0, 0\}$ and $\{\frac{m}{b}, \frac{r}{a}\}$, as we stated above. We now make the linearization of $V$ about $\{0, 0\}$ and about $\{\frac{m}{b}, \frac{r}{a}\}$:

```
MAPLE
> Jacobian(V,[x,y]);
> subs({x=0,y=0},%);
> subs({x=m/b,y=r/a},%%);
```

```
MATLAB
  % eps is matlab's smallest value; by divided difference
  % find the derivatives numerically; first at (0,0)
> M1=(predPrey44(0,[eps 0]) - predPrey44(0,[0 0]))/eps;
  % this is the first column of the Jacobian at x=y=0, i.e., derivatives with respect to x
> M2=(predPrey44(0,[0 eps]) - predPrey44(0,[0 0]))/eps;
  % the derivatives with respect to y
> M=[M1 M2]; % the Jacobian
  % calculate its eigenvalues
> eig(M) % get 1 and -1, +1 means unstable at (0,0)
```

Note that in matrix form,

$$\begin{pmatrix} \frac{dx}{dt} \\ \frac{dy}{dt} \end{pmatrix} = \begin{pmatrix} r & 0 \\ 0 & -m \end{pmatrix} \begin{pmatrix} x-0 \\ y-0 \end{pmatrix} + \begin{pmatrix} -a \\ b \end{pmatrix} (x-0)(y-0)$$

for linearization about $(0, 0)$ and

$$\begin{pmatrix} \frac{dx}{dt} \\ \frac{dy}{dt} \end{pmatrix} = \begin{pmatrix} 0 & -\frac{am}{b} \\ \frac{br}{a} & 0 \end{pmatrix} \begin{pmatrix} x-\frac{m}{b} \\ y-\frac{r}{a} \end{pmatrix} + \begin{pmatrix} -a \\ b \end{pmatrix} \left(x-\frac{m}{b}\right)\left(y-\frac{r}{a}\right)$$

for linearization about $(\frac{m}{b}, \frac{r}{a})$. Finally, we compute the eigenvalues for the linearization about each of the critical points:

```
MAPLE
> Eigenvalues(%%); Eigenvalues(%%);
```

```
MATLAB
  % now linearize at x=m/b=1, y=r/a=1
> M1=(predPrey44(0,[1+eps 1])-predPrey44(0,[1 1]))/eps;
> M2=(predPrey44(0,[1 1+eps])-predPrey44(0,[1 1]))/eps;
> M=[M1 M2];
> eig(M) % get +/-I (I=sqrt(-1)), so neutrally stable
```

The result is the same as that from (4.4.5).

## Exercises/Experiments

1. The following *competition model* is provided in [9]. Imagine rabbits and sheep competing for the same limited amount of grass. Assume a logistic growth for the two populations, that rabbits reproduce rapidly, and that the sheep will crowd out the rabbits. Assume that these conflicts occur at a rate proportional to the size of each population. Further, assume that the conflicts reduce the growth rate for each species, but make the effect more severe for the rabbits by increasing the coefficient for that term. A model that incorporates these assumptions is

$$\frac{dx}{dt} = x(3 - x - 2y),$$
$$\frac{dy}{dt} = y(2 - x - y),$$

where $x(t)$ is the rabbit population and $y$ is the sheep population. (Of course, the coefficients are not realistic but are chosen to illustrate the possibilities.) Find four stationary points and investigate the stability of each. Show that one of the two populations is driven to extinction.

2. Imagine a *three-species predator–prey problem* that we identify with grass, sheep, and wolves. The grass grows according to a logistic equation in the absence of sheep. The sheep eat the grass and the wolves eat the sheep. (See McLaren [10] for a three-species population under observation.) We model this with the equations that follow. Here $x$ represents the wolf population, $y$ represents the sheep population, and $z$ represents the area in grass:

$$\frac{dx}{dt} = -x + xy,$$
$$\frac{dy}{dz} = -y + 2yz - xy,$$
$$\frac{dz}{dt} = 2z - z^2 - yz.$$

What would be the steady state of grass with no sheep or wolves present? What would be the steady state of sheep and grass with no wolves present? What is the revised steady state with wolves present? Does the introduction of wolves benefit the grass? This study can be done as follows:

MAPLE
```
> restart:
> rsx:=-x(t)+x(t)*y(t);
> rsy:=-y(t)+2*y(t)*z(t)-x(t)*y(t);
> rsz:= 2*z(t)-z(t)^2-y(t)*z(t);
```

MATLAB
```
% make an m-file, exer442.m
% function Yprime=exer442(t,Y); % Y(1)=x, Y(2)=y, Y(3)=z;
% Yprime=[-Y(1)+Y(1).*Y(2); -Y(2)+2*Y(2).*Y(3)-Y(1).*Y(2); 2*Y(3)-Y(3).*Y(3)-Y(2).*Y(3)];
```

For just grass:

```
MAPLE
> sol:=dsolve({diff(x(t),t)=rsx,diff(y(t),t)=rsy,diff(z(t),t)=rsz,x(0)=0,y(0)=0,z(0)=1.5},{x(t),y(t),z(t)},
        type=numeric,output=listprocedure);
> zsol:=subs(sol,z(t)); zsol(1);
> plot(zsol,0..20,color=green);
```

```
MATLAB
% grass
> [t,Y]=ode23('exer442',[0 200],[0; 0; 1.5]);
> plot(t,Y(:,3))
```

For grass and sheep:

```
MAPLE
> sol:=dsolve({diff(x(t),t)=rsx,diff(y(t),t)=rsy,diff(z(t),t)=rsz,x(0)=0,y(0)=.5,z(0)=1.5},{x(t),y(t),z(t)},
        type=numeric,output=listprocedure);
> ysol:=subs(sol,y(t));zsol:=subs(sol,z(t));
> plot([ysol,zsol],0..20,color=[green,black]);
```

```
MATLAB
% grass and sheep
> [t,Y]=ode23('exer442',[0 200],[0; .5; 1.5]);
> plot(t,Y)
```

For grass, sheep, and wolves:

```
MAPLE
> sol:=dsolve({diff(x(t),t)=rsx,diff(y(t),t)=rsy,diff(z(t),t)=rsz,x(0)=.2,y(0)=.5,z(0)=1.5},{x(t),y(t),z(t)},
        type=numeric,output=listprocedure);
> xsol:=subs(sol,x(t));
> ysol:=subs(sol,y(t));
> zsol:=subs(sol,z(t));
> plot([xsol,ysol,zsol],0..20,color=[green,black,red]);
```

```
MATLAB
% all three
> [t,Y]=ode23('exer442',[0 200],[.2; .5; 1.5]);
> plot(t,Y(:,3),'g') % grass behavior
> hold on
> plot(t,Y(:,2),'b') % sheep behavior
> plot(t,Y(:,1),'r') % wolf behavior
```

**3.** J. M. A. Danby [11] has a collection of interesting population models in his delightful text. The following *predator–prey model with child care* is included. Suppose that the prey $x(t)$ is divided into two classes, $x_1(t)$ and $x_2(t)$, of young and adults. Suppose that the young are protected from predators $y(t)$. Assume that the young increase in proportion to the number of adults and decrease due to death or to moving into the adult class. Then

$$\frac{dx_1}{dt} = ax_2 - bx_1 - cx_1.$$

The number of adults is increased by the young growing up and decreased by natural death and predation, so that we model

$$\frac{dx_2}{dt} = bx_1 - dx_2 - ex_2y.$$

Finally, for the predators, we take

$$\frac{dy}{dt} = -fy + gx_2y.$$

Investigate the structure for the solutions of this model. Parameters that might be used are

$$a = 2, \qquad b = c = d = \frac{1}{2}, \quad \text{and} \quad e = f = g = 1.$$

**4.** Show that the linearization of the system

$$\frac{dx}{dt} = -y + ax(x^2 + y^2),$$
$$\frac{dy}{dt} = x + ay(x^2 + y^2)$$

predicts that the origin is a center for all values of $a$, whereas, in fact, the origin is a stable spiral if $a < 0$ and an unstable spiral if $a > 0$. Draw phase portraits for $a = 1$ and $a = -1$.

**5.** Suppose there is a small group of individuals who are infected with a contagious disease and who have come into a larger population. If the population is divided into three groups—the susceptible, the infected, and the recovered—we have what is known as a classical S–I–R problem. (We take up such problems again in Section 11.4.) The susceptible class consists of those who are not infected, but who are capable of catching the disease and becoming infected. The infected class consists of the individuals who are capable of transmitting the disease to others. The recovered class consists of those who have had the disease, but are no longer infectious.
A system of equations that is used to model such a situation is often described as follows:

$$\frac{dS}{dt} = -rS(t)I(t),$$
$$\frac{dI}{dt} = rS(t)I(t) - aI(t),$$
$$\frac{dR}{dt} = aI(t)$$

for positive constants $r$ and $a$. The proportionality constant $r$ is called the infection rate and the proportionality constant $a$ is called the removal rate.

(a) Rewrite this model as a matrix model and recognize that the problem forms a closed compartment model. Conclude that the total population remains constant.

(b) Draw graphs for solutions. Observe that the susceptible class decreases in size and that the infected size increases in size and later decreases.

```
MAPLE
> r:=1; a:=1;
> sol:=dsolve({diff(SU(t),t)=-r*SU(t)*IN(t),diff(IN(t),t)=r*SU(t)*IN(t)-a*IN(t),diff(R(t),t)=a*IN(t),
        SU(0)=2.8,IN(0)=0.2,R(0)=0},{SU(t),IN(t),R(t)},type=numeric,output=listprocedure):
> f:=subs(sol,SU(t)): g:=subs(sol,IN(t)): h:=subs(sol,R(t));
> plot({f,g,h},0..20,color=[green,red,black]);
```

MATLAB
```
% contents of the m-file exer445a.m:
% function SIRprime=exer445a(t,SIR); % S=SIR(1), I=SIR(2), R=SIR(3);
% r=1; a=1;
% SIRprime=[-r*SIR(1).*SIR(2); r*SIR(1).*SIR(2)-a*SIR(2);a*SIR(2)];
> r=1; a=1;
> [t,SIR]=ode45('exer445a',[0 20], [2.8; .2; 0]);
> plot(t,SIR)
```

(c) Suppose now that the recovered do not receive permanent immunity. Rather, we suppose that after a delay of one unit of time, those who have recovered lose immunity and move into the susceptible class. The system of equations changes to the following:

$$\frac{dS}{dt} = -rS(t)I(t) + R(t-1),$$
$$\frac{dI}{dt} = rS(t)I(t) - aI(t),$$
$$\frac{dR}{dt} = aI(t) - R(t-1).$$

Draw graphs for solutions to this system. Observe the possibility of oscillating solutions. How do you explain these oscillations from the perspective of an epidemiologist? (Note: The following has a long run time.)

MAPLE
```
> restart:with(plots):
> N:=5;
> f[0]:=t->2.8; g[0]:=t->0.2*exp(-t^2); h[0]:=t->0;
> P[0]:=plot([[t,f[0](t),t=-1..0],[t,g[0](t),t=-1..0],[t,h[0](t),t=-1..0]],color=[green,red,black]):
> for n from 1 to N do
> sol:=dsolve({diff(SU(t),t)=-SU(t)*IN(t)+h[n-1](t-1),diff(IN(t),t)=SU(t)*IN(t)-IN(t),
        diff(R(t),t)=IN(t)-h[n-1](t-1),SU(n-1)=f[n-1](n-1),IN(n-1)=g[n-1](n-1),
        R(n-1)=h[n-1](n-1)},{SU(t),IN(t),R(t)},numeric,output=listprocedure,known=h[n-1]):
> f[n]:=subs(sol,SU(t)); g[n]:=subs(sol,IN(t));
> h[n]:=subs(sol,R(t)):
> P[n]:=plot([[t,f[n](t),t=n-1..n],[t,g[n](t),t=n-1..n],[t,h[n](t),t=n-1..n]],color=[green,red,black]):
> od:
> n:='n';
> J:=plot([t,1,t=0..N],color=blue):
> display([J,seq(P[n],n=0..N)]);
> for n from 1 to N do
> Q[n]:=spacecurve([f[n](t),g[n](t),h[n](t)],t=n-1..n,axes=normal,color=black):
> od:
> PP:=pointplot3d([1,1,1],axes=normal,symbol=diamond,color=green):
> display([PP,seq(Q[n],n=1..N)]);
```

MATLAB
```
> N=100; % number steps per unit interval
> delT=1/N; % so delta t=0.01
  % t is now linked to index i by t=-1+(i-1)*delT, where i=1,2,...,nFinal
  % and the final index nFinal is given by solving tFinal = -1+(nFinal-1)*delT.
> tFinal=5; nFinal=(tFinal+1)*N+1;
  % set up the initial values of R on -1 to 0
> for i=1:N
> R(i)=0; S(i)=0; I(i)=0;
> end
  % work from t=0 in steps of delT
> S(N+1)=2.8; I(N+1)=0.2; R(N+1)=0;
> for i=N+1:nFinal-1
> delY=delT*[-r*S(i)*I(i)+R(i-N); r*S(i)*I(i)-a*I(i); a*I(i)-R(i-N)]; S(i+1)=S(i)+delY(1);...
  I(i+1)=I(i)+delY(2); R(i+1) = R(i)+delY(3);
```

```
> end
  % graph it
> t=-1:delT:tFinal;
> plot(t,S,t,I,t,R) % S blue, I green, R red
```

### Questions for Thought and Discussion

1. Name and discuss four factors that affect the carrying capacity of an environment for a given species.
2. Draw and explain the shape of survivorship and population growth curves for an $r$-strategist.
3. Draw and explain the shape of survivorship and population growth curves for a $K$-strategist.
4. Define carrying capacity and environmental resistance.
5. Discuss the concept of parental investment and its role in $r$- and $K$-strategies.

## References and Suggested Further Reading


[1] ENVIRONMENTAL RESISTANCE:
W. T. Keeton and J. L. Gould, *Biological Science*, 5th ed., Norton, New York, 1993.

[2] PARTITIONING OF RESOURCES:
J. L. Harper, *Population Biology of Plants*, Academic Press, New York, 1977.

[3] POPULATION ECOLOGY:
R. Brewer, *The Science of Ecology*, 2nd ed., Saunders College Publishing, Fort Worth, TX, 1988.

[4] ECOLOGY AND PUBLIC ISSUES:
B. Commoner, *The Closing Circle: Nature, Man, and Technology*, Knopf, New York, 1971.

[5] NATURAL POPULATION CONTROL:
H. N. Southern, The natural control of a population of tawny owls (*Strix aluco*), *J. Zool. London*, **162** (1970), 197–285.

[6] A DOOMSDAY MODEL:
D. A. Smith; Human population growth: Stability or explosion, *Math. Magazine*, **50**-4 (1977) 186–197.

[7] BUDWORM, BALSALM FIR, AND BIRDS:
D. Ludwig, D. D. Jones, and C. S. Holling, Qualitative analysis of insect outbreak systems: The spruce budworm and forests, *J. Animal Ecol.*, **47** (1978), 315–332.

[8] PREDATOR OR PREY:
J. D. Murray, Predator–prey models: Lotka–Volterra systems, in J. D. Murray, *Mathematical Biology*, Springer-Verlag, Berlin, 1990, Section 3.1.

[9] LINEARIZATION:
S. H. Strogatz, *Nonlinear Dynamics and Chaos, with Applications to Physics, Biology, Chemistry, and Engineering*, Addison–Wesley, New York, 1994.

[10] A MATTER OF WOLVES:
B. E. McLaren and R. O. Peterson, Wolves, moose, and tree rings on Isle Royale, *Science*, **266** (1994), 1555–1558.

[11] PREDATOR–PREY WITH CHILD CARE, CANNIBALISM, AND OTHER MODELS:
J. M. A. Danby, *Computing Applications to Differential Equations*, Reston Publishing Company, Reston, VA, 1985.

[12] CHAOS IN BIOLOGICAL POPULATIONS:
P. Raeburn, Chaos and the catch of the day, *Sci. Amer.*, **300**-2 (2009), 76–78.

# 5

# Age-Dependent Population Structures

## Introduction

This chapter presents an analysis of the distribution of ages in a population. We begin with a discussion of the aging process itself and then present some data on the age structures of actual populations. We finish with a mathematical description of age structures. Our primary interest is in humans, but the principles we present will apply to practically any mammal and perhaps to other animals as well.

## 5.1 Aging and Death

*The notion of aging is not simple. One must consider that oak trees, and perhaps some animals like tortoises, seem to have unlimited growth potential, that a Pacific salmon mates only once and then ages rapidly, and that humans can reproduce for many years. In each case a different concept of aging may apply.*

*The reason that aging occurs, at least in mammals, is uncertain. The idea that the old must die to make room for the new gene combinations of the young is in considerable doubt. An alternative hypothesis is that organisms must partition their resources between the maintenance of their own bodies and reproduction, and that the optimal partitioning for evolutionary fitness leaves much damage unrepaired. Eventually, the unrepaired damage kills the organism. We present several hypotheses about how and why damage can occur.*

*What is meant by "aging" in an organism?*

We will use a simple definition of aging, or *senescence*:[1] it is a series of changes that accelerate with age and eventually result in the death of an organism. This definition is a loose one because it does not specify the source of the changes—the

[1] There is much argument about definitions in the study of aging, and we wish to avoid being part of the dispute. Our simplification may have the opposite effect!

R.W. Shonkwiler and J. Herod, *Mathematical Biology: An Introduction with Maple and Matlab*, Undergraduate Texts in Mathematics, DOI: 10.1007/978-0-387-70984-0_5,

only requirement is that they accelerate. We will adopt a common approach and not regard predation, injury, and disease caused by parasites, e.g., microorganisms, as causes of aging, even though their incidence may increase with age.

The effect of aging on survival is demonstrated in Figure 5.1.1 for a simple model system of test tubes. Suppose that a laboratory technician buys 1000 test tubes and that 70% of all surviving test tubes are broken each month. Curve (a) of Figure 5.1.1 shows the specific rate of breakage of the tubes—a constant 70% per month.[2] Note that a test tube surviving for three months would have the same chance of breakage in the fourth month as would one at the outset of the experiment (because aging has not occurred). Alternatively, suppose that the test tubes broke more easily as time passed. A tube surviving for three months would have a much greater chance of breakage during the fourth month than would one at the outset of the experiment (because the older one has aged). Curve (b) shows the rate of breakage for these tubes (doubling each month in this example).

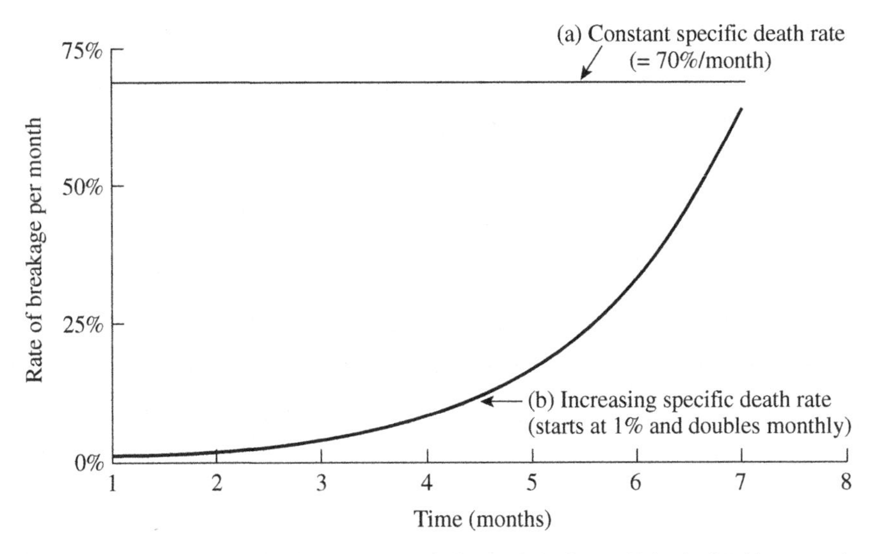


**Fig. 5.1.1.** Death rate, modeled on the breakage of test tubes. Curve (a) is obtained by assuming a specific death (breakage) rate of 70% of survivors per month of test tubes surviving to that point. This is equivalent to assuming that there is no aging, because the probability of death (breakage) is independent of time. The data of curve (b) is obtained by assuming that the specific death rate is 1% of the survivors in the first month and then doubles each month thereafter. This is equivalent to assuming that the test tubes age, because the probability of death (breakage) increases with time.

[2] The specific death (= breakage) rate is the number dying per unit time *among those of a specific age*. This is to be distinguished from the simple death rate, which is the death rate irrespective of age. In this experiment, of course, all the test tubes are of the same age.

Figure 5.1.2 shows survivorship curves for the two cases whose specific death rates are described by Figure 5.1.1. You should compare them to Figures 4.1.2 and 4.1.4, which are survivorship curves for $r$-strategists and $K$-strategists, respectively. It should be clear that $r$-strategists do not show aging (because they are held in check by climatic factors, which should kill a constant fraction of them, regardless of their ages).[3] The situation with regard to $K$-strategists is a bit more complex: Mammals, for instance, are held in check by density-dependent factors. If they live long enough, aging will also reduce their numbers. Both density-dependent factors and aging become more important as time passes. Thus the survivorship curve for a mammalian $K$-strategist should look somewhat like that shown in Figures 4.1.4 and 5.1.2(a).

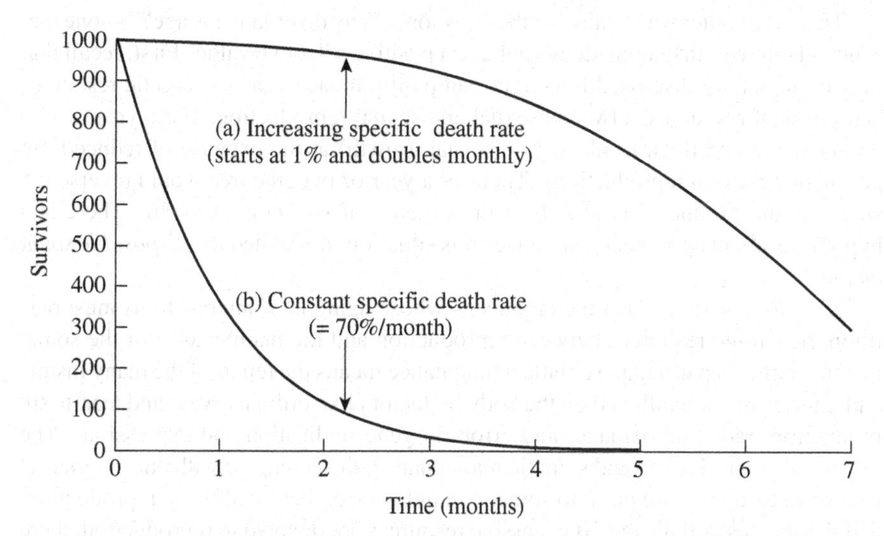


**Fig. 5.1.2.** (a) A survivorship curve for a nonaging system, using the data of Figure 5.1.1(b). (b) A survivorship curve for a system that exhibits aging, using the data of Figure 5.1.1(a). Both curves assume an initial cohort of 1000 test tubes at time $t = 0$. Note the similarity of curves (a) and (b) to Figures 4.1.2 and 4.1.4, which are survivorship curves for $r$-strategists and $K$-strategists, respectively.

*Why do organisms age and die?*

When asking "why" of any biological process as profound as senescence, we should immediately look to the Darwinian model of evolution for enlightenment and seek a positive selective value of aging to a species. A characteristic conferring a positive advantage is called an *adaptation*, and as we shall see, the adaptation we seek may not exist.

A simple adaptive explanation for senescence is that the Darwinian struggle for survival creates new organisms to fit into a changing environment. Thus the previous

[3] This is admittedly an approximation.

generation must die to make space and nutrients available for the new generation. Thomas Kirkwood has made two objections to this hypothesis [1]. The first objection is posed in the question, "How can aging have a positive selective value for a species when it can kill all the members of the species?" Besides, many organisms show the most evident aging only after their reproductive lives have ended. If the organism should show genetically programmed deterioration in its old age, that would have minimal (or no) selective value because the organism's reproductive life would have already ended anyway.

Kirkwood's second objection is that most organisms live in the wild and almost always die from disease and predation. Thus there is no need for selection based on aging in most organisms—they die too soon from other causes.

There is another way to answer the question, "Why do organisms age?"—one that is nonadaptive in that aging does not have a positive selective value. First, recall that in Section 4.1 we discussed how trees can partition each year's energetic resources and physical resources between asexual and sexual reproduction. For a year or two a tree would add thick trunk rings (asexual growth) at the expense of reduced nut production (sexual reproduction). Then for a year or two, the tree would reverse the situation and produce lots of nuts at the expense of vegetative growth. There is a hypothesis about aging that generalizes this situation; it is called the *disposable soma model*.[4]

Kirkwood assumes that the organisms whose aging is of interest to us must partition their finite resources between reproduction and the maintenance of the soma, i.e., the body. In particular, somatic maintenance means the repair of the many insults and injuries that are inflicted on the body by factors like ordinary wear and tear, toxin production, radiation damage, and errors in gene replication and expression. The two needs, reproduction and somatic maintenance, thus compete with one another. If excessive resources are put into somatic maintenance, there will be no reproduction, and the species will die out. If excessive resources are devoted to reproduction, there will be insufficient somatic maintenance, and the species will die out. We thus assume that there is an optimal partitioning of resources between somatic maintenance and reproduction. The disposable soma model postulates that this optimal partitioning is such that some somatic damage must go unrepaired and that the organism eventually dies because of it. Thus the organism has a finite lifetime, one marked by increasing rate of deterioration, i.e., aging.

The disposable soma model is nonadaptive in that aging is a harmful process. It is, however, an essential process because it is a measure of the resources diverted into reproduction. In a way, aging is a side effect, but, of course, it has powerful consequences to the organism.

*Aging of cells can provide insight into organismal aging.*

The death of the only cell comprising an amoeba has consequences that are quite different from those associated with the death of a single skin cell of a person; thus we will have to distinguish between aging in single-celled and multicellular organisms.

[4] "Soma" means "body."

It is fine to study the processes that lead to the death of a cell, but what if that cell is only one of many in an organ of a multicellular organism? To answer this question, we must first understand that cell death is a natural part of the life and development of organisms. Our hands are initially formed with interdigital webbing, perhaps suggesting our aquatic ancestry. This webbing is removed in utero by the death of the cells that comprise it in a process called *apoptosis*. There are many other examples of cell death as a natural consequence of living: our red blood cells live only about three months and our skin cells peel off constantly. Both are quickly replaced, of course.

We can now return to the question of what happens if one, or even a small fraction, of the cells in an organ die. Usually, nothing—we see that it happens all the time. But if that cell dies for a reason connected to the possible deaths of other cells, then the study of the one cell becomes very important. Thus the study of aging in cells can contribute greatly to our knowledge of aging in multicellular organisms.

*How do organisms become damaged?*

Whether we accept Kirkwood's disposable soma model or not, it is clear that our cells age, and we must suggest ways that the relevant damage occurs. Numerous mechanisms have been proposed, but no single one has been adequate, and in the end it may be that several will have to be accepted in concert. Some examples of damage mechanisms that have been proposed are the following:

(a) *Wear and tear*: A cell accumulates "insults," until it dies. Typical insults are the accumulation of wastes and toxins, as well as physical injuries like radiation damage and mechanical injury. These are all well known to cause cell death. Cells have several mechanisms by which insults can be repaired, but it may be that these repair systems themselves are subject to damage by insults.
(b) *Rate of living*: This is the "live fast, die young" hypothesis. In general, the higher a mammal's basal metabolic rate, the shorter its life span is. Perhaps some internal cellular resource is used up, or wastes accumulate, resulting in cell death.
(c) *Preprogrammed aging*: Our maximum life span is fixed by our genes. While the average life span of humans has increased over the past few decades, the maximum life span seems fixed at 100–110 years. Noncancerous mammalian cell lines in test tube culture seem capable of only a fixed number of divisions. If halfway through that fixed number of divisions, the cells are frozen in liquid nitrogen for ten years and then thawed, they will complete only the remaining half of their allotted divisions.

*Cell reproduction seems to have a rejuvenating effect on cells.*

It is a common observation that cells that reproduce often tend to age more slowly than cells that divide infrequently. This effect is seen in both asexual and sexual reproduction. Cancer cells divide rapidly and will do so in culture forever. Cells of our pancreas divide at a moderate rate, and our pancreas seems to maintain its function well into old age. Brain cells never divide and brain function deteriorates noticeably in old age. Even single-celled organisms can exhibit this effect: they may show obvious signs of senescence until they reproduce, at which point those signs disappear.

## 5.2 The Age Structure of Populations

*Age-structure diagrams show the frequency distribution of ages in a population. The data for males and females are shown separately. The shape of these diagrams can tell us about the future course of population changes: The existence of a large proportion of young people at any given time implies that there will be large proportions of individuals of childbearing age 20 years later and of retirees 60 years later. The shapes of age-structure diagrams are also dependent on migration into and out of a population. Comparison of data for males and females can tell us about the inherent differences between the genders and about the society's attitude toward the two genders.*

*Age-structure diagrams are determined by age-specific rates of birth, death, and migration.*

Figure 5.2.1 is a set of age-structure diagrams for the United States for 1955, 1985, 2015 (projected), and 2035 (projected) (see also [2]). They show how the population is, or will be, distributed into age groups. Data are included for males and females.

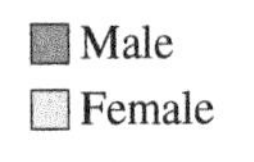


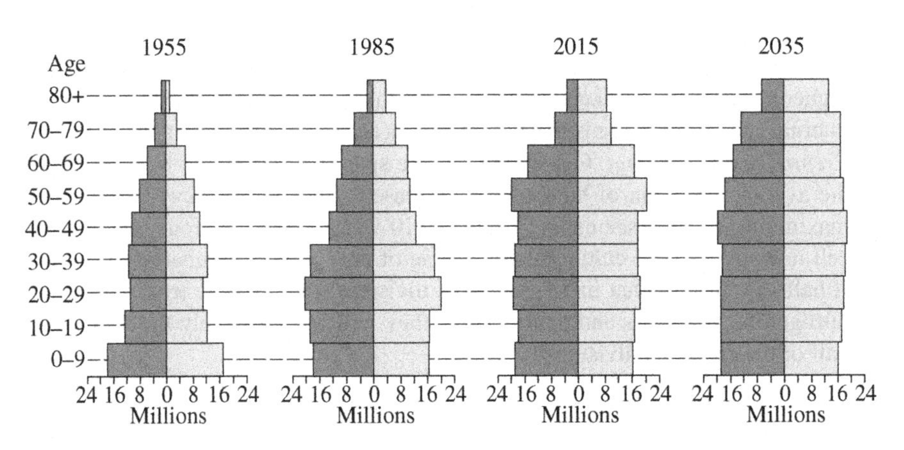


**Fig. 5.2.1.** Past and future (projected) age-structure diagrams for the United States. Note the growing proportion of elderly, compared to young, people. The cohort of "baby boomers" is evident at the base of the 1955 data. That group moves up in the 1985 and 2015 diagrams. (Redrawn from "Age and Sex Composition of the U.S. Population," in *U.S. Population: Charting the Change: Student Chart Book*, Population Reference Bureau, Washington, DC, 1988. Used with permission.)

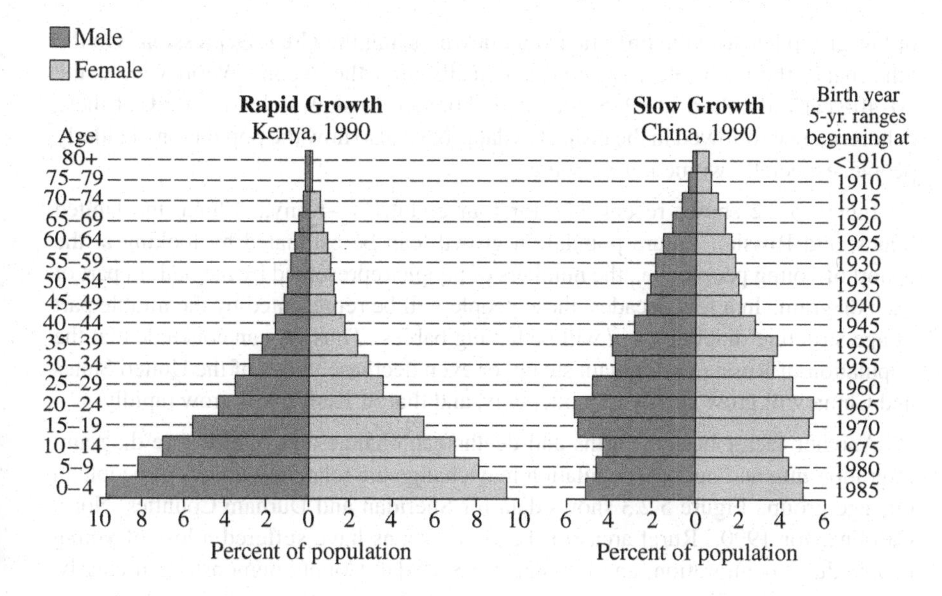


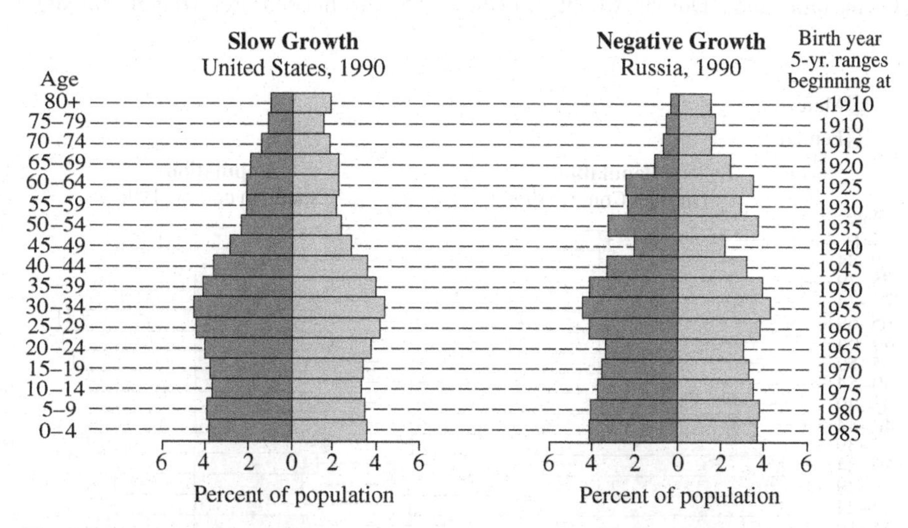


**Fig. 5.2.2.** Age-structure diagram for four countries for 1990. Each is labeled according to its expected future growth rate. For instance, Kenya has a high proportion of young people, so we expect its future growth rate to be high. (Redrawn from "Patterns of Population Change," in *World Population: Toward the Next Century*, Population Reference Bureau, Washington, DC, 1994, p. 5. Used with permission.)

These diagrams can convey a great deal of information. For example, look at the data for 1955 and note the 20–30-year-old *cohort*.[5] There are relatively fewer people

[5] A cohort is a group of people with a common characteristic. Here the characteristic they share is that they were born in the same decade.

in this group because the birth rate went down during the Great Depression. On the other hand, the birth rate went up dramatically after the Second World War, as the 20–40-year-old cohort in 1985 (the "baby boomers") shows clearly. Both of these cohorts can be followed in the projected data. Note also how the population of elderly people, especially women, is growing.

Figure 5.2.2 shows recent data for four countries—Kenya, China, the United States, and Russia. Future population growth can be estimated by looking at the cohort of young people, i.e., the numbers of people represented by the bottom part of each diagram. In a few decades, these people will be represented by the middle part of age-structure diagrams *and* will be having babies. Thus we can conclude that the population of Russia will remain steady or even decrease, those of the United States and China will grow slowly to moderately, and that of Kenya will grow rapidly.

Another factor besides births and deaths can change an age-structure diagram: migration into and out of a population may change the relative numbers of people in one age group. Figure 5.2.3 shows data for Sheridan and Durham Counties, North Carolina, for 1990. Rural areas of the Great Plains have suffered a loss of young people due to emigration, and the data for Sheridan County demonstrate it clearly. On the other hand, Durham County is in the North Carolina Research Triangle, the site

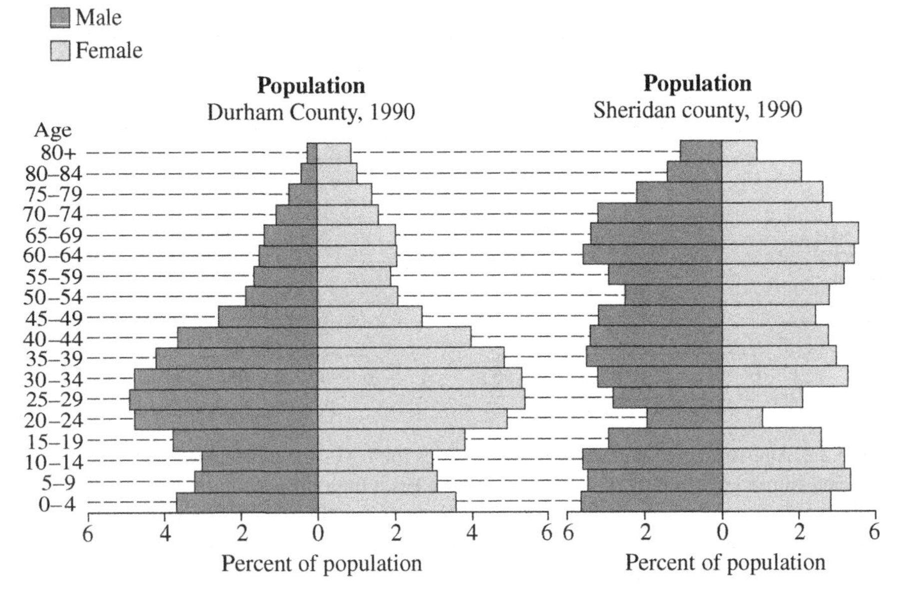


**Fig. 5.2.3.** An age-structure diagram showing the effects of migration. Many young people in the 20–45-year-old age group have moved into Durham County, North Carolina, and many young people in the 20–30-year-old age group have moved out of Sheridan County, North Carolina. (Redrawn from "Age and Sex Profiles of Sheridan and Durham Counties, 1990," in "Americans on the Move," *Population Bull.*, **48**-3 (1993), 25 (published by Population Reference Bureau, Washington, DC). Used with permission.)

of several major universities and many research industries. It is therefore a magnet for younger people, and its age-structure diagram reflects that fact.

*Some populations have more men than women.*

We are accustomed to the idea that there are more women than men in our country. That (true) fact can be misleading, however. While the sex ratio at conception is not known, there is evidence that a disproportionate number of female fetuses are spontaneously aborted in the first trimester of pregnancy. On the other hand, in the second and third trimesters, more male than female fetuses are lost. The ratio of sexes at birth in the United States is about 106 males to every 100 females. The specific death rate for males is higher than for women, and by early adolescence the sex ratio is 100:100. You can refer back to Figure 5.2.1 to see the effect of males' higher death rate on the relative numbers of males and females in later life.

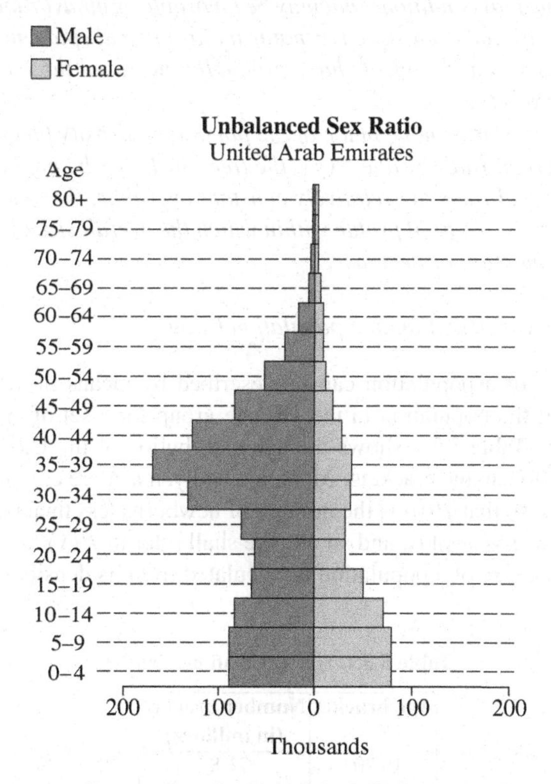


**Fig. 5.2.4.** Age-structure diagram from the United Arab Emirates showing an unbalanced sex ratio. The gender imbalance, males outnumbering females, is due to the importation of males to work in the oil fields: these males are not accompanied by their families. (Redrawn from "Unbalanced Sex Ratio: United Arab Emirates, 1985," in "Population: A Lively Introduction," *Population Bull*, **46**-2 (1991), 25 (published by Population Reference Bureau, Washington, DC). Used with permission.)

The fact that there are more females than males in the United States might lead us to be surprised by the data of Figure 5.2.4, an age-structure diagram for the United Arab Emirates. The unbalanced sex ratio, heavily tilted toward males, arises from immigration: U.A.E. has brought in many men from other countries to work in its oil fields, and the men seldom bring their families.

Another feature of gender ratios can be noted in age-structure diagrams of certain countries. In the late 1980s, the ratio of men to women in advanced countries was about 94:100; in developing countries, it was about 104:100.

## 5.3 Predicting the Age Structure of a Population

*A graph of population size $P$ as a function of age $y$ visually documents the age structure, or profile, of a population. Over time, a population profile can change due to periodic environmental conditions that may be favorable or unfavorable to the population, and to occasional events such as natural diasters and epidemics. For human populations, medical improvements have gradually increased the representation in the higher age brackets.*

*But much greater use can be made of the population density function $P$. With a knowledge of survival rates by age, $\ell(y)$, the trend in $P$ can be predicted. It can be shown that if survival rates are relatively constant over time, then the age structure of a population tends to a fixed profile within which the overall size of the population may nonetheless increase or decrease.*

*Age structure is the distribution of a population by age.*

The age structure of a population can be described by means of a function $P(y)$ giving the size of the population in the $y$th age group for a set of groups covering all possible ages. Table 5.3.1 shows the age distribution of the U.S. population in 1990 refined to 20-year age brackets. Mathematically, it is more common to use one-year age brackets, so that $P(0)$ is the number of newborns less than one year of age, $P(1)$ counts the one-year-olds, and so on. We shall refer to $P(y)$ as the *age density* function. The total size of a population is calculated from its density by summing,

**Table 5.3.1.** U.S. population, 1990.

| Age bracket | Number bracket (in millions) |
|---|---|
| 0–20 | 71.8 |
| 20–40 | 103.4 |
| 40–60 | 60.3 |
| 60–80 | 20.9 |
| 80–100 | .209 |
| 100 | .001 |

$$P = \sum_{n=0}^{\infty} P(n). \tag{5.3.1}$$

The use of infinity as the upper limit of this sum is a simplifying measure; for some age, maybe $y_{\max} = 115$, $P(y) = 0$ for $y > y_{\max}$, so the indicated infinite sum is in reality only from 0 to 115.

The age structure of the United States has gradually evolved over the last half of the twentieth century, as seen in Figure 5.2.1. On the other hand, any of several catastrophes can bring about rapid change to an age structure. We account for these possibilities by regarding the age-density function as dependent on calendar time $t$ as well as age $y$, and in deference to these dual dependencies we write $P(y, t)$. In addition, including the reference to time provides a mechanism for describing births year by year, namely, $P(0, t)$. This is the birth rate of a population in year $t$. If the birth rate is down in some year, say, $t = t_0$, this affects the population in subsequent years as well, as we have seen above. To begin with, the population of one-year-olds cannot exceed the population of newborns in the previous year,

$$P(1, t_0 + 1) \le P(0, t_0),$$

assuming no immigration into the population, of course. This is generally true for any age bracket; thus under the condition of no immigration,

$$P(y + 1, t + 1) \le P(y, t) \quad \text{for } y \ge 0 \text{ and for all } t. \tag{5.3.2}$$

While the population in an age bracket cannot increase in the following year, it can decrease due to deaths that occur during the year. Let $\mu(y)$ denote the death rate, or *mortality*, experienced by the population of age $y$. The *death rate* is dimensionless, being the fraction of deaths per individual, or since it is usually a number in the thousandths, it is frequently given as deaths per 1000 individuals. The actual number of deaths that occur among the segment of the population of age $y$ in year $t$ is the product of the death rate and the number of individuals at risk,

$$\mu(y)P(y, t)$$

($\mu$ must be deaths per individual here or $P$ must be population in thousands).

Virtually all natural populations experience very high preadult mortality rates. Insect populations and other unnurtured species ($r$-strategists; cf. Chapter 4) experience death rates similar to that shown in Figure 5.1.1(a). Notice that the newly hatched young suffer the highest mortality rates, with improvement as the animal ages. By contrast, nurtured species ($K$-strategists), such as mammals, experience much lower preadult mortality rates, as seen in Figure 5.1.1(b).

A mortality table for the United States is given in Table 5.3.2. In most species, mortality rates are lowest during the middle adult years.

Returning to (5.3.2), taking deaths into account yields the equality

$$P(y + 1, t + 1) = P(y, t) - \mu(y)P(y, t), \tag{5.3.3}$$

**Table 5.3.2.** U.S. mortality table for 1991. (Source: U.S. Department of Health and Human Services, Hyattsville, MD.)

| Age | Deaths (%) |
|---|---|
| 0–10 | 1.2 |
| 10–20 | .57 |
| 20–30 | 1.2 |
| 30–40 | 1.8 |
| 40–50 | 3.1 |
| 50–60 | 7.2 |
| 60–70 | 16.4 |
| 70+ | 100 |

**Table 5.3.3.** U.S. mortality rates; rates per 1,000 population. (Source: U.S. Department of Health and Human Services, Hyattsville, MD.)

| Year | Average mortality |
|---|---|
| 1920 | 13.0 |
| 1930 | 11.3 |
| 1940 | 10.6 |
| 1950 | 9.6 |
| 1960 | 9.5 |
| 1970 | 9.5 |
| 1980 | 8.6 |
| 1990 | 8.6 |

provided there is no immigration or emigration. But this equation ignores the effect of external events that may play havoc with death rates. For example, due to a catastrophic epidemic, death rates in the youth age groups may be high during the calendar year in which it strikes. On the other hand, the U.S. population has experienced a gradually decreasing death rate over this century as a result of improved medical care (see Table 5.3.3). To account for these and other factors unrelated to age, we must regard $\mu$ as a function of time as well as age. Thus (5.3.3) becomes

$$P(y+1,t+1) = P(y,t) - \mu(y,t)P(y,t) = \ell(y,t)P(y,t), \tag{5.3.4}$$

where $\ell(y,t) = 1 - \mu(y,t)$ is the fraction of the population of age $y$ that will live through year $t$. These factors $\ell(\cdot,\cdot)$ are called *survival rates*.

*In the absence of external events, populations evolve to a stable age distribution.*

While survival rates depend on calendar time in general, here we are interested in predicting the population structure in the absence of external events. Consequently, we will regard $\mu$ (and $\ell$) as a function of age only.

If we know yearly birth rates $P(0,t)$ and age-specific survival rates $\ell(y)$, (5.3.4) allows us to calculate the course of the population through time, including its age distribution and size. We also need to know the present age distribution, $P(y, 0)$, where we may regard the present time as $t = 0$. Usually the calculation is done for the female population of the species, since birth rates depend largely on the number of females while being somewhat independent of the number of males. The birth rates given will therefore pertain to the birth of females.

We illustrate this calculation for a $K$-strategist, specifically, for the gray seal, whose (female) fecundity and survival rates are given in Table 5.3.4.

**Table 5.3.4.** Gray seal fecundity and survival rates. (Source: D. Brown and P. Rothery, *Models in Biology: Mathematics, Statistics, and Computing*, Wiley, Chirchester, UK, 1993.)

| **Age** | 0 | 1 | 2 | 3 | 4 | 5 | 5+ |
|---|---|---|---|---|---|---|---|
| **Fecundity** | 0 | 0 | 0 | 0 | 0.08 | 0.28 | 0.42 |
| **Survival** | 0.657 | 0.930 | 0.930 | 0.930 | 0.935 | 0.935 | 0 |

To get it started, we make the assumption that the present population has uniform age density. Actually, this assumption about the starting population is not important in the long term, as we will see in the exercises. The key values are the birth and survival rates in the table. Since the survival rate for age 0 is 0.657, from (5.3.4) we have

$$P(1, t+1) = 0.657P(0, t) \quad \text{for all } t \geq 0.$$

Similarly, for $y = 1, 2, 3$,

$$P(y+1, t+1) = 0.930P(y, t) \quad \text{for all } t \geq 0.$$

And for $y = 4, 5$,

$$P(y+1, t+1) = 0.935P(y, t) \quad \text{for all } t \geq 0.$$

In this we take $5 + 1$ to be 5+. Since there is no category beyond "5+," the survival rate $\ell(5+)$ is 0. The birth-rate calculation uses the fecundity entries and is only slightly more complicated,

$$P(0, t+1) = 0.08P(4, t) + 0.28P(5, t) + 0.42P(5+, t).$$

It is convenient to write the calculation in matrix form. Let $\mathbf{p}(t)$ be the vector whose components are $P(y, t)$,

$$\mathbf{p}(t) = \begin{bmatrix} P(0, t) \\ P(1, t) \\ P(2, t) \\ P(3, t) \\ P(4, t) \\ P(5, t) \\ P(5+, t) \end{bmatrix}.$$

Then $\mathbf{p}(1)$ is given as the matrix product

$$\mathbf{p}(1) = \begin{bmatrix} 0 & 0 & 0 & 0 & 0.08 & 0.28 & 0.42 \\ 0.657 & 0 & 0 & 0 & 0 & 0 & 0 \\ 0 & 0.930 & 0 & 0 & 0 & 0 & 0 \\ 0 & 0 & 0.930 & 0 & 0 & 0 & 0 \\ 0 & 0 & 0 & 0.930 & 0 & 0 & 0 \\ 0 & 0 & 0 & 0 & 0.935 & 0 & 0 \\ 0 & 0 & 0 & 0 & 0 & 0.935 & 0 \end{bmatrix} \begin{bmatrix} P(0,0) \\ P(1,0) \\ P(2,0) \\ P(3,0) \\ P(4,0) \\ P(5,0) \\ P(5+,0) \end{bmatrix} = L\mathbf{p}(0). \tag{5.3.5}$$

Denote by $L$ the $7 \times 7$ matrix indicated. The first row reflects the births coming from various age groups and has nonzero terms indicated by them. Except for the first row, the only nonzero terms are the principal subdiagonal entries and those are the survival rates $\ell(y)$. This matrix is called the *Leslie matrix*, and it always has the same form:

$$L = \begin{pmatrix} a_1 & a_2 & a_3 & \cdots & a_n \\ b_1 & 0 & 0 & \cdots & 0 \\ 0 & b_2 & 0 & \cdots & 0 \\ \cdots & \cdots & \cdots & \cdots & 0 \\ 0 & 0 & 0 & \cdots & 0 \end{pmatrix}.$$

To be specific, assume a starting density $\mathbf{p}(0)$. The new density $\mathbf{p}(1)$ in (5.3.5) can be computed by inspection, or by using the computer:

```
MAPLE (symbolic calculation)
> with(LinearAlgebra):
> el:=Matrix(7,7); # Maple initializes the entries to 0
  # symbolic maple calculations require rational numbers,
  # .08 = 2/25, .28 = 7/25, and so on
> el[1,5]:=2/25: el[1,6]:=7/25: el[1,7]:=21/50: el[2,1]:=657/1000: el[3,2]:=93/100:
> el[4,3]:=93/100: el[5,4]:=93/100: el[6,5]:=935/1000: el[7,6]:=935/1000:
> el;
> evalm(el &* [P0,P1,P2,P3,P4,P5,P6]);
```

Either way, we get

$$\mathbf{p}(1) = \begin{pmatrix} 0.08P(4,0) + 0.28P(5,0) + 0.42P(5+,0), \\ 0.657P(0,0) \\ 0.930P(1,0) \\ 0.930P(2,0) \\ 0.930P(3,0) \\ 0.935P(4,0) \\ 0.935P(5,0) \end{pmatrix}.$$

Furthermore, the population size after one time period is simply the sum of the components of $\mathbf{p}(1)$.

The beauty of this formulation is that advancing to the next year is just another multiplication by $L$. Thus

$$\mathbf{p}(2) = L\mathbf{p}(1) = L^2\mathbf{p}(0), \qquad \mathbf{p}(3) = L\mathbf{p}(2) = L^3\mathbf{p}(0), \qquad \text{etc.}$$

The powers of a Leslie matrix have a special property, which we illustrate. For example, compute $L^{10}$:

MAPLE
```
> el10:=evalf(evalm(el^10)):
> Digits:=2; evalf(evalm(el10)); Digits:=10;
```

MATLAB
```
> L=[0 0 0 0 .08 .28 .42; .657 0 0 0 0 0 0; 0 .930 0 0 0 0 0; 0 0 .930 0 0 0 0; 0 0 0 .930 0 0 0;...
   0 0 0 0 .935 0 0; 0 0 0 0 0 .935 0]
> L^(10)
```

The result, accurate to three places, is

$$L^{10} = \begin{pmatrix} 0.0018 & 0.018 & 0.058 & 0.094 & 0.71 & 0 & 0 \\ 0 & 0.0018 & 0.013 & 0.041 & 0.067 & 0.050 & 0 \\ 0 & 0 & 0.0018 & 0.013 & 0.041 & 0.066 & 0.050 \\ 0.11 & 0 & 0 & 0.0018 & 0.013 & 0.031 & 0.033 \\ 0.073 & 0.16 & 0 & 0 & 0.0018 & 0.0063 & 0.0094 \\ 0.021 & 0.10 & 0.16 & 0 & 0 & 0 & 0 \\ 0 & 0.030 & 0.10 & 0.16 & 0 & 0 & 0 \end{pmatrix}. \tag{5.3.6}$$

Remarkably, the power $L^n$ can be easily approximated, as predicted by the Perron–Frobenius theorem [3], as we now describe. Letting $\lambda$ be the largest eigenvalue of $L$ (see Section 2.6) and letting $V$ be the corresponding normalized eigenvector, so $LV = \lambda V$, then

$$L^n p(0) \approx c\lambda^n \mathbf{V},$$

where $c$ is a constant determined by the choice of normalization; see (5.3.7). This approximation improves with increasing $n$. The importance of this result is that the long-range forecast for the population is predictable in form. That is, the ratios between the age classes are independent of the initial distribution and scale as powers of $\lambda$.

The number $\lambda$ is a real, positive eigenvalue of $L$. It can be found rather easily by a computer algebra system. The eigenvector can also be found numerically. It is shown in [4] that the eigenvector has the following simple form:

$$\mathbf{V} = \begin{pmatrix} 1 \\ \frac{b_1}{\lambda} \\ \frac{b_1 b_2}{\lambda^2} \\ \vdots \\ \frac{b_1 b_2 b_3 \cdots b_n}{\lambda^n} \end{pmatrix}. \tag{5.3.7}$$

To illustrate this property of Leslie matrices, we will find $\lambda$, $\mathbf{V}$, and $L^{10}$ for the gray seal example. Other models are explored in the exercises.

MAPLE
```
> vel:=Eigenvectors(fel);
> vals:=vel[1]; lambda:=vals[1] # only one real e-value, should be the first
  # grab the first e-vector and normalize it
> vects:=(Transpose(vel[2]): V:=vects[1]; V:=[seq(V[i]/V[1],i=1..7)]:
> V:=convert(V,Vector[column]);
```

MATLAB
```
> [evect,eval]=eig(L)
```

```
> lambda=eval(1)
> pf=evect(:,1)
  % get pf=0.8586 and eigenvector=[-0.3930 -0.3007 ... -0.4532],
  % multiply by a constant so leading term is 1
> pf=pf/pf(1)
```

The eigenvalue and eigenvector are given as

$$\lambda \approx 0.8586 \quad \text{and} \quad \mathbf{V} \approx \begin{pmatrix} 1.0 \\ 0.765 \\ 0.829 \\ 0.898 \\ 0.972 \\ 1.06 \\ 1.15 \end{pmatrix}, \tag{5.3.8}$$

which is normalized to have first component equal to 1. The alternative formula (5.3.7) for computing **V** can be used to check this result:

```
MAPLE
> chk:=[1,el[2,1]/lambda, el[2,1]*el[3,2]/lambda^2, el[2,1]*el[3,2]*el[4,3]/lambda^3,
      el[2,1]*el[3,2]*el[4,3]*el[5,4]/lambda^4, el[2,1]*el[3,2]*el[4,3]*el[5,4]*el[6,5]/lambda^5,
      el[2,1]*el[3,2]*el[4,3]*el[5,4]*el[6,5]*el[7,6]/lambda^6];
```

```
MATLAB
> V=[1; L(2,1)/lambda; L(2,1)*L(3,2)/lambda^2; L(2,1)*L(3,2)*L(4,3)/lambda^3;...
    L(2,1)*L(3,2)*L(4,3)*L(5,4)/lambda^4; L(2,1)*L(3,2)*L(4,3)*L(5,4)*L(6,5)/lambda^5;...
    L(2,1)*L(3,2)*L(4,3)*L(5,4)*L(6,5)*L(7,6)/lambda^6]
```

Evidently, we get the same vector **V** as (5.3.8). Next, we illustrate the approximation of the iterates for this example. Take the intial value to be uniform, say, 1; then make the following calculations:

```
MAPLE
> evalf(evalm(el10 &* [1,1,1,1,1,1,1]));
> evalm(lambda^10*V);
```

```
MATLAB
> p=ones(7,1) % column vector of 1s
> (L^10)*p
> lambda^10*V
```

$$p(0) = \begin{pmatrix} 1 \\ 1 \\ 1 \\ 1 \\ 1 \\ 1 \\ 1 \end{pmatrix}, \qquad L^{10}p(0) = \alpha \begin{pmatrix} .24 \\ .17 \\ .17 \\ .19 \\ .25 \\ .28 \\ .29 \end{pmatrix} \approx c\lambda^{10}\mathbf{V} = c \begin{pmatrix} .22 \\ .17 \\ .18 \\ .19 \\ .21 \\ .23 \\ .25 \end{pmatrix}.$$

One implication of this structure is that the total population is stable if $\lambda = 1$, and it increases or decreases depending on the comparative size of $\lambda$ to 1.

*Continuous population densities provide exact population calculations.*

Any table of population densities, such as $P(y, t)$ for $n = 0, 1, \ldots$ as above, will have limited resolution, in this case one-year brackets. Alternatively, an age distribution can be described with unlimited resolution by a continuous age-density function, which we also denote by $P(y,t)$, such as we have shown in Figures 4.1.2 and 4.1.4.

Given a continuous age density $P(y,t)$, to find the population size in any age group, just integrate. For instance, the number in the group 17.6 to 21.25 is

$$\text{number between age 17.6 and 21.25} = \int_{17.6}^{21.25} P(y,t)dy.$$

This is the area under the density curve between $y = 17.6$ and $y = 21.25$. The total population at time $t$ is

$$P = \int_0^{\infty} P(y,t)dy,$$

which is the analogue of (5.3.1). For a narrow range of ages at age $y$, for example, $y$ to $y+\Delta y$ with $\Delta y$ small, there is a simpler formula: Population size is approximately given by the product

$$P(y,t) \cdot \Delta y$$

because density is approximately constant over a narrow age bracket.

The variable $y$ in an age density function is a continuous variable. The period of time an individual is exactly 20, for instance, is infinitesimal; so what does $P(20,t)$ mean? In general, $P(y,t)$ is the limit as $\Delta y \to 0$ of the number of individuals in an age bracket of size $\Delta y$ that includes $y$, divided by $\Delta y$,

$$P(y,t) = \lim_{\Delta y \to 0} \frac{\text{population size between } y \text{ and } y + \Delta y}{\Delta y}.$$

As above, the density is generally a function of time as well as age, and it is written $P(y,t)$ to reflect this dependence.

Table 2.7.3 gives the mortality rate for Alabama in 1990. From the table, the death rate for 70-year-olds, i.e., someone between 70.0 and 70.999..., is approximately 40 per 1000 individuals over the course of the year. Over one-half of the year it is approximately 20 per 1000, and over $\Delta t$ fraction of the year the death rate is approximately $\mu(70, 1990) \cdot \Delta t$ in deaths per 1000, where $\mu(70, 1990)$ is 40. To calculate the actual number of deaths, we must multiply by the population size of the 70-year-olds in thousands. On January 1, 1990, the number of such individuals was $\int_{70}^{71} P(y, 1990)dy/1000$. Thus the number of deaths among 70-year-olds over a small fraction $\Delta t$ of time at the beginning of the year 1990 is given by

$$\mu(70, 1990)\Delta t \int_{70}^{71} P(y, 1990)dy/1000. \qquad (5.3.9)$$

A calculation such as (5.3.9) works, provided the death rate is constant over the year and the time interval $\Delta t$ is less than one year. But in general, death rates vary continuously with age. In Figure 5.3.1, we show an exponential fit to the data of Table 2.7.3. The approximate continuously varying death rate is

$$\mu(y,t) = Ae^{by},$$

which is drawn using the methods of Exercise 1 in Section 2.7. This equation assumes that the death rate is independent of time; but as we have seen, it can depend on time as well as age.

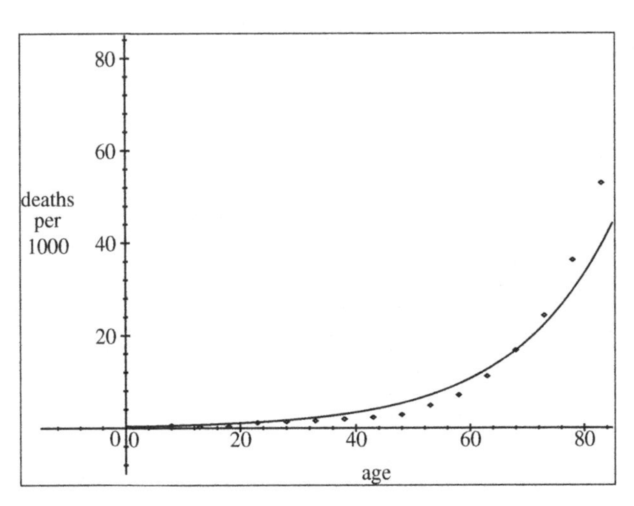


**Fig. 5.3.1.** Least squares fit to the death rate table, Table 2.7.3.

To calculate a number of deaths accurately, we must account for the changing death rate as well as the changing density. The term that calculates the number of deaths to individuals of exact age $y$ at time $t$ over the interval of time $\Delta t$ is

$$P(y,t)\mu(y,t)\Delta t. \tag{5.3.10}$$

The number of deaths among those individuals who are between $y$ and $y + \Delta y$ years old over this same period of time is

$$[P(y,t)\Delta y]\mu(y,t)\Delta t.$$

Suppose we want to do the calculation for those between the ages of $a_1$ to $a_2$ over the calendar time $t_t$ to $t_2$. The approximate answer is given by the double sum of such terms,

$$\sum\sum \mu(y,t)P(y,t)\Delta y\Delta t,$$

over a grid of small rectangles $\Delta y\Delta t$ covering the range of ages and times desired. In the limit as the grid becomes finer, this double sum converges to the double integral

$$\int_{t_1}^{t_2}\int_{a_1}^{a_2} \mu(y,t)P(y,t)dydt. \tag{5.3.11}$$

Return to (5.3.10), which calculates the loss of population, $\Delta P$, in the exact age group $y$ over the time interval $\Delta t$,

$$\Delta P = -\mu(y,t)P(y,t)\Delta t.$$

But by definition, the change in population is

$$\Delta P = P(y+\Delta y, t+\Delta t) - P(y,t).$$

Equate these two expressions for $\Delta P$, incorporating the fact that as time passes, the population ages at the same rate, that is, $\Delta y = \Delta t$. Therefore, we have the continuous analogue of (5.3.3),

$$P(y+\Delta t, t+\Delta t) - P(y,t) = -\mu(y,t)P(y,t)\Delta t.$$

Subtract the term $P(y, t+\Delta t)$ from both sides, transpose $P(y,t)$, and divide by $\Delta t$:

$$\frac{P(y+\Delta t, t+\Delta t) - P(y, t+\Delta t)}{\Delta t} + \frac{P(y, t+\Delta t) - P(y,t)}{\Delta t} = -\mu(y,t)P.$$

Finally, take the limit as $\Delta t \to 0$ to get

$$\frac{\partial P}{\partial y} + \frac{\partial P}{\partial t} = -\mu(y,t)P. \tag{5.3.12}$$

This is referred to as the *Von Foerster equation.* Its solution for $y > t$ is

$$P(y,t) = P(y-t,0)e^{-\int_0^t \mu(y-t+u,u)du},$$

as can be verified by direct substitution.

```
MAPLE (symbolic, no MATLAB)
> P:=(n,t)->h(n-t)*exp(-int(mu(n-t+u,u),u=0..t));
> diff(P(n,t),t)+diff(P(n,t),n)+mu(n,t)*P(n,t);
> simplify(%);
```

This solution does not incorporate new births, however. Just as in the discrete case, we must use experimental data to determine $P(0,t)$ as a function of $P(y,t)$, $y > 0$.

### Exercises/Experiments

**1.** Consider the following discrete population model based on (5.3.1). Suppose the initial population distribution (year $t = 0$) is given by

$$P(n,0) = (100-n)\cdot(25+n), \quad n = 0,\ldots,100.$$

Take the birth rate to be 1.9 children per couple per 10 years in the ten-year age bracket from 21 to 30 years of age. Thus over the year $t$, the number of births (number of people aged 0 in year $t+1$) is

$$P(0,t+1) = \frac{1.9}{2}\sum_{i=21}^{30}\frac{P(i,t)}{10}.$$

(Assume that this formulation accounts for the complication of a $\frac{3}{4}$-year gestation period.) Take the death rate for people of age $n$ to be given by the exponential

$$\mu(n) = 0.0524(\exp(0.03n) - 1), \quad n > 0.$$

The problem is to advance the population for three years, keeping track of the total population:

$$\text{Total}(t) = \sum_{n=0}^{100} P(n, t).$$

Does the total population increase? (One can use the Leslie matrix approach or (5.3.3) directly.)

```
MAPLE
> restart;
> for n from 0 to 100 do
   P[n,0]:=(100-n)*(25+n); mu[n]:=.0524*(exp(.03*n)-1);
   od:
> plot([seq([i,P[i,0]],i=0..100)]);
> plot([seq([i,mu[i]],i=0..100)]);
> for t from 1 to 3 do
   P[0,t]:=(1.9/20)*sum(P[i,t-1],i=21..30);
   for k from 1 to 100 do
     P[k,t]:=(1-mu[k-1])*P[k-1,t-1];
   od: od:
> for t from 0 to 3 do
   total[t]:=sum(P[i,t],i=0..100);
> od;
```

```
MATLAB
> n=0:1:100;
> P0=(100-n).*(25+n);
> plot(n,P0);
> P=P0'; % rows=age, columns=time
  % no base 0 indexing so P(n,t) = number aged n-1 in year t-1
> mu=.0524*(exp(.03.*n)-1); % mu(n) applies to age n-1
> plot(n,mu);
> for t=1:3
   total(t)=sum(P(:,t));
   P(1,t+1)=(1.9/20)*sum(P(22:31,t));
   for n=2:101
     P(n,t+1)=(1-mu(n-1))*P(n-1,t);
   end; end;
> total(1) %starting year
> total(4)=sum(P(:,4)) % 3 years later
```

**2.** For the following two Leslie matrices find $\lambda$ and $\mathbf{V}$ as given in (5.3.4). What is the ratio of the ages of associated populations?

$$L_1 = \begin{pmatrix} 1 & \frac{2}{3} \\ \frac{1}{2} & 0 \end{pmatrix}, \qquad L_2 = \begin{pmatrix} 0 & 4 & 3 \\ \frac{1}{2} & 0 & 0 \\ 0 & \frac{1}{4} & 0 \end{pmatrix}.$$

## Questions for Thought and Discussion

**1.** Draw age-structure diagrams for the three cases of populations whose maximum numbers are young, middle-aged, and elderly people. In each case, draw the age-structure diagram to be expected 30 years later if birth and death rates are equal and constant and if there is no migration.

2. Repeat Question 1 for the situation in which the birth rate is larger than the death rate and there is no migration.
3. Repeat Question 1 for the situation in which the birth and death rates are constant, but there is a short but extensive incoming migration of middle-aged women at the outset.

## References and Suggested Further Reading

[1] AGING:
T. B. L. Kirkwood, The nature and causes of ageing, in D. Evered and J. Whelan, eds., *Research and the Ageing Population*, Ciba Foundation Symposium, Vol. 134, Wiley, Chichester, UK, 1988, 193–202.

[2] AGING IN HUMANS:
R. L. Rusting, Why do we age?, *Sci. Amer.*, **267**-6 (1992), 130–141.

[3] PERRON–FROBENIUS THEOREM:
E. Senata, *Non-Negative Matrices and Markov Chains*, Springer-Verlag, New York, 1973.

[4] LESLIE MATRICES:
H. Anton and C. Rorres, *Elementary Linear Algebra*, Wiley, New York, 1973, 653.

# 6

# Random Movements in Space and Time

## Introduction

Many biological phenomena, at all levels of organization, can be modeled by treating them as random processes, behaving much like the diffusion of ink in a container of water. In this chapter, we discuss some biological aspects of random processes, namely, the movement of oxygen across a human placenta. While these processes might seem to be quite different at first glance, they actually act according to very similar models.

We begin with a description of biological membranes, structures that regulate the movement of material into, out of, and within the functional compartments of a cell. At the core of a membrane is a layer of water-repelling molecules. This layer has the effect of restricting the free transmembrane movement of any substance that is water soluble, although water itself can get past the layer. The transmembrane movement of the normal water-soluble compounds of cellular metabolism is regulated by large biochemical molecules that span the membrane. They are called *permeases*, or *transport proteins*. Permeases have the ability to select the materials that cross a membrane. Other membranes anchor critical cellular components that promote chemical reactions through catalysis.

A human fetus requires oxygen for its metabolic needs. This oxygen is obtained from its mother, who breathes it and transfers it via her blood to the placenta, an organ that serves as the maternal–fetal interface. Because the blood of mother and child do not mix, material exchange between them must take place across a group of membranes. The chemical that transports the oxygen is hemoglobin, of which there are at least two kinds, each exhibiting a different strength of attachment to oxygen molecules. Further, chemical conditions around the hemoglobin also affect its attachment to oxygen. The conditions at the placenta are such that there is a net transmembrane movement of oxygen from maternal hemoglobin to fetal hemoglobin.

This chapter also serves as an introduction to the discussions of the blood vascular system of Chapter 9, of biomolecular structure of Chapter 8, and of HIV in Chapter 10.

R.W. Shonkwiler and J. Herod, *Mathematical Biology: An Introduction with Maple and Matlab*, Undergraduate Texts in Mathematics, DOI: 10.1007/978-0-387-70984-0_6,

## 6.1 Biological Membranes

*Biological membranes do much more than physically separate the interior of cells from the outside world. They provide organisms with control over the substances that enter, leave, and move around their cells. This is accomplished by selective molecules that can recognize certain smaller molecules whose transmembrane movement is required by the cell. A waterproof layer in the membrane otherwise restricts the movement of these smaller compounds. In addition, membranes maintain compartments inside a cell, allowing the formation of specific chemical environments in which specialized reactions can take place.*

*The molecular structure of a substance determines its solubility in water.*

Distinctions between oil and water are everywhere. We have all seen how salad oil forms spherical globules when it is mixed with watery vinegar. Likewise, we say that two hostile people "get along like oil and water." On the other hand, a drop of ink or food coloring dissolves immediately in a glass of water. These experiences seem to suggest that all materials are either water soluble or not. This is an oversimplification: ethyl alcohol is infinitely soluble in water (gin is about half alcohol, half water), but isopropanol (rubbing alcohol), table salt, and table sugar all have moderate water-solubility. Salt and sugar have very low solubility in gasoline and benzene (erstwhile dry-cleaning fluid). On the other hand, benzene will easily dissolve in gasoline and in fatty substances.

The electronic basis for water-solubility will be described in Chapter 8, but for now it is sufficient that we recognize that the ability of a substance to dissolve in water is determined by its electronic structure. Further, an appropriate structure is found in ions (like sodium and chlorine from salt) and in molecules with oxygen and nitrogen atoms (like sugars and ammonia). Such substances are said to be *hydrophilic*, or *polar*. Hydrophilic structures are not found in electrically neutral atoms, nor in most molecules lacking oxygen and nitrogen. This is especially true when the latter molecules have a very high proportion of carbon and hydrogen (e.g., benzene, gasoline and fatty substances). These latter materials are said to be *hydrophobic*, or *nonpolar*.

*Both faces of a membrane are attracted to water, but the interior of the membrane repels water.*

The biological world is water based.[1] Therefore, cells face a bit of a problem in that water is a major component of the external world, which could lead to too much interaction between a cell's contents and its environment. To deal with this problem, cells are surrounded by a water-proofing, or hydrophobic, membrane layer. We should be glad for this structural feature of our bodies—it keeps us from dissolving in the shower!

[1] Our bodies must resort to special tricks to solubilize fats that we eat. Our livers produce a detergent-like substance, called bile, that allows water to get close to the fats. The hydrocarbon-metabolizing microorganisms that are useful in dealing with oil spills often use similar methods.

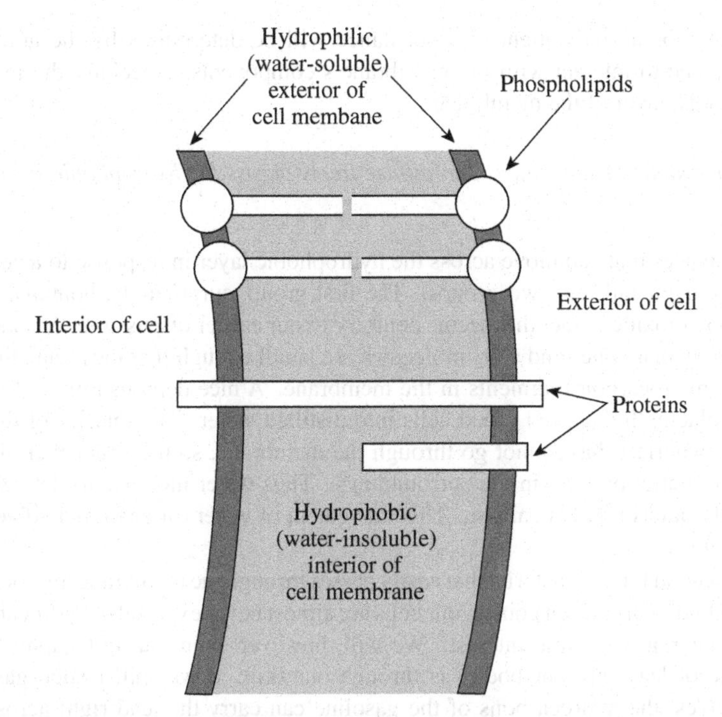


**Fig. 6.1.1.** A model of a cell membrane, showing the hydrocarbon (hydrophobic, water-insoluble) interior and hydrophilic (water-soluble) exterior of the membrane. This dual nature of the membrane is the result of the orientation of many phospholipid molecules, only four of which are actually shown in the figure. Figure 8.2.5 will show how the chemical nature of a phospholipid leads to hydrophobic and hydrophilic parts of the membrane. Two proteins in the figure are also shown to demonstrate that some span the membrane completely and others only pierce the outside halfway through (on either side of the membrane).

Figure 6.1.1 shows a model of a cell membrane. The side facing the cellular interior is hydrophilic because it must interact with the cell's internal environment; the outside is also hydrophilic because it interacts with the external world.[2] The interior of the membrane, however, is strongly hydrophobic, being a kind of *hydrocarbon* (constructed from hydrogen and carbon only). This arrangement is thermodynamically favorable because there are no direct interactions between hydrophilic and hydrophobic groups.[3] Attached to, and sometimes piercing, the membrane are complicated biological molecules called proteins, which will be described in more detail in Chapter 8.

No material can enter or leave the cell unless it negotiates its way past this membrane, because the membrane completely envelope the cell. Clearly, the efficiency

[2] It might be easiest here to picture a single-celled organism in a pond.

[3] The arrangement of molecules in the membrane of Figure 6.1.1 is called a *bilayer* because it consists of two leaflets of molecules, arranged back to back.

of transmembrane movement of a substance will be determined by the ability of the substance to interact with the membrane's components, especially the interior, hydrophobic, layer of the membrane.

*Only a few kinds of substances can diffuse freely across the hydrophobic layer of a membrane.*

The substances that can move across the hydrophobic layer in response to a concentration gradient fall into two groups. The first group, surprisingly, contains water and carbon dioxide, a fact that seems contrary to our earlier discussion. What seems to happen is that water and $CO_2$ molecules are small enough that they can slip past the large hydrocarbon fragments in the membrane. A nice demonstration of this is seen by placing human red blood cells into distilled water. The interior of the cell contains materials that cannot go through the membrane, so the water there is at a *lower* concentration than in the surroundings. Thus water moves into the cell and eventually bursts it like a balloon. This movement of water (or any other solvent) is called *osmosis*.

The second kind of material that easily passes through the membrane hydrocarbon layer is a hydrocarbon. Of course, our cells are almost never exposed to hydrocarbons, so this material is of little interest. We will, however, point out in Chapter 9 that one route of lead into our bodies is through our skin: if we spill leaded gasoline on ourselves, the hydrocarbons of the gasoline can carry the lead right across our hydrophobic membrane barriers and into our bloodstream.

*Selective channels control the passive and active movements of ions and large molecules across membranes.*

Many relatively large hydrophilic particles, such as ions and sugars, can pass through membranes—after all, these particles are essential components of cellular systems. They do not move directly through the bulk of the membrane. Rather, their movement is regulated by large proteins that completely penetrate the membrane and act like specialized channels, choosing which substances get past the membrane (see Figure 6.1.1). These proteins are called *permeases*, or *transport proteins*, and they are very selective: The substitution of a single atom in a transportable molecule with molecular weight of several hundred can cause the molecule to be excluded by its normal permease. Permeases thus act like selective gates to control material transport into and out of a cell.

Materials can move across membranes via permeases by two different mechanisms, both often called *facilitated transport*. First, the *passive* movement of a material in response to a concentration gradient (diffusion) is usually facilitated by permeases. The point is that only those substances recognized by a permease will behave in this way. Any other substances will diffuse up to the membrane and then be stopped by the hydrophobic layer of the membrane.

Second, many materials are pumped *against* a concentration gradient past a membrane. This process, called *active transport*, requires energy because it is in the opposite direction to the usual spontaneous movement of particles. Active transport also requires a facilitating permease.

Facilitated transport is discussed further in Chapter 8, in the text by Beck et al. [1], and in the reference by Yeargers [2].

*Some cellular membranes face the outside world and regulate intercellular material movement.*

The day-to-day processes that a cell must perform require that nutrients and oxygen move into the cell and that wastes and carbon dioxide move out. In other words, the cell must maintain constant, intimate communication with its external environment. The cell membrane provides the interface between the cell and the outside world, and membrane permeases, because of their selectivity, control the transmembrane movement of most of the substances whose intercellular transport is required.

What about water? It moves across membranes irrespective of permeases and would therefore seem to be uncontrollable. In fact, cells can regulate water movement, albeit by indirect means. They accomplish this by regulating other substances and that, in turn, affects water. For example, a cell might pump sodium ions across a membrane to a high concentration. Water molecules will then follow the sodium ions across the membrane by simple osmosis, to dilute the sodium.

*Some cellular membranes are inside the cell and regulate intracellular material movements.*

Students are sometimes surprised to learn that the interior of a cell, exclusive of the nucleus, is a labyrinth of membranes. A mechanical analogue can be obtained by combining some confetti and a sheet of paper, crumpling up the whole mess into a wad, and then stuffing it into a paper bag. The analogy cannot be pushed too far; membranes inside the cell often have very regular structures, lying in parallel sheets or forming globular structures (like the nucleus). In short, the interior of a cell is a very complicated place.

Many thousands of different biochemical reactions occur in a mammalian cell. If these reactions were not coordinated in space and time the result would be chaos. Membranes provide coordinating mechanisms in several ways: First, large biochemical molecules are always assembled stepwise, beginning with small structures and ending up with large ones. Each of the fragments to be added must be close to the nascent biomolecule so that it can be added at the right time. Intracellular membranes provide compartmentalization to keep the reactants and products of related reactions in close proximity to one another. Second, the efficiencies of different cellular biochemical reactions are dependent on environmental conditions, e.g., pH and salt concentration. The specialized environmental needs of each reaction, or set of reactions, are maintained by membrane compartmentalization. Thus a cell is partitioned into many small chambers, each with a special set of chemical conditions. A third point, related to the first two, is that virtually all chemical reactions in a cell are catalyzed by special proteins, and these catalysts often work only when they are attached to a membrane. Refer back to Figure 6.1.1 and note that many of the proteins do not pierce the membrane, but rather are attached to one side or the other. These proteins represent several of the membrane-bound protein catalysts of a cell. You will read more about these catalysts in Chapter 8.

*Large objects move into and out of a cell by special means.*

Neither simple diffusion nor facilitated transport can move particulate objects such as cellular debris, food, bacteria, and viruses into or out of a cell; those require completely different routes. If a cell is capable of amoeboid movement, it can surround the particle with *pseudopods* and draw it in by *phagocytosis*. If the cell is not amoeboid, it can form small pockets in its surface to enclose the particle; this process is *pinocytosis*, as shown in Figure 6.1.2. Both phagocytosis and pinocytosis can be reversed to rid the cell of particulate waste matter.

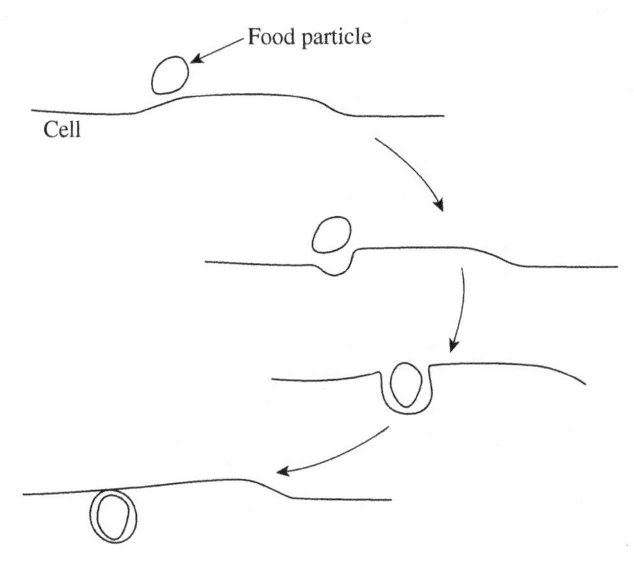


**Fig. 6.1.2.** A schematic diagram showing the process of pinocytosis. A small indentation forms at the particle's point of contact, and the particle is then drawn into the cell's interior.

## 6.2 The Mathematics of Diffusion

*In this section, we derive Fick's laws of diffusion by studying a random walk model. Using the normal approximation to the binomial distribution, we obtain the Gaussian solution of the diffusion equation for a point-source concentration. It is seen that particles disperse on average in proportion to the square root of time.*

*Fick's laws are applied to investigate one aspect of diffusion through biological membranes. It is shown that the rate of mass transport is proportional to the concentration difference across the membrane and inversely proportional to the thickness of the membrane.*

*Random processes in the biosphere play a major role in life.*

In Section 6.1, we described the membrane system that surrounds and pervades a cell. In this section, we show how the random motion of substances can carry materials across these membranes and through the bulk of a cell.

Chance plays a major role in the processes of life. On the microscopic scale, molecules are in constant random motion corresponding to their temperature. Consequently, chance guides the fundamental chemistry of life. On a larger scale, genes mutate and recombine by random processes. Thus chance is a fundamental component of evolution. Macroscopically, unpredictable events such as intraspecies encounter lead to matings or maybe the transmission of disease, which interspecies encounter claims many a prey victim, but not with certainty. The weather can affect living things throughout an entire region and even an entire continent. And on a truly grand scale, random astronomical impacts can cause mass extinction.

*Diffusion can be modeled as a random walk.*

Molecules are in a constant state of motion as a consequence of their temperature. According to the kinetic theory of matter, there is a fundamental relationship between molecular motion and temperature, which is simplified by measuring the latter on an absolute scale, degrees Kelvin. Zero degrees Kelvin, or absolute zero, is −273.15°C. Moreover, Albert Einstein showed in 1905 that the principle extends to particles of any size, for instance, to pollen grains suspended in water. Einstein's revelation explained an observation made in 1828 by the Scottish botanist Robert Brown, who reported on seeing a jittery, undirected motion of pollen grains in the water of his microscope plate. We now refer to this phenomenon as *Brownian motion*. It is a visible form of diffusion.

The relationship between temperature and particle motion can be precisely stated: *The average kinetic energy of a particle along a given axis is* $\frac{kT}{2}$, where $T$ is in degrees Kelvin and $k$ is the *universal Boltzmann's constant*, $k = 1.38 \times 10^{-16}$ ergs per degree [4]. The principle is stated in terms of the *time average* of a single particle, but we will assume that it applies equally well to the average of an ensemble or collection of identical particles taken at the same time, the *ensemble average*.

The kinetic energy of an object of mass $m$ and velocity $v$ is $\frac{1}{2}mv^2$. And so the average kinetic energy of $N$ particles of the same mass $m$ but possibly different velocities is

$$\overline{\frac{mv^2}{2}} = \frac{\sum_{i=1}^{N} \frac{mv_i^2}{2}}{N} = \frac{m}{2N} \sum_{i=1}^{N} v_i^2 = \frac{m\overline{v^2}}{2}.$$

In this we have used an overline to denote average ensemble value.

Therefore, for a collection of particles of mass $m$, the kinetic theory gives

$$\frac{m\overline{v^2}}{2} = \frac{kT}{2},$$

or

**Table 6.2.1.** Root mean square (RMS) velocities at body temperature.

| Molecule | Molecular weight | Mass (g) | RMS speed at 36°C (m/sec) |
|---|---|---|---|
| $H_2O$ | 18 | $3 \times 10^{-26}$ | 652 |
| $O_2$ | 32 | $5.4 \times 10^{-26}$ | 487 |
| glucose | 180 | $3 \times 10^{-25}$ | 200 |
| lysozyme | 1,400 | $2.4 \times 10^{-23}$ | 23 |
| hemoglobin | 65,000 | $1 \times 10^{-22}$ | 11 |
| bacteriophage | $6.2 \times 10^{6}$ | $1 \times 10^{-20}$ | 1.1 |
| *E. coli* | $\approx 2.9 \times 10^{11}$ | $2 \times 10^{-15}$ | 0.0025 |

$$\overline{v^2} = \frac{kT}{m}. \tag{6.2.1}$$

Table 6.2.1 gives the average thermal velocity of some biological molecules at body temperature predicted by this equation.

A particle does not go very far at these speeds before undergoing a collision with another particle or the walls of its container and careers off in some new direction. With a new direction and velocity, the process begins anew, destined to undergo the same fate. With all the collisions and rebounds, the particle executes what can be described as a random walk through space. To analyze this process, we model it by stripping away as much unnecessary complication as possible while still retaining the essence of the phenomenon.

For simplicity, assume that time is divided into discrete periods $\Delta t$ and in each such period a particle moves one step $\Delta x$ to the left or right along a line, the choice being random. After $n$ time periods the particle lies somewhere in the interval from $-n(\Delta x)$ to $n(\Delta x)$ relative to its starting point, taken as the origin 0.

For example, suppose that $n = 4$. If all four choices are to the left, the particle will be at $-4$; if three are left and one right, it will be at $-2$. The other possible outcomes are $0, 2$, and 4. Notice the outcomes are separated by two steps. Also notice that there are several ways most of the outcomes can arise, the outcome 2, for instance. We can see this as follows. Let $R$ denote a step right and $L$ a step left. Then a path of four steps can be coded as a string of four choices of the symbols $R$ or $L$. For example, $LRRR$ means that the first step is to the left and the next three are to the right. For an outcome of four steps to be a net two to the right, three steps must be taken to the right and one to the left, but the order does not matter. There are four possibilities that do it; they are $LRRR$, $RLRR$, $RRLR$, and $RRRL$.

In general, let $p(m, n)$ denote the probability that the particle is at position $x = m(\Delta x)$, $m$ steps right of the origin, after $n$ time periods, $t = n(\Delta t)$. We wish to calculate $p(m, n)$. It will help to recognize that our random walk with $n$ steps is something like tossing $n$ coins. For every coin that shows heads, we step right, and for tails we step left. Let $r$ be the number of steps taken to the right and $l$ the number of steps taken to the left; then to be at position $m(\Delta x)$, it must be that their difference is $m$:

$$m = r - l, \quad \text{where } n = r + l.$$

Thus $r$ can be given in terms of $m$ and $n$ by adding these two equations, and $l$ is given by subtracting:

$$r = \frac{1}{2}(n + m) \quad \text{and} \quad l = \frac{1}{2}(n - m). \tag{6.2.2}$$

As in a coin toss experiment, the number of ways of selecting $r$ moves to the right out of $n$ possibilities is the problem of counting combinations and is given by (see Section 2.8)

$$C(n, r) = \frac{n!}{r!(n - r)!}.$$

For example, three moves right out of four possible moves can happen in $\frac{4!}{3!1!} = 4$ ways, in agreement with the explicitly written-out $L$ $R$ possibilities noted above. Therefore, if the probabilities of going left and going right are equal, then

$$p(m, n) = \text{probability of } r \text{ steps right} = \frac{C(n, r)}{2^n}, \quad r = \frac{1}{2}(n + m). \tag{6.2.3}$$

This is the *binomial* distribution with $p = q = \frac{1}{2}$. The solid curve in Figure 6.2.1 is a graph of $p(m, 40)$. If the random walk experiment with $n = 40$ steps is conducted a large number of times, then a histogram of the resulting particle positions will closely approximate this figure. This histogram is also shown in Figure 6.2.1. Equivalently, the same picture pertains to the fate of a large number of particles randomly walking at the *same* time, each taking 40 steps, provided they may slide right past each other without collisions.

*Particles disperse in proportion to the square root of time.*

The average, or *mean*, position, $\overline{m}$, of a large number of particles after a random walk

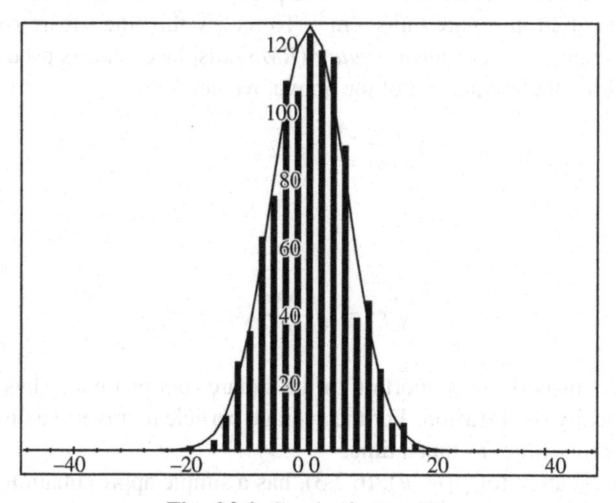


**Fig. 6.2.1.** Graph of $p(m, 40)$.

of $n$ steps with equal probabilities of stepping left or right is 0. To show this, start with (6.2.2) to get $m = 2r - n$. Then since $r$ has a binomial distribution, we can write down its mean and variance from Section 2.8. Equations (2.8.12) and (2.8.13) with $p = \frac{1}{2}$ and $q = 1 - p = \frac{1}{2}$ give

$$\begin{aligned} \overline{r} &= np = \frac{n}{2}, \\ \operatorname{var}(r) &= \overline{(r - \overline{r})^2} = npq = \frac{n}{4}. \end{aligned} \tag{6.2.4}$$

Hence

$$\overline{m} = \overline{2r - n} = 2\overline{r} - \overline{n} = 2\frac{n}{2} - n = 0,$$

since the average value of the constant $n$ is $n$. Unfortunately, knowing that the average displacement of particles is 0 does not help in expressing how quickly particles are moving away from the origin. The negative values of those that have moved to the left cancel the positive values of those that have gone right.

We can avoid the left vs. right cancellation by using the squares of displacements; we will get thereby the *mean square* displacement, $\overline{m^2}$. Since the mean position is 0, the mean square displacement is equal to the variance here; hence from (6.2.4),

$$\overline{m^2} = \overline{(m - \overline{m})^2} = \overline{(2r - n)^2} = 4\overline{\left(r - \frac{n}{2}\right)^2} = 4\frac{n}{4} = n.$$

Since $m = \frac{x}{\Delta x}$ and $n = \frac{t}{\Delta t}$, we can convert this into statements about $x$ and $t$:

$$\overline{x^2} = \frac{\Delta x^2}{\Delta t} t. \tag{6.2.5}$$

But mean square displacement is not a position, a distance from 0. For one thing, it is measured in square units, $cm^2$. To rectify this, the square root of mean square displacement, or *root mean square* (*RMS*) displacement, is used to quantify dispersion; taking the square root of the above, we get

$$\sqrt{\overline{m^2}} = \sqrt{n}$$

and

$$\sqrt{\overline{x^2}} = \sqrt{\frac{\Delta x^2}{\Delta t}}\sqrt{t}. \tag{6.2.6}$$

Hence particles disperse in proportion to the square root of time. Thus there is no concept of velocity for diffusion. For the average particle to traverse a distance twice as far requires four times as much time.

The exact equation for $p(m, n)$, (6.2.3), has a simple approximation. There is a real need for such an approximation because it is difficult to compute the combinatorial factor $C(n, r)$ for large values of $n$. Moreover, the approximation improves with an error that tends to 0 as $n \to \infty$. The binomial distribution (see (6.2.3) and Figure 6.2.1) looks very much like that of a normal distribution, as discussed in Chapter 2. Although *Stirling's formula* for approximating $n!$,

$$n! \approx \sqrt{2\pi n} n^n e^{-n},$$

may be used to prove it, we will not do this. Instead, we will match the means and standard deviations of the two distributions. First, recall that the probability that a normally distributed observation will fall within an interval of width $dm$ centered at $m$ is approximately (see Section 2.8)

$$\frac{1}{\sqrt{2\pi\sigma^2}} e^{-\frac{(m-\mu)^2}{2\sigma^2}} dm,$$

where $\mu$ is the mean and $\sigma$ is the standard deviation of the distribution. On the other hand, $p(m, n)$ is the probability the walk will end between $m-1$ and $m+1$, an interval of width 2, and from above, its mean is 0 and standard deviation is $\sqrt{n}$. Hence

$$p(m, n) \approx \frac{1}{\sqrt{2\pi n}} e^{-\frac{m^2}{2n}} (2) \approx \sqrt{\frac{2}{\pi n}} e^{-\frac{m^2}{2n}}. \tag{6.2.7}$$

Our last refinement is to let $\Delta x$ and $\Delta t$ tend to 0 to obtain a continuous version of $p(m, n)$. Of course, without care, $p(m, n)$ will approach zero too because the probability will spread out over more and more values of $m$. But since each value of $m$ corresponds to a probability over a width of $2\Delta x$, we take the quotient of $p(m, n)$ by this width. That is, let $u(x, t)$ denote the probability that the particle lies in an interval of width $2(\Delta x)$ centered at $x$ at time $t$. Then

$$u(x, t) = \frac{P\left(\frac{x}{\Delta x}, \frac{t}{\Delta t}\right)}{2(\Delta x)} = \frac{1}{2(\Delta x)} \sqrt{\frac{2}{\pi \left(\frac{t}{\Delta t}\right)}} e^{-\frac{\left(\frac{x}{\Delta x}\right)^2}{2\left(\frac{t}{\Delta t}\right)}}.$$

And upon simplification,[4]

$$u(x, t) = \frac{e^{-\left(\frac{x^2}{4\left(\frac{\Delta x^2}{2\Delta t}\right)t}\right)}}{\sqrt{4\pi \left(\frac{\Delta x^2}{2(\Delta t)}\right) t}}.$$

Now keeping the ratio

$$D = \frac{\Delta x^2}{2(\Delta t)} \tag{6.2.8}$$

[4] For more on this and an alternative derivation, see C. W. Gardiner, *Handbook of Stochastic Methods*, Springer-Verlag, Berlin, 1983.

fixed as $\Delta x$ and $\Delta t$ tend to 0, we obtain the Gaussian distribution

$$u(x,t) = \frac{e^{-\frac{x^2}{4Dt}}}{\sqrt{4\pi Dt}}. \tag{6.2.9}$$

The parameter $D$ is called the *diffusion coefficient* or *diffusivity* and has units of area divided by time. Diffusivity depends on the solute, the solvent, and the temperature, among other things. See Table 6.2.2 for some pertinent values.

**Table 6.2.2.** Diffusion coefficients in solution.

| | | | | Seconds to cross | |
|---|---|---|---|---|---|
| **Molecule** | **Solvent** | **T, °C** | **$D$ ($10^{-6}$ cm$^2$/sec)** | **0.01 mm** | **1 mm** |
| $O_2$ | blood | 20 | 10.0 | 0.05 | 500 |
| acetic acid | water | 25 | 12.9 | 0.04 | 387 |
| ethanol | water | 25 | 12.4 | 0.04 | 403 |
| glucose | water | 25 | 6.7 | 0.07 | 746 |
| glycine | water | 25 | 10.5 | 0.05 | 476 |
| sucrose | water | 25 | 5.2 | 0.10 | 961 |
| urea | water | 25 | 13.8 | 0.04 | 362 |
| ribonuclease | water | 20 | 1.07 | 0.46 | 4671 |
| fibrinogen | water | 20 | 2.0 | 0.25 | 2500 |
| myosin | water | 20 | 1.1 | 0.45 | 4545 |

Diffusivity quantifies how rapidly particles diffuse through a medium. In fact, from (6.2.6), the rate at which particles wander through the medium in terms of root mean square distance is

$$\text{RMS distance} = \sqrt{\frac{\Delta x^2}{\Delta t}t} = \sqrt{2Dt}. \tag{6.2.10}$$

In Table 6.2.2, we give some times required for particles to diffuse the given distances. As seen, the times involved become prohibitively long for distances over 1 mm. This explains why organisms whose oxygen transport is limited to diffusion cannot grow very large in size.

The function $u$ has been derived as the probability for the ending point, after time $t$, of the random walk for a single particle. But as noted above, it applies equally well to an ensemble of particles if we assume that they "walk" independently of each other. In terms of a large number of particles, $u$ describes their concentration as a function of time and position. Starting them all at the origin corresponds to an infinite concentration at that point, for which (6.2.9) does not apply. However, for any positive time, $u(x,t)$ describes the concentration profile (in number of particles per unit length); see Figure 6.2.2 for the times $1, 2, 4$. Evidently, diffusion transports particles from regions of high concentration to places of low concentration. Fick's first law, derived below, makes this precise.

MAPLE
```
> plot({exp(-x^2/4)/sqrt(4*Pi), exp(-x^2/(4*2))/sqrt(4*Pi*2), exp(-x^2/(4*4))/sqrt(4*Pi*4)},
      x=-10..10,color=BLACK);
```

MATLAB
```
% make an m-file, gaussian.m containing
% function y=gaussian(x,m,s); % m=mean, s=stddev
% note 1/sqrt(2*pi) = .3989422803
% y=(.3989422803/s)*exp(-0.5*((x-m)./s).^2);
> x=[-10:.1:10]; y=gaussian(x,0,1);
> plot(x,y); hold on;
> y=gaussian(x,0,2); plot(x,y);
> y=gaussian(x,0,4); plot(x,y);
```

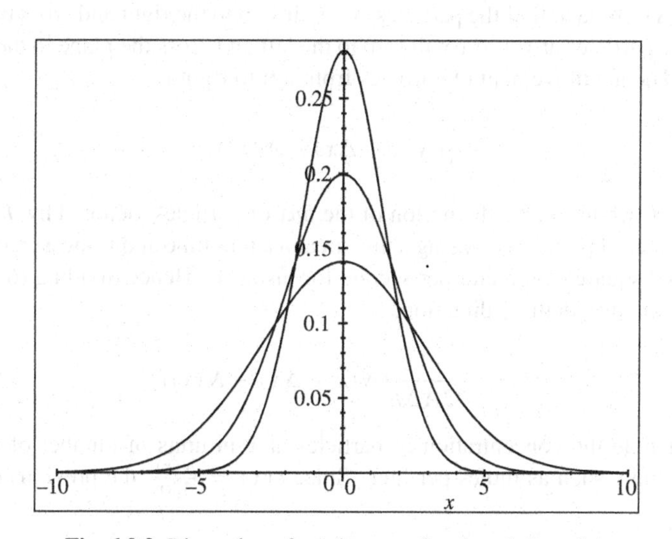


**Fig. 6.2.2.** Dispersion of a unit mass after time 1, 2, and 4.

To treat diffusion in three dimensions, it is postulated that the random walk proceeds independently in each dimension. Hence the mean transport in the $x$, $y$, and $z$ directions is each given by (6.2.10), $\overline{x^2} = 2Dt$, $\overline{y^2} = 2Dt$, and $\overline{z^2} = 2Dt$. In two dimensions, if $r^2 = x^2 + y^2$, then

$$\overline{r^2} = 4Dt,$$

and in three dimensions, if $r^2 = x^2 + y^2 + z^2$, we get

$$\overline{r^2} = 6Dt.$$

*Fick's laws describe diffusion quantitatively.*

Again consider a one-dimensional random walk, but now in three-dimensional space, for example, along a channel of cross-sectional area $A$; see Figure 6.2.3.

Let $N(x)$ denote the number of particles at position $x$. We calculate the net movement of particles across an imaginary plane perpendicular to the channel between

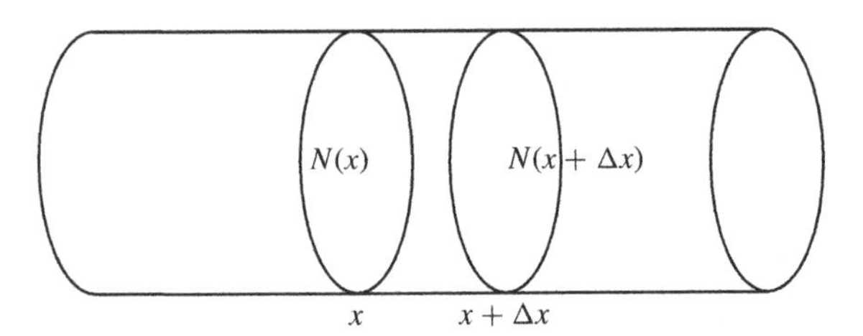


**Fig. 6.2.3.** One-dimensional diffusion along a channel.

$x$ and $x + \Delta x$. In fact, half the particles at $x$ will step to the right and cross the plane, and half the particles at $x + \Delta x$ will step to the left and cross the plane in the reverse direction. The net movement of particles from left to right is

$$-\frac{1}{2}(N(x + \Delta x) - N(x)). \tag{6.2.11}$$

At this point, we introduce the notion of the *flux* of particles, denoted by $J$. This is the net number of particles crossing a unit area in a unit time and is measured in units of moles per square centimeter per second for instance. Hence dividing (6.2.11) by $A\Delta t$ gives the flux in the $x$ direction,

$$J_x = -\frac{1}{2A\Delta t}(N(x + \Delta x) - N(x)).$$

Let $c(x)$ denote the concentration of particles at $x$ in units of number of particles per unit volume such as moles per liter. Since $c(x) = \frac{N(x)}{A\Delta x}$, the previous equation becomes

$$J_x = -\frac{\Delta x}{2\Delta t}(c(x + \Delta x) - c(x)) = -\frac{\Delta x^2}{2\Delta t}\frac{c(x + \Delta x) - c(x)}{\Delta x}.$$

Now let $\Delta x \to 0$ and recall the definition of diffusivity, (6.2.8); we get *Fick's first law*:

$$J = -D\frac{\partial c}{\partial x}. \tag{6.2.12}$$

A partial derivative is used here because $c$ can vary with time as well as location. The sign is negative because the flow of particles is from high concentration to low, i.e., if the concentration increases from left to right, then the flow is from right to left. Multiply (6.2.12) by $A$. and we have

$$\text{net number of particles crossing area } A \text{ per unit time} = -DA\frac{\partial c}{\partial x}. \tag{6.2.13}$$

To obtain Fick's second law, consider the channel again. The number of particles $N(x)$ in the section running from $x$ to $x + \Delta x$ is $c(x, t)A\Delta x$. If concentration is not constant, then particles will diffuse into (or out of) this section according to Fick's first law:

the decrease in the number of particles in the section
= (the flux out at $x + \Delta x$ − the flux in at $x$)$A$.

More precisely,

$$-\frac{\partial}{\partial t}(c(x,t)A\Delta x) = (J(x+\Delta x,t) - J(x,t))A.$$

Dividing by $\Delta x$ and letting $\Delta x \to 0$ gives

$$-A\frac{\partial c}{\partial t} = A\frac{\partial J}{\partial x}.$$

Canceling the $A$ on each side, we obtain the *continuity equation*,

$$\frac{\partial c}{\partial t} = -\frac{\partial J}{\partial x}. \tag{6.2.14}$$

Differentiating $J$ in Fick's first law and substituting into this gives *Fick's second law* of diffusion, also known as the *diffusion equation*,

$$\frac{\partial c}{\partial t} = D\frac{\partial^2 c}{\partial x^2}. \tag{6.2.15}$$

Direct substitution shows that the Gaussian distribution $u$, (6.2.9), satisfies the diffusion equation.

*Oxygen transfer by diffusion alone limits organisms in size to about one-half millimeter.*

As an application of Fick's laws, we may calculate how large an organism can be if it has no circulatory system. Measurements taken for many organisms show that the rate of oxygen consumption by biological tissues is on the order of $R_{O_2} = 0.3$ microliters of $O_2$ per gram of tissue per second. Also note that the concentration of oxygen in water at physiological temperatures is 7 microliters of $O_2$ per $cm^3$ of water. Assuming an organism of spherical shape, balancing the rate of oxygen diffusion through the surface with that consumed by interior tissue, we get, using Fick's first law, (6.2.12),

$$AJ = DA\frac{dc}{dr} = VR_{O_2},$$
$$D(4\pi r^2)\frac{dc}{dr} = \frac{4}{3}\pi r^3 R_{O_2}.$$

Isolate $\frac{dc}{dr}$ and integrate; use the boundary condition that at the center of the sphere the oxygen concentration is zero, and at the surface of the sphere, where $r = r_m$, the concentration is $C_{O_2}$. We get

$$\frac{dc}{dr} = \frac{R_{O_2}}{3D}r,$$

$$\int_0^{C_{O_2}} dc = \int_0^{r_m} \frac{R_{O_2}}{3D} r dr,$$
$$C_{O_2} = \frac{R_{O_2}}{6D} r_m^2.$$

Using the values for $C_{O_2}$ and $R_{O_2}$ above and the value $2 \times 10^{-5}$ cm/sec for a typical value of $D$, we get

$$\begin{aligned} r_m^2 &= \frac{6DC_{O_2}}{R_{O_2}} \\ &= \frac{6 \times 2 \times 10^{-5}(\text{cm}^2/\text{sec}) \times 7(\mu\text{l}/\text{cm}^3)}{0.3(\mu\text{l}/(\text{gm} \times \text{sec})} \\ &= 0.0028(\text{cm}^2), \end{aligned}$$

where we have assumed that one gram of tissue is about 1 $\text{cm}^3$ water. Taking the square root gives the result

$$r_m = 0.53 \text{ mm}.$$

We will use this value in Chapter 9 as a limitation on the size of certain organisms.

*Resistance to fluid flow is inversely proportional to the fourth power of the radius of the vessel.*

From the discussion above, it is clear that large organisms must actively move oxygen to the site of its use, possibly dissolved in a fluid. In this section, we derive the equation governing resistance to flow imposed by the walls of the vessel through which the fluid passes. In Chapter 9, we will discuss the anatomical and physiological consequences of this resistance to flow.

As a fluid flows through a circular vessel, say of radius $a$, resistance to the flow originates at the walls of the vessel. In fact, at the wall itself, the fluid velocity $u(a)$ is barely perceptible; thus $u(a) = 0$. A little further in from the wall the velocity picks up, and at the center of the vessel the velocity is largest. By radial symmetry, we need only one parameter to describe the velocity profile, namely, the radius $r$ from the center of the vessel. The fluid travels downstream in the form of concentric cylinders, the cylinders nearer the center moving fastest. This results in a shearing effect between the fluid at radius $r$ and the fluid just a little farther out, at radius $r+\delta r$. Shear stress, $\tau$, is defined as the force required to make two sheets of fluid slide past each other, divided by their contact area. It is easy to imagine that the shear stress depends on the difference in velocity of the two sheets, and, in fact, the two quantities are proportional.

Consider a portion of the vessel of length $\ell$ and let $\Delta p$ denote the difference in fluid pressure over this length. This pressure, acting on the cylinder of fluid of radius $r$, is opposed by the shear stress mentioned above. The force on the cylinder is being applied by the difference in pressure acting on its end; thus

$$\text{force} = \Delta p \pi r^2,$$

while the equal opposing force is due to the shear stress acting on the circular side of the cylinder,

$$\text{force} = \tau(2\pi r\ell) = \mu \frac{du}{dr}(2\pi r\ell).$$

Since these forces are equal and opposite,

$$\Delta p \pi r^2 = -\mu \frac{du}{dr}(2\pi r\ell).$$

This simple differential equation can be solved by integrating to obtain

$$u = -\frac{r^2 \Delta p}{4\ell\mu} + C,$$

where $C$ is the constant of integration. Using the zero velocity condition at the vessel walls gives the value of $C$ to be

$$C = \frac{a^2 \Delta p}{4\ell\mu}.$$

Hence for any radius, the velocity is given by

$$u(r) = \frac{\Delta p}{4\ell\mu}(a^2 - r^2).$$

We see that the velocity profile is parabolic (see Figure 6.2.4).

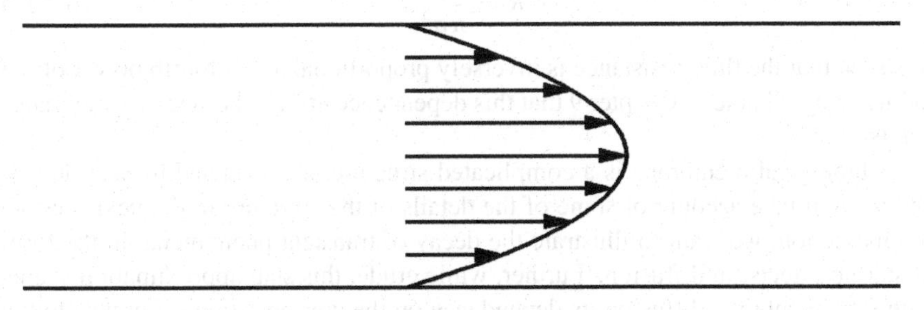

**Fig. 6.2.4.** Parabolic velocity profile of flow in a tube.

Now we can calculate the total flow rate, $Q$, of a volume of fluid through a cross-section of the vessel per unit time. Since a thin ring of fluid all of whose molecules are at the same distance $r$ from the axis of the vessel moves as a unit (see Figure 6.2.5), we consider all these molecules together. The volume of fluid per unit time that passes through a given cross-section of the vessel arising from such a ring, $dQ$, is given as the product of its velocity $u(r)$ and its area $dS$,

$$dQ = u(r)dS = u(r)(2\pi r)dr.$$

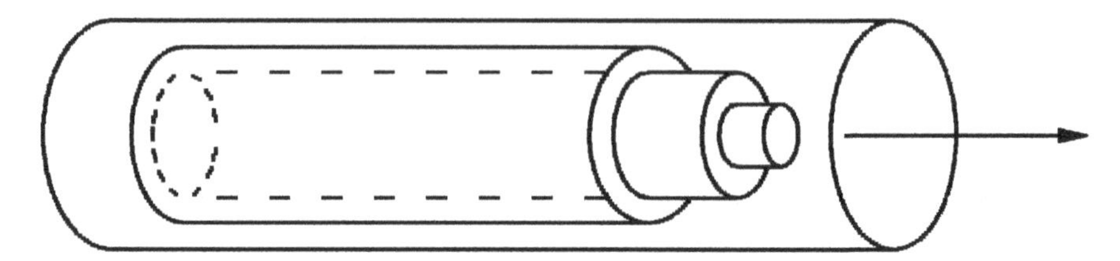

**Fig. 6.2.5.** Circular sheet of fluid all of whose molecules are moving together.

Substituting the velocity profile from above and integrating gives

$$\begin{aligned} Q &= \int_0^a \frac{\Delta p}{4\ell\mu}(a^2 - r^2)2\pi r dr \\ &= \frac{\pi \Delta p}{2\ell\mu}\left[\frac{a^2 r^2}{2} - \frac{r^4}{4}\right]_0^a \\ &= \frac{\pi \Delta p a^4}{8\mu\ell}. \end{aligned} \tag{6.2.16}$$

This expression is known as *Poiseuille's equation.* It shows that the flow rate increases as the fourth power of the vessel radius. It means that a vessel of twice the radius can carry 16 times the fluid volume for the same pressure drop per unit length.

It is natural to think of the shear stress in the moving fluid due to its contact with the walls as a *resistance* to flow. Fluid resistance $R$ is defined as $R = \frac{\Delta p}{Q}$ and hence is given by

$$R = \frac{8\mu\ell}{\pi a^4}. \tag{6.2.17}$$

It is seen that the fluid resistance is inversely proportional to the fourth power of the radius. We will note in Chapter 9 that this dependence affects the size of vertebrates' hearts.

A biological membrane is a complicated structure, as explained in Section 6.1, and we will take account of some of the details of the structure in the next section. In this section, we want to illustrate the decay of transient phenomena, in the form of startup effects, in diffusion. Further, while crude, this slab approximation shares with real membrane diffusion its dependence on the concentration difference to the first power as the driving force behind the transport of solute particles.

By a slab we mean a solid homogeneous material throughout which the diffusivity of solute particles is $D$. The slab has thickness $h$ but is infinite in its other two dimensions, and so diffusion through it takes place one-dimensionally.

To complete the statement of the problem, additional information, referred to as *boundary conditions* or *initial conditions*, must be specified. We will assume that the concentrations of solute on the sides of the slab are maintained at the constant values of $C_0$ at $x = 0$ and $C_h$ at $x = h$; assume that $C_0 > C_h$:

$$c(0,t) = C_0, \qquad c(h,t) = C_h \quad \text{for all } t \geq 0.$$

This could happen if the solvent reservoirs on either side of the slab were so large that the transport of solute is negligible. Or it could happen if solute particles are whisked away as soon as they appear at $x = h$ or are immediately replenished at $x = 0$ as they plunge into the slab. Or, of course, any combination of these. Further, we assume the startup condition that the concentration in the slab is 0:

$$c(x, 0) = 0, \quad 0 \le x \le h.$$

We begin by assuming that the solution, $c(x, t)$, can be written in the form of a product of a function of $x$ only with a function of $t$ only:

$$c(x, t) = X(x)T(t).$$

Then $\frac{\partial c}{\partial t} = X(x)T'(t)$ and $\frac{\partial^2 c}{\partial x^2} = X''(x)T(t)$. Substituting into the diffusion equation, (6.2.15), and dividing, we get

$$\frac{1}{T} dT dt = \frac{D}{X}\frac{d^2X}{dx^2}. \tag{6.2.18}$$

Now the left-hand side of this equation is a function of $t$ only and the right-hand side is a function of $x$ only, and the equality is maintained over all values of $x$ and $t$. This is possible only if both are equal to a constant, which we may write as $-\lambda^2 D$. Then (6.2.18) yields the two equations

$$\frac{1}{T}\frac{dT}{dt} = -\lambda^2 D \tag{6.2.19}$$

and

$$\frac{1}{X}\frac{d^2X}{dx^2} = -\lambda^2. \tag{6.2.20}$$

It is easy to verify that the solution of (6.2.19) is

$$T = Ae^{-\lambda^2 Dt},$$

and the solution of (6.2.20) is

$$X = \begin{cases} ax + b & \text{if } \lambda = 0, \\ c \sin \lambda x + d \cos \lambda x & \text{if } \lambda \ne 0, \end{cases}$$

where $A$, $a$, $b$, $c$, and $d$ are constants.

The solution so far is

$$c(x, t) = \begin{cases} ax + b & \text{if } \lambda = 0, \\ (c \sin \lambda x + d \cos \lambda x)e^{-\lambda^2 Dt} & \text{if } \lambda \ne 0. \end{cases}$$

The constant $A$ has been absorbed into the other constants, all of which have yet to be determined using the boundary conditions. For $\lambda = 0$, the conditions on either side of the slab give

$$a0 + b = C_0 \quad \text{and} \quad ah + b = C_h.$$

Hence it must be that $b = C_0$ and $a = -\frac{C_0 - C_h}{h}$. But this solution cannot satisfy the initial condition, we will use the $\lambda \neq 0$ case for that.

First, note that if two functions $c_1(x,t)$ and $c_2(x,t)$ both satisfy the diffusion equation, then so does their sum, as follows:

$$\frac{\partial(c_1 + c_2)}{\partial t} = \frac{\partial c_1}{\partial t} + \frac{\partial c_2}{\partial t}$$

and

$$-D\frac{\partial^2(c_1 + c_2)}{\partial x^2} = -D\frac{\partial^2 c_1}{\partial x^2} - D\frac{\partial^2 c_2}{\partial x^2}.$$

And so

$$\frac{\partial(c_1 + c_2)}{\partial t} = -D\frac{\partial^2(c_1 + c_2)}{\partial x^2}.$$

In particular, the sum of the $\lambda = 0$ and $\lambda \neq 0$ solutions,

$$c(x,t) = -\frac{C_0 - C_h}{h}x + C_0 + (c \sin \lambda x + d \cos \lambda x)e^{-\lambda^2 Dt}, \tag{6.2.21}$$

will satisfy the diffusion equation. It remains to satisfy the boundary conditions.

At $x = 0$ in (6.2.21),

$$C_0 = -\frac{C_0 - C_h}{h}0 + C_0 + (c \sin \lambda 0 + d \cos \lambda 0)e^{-\lambda^2 Dt}.$$

Upon simplifying, this becomes

$$0 = de^{-\lambda^2 Dt},$$

which must be valid for all $t$; thus $d = 0$. Continuing with the $x = h$ boundary condition in (6.2.21), we have

$$C_h = -\frac{C_0 - C_h}{h}h + C_0 + (c \sin \lambda h)e^{-\lambda^2 Dt},$$

or

$$0 = (c \sin \lambda h)e^{-\lambda^2 Dt}.$$

As before, this must hold for all $t \geq 0$. We cannot take $c = 0$, for then the initial condition cannot be satisfied. Instead, we solve $\sin \lambda h = 0$ for $\lambda$, which is as yet unspecified. The permissible values of $\lambda$ are

$$\lambda = \frac{\pm n\pi}{h}, \quad n = 1, 2, \ldots, \tag{6.2.22}$$

known as the *eigenvalues* of the problem. The negative values of $n$ may be absorbed into the positive ones, since $\sin \lambda_{-n} x = -\sin \lambda_n x$. Remembering that solutions of the diffusion equation may be added, we can form an infinite series solution with a term for each eigenvalue, and, possibly, each with a different coefficient, $c_n$,

$$c(x,t) = -\frac{C_0 - C_h}{h}x + C_0 + \sum_{n=1}^{\infty}(c_n \sin \lambda_n x)e^{-\lambda_n^2 Dt}. \tag{6.2.23}$$

Finally, in order to fulfill the initial condition, the coefficients $c_n$ must be chosen to satisfy the initial condition

$$c(x,0) = 0 = -\frac{C_0 - C_h}{h}x + C_0 + \sum_{n=1}^{\infty}(c_n \sin \lambda_n x)e^{-\lambda_n D0},$$

or upon simplifying,

$$\sum_{n=1}^{\infty} c_n \sin \lambda_n x = \frac{C_0 - C_h}{h}x - C_0.$$

We will not show how to calculate the $c_n$s; we note only that it can be done [7]. The infinite series is referred to as the *Fourier series* representation of the function on the right.

Thus the solution occurs in two parts; in one part, every term contains the decaying exponential $e^{-\lambda_n Dt}$ for constants $\lambda_n$ given above. These terms tend to zero and, in time, become negligible. That leaves the *steady-state* part of the solution,

$$c(x,t) = -\frac{C_0 - C_h}{h}x + C_0,$$

a linear concentration gradient. The amount of solute delivered in the steady state is the flux given by Fick's first law,

$$J = -D\frac{\partial c}{\partial x} = \frac{D}{h}(C_0 - C_h). \tag{6.2.24}$$

In summary, the rate of material crossing a membrane is directly proportional to the concentration difference and inversely proportional to the thickness of the membrane.

*Membrane diffusion decays exponentially as particles accumulate.*

The structure of cell membranes was described in Section 6.1. It consists of a double layer of lipid molecules studded with proteins. Some of the latter penetrate entirely through the lipid bilayer and serve to mediate the movement of various substances into and out of the cell's interior. This section is not about the transport of such substances. Rather, we describe the transport of those molecules that pass through the lipid bilayer itself by diffusion. These are mainly lipid-soluble molecules.[5]

However, the membrane molecules themselves have two ends: a hydrophilic head and lipid tail. Functionally, the head end of one layer faces outward in contact with the aqueous environment of the cell, while the head end of the other layer faces inward in contact with the aqueous interior of the cell. The lipid tails of both layers face together and constitute the interior of the membrane. Thus the concentration of solute just under the head of the membrane molecule is not necessarily the same as in the aqueous phase. Denote by $C'$ the solute concentration just inside the membrane on the environmental side, and by $c'$ the concentration just inside the membrane on the cell interior side. As we saw in the derivation of Fick's first law, (6.2.12), the flux of solute through the lipid part of the membrane is proportional to the concentration difference and inversely proportional to the separation distance, so we have

$$J = \frac{D}{h}(C' - c').$$

We next assume a linear relation between the concentrations across the molecular head of the membrane molecule; thus

$$C' = \Gamma C \quad \text{and} \quad c' = \Gamma c,$$

where $C$ is the environmental concentration of the solute and $c$ is the concentration inside the cell. The constant $\Gamma$ is called the *partition coefficient*. With this model, the partition coefficient acts as a diffusivity divided by thickness ratio for the diffusion of solute across the head of the membrane molecule. The partition coefficient is less than 1 for most substances, but can be greater than 1 if the solute is more soluble in lipid than in water.

Combining the development above, we calculate flux in terms of exterior and interior concentrations as

$$J = \frac{\Gamma D}{h}(C - c); \tag{6.2.25}$$

this is in moles/cm$^2$/sec, for instance.

As solute molecules accumulate inside the cell, the concentration difference in (6.2.25) diminishes, eventually shutting off the transport. Denote the volume of the cell by $V$ and the surface area by $S$. The quantity $SJ$ is the rate of mass transport across the membrane in moles/sec, that is,

[5] In Section 6.1, we noted that water and carbon dioxide, although polar molecules, can move through the lipid part of a membrane.

$$SJ = \frac{dm}{dt} = V\frac{dc}{dt},$$

since concentration is mass per unit volume. Therefore, multiplying (6.2.25) by $S$ and using this relation, we get

$$V\frac{dc}{dt} = \frac{S\Gamma D}{h}(C - c),$$

or

$$\frac{dc}{dt} = k(C - c),$$

where $k = \frac{S\Gamma D}{Vh}$ is a constant. The solution, given by integration, is

$$c = C - c_0 e^{-kt},$$

where $c_0$ is the initial concentration inside the membrane. In summary, the interior concentration becomes exponentially asymptotic to that of the environment.

## Exercises/Experiments

**1.** Instead of presenting a theoretical distribution of the position of particles after 40 steps, we simulate the random movement using the built-in random number generator of the computer system. Simulate a random walk of 40 steps for each of 500 particles and histogram the place they end up.

```
MAPLE
> with(stats): with(plots):
> for i from 1 to 100 do
    count[i]:=0; od: # initialize a counter
> N:=rand(0..1): #random integer 0/1
> particles:=500; steps:=40;
> for m from 1 to particles do
  place:=sum('2*N(p)-1','p'=1..steps)+steps:
  count[place]:=count[place]+1: # record endpt
  od:
  # histogram the endpoints
> ranges:=[seq(-steps/2+2*(i-1)..-steps/2+2*i,i=1..steps/2)];
> movement:=[seq(count[20+2*j],j=1..20)];
> diffusion:=[seq(Weight(ranges[i],movement[i]),i=1..20)];
> statplots[histogram](diffusion);
```

```
MATLAB
> particles=500; steps=40;
> for k=1:particles
    steplog=fix(2*rand(1,steps)); % random vectors of 0s/1s
    steplog = 2*steplog-1; % random vectors of -1/+1
    place(k) = sum(steplog); % endpt for this 40 step walk
  end
> x=-20:2:20;
> hist(place,x)
```

(a) Plot the Gaussian distribution of (6.2.9) with $D = 1$ and for $t = 1, 2$, and 3.

```
MAPLE
> plot({exp(-x^2/4)/sqrt(4*Pi), exp(-x^2/(4*2))/sqrt(4*Pi*2), exp(-x^2/(4*3))/sqrt(4*Pi*3)},x=-10..10);
```

```
MATLAB
% Equation (6.2.9) is the gaussian with mean = 0 and stddev = sqrt(2Dt).
% Therefore make an m-file gaussian.m (already done in Section 2.6, repeated here):
% function y=gaussian(x,m,s); y=(.3989422803/s)*exp(-0.5*((x-m)./s).^2);
% Part (a)
> D=1; t=1; s=sqrt(2*D*t);
> x=[-10:.1:10]; y=gaussian(x,0,s);
> plot(x,y); hold on;
> D=1; t=2; s=sqrt(2*D*t);
> y=gaussian(x,0,s); plot(x,y);
> D=1; t=3; s=sqrt(2*D*t);
> y=gaussian(x,0,s); plot(x,y);
```

(b) Verify that (6.2.9) satisfies the partial differential equation (6.2.15) with $D = 1$.

```
MAPLE (symbolic)
> u:=(t,x)->exp(-x^2/(4*t))/sqrt(4*Pi*t);
> diff(u(t,x),t)-diff(u(t,x),x,x);
> simplify(%);
```

(c) The analogue of (6.2.15) for diffusion in a plane is

```
MAPLE (symbolic)
> diff(U(t,x,y),t) = diff(U(t,x,y),x,x)+diff(U(t,x,y),y,y);
```

Show that the function $U$ given below satisfies this two-dimensional diffusion equation:

```
MAPLE (symbolic)
> U:=(t,x,y)->exp(-(x^2+y^2)/(4*t))/t;
> diff(U(t,x,y),t) - diff(U(t,x,y),x,x)-diff(U(t,x,y),y,y);
> simplify(%);
```

(d) Give visualization to these two diffusions by animation of (6.2.9) and of the two-dimensional diffusion.

```
MAPLE (animation)
> plot({exp(-x^2/4)/sqrt(4*Pi),exp(-x^2 /(4*2))/sqrt(4*Pi*2), exp(-x^2/(4*3))/sqrt(4*Pi*3)},x=-10..10);
> with(plots):
> animate(exp(-x^2/(4*t))/sqrt(4*Pi*t),x=-10..10,t=0.1..5);
> animate3d(exp(-(x^2+y^2)/ (4*t))/t,x=-1..1,y=-1..1,t=0.1..0.5);
```

**2.** A moment's reflection on the form of (6.2.15) suggests a geometric understanding. The left side is the rate of change in time of $c(t, x)$. The equation asserts that this rate of change is proportional to the curvature of the function $c(t, x)$ as a graph in $x$ and as measured by the second derivative. That is, if the second derivative in $x$ is positive and the curve is concave up, expect $c(t, x)$ to increase in time. If the second derivative is negative and the curve is concave down, expect $c(t, x)$ to decrease in time. We illustrate this with a single function. Note that the function $\sin(x)$ is concave down on $[0, \pi]$ and concave up on $[\pi, 2\pi]$. We produce a function such that with $t = 0$, $c(0, x) = \sin(x)$, and for arbitrary $t$, $c(t, x)$ satisfies (6.2.15).

```
> MAPLE (symbolic)
> c:=(t,x)->exp(-t)*sin(x);
```

Here we verify that this is a solution of (6.2.15).

```
> MAPLE (symbolic)
> diff(c(t,x),t)-diff(c(t,x),x,x);
```

Now we animate the graph. Observe where $c(t, x)$ is increasing and where it is decreasing.

```
> MAPLE (animation)
> with(plots): animate(c(t,x),x=0..2*Pi,t=0..2);
```

## 6.3 Transplacental Transfer of Oxygen: Biological and Biochemical Considerations

*A fetus must obtain oxygen from its mother. Oxygen in the mother's blood is attached to a blood pigment called hemoglobin and is carried to the placenta, where it diffuses across a system of membranes to the fetus's hemoglobin. A number of physical factors cause the fetal hemoglobin to bind the* $O_2$ *more securely than does the maternal, or adult, hemoglobin, thus ensuring a net* $O_2$ *movement from mother to fetus.*

*The blood of a mother and her unborn child do not normally mix.*

The circulatory systems of a mother and her unborn child face one another across a platelike organ called the *placenta*. The placenta has a maternal side and a fetal side, each side being fed by an artery and drained by a large vein, the two vessels being connected by a dense network of fine capillaries. The two sides of the placenta are separated by membranes, and the blood of the mother and that of the child do not mix. All material exchange between mother and child is regulated by these placental membranes, which can pass ions and small-to-medium biochemical molecules. Large molecules, however, do not usually transit the placental membranes.

*Hemoglobin carries oxygen in blood.*

The chemical hemoglobin is found in anucleate cells called red blood cells or *erythrocytes*. Hemoglobin picks up $O_2$ at the mother's lungs and takes it to the placenta, where the $O_2$ crosses the placenta to the hemoglobin of fetal red blood cells for distribution to metabolizing fetal tissues.

*Oxygen affinity measures the strength with which hemoglobin binds oxygen.*

A fixed amount of hemoglobin can hold a fixed maximum amount of oxygen, at which point the hemoglobin is said to be saturated. Figure 6.3.1 is an *oxygen dissociation curve*; it shows the extent to which saturation is approached as determined by the *partial pressure* of the oxygen.[6] The partial pressure of $O_2$ at which the hemoglobin is half-saturated is a measure of the *oxygen affinity* of the hemoglobin. Thus hemoglobin that reaches half-saturation at low $O_2$ partial pressure has a high oxygen affinity (see [1] and [3] for further discussion).

---

[6] The partial pressure of a gas is the pressure exerted by that specific gas in a mixture of gases. The partial pressure is proportional to the concentration of the gas. The total pressure exerted by the gaseous mixture is the sum of the partial pressures of the various constituent gases.

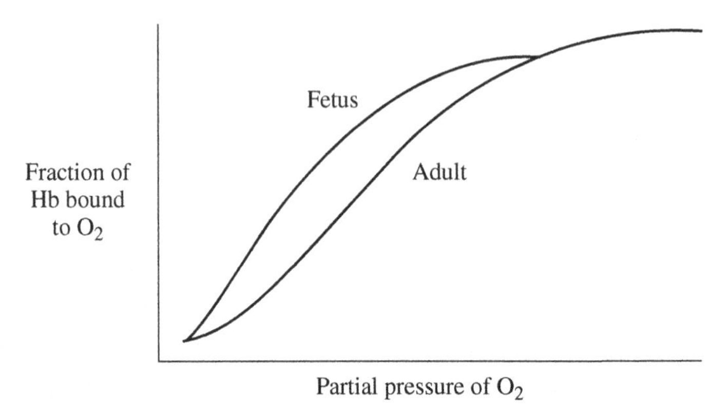


**Fig. 6.3.1.** Oxygen dissociation curves for adult and fetal hemoglobin. Note that for a given partial pressure (concentration) of oxygen, the fetal hemoglobin has a greater fraction of its hemoglobin bound to oxygen than does the adult hemoglobin.

The reversible attachment of $O_2$ to hemoglobin is represented by

$$\underset{\text{hemoglobin}}{Hb + O_2} \rightleftarrows \underset{\text{oxyhemoglobin}}{O_2\text{–}Hb} \ .$$

At equilibrium the relative amounts of hemoglobin and oxyhemoglobin are fixed, the reaction going to the right as often as it goes to the left. The relative amounts of oxyhemoglobin, hemoglobin, and oxygen at equilibrium are determined by the oxygen affinity of the hemoglobin. The greater the oxygen affinity, the more oxyhemoglobin there will be relative to the amounts of hemoglobin and oxygen, i.e., the more the equilibrium will move toward the right in the above reaction scheme.

*Oxygen affinity depends on a variety of factors.*

In practice, the oxygen affinity is determined by multiple factors: First, we would surely expect that the structure of hemoglobin would be important, and that will be discussed below. Second, oxygen affinity is affected by the extent to which oxygen molecules are already attached. Hemoglobin can bind to as many as four $O_2$ molecules. The second, third, and fourth are progressively easier to attach because the oxygen affinity of the hemoglobin increases as more $O_2$ molecules are added. Third, blood pH affects its oxygen affinity. The pH of the blood and the presence of $CO_2$ are related; this will be discussed in Section 9.6. Finally, a chemical constituent of red blood cells, called D-2, 3-*biphosphoglycerate* (*BPG*), plays an important role in the oxygen-binding properties of hemoglobin by binding to it and thereby decreasing its $O_2$ affinity. The role of BPG is a crucial one because the more BPG is bound to hemoglobin, the less tightly the hemoglobin binds oxygen. Therefore, the oxygen will be released more easily and will be provided to metabolizing tissues in higher concentration. In terms of the chemical reaction above, BPG moves the equilibrium toward the left.

*Fetal hemoglobin has a greater affinity for oxygen than does adult hemoglobin.*

Adult and fetal hemoglobins have somewhat different structures. The result is that fetal hemoglobin binds less BPG than does adult hemoglobin, and therefore fetal hemoglobin has the higher oxygen affinity of the two. Figure 6.3.1 shows oxygen dissociation curves for the hemoglobin of an adult and for that of a fetus. Note that at a given partial pressure of $O_2$, the fetal hemoglobin has a greater $O_2$ affinity than does maternal hemoglobin. Thus there is a net movement of oxygen from the mother to the fetus.

We must be very careful here: We must not think that the fetal hemoglobin somehow drags $O_2$ away from that of the mother. This would require some sort of "magnetism" on the part of the fetal hemoglobin, and such "magnetism" does not exist. What does happen is represented by the following diagram:

$$O_2\text{–}Hb_{adult} \rightleftarrows O_2 + Hb_{adult}$$

$$\uparrow\downarrow \quad \text{(placenta)}$$

$$O_2 + Hb_{fetal} \rightleftarrows O_2\text{–}Hb_{fetal}.$$

Both kinds of hemoglobin are constantly attaching to, and detaching from, oxygen—consistent with their oxygen affinities. The mother's breathing gives her blood a high concentration of oxyhemoglobin, and that leads to a high concentration of *free oxygen* on her side of the placenta. On the fetal side of the placenta, the fetus, which does not breathe, has a low $O_2$ concentration. Therefore, $O_2$, once released from maternal oxyhemoglobin, moves by simple diffusion across the placenta, in response to the concentration gradient. On the fetal side, fetal hemoglobin attaches to the oxygen and holds it tightly because of its high oxygen affinity. Some oxygen will dissociate from the fetal hemoglobin, but little will diffuse back to the maternal side because the concentration gradient of the oxygen across the placenta is in the other direction.[7] In summary, oxygen diffuses across the placenta from mother to fetus, where it tends to stay because of its concentration gradient and the high oxygen affinity of fetal hemoglobin compared with that of adult hemoglobin.

## 6.4 Oxygen Diffusion Across the Placenta: Physical Considerations

*The delivery of fetal oxygen typifies the function of the placenta. In this organ, fetal blood flow approaches maternal blood flow, but the two are separated by membranes. Possible mechanisms for oxygen transfer are simple diffusion, diffusion facilitated by some carrier substance, or active transport requiring metabolic energy. No evidence for facilitated diffusion or active transport has been found. We will see that simple diffusion can account for the required fetal oxygen consumption.*

[7] In Chapter 9, we will see that the concentration of $CO_2$ in the blood also affects the oxygen affinity of hemoglobin.

*The oxygen dissociation curve is sigmoid in shape.*

Oxygen in blood exists in one of two forms, either dissolved in the plasma or bound to hemoglobin as oxyhemoglobin. Only the dissolved oxygen diffuses; oxyhemoglobin is carried by the moving red blood cells. The binding of oxygen to hemoglobin depends mostly on $O_2$ partial pressure but also on blood acidity. The relationship, given by a dissociation curve, possesses a characteristic sigmoid shape as a function of partial pressure (see Figure 6.4.1 and Table 6.4.1).

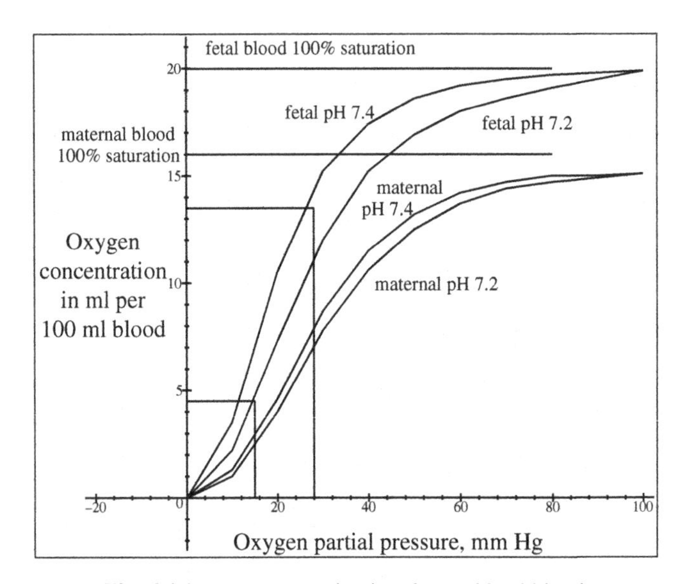


**Fig. 6.4.1.** $O_2$ concentration in ml per 100 ml blood.

**Table 6.4.1.** $O_2$ concentration in ml per 100 ml blood (see [16]).

| $pO_2$ mm Hg ⟶ | 10 | 20 | 30 | 40 | 50 | 60 | 70 | 80 | 90 | 100 |
|---|---|---|---|---|---|---|---|---|---|---|
| fetal (pH 7.4) | 3.5 | 10.5 | 15.2 | 17.4 | 18.6 | 19.2 | 19.5 | 19.7 | 19.8 | 19.9 |
| fetal (pH 7.2) | 2.2 | 7.3 | 12.0 | 15.2 | 16.9 | 18.0 | 18.6 | 19.1 | 19.5 | 19.8 |
| maternal (pH 7.4) | 1.3 | 4.6 | 8.7 | 11.5 | 13.2 | 14.2 | 14.7 | 14.9 | 15.0 | 15.1 |
| maternal (pH 7.2) | 1.0 | 4.0 | 7.8 | 10.6 | 12.5 | 13.7 | 14.4 | 14.7 | 14.9 | 15.1 |

The effect of increasing acidity is to shift the curve rightward. (The dissociation curves can be constructed using the data in Table 6.4.1.)

```
MAPLE
> with(plots):
> ppo:=[seq(10*(i-1),i=1..11)]:
> fetal74:=array([0,3.5,10.5,15.2,17.4,18.6,19.2,19.5,19.7,19.8,19.9]):
> fetal72:=array([0,2.2,7.3,12.0,15.2,16.9,18.0,18.6,19.1,19.5,19.9]):
> maternal74:=array([0,1.3,4.6,8.7,11.5,13.2,14.2,14.7,15.0,15.0,15.1]):
```

```
> maternal72:=array([0,1.0,4.0,7.8,10.6,12.5,13.7,14.4,14.7,14.9,15.1]):
> f74plot:=plot([seq([ppo[i],fetal74[i]],i=1..11)]):
> f72plot:=plot([seq([ppo[i],fetal72[i]],i=1..11)]):
> m74plot:=plot([seq([ppo[i],maternal74[i]],i=1..11)]):
> m72plot:=plot([seq([ppo[i],maternal72[i]],i=1..11)]):
> with(plots):
> display({f74plot,f72plot,m74plot,m72plot});
```

MATLAB

```
> ppo=[0:10:100];
> fetal74=[0 3.5 10.5 15.2 17.4 18.6 19.2 19.5 19.7 19.8 19.9];
> plot(ppo,fetal74); hold on
> fetal72=[0 2.2 7.3 12.0 15.2 16.9 18.0 18.6 19.1 19.5 19.9];
> plot(ppo,fetal72)
> maternal74=[0 1.3 4.6 8.7 11.5 13.2 14.2 14.7 15.0 15.0 15.1];
> plot(ppo,maternal74)
> maternal72=[0 1.0 4.0 7.8 10.6 12.5 13.7 14.4 14.7 14.9 15.1];
> plot(ppo,maternal72)
```

When maximally saturated, hemoglobin (Hb) holds about 1.34 ml $O_2$ per gm. Fetal blood contains about 15 gm Hb per 100 ml, while maternal blood has 12 gm per 100 ml. Therefore, maternal blood is 100% saturated at a concentration of $1.34 \times 12 \approx 16$ ml of $O_2$ per 100 ml of blood, and fetal blood is 100% saturated at 20 ml per 100 ml blood.

Although only the dissolved oxygen diffuses, hemoglobin acts like a moving reservoir on both the maternal and fetal sides of the placenta. On the maternal side, $O_2$ diffuses across the placental membrane from the maternal blood plasma, causing a decrease in the partial pressure of $O_2$, symbolized $pO_2$. But a lower oxygen partial pressure dissociates oxygen out of the hemoglobin to replace what was lost. This chemical reaction is very fast. Consequently, hemoglobin acts to preserve the partial pressure while its oxygen, in effect, is delivered to the fetal side. Of course as more and more oxygen dissociates, $pO_2$ gradually decreases.

On the fetal side, the opposite occurs. The incoming oxygen raises the partial pressure, with the result that oxygen associates with fetal hemoglobin with gradual increase of $pO_2$.

*Fetal oxygen consumption rate at term is* 23 *ml per minute.*

The first step in showing that simple diffusion suffices for oxygen delivery is to determine how much oxygen is consumed by the fetus. By direct measurement, oxygen partial pressure and blood pH at the umbilical cord are as shown in Table 6.4.2.

It follows from Figure 6.4.1 that each 100 ml of venous blood in the fetus contains approximately 13.5 ml $O_2$, while for arterial blood it is about 4.5 ml.

Evidently an $O_2$ balance for fetal circulation measured at the umbilical cord is given by

$$O_2 \text{ in} - O_2 \text{ out} = O_2 \text{ consumed}.$$

For each minute, this gives

$$\begin{aligned}\text{rate } O_2 \text{ consumed} &= 250\frac{\text{ml blood}}{\text{min}} \times (13.5 - 4.5)\frac{\text{ml } O_2}{100 \text{ ml blood}} \\ &= 22.5 \text{ ml } O_2/\text{min}.\end{aligned}$$

**Table 6.4.2.** Placental oxygen and flow rate data.

| | |
|---|---|
| umbilical artery | $pO_2$: 15 mm Hg, pH: 7.24 [15] |
| umbilical vein | $pO_2$: 28 mm Hg, pH: 7.32 [15] |
| umbilical flowrate | 250 ml per minute [15] |
| maternal artery | $pO_2$: 40 mm Hg [15] |
| maternal vein | $pO_2$: 33 mm Hg [15] |
| maternal flowrate | 400 ml per minute [15] |
| placental membrane surface | 12 square meters [16] |
| placental membrane thickness | $3.5 \times 10^{-4}$ cm [16] |
| $pO_2$ diffusivity (see text) | $3.09 \times 10^{-8}$ cm$^2$/min/mm Hg [15, Figure 5] |

*Maximal oxygen diffusion rate is* 160 *ml per minute.*

Next, we estimate the maximum diffusion possible across the placenta. Recall the membrane transport (6.2.25),

$$J = -\frac{D}{w}\Delta c, \tag{6.4.1}$$

where we have taken the partition coefficient $\Gamma = 1$ and the membrane thickness to be $w$. This holds for those sections of the membrane that happen to have thickness $w$ and concentration difference $\Delta c$. Normally, both these attributes will vary throughout the placenta. However, since we are interested in the maximal diffusion rate, we assume them constant for this calculation. Placental membrane thickness has been measured to be about 3.5 microns ($3.5 \times 10^{-4}$ cm). Since flux is the diffusion rate per unit area, we must multiply it by the surface area, $S$, of the membrane. Careful measurements show this to be about 12 square meters at term [15].

Actually, taking a constant average value for $w$ is a reasonable assumption. But taking $O_2$ concentrations to be constant is somewhat questionable. Mainly, doing so ignores the effect of the blood flow. We will treat this topic below in connection with the countercurrent flow model. For this derivation, we assume that $O_2$ dissociates out of maternal blood in response to diffusion, all the while maintaining the concentration constant on the maternal side. On the fetal side, $O_2$ associates with fetal blood and likewise maintains a constant concentration there. In the countercurrent flow these assumptions tend to be realized.

A lesser difficulty in applying Fick's law has to do with the way an oxygen concentration is normally measured, namely, in terms of partial pressure. By Henry's law [15, p. 1714], there is a simple relationship between them: The concentration of a dissolved gas is proportional to its partial pressure, in this case $pO_2$. Hence

$$c = \delta(pO_2),$$

for some constant $\delta$. Incorporating surface area and Henry's law, (6.4.1) takes the form

$$SJ = -\delta D\frac{S}{w}\Delta(p0_2).$$

The product $\delta D$ has been calculated to be about $3.09 \times 10^{-8}$ cm$^2$/min/mm Hg (derived from data in [15] and investigated in the problems).

For the constant fetal partial pressure, we take the average of the range 15 to 28 mm Hg noted in Table 6.4.2, so about 21.5 mm Hg. Maternal arterial $pO_2$ is 40 mm Hg, while venous $pO_2$ is 33 mm Hg for an average of 36.5 mm Hg. Hence using the values in Table 6.4.2,

$$\begin{aligned} O_2 \text{ diffusion rate} &= 3.09 \cdot 10^{-8} \frac{\text{cm}^2}{\text{min-mm Hg}} \\ &\quad \cdot \frac{12\ \text{m}^2 \cdot 10^4 \frac{\text{cm}^2}{\text{m}^2}}{3.5 \cdot 10^{-4}\ \text{cm}} (36.5 - 21.5)\ \text{mm Hg} \\ &= 159 \frac{\text{cm}^3}{\text{min}}. \end{aligned}$$

Recalling that only 22.5 ml of oxygen per minute are required, the placenta, in its role of transferring oxygen, need only be about $\frac{22.5}{159} = 14\%$ efficient.

*The fetal flow rate limits placental transport efficiency.*

The placenta as an exchanger is not 100% efficient (it only needs to be 14% efficient) due to (1) maternal and fetal shunts, (2) imperfect mixing, and, (3) most importantly, flow of the working material, which we have not taken into consideration. Let $F$ be the maternal flow rate, $f$ the fetal flow rate, $C$ maternal $O_2$ concentration, and $c$ fetal $O_2$ concentration. Use the subscript $i$ for in and $o$ for out (of the placenta). Let $r$ denote the transfer rate across the placenta ($r = SJ$). From the mass balance equation,

$$O_2\text{/min in} \pm O_2 \text{ gained or lost per min} = O_2\text{/min out},$$

we get

$$fc_i + r = fc_o; \qquad FC_i - r = FC_o, \tag{6.4.2}$$

since the oxygen rates in or out of the placenta are the product of conentration times flow rate. From the membrane equation (6.2.25),

$$r = K(\Delta\text{concentration across membrane}), \quad \text{where } K = \frac{\Gamma DS}{w}.$$

For the fetal and maternal concentrations, we use $C_o$ and $c_o$. From (6.4.2),

$$r = K(C_o - c_o) = K\left(C_i - \frac{r}{F} - \frac{r}{f} - c_i\right).$$

Solve this for $r$ and get

$$r = \frac{C_i - c_i}{\frac{1}{K} + \frac{1}{F} + \frac{1}{f}}. \tag{6.4.3}$$

Now consider the magnitude of the three terms in the denominator. If $F$ and $f$ are infinite, then the denominator reduces to $\frac{1}{K}$ and the transfer rate becomes

$$r = K(C_i - c_i)$$

as before. In this case, diffusion is the limiting factor.

Since the flow terms are not infinite, their effect is to increase the denominator and consequently reduce the transfer coefficient, that is,

$$\frac{1}{\frac{1}{K} + \frac{1}{F} + \frac{1}{f}} < \frac{1}{\frac{1}{K}} = K.$$

Moreover, depending on the relative size of the three terms in the denominator of (6.4.3), diffusion may not be the limiting factor. In particular, the smallest of the quantities $K$, $F$, and $f$, corresponds to the largest of the reciprocals $\frac{1}{K}$, $\frac{1}{F}$, and $\frac{1}{f}$. Using the values of $S$ and $w$ from Table 6.4.2, and taking $\Gamma D = 4 \times 10^{-7}$ cm$^2$/sec (see [15]), gives

$$K = \frac{4 \cdot 10^{-7} \cdot 12 \cdot 10^4}{3.5 \cdot 10^{-4}} = 137 \frac{\text{cm}^3}{\text{sec}}.$$

Compare this with a maternal flow rate $F$ of 400 ml/min, or 6.7 ml/sec, and a fetal flow rate $f$ of 250 ml/min, or 4.2 ml/sec. Thus fetal flow rate is the smallest term and so is the limiting factor. Furthermore, from (6.4.3) we can see that diffusion is a relatively minor factor compared to the maternal and fetal flow rates. That is,

$$\frac{1}{\frac{1}{137} + \frac{1}{6.7} + \frac{1}{4.2}} = 2.53,$$

while

$$\frac{1}{\frac{1}{6.7} + \frac{1}{4.2}} = 2.58.$$

*Countercurent flow is more efficient than concurrent flow.*

In this section, we will compare the diffusion properties of the placenta depending on whether the maternal and fetal blood flow in the same or the opposite directions through the placenta. For this we assume the placenta to be a channel separated by the placental membrane. Assume first that fetal and maternal blood flow in the same direction. As shown in Figure 6.4.2, on the maternal side we take the channel height to be $H$ and the velocity to be $v_m$, while these will be $h$ and $v_f$ respectively on the fetal side. Let the channel width be $b$. Assume that steady state has been reached and take the oxygen concentrations at position $x$ along the maternal and fetal sides to be $C(x)$ and $c(x)$, respectively.

On the fetal side, a block $\Delta V$ of blood at position $x$ gains in concentration in moving distance $\Delta x$ due to the flux $J(x)$ at $x$. Let $\Delta S$ denote the area of contact of the block with the placental membrane. Since the time required to move this distance is $\Delta t = \frac{\Delta x}{v_f}$, we have

$$c(x + \Delta x) = \frac{c(x)\Delta V + J(x)\Delta S\left(\frac{\Delta x}{v_f}\right)}{\Delta V}.$$

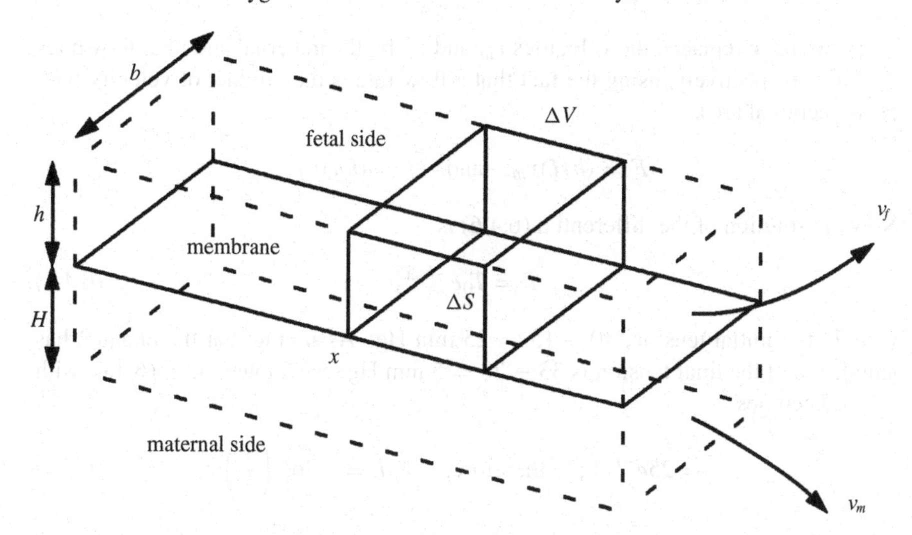


**Fig. 6.4.2.** Diffusion of oxygen into a moving incremental volume.

But from the membrane (6.2.25) (with $\Gamma = 1$),

$$J(x) = \left(\frac{D}{w}\right)(C(x) - c(x)),$$

and so

$$\frac{c(x + \Delta x) - c(x)}{\Delta x} = \frac{D}{w v_f}\frac{\Delta S}{\Delta V}(C(x) - c(x)).$$

Now $\frac{\Delta S}{\Delta V} = h$, and so in the limit we have

$$\frac{dc}{dx} = \frac{\left(\frac{D}{w}\right)}{h v_f}(C - c). \tag{6.4.4}$$

A similar calculation on the maternal side gives

$$\frac{dC}{dx} = -\frac{\left(\frac{D}{w}\right)}{H v_m}(C - c). \tag{6.4.5}$$

Denote by $T$ the *tension* or concentration difference $C(x) - c(x)$. By subtracting the first equation from the second, we get

$$\frac{dT}{dx} = -\frac{D}{w}\left(\frac{1}{H v_m} + \frac{1}{h v_f}\right)T. \tag{6.4.6}$$

For this parallel flow, denote by $k_p$ the constant coefficient,

$$k_p = \frac{D}{w}\left(\frac{1}{H v_m} + \frac{1}{h v_f}\right) = \frac{D}{w}\left(\frac{b}{F} + \frac{b}{f}\right), \tag{6.4.7}$$

where we have replaced the velocities $v_m$ and $v_f$ by the maternal and fetal flow rates $F$ and $f$, respectively, using the fact that a flow rate is the product of velocity with cross-sectional area,

$$F = (bH)v_m \quad \text{and} \quad f = (bh)v_f.$$

Now the solution of the differential (6.4.6) is

$$T = T_0 e^{-k_p x}, \tag{6.4.8}$$

with $T_0$ the initial tension, $40 - 15 = 25$ mm Hg. Assuming that the channel has length $L$ and the final tension is $33 - 28 = 5$ mm Hg (see Table 6.4.2), (6.4.8) with $x = L$ becomes

$$5 = 25e^{-k_p L}; \quad \text{therefore,} \quad k_p L = -\log\left(\frac{1}{5}\right). \tag{6.4.9}$$

Next, we calculate the average tension $\bar{T}$ over the run of the channel. The average of (6.4.8) is given by the integral

$$\begin{aligned}\bar{T} &= \frac{1}{L}\int_0^L T_0 e^{-k_p x}\,dx = -\frac{T_0}{k_p L}e^{-k_p x}\Big|_0^L \\ &= \frac{T_0 - T_0 e^{-k_p L}}{k_p L} = \frac{25-5}{\log(5)} = 12.4 \text{ mm Hg},\end{aligned}$$

where (6.4.9) was used to substitute for $k_p L$.

Next, consider countercurrent flow. Arguing in the same manner as above, we see that the differential equations for countercurrent flow are similar to (6.4.4) and (6.4.5); the sign of the second is reversed because the flow is reversed here:

$$\frac{dc}{dx} = \frac{\left(\frac{D}{w}\right)}{hv_f}(C - c)$$

on the fetal side and

$$\frac{dC}{dx} = \frac{\left(\frac{D}{w}\right)}{Hv_m}(C - c)$$

on the maternal side. Subtracting, we get

$$\frac{dT}{dx} = -\frac{D}{w}\left(\frac{1}{hv_f} - \frac{1}{Hv_m}\right)T.$$

For this countercurrent flow, denote by $k_c$ the coefficient

$$k_c = -\frac{D}{w}\left(\frac{b}{f} - \frac{b}{F}\right). \tag{6.4.10}$$

As before, the solution is

$$T = T_0 e^{-k_c x}. \qquad (6.4.11)$$

Now, however, that $T_0 = 33 - 15 = 18$ mm Hg. And for $x = L$, $T = 40 - 28 = 12$ mm Hg; therefore, $12 = 18e^{-k_c L}$, from which it follows that $k_c L = \log(1.5)$. Hence the average tension in this case is

$$\bar{T} = \frac{T_0 - T_0 e^{-k_c L}}{k_c L} = \frac{6}{\log(1.5)} = 14.8 \text{ mm Hg}.$$

Therefore, countercurrent flow is somewhat more efficient than concurrent flow, by these calculations, $\frac{14.8}{12.4} = 1.2$ times more efficient. Note that the average tension for countercurrent flow is approximately equal to the numerical average, $\frac{18+12}{2} = 15$.

## Exercises/Experiments

1. Suppose the placenta becomes injured or impaired. How much of it is necessary in order to deliver adequate amounts of oxygen?
2. From Table 6.4.2, maternal $pO_2$ falls from 40 mm Hg to 33 mm Hg in its course through the placenta traveling at 400 ml/min. How much $O_2$ is delivered? (Compare with the text calculation.) If the flow rate falls to 300 ml/min, what must be the corresponding $pO_2$ difference to maintain this rate?
3. For fetal blood at 28 mm Hg $pO_2$, what is the amount of dissolved oxygen for a pH of 7.4? If the pH shifts to 7.2, what must be $pO_2$ so that the blood contains the same amount? Extrapolate to a pH of 7.0.
4. For maternal blood at 40 mm Hg $pO_2$, what is the amount of dissolved oxygen for a pH of 7.4? If the pH shifts to 7.2, what must be $pO_2$ so that the blood contains the same amount? Extrapolate to a pH of 7.0.
5. Modify the calculation to account for the diffusion of $O_2$ through 1 micron of plasma before reaching the placental membrane on the maternal side and 1 micron of plasma upon leaving the placental membrane on the fetal side before entering an erythrocyte.
6. Assume that carbon monoxide, CO, in the maternal blood reaches 5% (as is typical for smokers). Also assume that CO binds 220 times more readily than $O_2$ to hemoglobin. Recalculate diffusion across the placenta under these conditions.

## Questions for Thought and Discussion

1. Discuss why the evolution of small animals into large animals required the evolution of a closed circulatory system and a concomitant coelom.
2. Starting with the number 2, number the following events in the order in which they occur:

| | |
|---|---|
| blood enters right atrium | 1 |
| fluid from blood enters lymphatic system | ______ |
| blood gives up $CO_2$ at alveoli | ______ |
| blood enters systemic capillaries | ______ |

blood enters pulmonary artery ____________
blood enters aorta ____________

**3.** What factors determine how large the placenta has to be? By weight (including its blood supply), how large is the placenta relative to its fetus? Does this change over gestation?

## References and Suggested Further Reading


[1] MEMBRANE STRUCTURE:
W. S. Beck, K. F. Liem, and G. G. Simpson, *Life: An Introduction to Biology*, 3rd ed., Harper–Collins, New York, 1991.

[2] MEMBRANE TRANSPORT:
E. K. Yeargers, *Basic Biophysics for Biology*, CRC Press, Boca Raton, FL, 1992.

[3] DIFFUSION:
H. C. Berg, *Random Walks in Biology*, Princeton University Press, Princeton, NJ, 1993.

[4] DIFFUSION:
R. K. Hobbie, *Intermediate Physics for Medicine and Biology*, 2nd ed., Wiley, New York, 1988, 65.

[5] DIFFUSION IN BIOLOGY:
J. D. Murray, *Mathematical Biology*, Springer-Verlag, New York, 1989.

[6] FLUID RESISTANCE:
S. I. Rubinow, *Introduction to Mathematical Biology*, Wiley, New York, 1975.

[7] DIFFUSION ACROSS A SLAB:
D. L. Powers, *Boundary Value Problems*, Academic Press, New York, 1979.

[8] OXYGEN DISSOCIATION CURVES, FETAL BLOOD:
W. T. Keeton and J. L. Gould, *Biological Science*, 5th ed., Norton, New York, 1993.

[9] EPIDEMIOLOGY:
J. P. Fox, C. E. Hall, and L. R. Elveback, *Epidemiology: Man and Disease*, Macmillan, New York, 1970.

[10] EPIDEMIOLOGY AND DISEASE:
J. P. Krier and R. F. Mortenson, *Infection, Resistance, and Immunity*, Harper and Row, New York, 1990.

[11] PLACENTA:
F. C. Battaglia and G. Meschia, *An Introduction to Fetal Physiology*, Academic Press/Harcourt Brace Jovannovich, New York, 1986.

[12] PLACENTA:
K. S. Comline, K. W. Cross, G. S. Dawes, and P. W. Nathanielsz, *Foetal and Neonatal Physiology*, Cambridge University Press, Cambridge, UK, 1973.

[13] PLACENTA:
R. E. Forster II, Some principles governing maternal–foetal transfer in the placenta, in *Foetal and Neonatal Physiology*, Cambridge University Press, Cambridge, UK, 1973, 223–237.

[14] PLACENTA:
A. Costa, M. L. Costantino, and R. Fumero, Oxygen exchange mechanisms in the human placenta: Mathematical modelling and simulation, *J. Biomed. Engrg.*, **14** (1992), 85–389.

[15] PLACENTA:
H. Bartels, W. Moll, and J. Metcalfe, Physiology of gas exchange in the human placenta, *Amer. J. Obstetrics Gynecol.*, **84** (1962), 1714–1730.

[16] PLACENTA:
J. Metcalfe, H. Bartels, and W. Moll, Gas exchange in the pregnant uterus, *Physiol. Rev.*, **47** (1967), 782–838.

[17] PLACENTA:
A. Guettouche, J. C. Challier, Y. Ito, Y. Papapanayotou Cherruault, and A. Azancot-Benisty, Mathematical modeling of the human fetal arterial blood circulation, *Internat. J. Biomed. Comput.*, **31** (1992), 127–139.

# 7

# Neurophysiology

## Introduction

This chapter presents a discussion of the means, primarily electrical, by which the parts of an organism communicate with each other. We will see that this communication is not like that of a conducting wire; rather, it involves a self-propagating change in the ionic conductance of the cell membrane.

The nerve cell, or neuron, has an energy-requiring, steady-state condition in which the interior of the cell is at a negative potential relative to the exterior. Information transfer takes the form of a disruption of this steady-state condition, in which the polarity of a local region of the membrane is transiently reversed. This reversal is self-propagating, and is called an action potential. It is an all-or-none phenomenon: Either it occurs in full form or it doesn't occur at all.

Neurons are separated by a synaptic cleft, and interneuronal transmission of information is chemically mediated. An action potential in a presynaptic neuron triggers the release of a neurotransmitter chemical that diffuses to the postsynaptic cell. The sum of all the excitatory and inhibitory neurotransmitters that reach a postsynaptic cell in a short period of time determines whether a new action potential is generated.

## 7.1 Communication Between Parts of an Organism

*Specialization of structure and function in all organisms necessitates some means of communication among the various parts. Diffusive or convective flow of chemicals provides methods of communication, but they are very slow compared to the speed with which many intraorganismal needs must be conveyed. The high-speed alternative is electrical communication, for which complex nervous systems have evolved.*

*Communication is necessary at all levels of biological organization.*

The wide variety of molecular structures available in living systems is necessitated by the wide variety of physicochemical tasks required. Each kind of molecule,

R.W. Shonkwiler and J. Herod, *Mathematical Biology: An Introduction with Maple and Matlab*, Undergraduate Texts in Mathematics, DOI: 10.1007/978-0-387-70984-0_7,

supramolecular structure, organelle, cell, tissue, and organ is usually suited to just one or a few tasks. This kind of specialization ensures that each task is performed by the structure best adapted to it, one that Darwinian selection has favored above all others. The down side to such specialization is that the resultant structures are often localized into one region, well isolated from all others. If important parts are separated, clustered, there must be some means of communication between them to allow the entire organism to behave as a single integrated unit.

We find such specialization at all biological levels of organization. For example, there are many microscopic organisms that seem to be single cells, but close examination reveals that they possess special anatomical structures dedicated to quite different functions. Because of these differentiated structures, these organisms are often said to be "*acellular*," which is a simple admission that they do not fit into classical descriptive categories. Good examples are found in the organisms called *protozoans*: many have light-sensitive spots to help them orient. Others have elementary digestive tracts, with an opening to the outside and a tube leading into the body of the "cell." Of special interest are the primitive neural structures of the protozoan *ciliates*. These one-celled (or acellular) organisms move from place to place via the rapid beating of many small hairlike *cilia*. If these cilia were to beat at random there would be as many pushing in one direction as in the other, and the ciliate would not move. Observation of the cilia shows that they beat in synchronized waves, pushing the organism in a particular direction. If a small needle is inserted into the ciliate and then moved around to cause mechanical damage, the cilia will continue to beat—but not synchronously. Evidently there is some kind of primitive neural system to coordinate the movements of the cilia, and the needle damages that system, thus desynchronizing the cilia.

In multicellular animals, the need for a quick coordination between the perception of stimuli and consequent responses has led to the evolution of an endocrine system and a specialized nervous system. The endocrine system, facilitated by blood flow, provides chemical communication. The nervous system, the most complicated system in our bodies, provides the high-speed network that allows the other organs to work in harmony with each other.

*Communication in multicellular organisms can be chemical or electrical.*

*Hormones* are chemicals secreted by one kind of tissue in a multicellular organism and transported by the circulatory system to *target organs*. At its target organ a hormone exerts powerful chemical effects that change the basic physiology of cells. For example, sex hormones manufactured in the reproductive system cause changes in the skeletal and muscular structures of mammals, preparing them for the physical processes of mate attraction and reproduction. Insulin, a pancreatic hormone, affects the way cells in various tissues metabolize sugars.

The target organs of hormones are specifically prepared for hormone recognition. Protein receptors on the surfaces of cells in those organs can recognize certain hormones and not others. For instance, both men and women produce the hormone called *follicle stimulating hormone* (FSH), but it stimulates sperm production in males and egg production in females. The different effects are attributable to the target organs, not to the hormone, which is the same in both sexes. On the other hand, FSH has no effect on the voluntary muscles, which evidently have no FSH receptors. In the unusual phenomenon called *testicular feminization*, a person with both a man's chromosomal complement and *testosterone* levels has a woman's body, including external genitalia. The reason is that the person has no testosterone receptors on their target organ cells; thus the testosterone in the blood is not recognized by the organs that manifest secondary sexual characteristics. In humans, the default gender is evidently female, the generation of male characteristics requiring the interaction of testosterone and its appropriate receptors.

A second form of communication in an animal is electrical. We have a highly developed nervous system that provides us with a means of perceiving the outside world and then reacting to it. Indeed, the elaborate nervous system of primates and the complex behaviors it supports are defining characteristics.

Stimuli such as light, salt, pressure, and heat generate electrical signals in receptors. These signals are relayed to the *central nervous system*, consisting of the brain and spinal cord, where they are processed and where an appropriate response is formulated. The information for the response is then sent out to muscles or other organs where the response actually takes place. Examples of responses are muscular recoil, glandular secretion, and sensations of pleasure. There are special cells along which the electrical signals are passed. They are called *neurons*.

## 7.2 The Neuron

*Neurons, and other cells as well, are electrically polarized, the interior being negative with respect to the exterior. This polarization is due to the differential permeability of the neuron's plasma membrane to various ions, of which potassium is the most important. Active transport by sodium/potassium pumps maintains the interior concentration of sodium low and potassium high.*

*Neurons are highly specialized cells for conveying electrical information.*

Figure 7.2.1 shows a model of a typical mammalian neuron. The cell's long, narrow shape suggests its role: it is specifically adapted to the task of conveying information from one location in the body to another. The direction of information flow is from the dendrites, through the cell body, to the axon and on to other neurons, muscles, and glands. Neurons also exhibit unusual electrical behavior, variations of which permit these cells to pass information from dendrite to axon. These two properties of neurons, shape and electrical behavior, suggest an analogy with the conduction of electric signals by a copper wire. That analogy is incorrect, however, and we will spend part of this chapter discussing how that is so.

*A membrane can achieve electrical polarization passively.*

Under certain conditions a voltage, or *potential difference*, can be maintained across

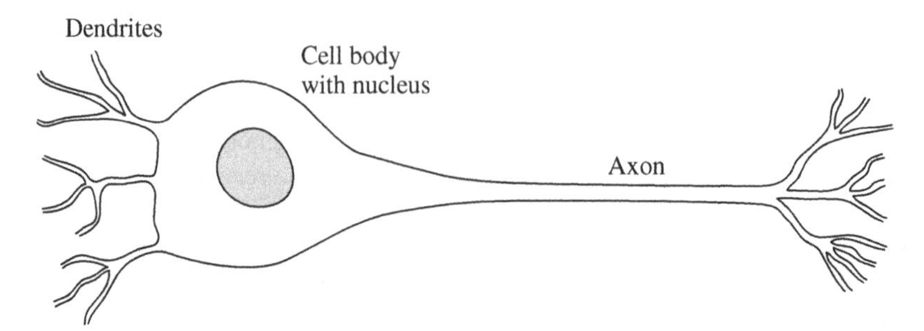


**Fig. 7.2.1.** A model of a neuron. The direction of information transfer is from left to right.

a membrane without the expenditure of energy.[1] We will use Figures 7.2.2(a)–(c) to show how this can happen in a model system of two water-filled compartments that are separated by a membrane. In each case, we begin by dissolving a compound KQ in the water on the left of the membrane, where K is a positive ion, perhaps potassium, and Q is any large organic group. This compound will immediately dissociate into $K^+$ and $Q^-$, such that $[K^+] = [Q^-]$, the braces indicating concentrations. (We assume that KQ dissociates completely.)

In Figure 7.2.2(a), we imagine that the membrane is completely permeable to both ions. They will both move to the right by passive diffusion, eventually reaching an equilibrium state in which the concentrations of each of the two ions will be the same on both sides of the membrane. The net change will be zero on each side of the membrane, and therefore no *potential difference* will exist across the membrane.

In Figure 7.2.2(b), we imagine that the membrane is completely impermeable to both ions. Neither will therefore move across the membrane, and the electrical charges on each side of the membrane will total zero. Thus there will again be no transmembrane potential difference.

In Figure7.2.2(c) we will finally see how a potential difference can be generated. We imagine that the membrane is permeable to $K^+$, but is impermeable to $Q^-$, perhaps because $Q^-$ is too big to pass through the membrane. $K^+$ will then move across the membrane in response to its own *concentration* gradient. Very quickly, however, the movement of $K^+$ away from the $Q^-$ will establish a transmembrane *electrical charge* gradient that opposes the concentration gradient. The concentration gradient pushes potassium to the right and the electrical gradient pushes it to the left. The system will then quickly reach *electrochemical equilibrium*, because there will be no further net change in $[K^+]$ on either side of the membrane. Note that all of $Q^-$ is on the left, but the $K^+$ is divided between the right and left sides. The right side thus will be at a positive electrical potential with respect to the left.

Note that the potential difference in the system of Figure 7.2.2(c) is achieved spontaneously, i.e., without the expenditure of energy. Real biological systems can

[1] This voltage difference across the neuron's plasma membrane is called a potential difference because it represents a form of potential energy, or energy conferred by virtue of the positions of electrical charges.

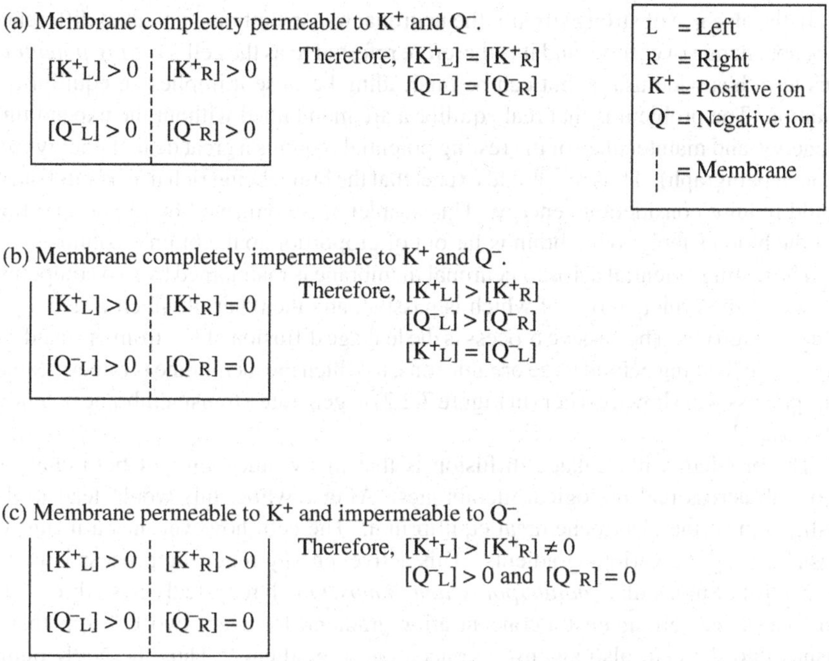

(a) Membrane completely permeable to $K^+$ and $Q^-$.

| | |
|---|---|
| $[K^+_L] > 0$ | $[K^+_R] > 0$ |
| $[Q^-_L] > 0$ | $[Q^-_R] > 0$ |

Therefore, $[K^+_L] = [K^+_R]$
$[Q^-_L] = [Q^-_R]$

L = Left
R = Right
$K^+$ = Positive ion
$Q^-$ = Negative ion
¦ = Membrane

(b) Membrane completely impermeable to $K^+$ and $Q^-$.

| | |
|---|---|
| $[K^+_L] > 0$ | $[K^+_R] = 0$ |
| $[Q^-_L] > 0$ | $[Q^-_R] = 0$ |

Therefore, $[K^+_L] > [K^+_R]$
$[Q^-_L] > [Q^-_R]$
$[K^+_L] = [Q^-_L]$

(c) Membrane permeable to $K^+$ and impermeable to $Q^-$.

| | |
|---|---|
| $[K^+_L] > 0$ | $[K^+_R] > 0$ |
| $[Q^-_L] > 0$ | $[Q^-_R] = 0$ |

Therefore, $[K^+_L] > [K^+_R] \neq 0$
$[Q^-_L] > 0$ and $[Q^-_R] = 0$

**Fig. 7.2.2.** This figure shows how a potential difference can be generated across a selectively permeable membrane without the expenditure of energy. See the text for a detailed discussion. Parts (a) and (b) depict extreme situations, in which no potential difference is generated across the membrane: In (a), the membrane is equally permeable to both the positive and the negative ions. In (b), the membrane is impermeable to both ions. In (c), however, the membrane is permeable to the positive ion and impermeable to the negative ion. This generates an equilibrium that is a compromise between electrical and mechanical diffusion properties, and results in an imbalance of charges across the membrane.

establish potential differences in much the same way: They possess membranes that are permeable to some materials and not to others, a feature that Figures 7.2.2(a)–(c) show to be essential to the establishment of a potential difference. Our model system, however, lacks some realistic features: For example, real systems have many other ions, such as $Cl^-$ and $Na^+$, that must be considered. Further, consideration of Figure 7.2.2 suggests that the *degree* of permeability will be important in the establishment of transmembrane potentials (see [1] and [2]).

*The membrane of a "resting" neuron is electrically polarized.*

If one probe of a voltmeter is inserted into a neuron and the other probe is placed on the outside, the voltmeter will show a potential difference of about −70 millivolts.[2] In other words, the interior of the cell is 70 mv more negative than the outside, and a positive charge would tend to be attracted from the outside to the inside.

[2] For various kinds of cells, the potential may vary from about −40 mv to more than −100 mv.

In the absence of strong stimuli, the neuronal transmembrane potential difference does not change over time, and it is therefore referred to as the cell's *resting potential*. This term has wide usage, but can be misleading because it implies an equilibrium situation. The problem is that real equilibria are maintained without the expenditure of energy, and maintenance of the resting potential requires a great deal of energy (see the next paragraph). Thus we should expect that the brain, being rich in nervous tissue, would require considerable energy. This suspicion is confirmed by the observation that the blood supply to the brain is far out of proportion to the brain's volume.

The resting potential across a neuronal membrane is maintained by two competing processes, the principal one of which is passive, and the other of which is active, or energy-requiring. The passive process is the leakage diffusion of $K^+$ from the inside to the outside, leaving behind large organic ions, to which the membrane is impermeable. This process was shown earlier (in Figure 7.2.2) to generate a transmembrane potential difference.[3]

The problem with leakage diffusion is that many other ions, of both charges, also leak across real biological membranes. After a while, this would lead to the destruction of the electrochemical equilibrium. The cell, however, has a means of reestablishing the various gradients. This active, energy-requiring process is under the control of molecular *sodium/potassium pumps*, which repeatedly push three $Na^+$ ions out of the cell, against a concentration gradient, for every two $K^+$ ions that it pushes into the cell, also against a concentration gradient.[4] Thus the Na/K pump and diffusion work against each other in a nonequilibrium, steady-state way. Such nonequilibrium, steady-state processes are very common in living cells.

Consideration of Figure 7.2.2 suggests that the low permeability of a membrane to sodium eliminates that ion as a contributor to the resting potential, and that the high permeability to potassium implicates that ion in the resting potential. This suspicion is confirmed by experiment: The neuronal resting potential is relatively insensitive to changes in the extracellular concentration of sodium, but highly sensitive to changes in the extracellular potassium concentration.

The concentrations of $Na^+$, $K^+$, and $Cl^-$ inside and outside a typical neuron are given later, in Table 7.5.1. The asymmetric ionic concentrations are maintained by the two factors mentioned above: the Na/K pump and the differential permeabilities of the cell's plasma membrane to the different ions. In particular, we note that $Na^+$, to which the membrane is poorly permeable, might rush across the membrane if that permeability increased.

## 7.3 The Action Potential

*The neuronal plasma membrane contains voltage-controlled gates for sodium and for potassium. The gates are closed at the usual resting potential. When a stimulus*

[3] The fraction of potassium ions that must leak out of a neuron to establish a potential difference of 100 mv is estimated to be only about $10^{-5}$.

[4] The molecular basis for the Na/K pump is not known. What is known is its effect—pumping $Na^+$ outward and $K^+$ inward.

*depolarizes the membrane beyond a threshold value, the sodium gates open and sodium rushes into the cell. Shortly thereafter, potassium rushes out. The Na/K pump then restores the resting concentrations and potential. The change in potential associated with the stimulus is called an action potential, and it propagates itself without attenuation down the neuronal axon.*

*The resting potential can be changed.*

The usual resting potential of a neuron is about −70 mv, but this value can be changed either by changing the permeability of the membrane to any ion or by changing the external concentration of various ions. If the potential is increased, say to −100 mv, the neuron is said to be *hyperpolarized*. If the potential is decreased, say to −40 mv, the neuron is said to be *depolarized* (see [1, 2, 3, 4]).

*The membranes of neurons contain gated ion channels.*

In Sections 6.1 and 7.2, we described several ways that materials could move across membranes. They were the following:

(a) Simple passive diffusion: Material moves directly through the bulk part of the membrane, including the hydrophobic layer. Water, carbon dioxide, and hydrocarbons move in this fashion.
(b) Facilitated passive diffusion: Ions and neutral materials move through special selective channels in the membrane. The selectivity of these channels resides in the recognition of the moving material by transport proteins, or permeases, in the membrane. Nevertheless, the process is spontaneous in that the material moves from regions of high concentration to regions of low concentration, and no energy is expended.
(c) Facilitated active transport: Ions and neutral materials move through selective channels (determined by transport proteins) but they move from regions of low concentration to regions of high concentration. Thus energy is expended in the process.

Here we are interested in certain aspects of facilitated, passive diffusion. Some channels through which facilitated, passive diffusion occurs seem to function all the time. Others, however, are controlled by the electrical or chemical properties of the membrane in the area near the channel. Such channels can open and close, analogously to fence gates, and are therefore called *voltage-gated channels* and *chemically gated channels*, respectively.

We now look more closely at voltage-gated channels; we will return to chemically gated channels a bit later. The following narrative corresponds to Figure 7.3.1. Voltage-gated sodium channels and potassium channels are closed when the potential across the membrane is −70 mv, i.e., the usual resting potential. If a small region of the axonal membrane is depolarized with an electrode to about −50 mv, the voltage-gated sodium channels in that area will open. Sodium ions will then rush into the cell (because the $Na^+$ concentration outside the cell is so much higher than it is inside the cell, as shown in Table 7.5.1).

(a) Unstimulated neuron

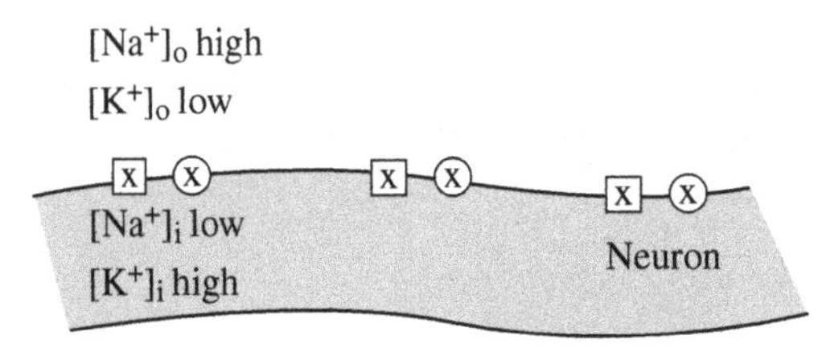


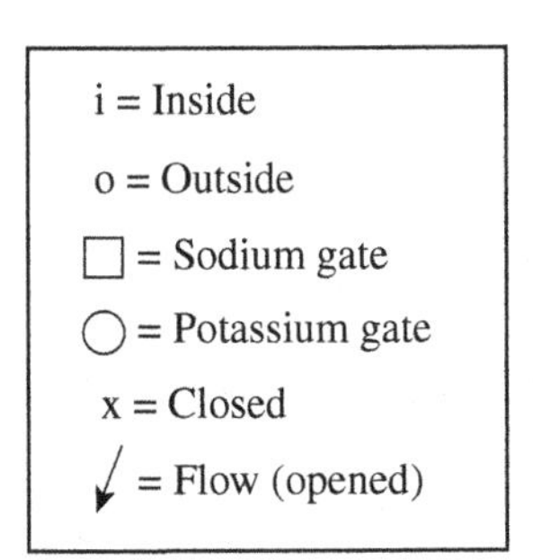


(b) Part of membrane depolarized; depolarization causes $Na^+$ gate to open

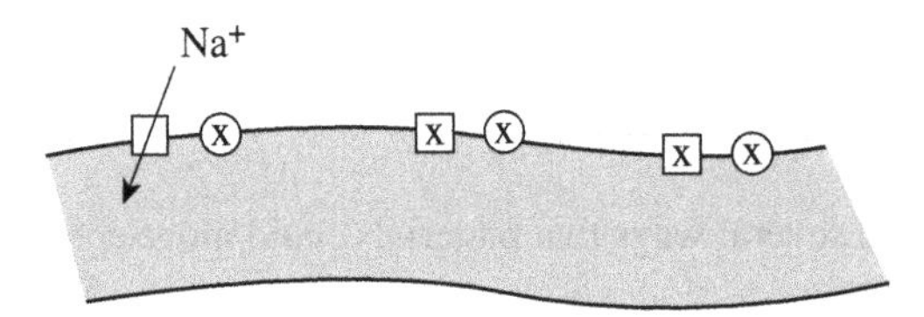


(c) Original $Na^+$ gate closes; $K^+$ gate opens, then nearby $Na^+$ gate opens

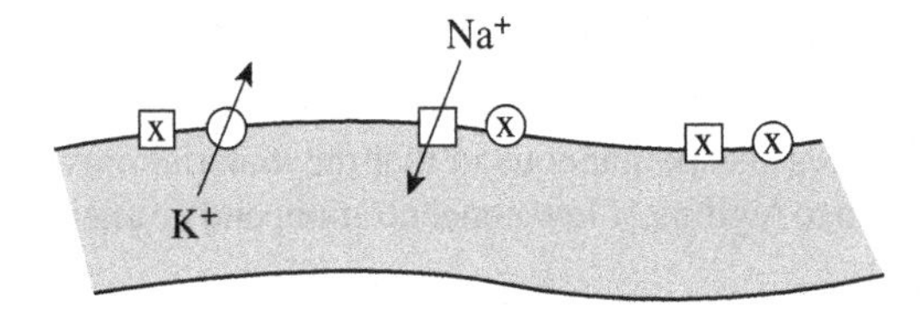


(d) Disturbance (action potential) propagated down axon

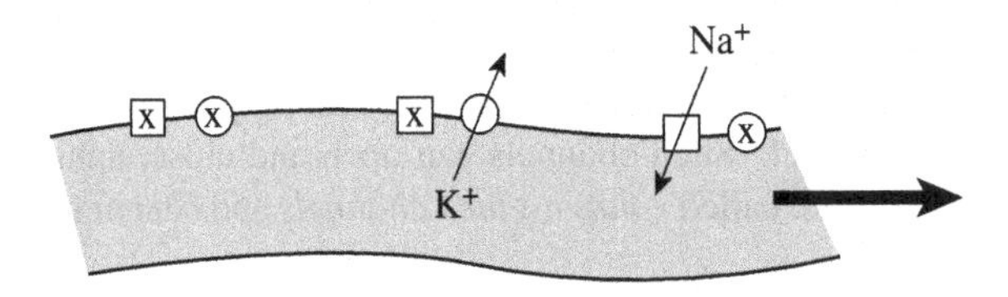


(e) $Na^+$ pumped back out and $K^+$ pumped back in after action potential passes

**Fig. 7.3.1.** The generation and movement of an action potential. This figure shows how an action potential is generated by an initial, localized depolarization of a neuron. This depolarization causes a depolarization at a neighboring site. Thus the disturbance, or action potential, propagates itself down the neuron.

The inward rush of sodium will *further* depolarize the membrane, opening even more sodium channels, suggesting an avalanche. This is prevented at the site of the *original* depolarization, however, by the spontaneous closing of the sodium gates about a millisecond after they open. During that millisecond the membrane potential difference not only goes to zero, but becomes positive. At that time, the potassium gates open, and potassium rushes out of the neuron in response to the positive charges that have accumulated in the interior. This makes the transmembrane potential drop rapidly, even hyperpolarizing to below $-70$ mv. The Na/K pump now restores $Na^+$ and $K^+$ to resting levels inside and outside the neuron.

The entire process of action potential generation, from initiation to complete recovery, takes about two milliseconds. The neuron can be restimulated before it has recovered completely, i.e., during the period of hyperpolarization, but such stimulation naturally takes a greater depolarizing stimulus than the original one. Finally, note the "*all-or-none*" nature of action potential generation: No action potential is generated until the threshold degree of depolarization is reached, and then the same-size action potential is generated, no matter how far past the threshold the stimulatory depolarization is carried.

*The action potential propagates itself down the neuron.*

We saw just above how sufficient depolarization of the axonal membrane at a localized point can cause the voltage-controlled sodium gates to open, permitting sodium ions to rush into the cell at that point. These sodium ions further depolarize the membrane at that point and permit even more sodium to enter the cell. Figure 7.3.1 shows that the inrushing sodium ions now spread laterally along the inside of the membrane, attracting negative ions from the extracellular fluid to the outside of the membrane. As the numbers of these ions increase, they will eventually depolarize the *neighborhood* of the site of the original stimulus and thus open the sodium gates there. Meanwhile, the site of the original stimulus cannot be restimulated during a several millisecond *latent period*. The temporary latency at the site of the original stimulus and the depolarization of the membrane in the neighborhood of that site combine to cause the action potential to *spread away* from the site of the original stimulus.[5]

The decay of the action potential at its original site and its lateral spread from that site causes the disturbance to move along the neuron. The movement of the action potential is often compared to the burning of a fuse: Action at one site triggers action at the adjacent site and then the original action is extinguished. The size of the disturbance is the same as it passes any point on the fuse.

In closing this section, we bring up a point made at the beginning of this section: The electrical propagation of an action potential is not like the propagation of an electrical current down a copper wire. The latter involves the lengthwise motion of electrons in the wire; the former involves radial and lengthwise motions of atomic ions.

[5] The effect of some anaesthetics, e.g., ether and ethyl alcohol, is to reduce the electrical resistance of the neuronal membrane. This causes the action potential to be extinguished. The effect of membrane resistance on action potential velocity is discussed in Section 7.5.

## 7.4 Synapses: Interneuronal Connections

*When an action potential reaches the junction, or synaptic gap, between two neurons, it triggers the release of a neurotransmitter chemical from the presynaptic cell. The neurotransmitter then diffuses across the synaptic gap and, combining with other incoming signals, may depolarize the postsynaptic cell enough to trigger a new action potential there.*

*Besides the excitatory neurotransmitters, there are inhibitory neurotransmitters. The latter hyperpolarize the* postsynaptic neuron, *making it more difficult for a new action potential to be generated. Thus whether an action potential is generated is determined by the sum of all incoming signals, both excitatory and inhibitory, over a short time period. The nervous system can use this summation to integrate the signals induced by a complex environment, and thereby generate complex responses.*

*In some cases, accessory cells wrap around neuronal axons to increase the electrical resistance between the cell's interior and its exterior. The action potential seems to jump between gaps in this sheath, thus greatly increasing the velocity with which the action potential is propagated.*

*Most communication between neurons is chemical.*

The simplest "loop" from stimulus to response involves three neurons: one to detect the stimulus and carry the message to the central nervous system, one in the central nervous system, and one to carry the message to the responding organ. Most neural processing is much more complicated than that, however. In any case, some means of cell-to-cell communication is necessary. It surprises many biology students to learn that the mode of communication between such neurons is almost always chemical, not electrical.

Figure 7.4.1 shows the junction of two typical neurons, There is a gap of about 30 nm between the axon of the neuron carrying the incoming action potential and the dendrite of the neuron in which a new action potential will be generated.[6] The gap is call a *synapse*. The arriving action potential opens certain voltage-gated ion channels, which causes small packets of a *neurotransmitter chemical* to be released from the *presynaptic membrane* of the *presynaptic neuron*. This neurotransmitter then diffuses to the *postsynaptic membrane* of the postsynaptic neuron, where it opens chemically gated ion channels. The opening of these chemically gated channels causes a local depolarization of the dendrites and cell body of the postsynaptic neuron. If the depolarization is intense enough, a new action potential will be created in the area between the postsynaptic cell body and its axon.

More than a hundred neurotransmitters have been identified, but *acetylcholine* is a very common one, especially in the part of our nervous system that controls voluntary movement. If acetylcholine were to remain at the postsynaptic membrane, it would continue to trigger action potentials, resulting in neuronal chaos. There is, however, a protein catalyst called *acetylcholine esterase* that breaks down acetylcholine soon after it performs its work. Most *nerve gases*, including many insecticides, work

[6] 1 nm = 1 nanometer = $10^{-9}$ meter.

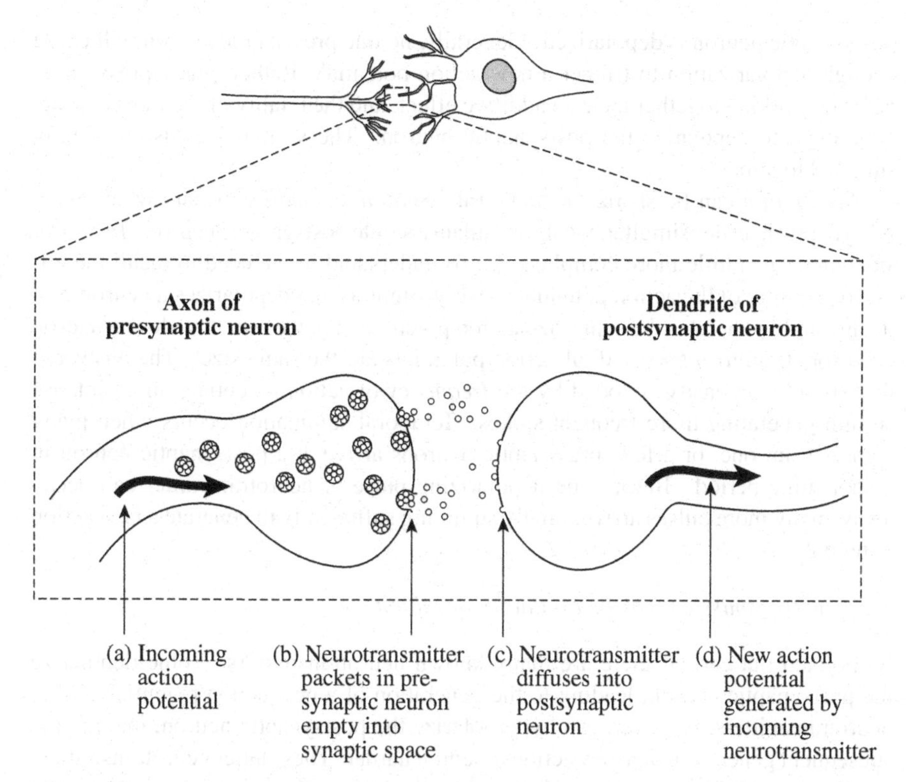


**Fig. 7.4.1.** Synaptic information transfer between neurons. The incoming action potential causes the presynaptic neuron to release a neurotransmitter chemical, which diffuses across the synaptic space, or cleft. At the postsynaptic neuron the neurotransmitter causes a new action potential to be generated.

by inactivating acetylcholine esterase, leading to uncontrolled noise in the animal's nervous system, and therefore death.

Note that synaptic transmission causes information flow to be one-way, because only the end of an axon can release a neurotransmitter and only the end of a dendrite can be stimulated by a neurotransmitter.

*Occasionally, synaptic transmission is electrical.*

At times, chemical transmission by synapses is too slow because diffusion may result in a delay of more than a millisecond. For these situations, there are direct cytoplasmic connections that allow electrical communication between cells. For example, parts of the central nervous system controlling jerky eye movements have electric synapses, as do parts of the heart muscle (which, of course, is not nervous tissue).

*Summation of incoming signals occurs at the postsynaptic neuron.*

Whether or not a new action potential is generated depends on the degree to which the postsynaptic neuron is depolarized. Generally, no one presynaptic neuron will cause enough depolarization to trigger a new action potential. Rather, many presynaptic neurons working together have an additive effect, and the totality of their effects may be enough to depolarize the postsynaptic neuron. The various signals are simply summed together.

*Summation* can be spatial or temporal. *Spatial summation* occurs when many presynaptic neurons simultaneously stimulate a single postsynaptic neuron. *Temporal summation* is a little more complicated. To understand it, we need to recall the all-or-none nature of the action potential—every stimulus that depolarizes a neuron past its threshold generates the same size action potential. So, we ask, how does a neuron code for stimulus *intensity* if all action potentials are the same size? The answer is that stimulus intensity is coded by the *frequency* of action potentials, more intense stimuli generating more frequent spikes. Temporal summation occurs when many signals from one, or a few, presynaptic neurons arrive at a postsynaptic neuron in a short time period. Before one depolarizing pulse of neurotransmitter can decay away, many more pulses arrive, finally summing sufficiently to generate a new action potential.

*Synaptic transmission may be excitatory or inhibitory.*

As pointed out earlier, there are many known neurotransmitters. Some depolarize the postsynaptic neuron, leading to the generation of a new action potential. Other neurotransmitters, however, can hyperpolarize the postsynaptic neuron, making the subsequent generation of a new action potential harder. These latter neurotransmitters are therefore *inhibitory*. In general, therefore, the information reaching a postsynaptic cell will be a mixture of excitatory and inhibitory signals. These signals are summed, and the resultant degree of depolarization determines whether a new action potential is generated.

We can now see an important way that the nervous system integrates signals. The actual response to a stimulus will depend on the pattern of excitatory and inhibitory signals that pass through the network of neurons. At each synapse, spatial and temporal summation of excitatory and inhibitory signals determines the path of information flow.

*Myelinated neurons transmit information very rapidly.*

As will be demonstrated in the next section, the velocity of conduction of an action potential down an axon depends on the diameter of the axon and on the electrical resistance of its outer membrane. Figure 7.4.2 shows how special accessory cells, called *Schwann cells*, repeatedly wrap around neuronal axons, thus greatly increasing the electrical resistance between the axon and the extracellular fluid. This resistance is so high, and the density of voltage-gated sodium channels so low, that no ions enter or leave the cell in these regions. On the other hand, there is a very high sodium channel density in the unwrapped regions, or *nodes*, between neighboring Schwann cells. The action potential exists only at the nodes, internodal information flow occurring by ion flow *under* the myelinated sections. Thus the action potential seems to skip from one

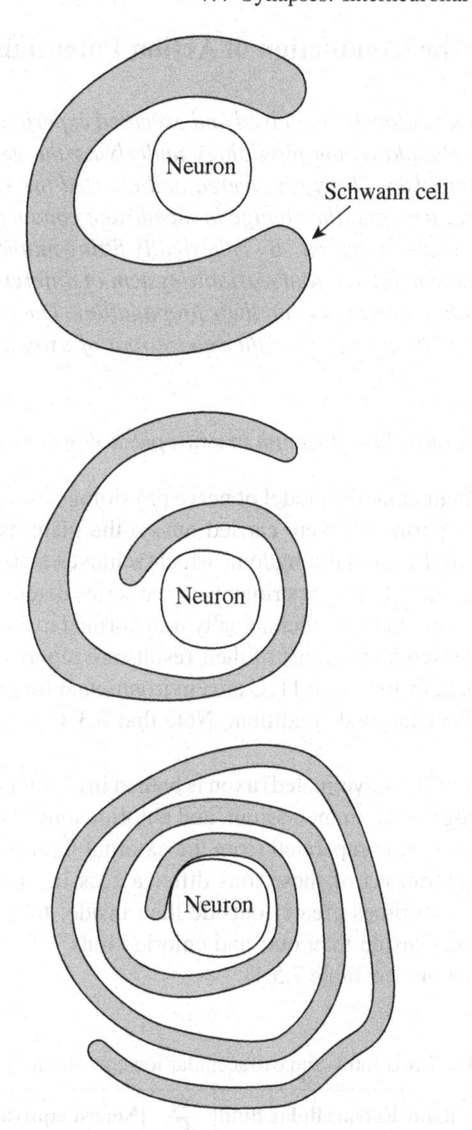


**Fig. 7.4.2.** This picture shows how a Schwann cell wraps around a neuron, resulting in many layers of membrane between the neuron and the outside world. This many-layered structure has a very high electrical resistance, which radically alters the action-potential–transmitting properties of the neuron.

node to another, in a process called *saltatory conduction*. Saltatory conduction in a myelinated nerve is about one hundred times faster than the movement of an action potential in a nonmyelinated neuron. We would (correctly) expect to see it in neurons that control rapid-response behavior.

## 7.5 A Model for the Conduction of Action Potentials

*Through a series of meticulously conceived and executed experiments, Alan Hodgkin and Andrew Huxley elucidated the physiology underlying the generation and conduction of nervous impulses. They discovered and charted the key roles played by sodium and potassium ions and the change in membrane conductance to those ions as a function of membrane potential. By empirically fitting mathematical equations to their graphs, they formulated a four-variable system of differential equations that accurately models action potentials and their propagation. The equations show that propagation velocity is proportional to the square root of axon diameter for an unmyelinated nerve.*

*The Hodgkin–Huxley model is a triumph in neurophysiology.*

We present here the mathematical model of nerve physiology as reported by Hodgkin and Huxley. Their experiments were carried out on the giant axon of a squid, the largest axon known in the animal kingdom, which achieves a size sufficient for the implantation of electrodes. Early experiments in the series determined that results at different temperatures could be mathematically transformed to any other (physiological) temperature. Consequently, most of their results are reported for a temperature of 6.3°C. Temperatures from 5°C to 11°C are environmental for the animal, and help maintain the nerve fiber in good condition. Note that 6.3°C is approximately 300° Kelvin.

In the resting state a (nonmyelinated) axon is bathed in an interstitial fluid containing, among other things, sodium, potassium, and chloride ions. The interior material of the axon, the *axoplasm*, is separated from the external fluid by a thin lipid membrane, and the concentrations of these ions differ across it. The concentration of sodium ions, $Na^+$, is 10 times greater outside than inside; that of potassium ions, $K^+$, is 30 times greater inside than out; and chloride ions, $Cl^-$, are 15 times more prevalent outside than in (see Table 7.5.1).

**Table 7.5.1.** Intra- and extracellular ion concentrations.

| Inside axon | Extracellular fluid | $\frac{C_i}{C_o}$ | Nernst equivalent |
|---|---|---|---|
| $Na^+$ 15 | $Na^+$ 145 | 0.10 | −55 mv |
| $K^+$ 150 | $K^+$ 5 | 30 | 82 mv |
| $Cl^-$ 7.5 | $Cl^-$ 110 | 0.068 | −68 mv |

Diffusion normally takes place down a concentration gradient, but when the particles are electrically charged, electrical potential becomes a factor as well. In particular, if the interior of the axon is electrically negative relative to the exterior, then the electrically negative chloride ions will tend to be driven out against its own concentration gradient until a balance between the electrical pressure and concentration gradient is reached (see Section 7.2). The balancing voltage difference, $V$, is given by the *Nernst equation*,

$$V = \frac{RT}{F} \ln \frac{C_i}{C_o}, \tag{7.5.1}$$

where $R$ is the gas constant, $R = 8.314$ joules/°C/mole, $T$ is absolute temperature, and $F$ is Faraday's constant, 96,485 coulombs, the product of Avogadro's number with the charge on an electron. The concentrations inside and out are denoted by $C_i$ and $C_o$, respectively.

For $Cl^-$, this potential difference is $-68$ mv. But in the resting state (see Section 7.2), the interior of an axon is at $-70$ mv relative to the exterior (see Figure 7.5.1). Thus there is little tendency for chloride ions to migrate. The same holds for potassium ions, which are electrically positive. Their Nernst equivalent is 82 mv, that is, the outside needs to be 82 mv more positive than the inside for a balance of the two effects. Since the outside is 70 mv more positive, the tendency for $K^+$ to migrate outward, expressed in potential, is only 12 mv.

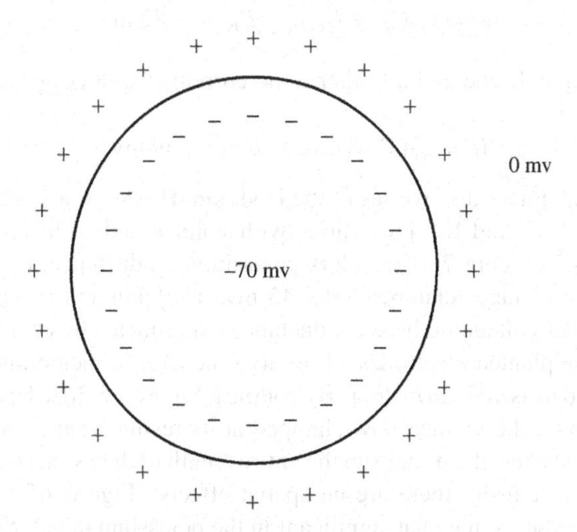


**Fig. 7.5.1.** Charges inside and outside an axon.

The situation is completely different for sodium ions, however. Their Nernst equivalent is $-55$ mv, but since sodium ions are positive, the interior would have to be 55 mv electrically positive to balance the inward flow due to the concentration gradient. Since it is $-70$ mv instead, there is an equivalent of 125 mv ($= 70 + 55$) of electrical potential for sodium ions to migrate inward. But in fact, there is no such inward flow; we must therefore conclude that the membrane is relatively impermeable to sodium ions. In electrical terms, the conductance of the membrane to sodium ions, $g_{Na}$, is small.

Electrical conductance is defined as the reciprocal of electrical resistance,

$$g = \frac{1}{R}.$$

The familiar Ohm's law,

$$V = iR,$$

then takes the form

$$i = gV. \tag{7.5.2}$$

In discrete-component electronics, resistance is measured in ohms and conductance in mhos. However, for an axon, conductance and current will depend on surface area. Therefore, conductance here is measured in mhos per square centimeter. Letting $E_{\mathrm{Na}} = 55$ mv denote the equilibrium potential given by the Nernst equation for sodium ions, the current per square centimeter of those ions can be calculated as

$$i_{\mathrm{Na}} = g_{\mathrm{Na}}(V - E_{\mathrm{Na}}), \quad E_{\mathrm{Na}} = 55 \text{ mv}, \tag{7.5.3}$$

where $V$ is the voltage inside the axon. A negative current corresponds to inward flow. Similarly, for potassium ions,

$$i_{\mathrm{K}} = g_{\mathrm{K}}(V - E_{\mathrm{K}}), \quad E_{\mathrm{K}} = -82 \text{ mv}. \tag{7.5.4}$$

Finally, grouping chlorine and all other ionic currents together as *leakage currents* gives us

$$i_{\ell} = g_{\ell}(V - E_{\ell}), \quad E_{\ell} = -59 \text{ mv}. \tag{7.5.5}$$

One of the landmark discoveries of the Hodgkin–Huxley study is that membrane permeability to $\mathrm{Na}^{+}$ and $\mathrm{K}^{+}$ ions varies with voltage and with time as an action potential occurs.[7] Figure 7.5.2(a) plots potassium conductance against time with the interior axon voltage "clamped" at −45 mv. Hodgkin and Huxley's apparatus maintained a fixed voltage in these conductance experiments by controlling the current fed to the implanted electrodes. The large increase in membrane conductance shown is referred to as *depolarization*. By contrast, Figure 7.5.2(b) depicts membrane *repolarization* with the voltage now clamped at its resting value. In these *voltage clamp* experiments, the electrodes run the entire length of the axon, so the entire axon membrane acts in unison; there are no spatial effects. Figure 7.5.3 shows a $\mathrm{Na}^{+}$ conduction response. Somewhat significant in the potassium depolarization curve is its sigmoid shape. On the other hand, the sodium curve shows an immediate rise in conductance. We conclude that the first event in a depolarization is the inflow of $\mathrm{Na}^{+}$ ions. An outflow of $\mathrm{K}^{+}$ ions follows shortly thereafter. The leakage conductance is 0.3 m-mhos/cm$^2$ and, unlike the sodium and potassium conductances, is constant.

*Potassium conductance is interpolated by a quartic.*

Hodgkin and Huxley went beyond just measuring these conductance variations; they modeled them mathematically. To capture the inflexion in the potassium conductance curve, it was empirically modeled as the fourth power of an exponential rise; thus

$$g_{\mathrm{K}} = \bar{g}_{\mathrm{K}} n^4, \tag{7.5.6}$$

[7] The biological aspects of this variation were discussed in Section 7.3.

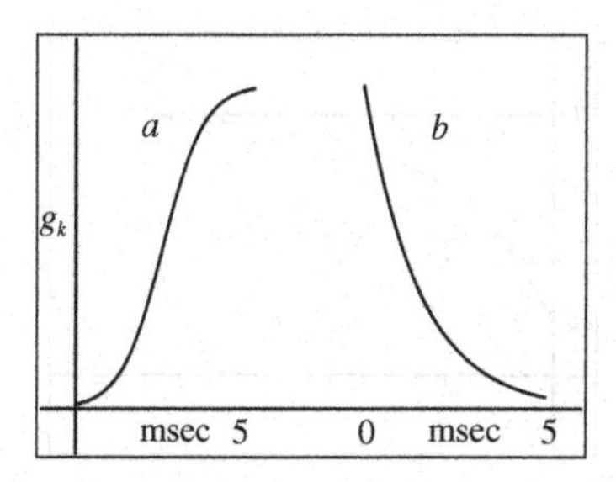


**Fig. 7.5.2.** Potassium conductance vs. time.

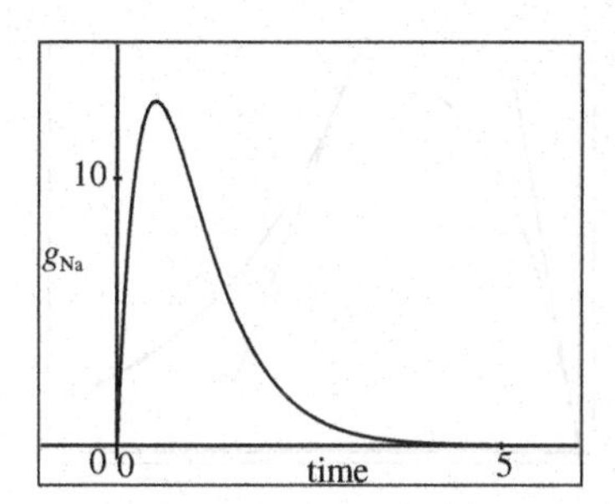


**Fig. 7.5.3.** Sodium conductance vs. time.

where $\bar{g}_K$ is a constant with the same units as $g_K$ and $n$ is a dimensionless exponential variable,

$$n = n_\infty - (n_\infty - n_0)e^{-\frac{t}{\tau_n}},$$

increasing from $n_0$ to the asymptote $n_\infty$; see Figure 7.5.4. If $n_0 = 0$, then for small $t$, $n$ behaves like $t^4$ as seen from a Taylor expansion,

$$n^4 = (n_\infty - n_\infty e^{-\frac{t}{\tau_n}})^4 = \left(\frac{n_\infty}{\tau_n}\right)^4 t^4 - 4\left(\frac{n_\infty}{\tau_n}\right)^3 \left(\frac{n_\infty}{2!\tau_n^2}\right) t^5 + \cdots .$$

Therefore, the $n^4$ curve has a horizontal tangent at 0, as desired.

In order that $n$ become no larger than 1 in the fit to the experimental data, $\bar{g}_K$ is taken as the maximum conductance attainable over all depolarization voltages. This value is

$$\bar{g}_K = 24.34 \text{ m-mhos/cm}^2.$$

The function $n$ may also be cast as the solution of a differential equation; thus

$$\frac{dn}{dt} = \alpha_n(1 - n) - \beta_n n, \tag{7.5.7}$$

where the coefficients $\alpha_n$ and $\beta_n$ are related to the rising and falling slopes of $n$, respectively; see Figure 7.5.5. Their relationship with the parameters of the functional form is

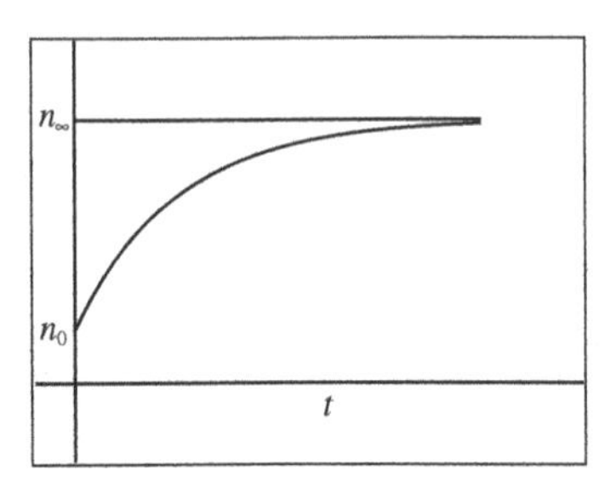


**Fig. 7.5.4.** Dimensionless variable $n$ vs. time.

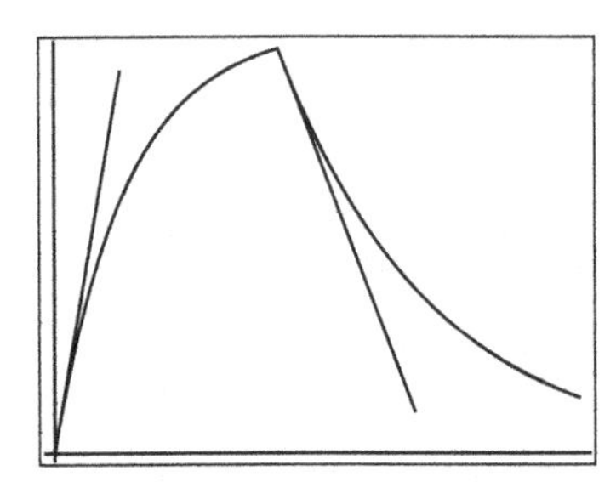

**Fig. 7.5.5.** Slope of the differential form of dimensionless variable $n$.

$$\tau_n = \frac{1}{\alpha_n + \beta_n} \quad \text{and} \quad n_\infty = \frac{\alpha_n}{\alpha_n + \beta_n}.$$

Experimentally, these coefficients vary with voltage. Hodgkin and Huxley fit the experimental relationship by the functions

$$\alpha_n = \frac{0.01(10 - (V - V_r))}{e^{1-0.1(V-V_r)} - 1} \tag{7.5.8}$$

and

$$\beta_n = 0.125e^{-\frac{V-V_r}{80}}, \tag{7.5.9}$$

where $V_r = -70$ mv is resting potential. Note that (7.5.8) is singular at $V = -60$; however, the singularity is removable with limiting value 0.1.

Substituting (7.5.6) into (7.5.4) gives

$$i_K = \bar{g}_K n^4 (V - E_K). \tag{7.5.10}$$

*Sodium conductance is interpolated by a cubic.*

Sodium conductance is modeled as

$$g_{Na} = \bar{g}_{Na} m^3 h, \tag{7.5.11}$$

where $\bar{g}_{Na} = 70.7$ m-mhos/cm$^2$ is a fixed parameter carrying the units and $m$ and $h$ are dimensionless. The use of two dimensionless variables helps smooth the transition between the ascending portion of the curve and the descending portion. As above, $m$ and $h$ are exponential and satisfy similar differential equations,

$$\frac{dm}{dt} = \alpha_m(1 - m) - \beta_m m, \tag{7.5.12}$$

$$\frac{dh}{dt} = \alpha_h(1 - h) - \beta_h h. \tag{7.5.13}$$

Further, the coefficients are functions of $V$ interpolated as follows:

$$\alpha_m = \frac{0.1(25 - (V - V_r))}{e^{0.1(25-(V-V_r))} - 1}, \tag{7.5.14a}$$

$$\beta_m = 4e^{-\frac{V-V_r}{18}}, \tag{7.5.14b}$$

$$\alpha_h = 0.07e^{-0.05(V-V_r)}, \tag{7.5.14c}$$

$$\beta_h = \frac{1}{e^{0.1(30-(V-V_r))} + 1}. \tag{7.5.14d}$$

Substituting (7.5.11) into (7.5.3) gives

$$i_{\text{Na}} = \bar{g}_{\text{Na}} m^3 h(V - E_{\text{Na}}). \tag{7.5.15}$$

*The Hodgkin–Huxley space-clamped-axon equations produce action potentials.*

In another series of experiments, Hodgkin and Huxley fixed electrodes along the entire length of the axon as before, but now the electrodes were used to measure the voltage as it varied during a depolarization event. These are called the *space clamp* experiments. In addition to its role as a variable conductor of electrically charged particles, the axon membrane also functions as the dielectric of an electrical capacitor in that charged particles accumulate on either side of it. The capacitance of the membrane was determined by Hodgkin and Huxley to be about 1 microfarad per square centimeter:

$$C_{\text{m}} = 1 \times 10^6 \text{ farad/cm}^2.$$

Electrically, the membrane may be depicted as in Figure 7.5.6. When space clamped, the sum of the membrane ionic currents serves to deposit charge on or remove charge from this membrane capacitor. Said differently, the effective current "through" the capacitor balances the membrane current made up of the sodium, potassium, and leakage components; the sum of these currents must be 0 by Kirchhoff's law,

$$i_C + i_{\text{Na}} + i_{\text{K}} + i_\ell = 0. \tag{7.5.16}$$

The effective current through a capacitor is given by

$$i_C = C_{\text{m}} \frac{dV}{dt},$$

where membrane capacitance $C_{\text{m}}$ is measured in farads per square centimeter, $\frac{dV}{dt}$ is in volts/second, and $i$ is in amperes per square centimeter. Substituting from (7.5.16) gives

$$C_{\rm m}\frac{dV}{dt} = -(i_{\rm Na} + i_{\rm K} + i_\ell). \tag{7.5.17}$$

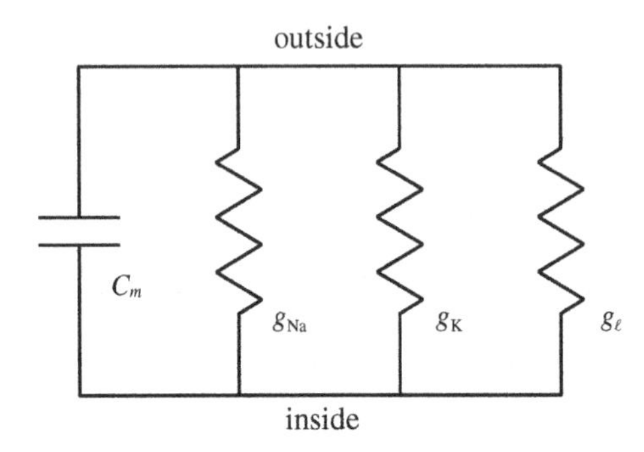


**Fig. 7.5.6.** Axon membrane circuit equivalent.

We now collect the various equations to give the Hodgkin–Huxley space-clamped equations. Substituting (7.5.5), (7.5.10), and (7.5.15) into (7.5.17) and recalling (7.5.7), (7.5.12), and (7.5.13) gives

$$\frac{dV}{dt} = \frac{-1}{C_{\rm m}}(\bar{g}_{\rm Na}m^3h(V - E_{\rm Na}) + \bar{g}_{\rm K}n^4(V - E_{\rm K}) + g_\ell(V - E_\ell)), \tag{7.5.18a}$$

$$\frac{dn}{dt} = \alpha_n(1 - n) - \beta_n n, \tag{7.5.18b}$$

$$\frac{dm}{dt} = \alpha_m(1 - m) - \beta_m m, \tag{7.5.18c}$$

$$\frac{dh}{dt} = \alpha_h(1 - h) - \beta_h h, \tag{7.5.18d}$$

where the alphas and betas are given functions of $V$ according to (7.5.8), (7.5.9), and (7.5.14).

The following code solves this system of differential equations and produces an action potential. The action potential is initiated by a pulse of current lasting 0.001 second. Figure 7.5.7 shows the result.

MAPLE

```
> Ena:=55: Ek:=-82: El:=-59: gkbar:=24.34: gnabar:=70.7:
> gl:=0.3: vrest:= -70: Cm:=0.001:
> alphan:=v->0.01*(10-(v-vrest))/(exp(0.1*(10-(v-vrest)))-1);
> betan:=v->0.125*exp(-(v-vrest)/80);
> alpham:=v->0.1*(25-(v-vrest))/(exp(0.1*(25-(v-vrest)))-1);
> betam:=v->4*exp(-(v-vrest)/18);
> alphah:=v->0.07*exp(-0.05*(v-vrest));
> betah:=v->1/(exp(0.1*(30-(v-vrest)))+1);
> pulse:=t->-20*(Heaviside(t-0.001)-Heaviside(t-0.002));
> rhsV:=(t,V,n,m,h)->-(gnabar*m^3*h*(V - Ena)+gkbar*n^4*(V - Ek)+gl*(V - El)+pulse(t))/Cm;
> rhsn:=(t,V,n,m,h)->1000*(alphan(V)*(1-n)-betan(V)*n);
> rhsm:=(t,V,n,m,h)->1000*(alpham(V)*(1-m)-betam(V)*m);
> rhsh:=(t,V,n,m,h)->1000*(alphah(V)*(1-h)-betah(V)*h);
> inits:=V(0)=vrest,n(0)=0.315,m(0)=0.042, h(0)=0.608;
```

```
> sol:=dsolve({diff(V(t),t)=rhsV(t,V(t),n(t),m(t),h(t)), diff(n(t),t)=rhsn(t,V(t),n(t),m(t),h(t)),
      diff(m(t),t)=rhsm(t,V(t),n(t),m(t),h(t)), diff(h(t),t)=rhsh(t,V(t),n(t),m(t),h(t)),inits},
      {V(t),n(t),m(t),h(t)},type=numeric, output=listprocedure);
> Vs:=subs(sol,V(t));
> plot(Vs,0..0.02,tickmarks=[4,7]);
```

```
MATLAB
% first create 12 m-files
%  function y=alphan(v); % (1) alphan.m
%  vrest=-69;
%  y=0.01*(10-(v-vrest))/(exp(0.1*(10-(v-vrest)))-1);
%  function y=betan(v); % (2) betan.m
%  vrest=-69; y=0.125*exp(-(v-vrest)/80);
%  function y=alpham(v); % (3) alpham.m
%  vrest=-69;
%  y=0.1*(25-(v-vrest))/(exp(0.1*(25-(v-vrest)))-1);
%  function y=betam(v); % (4) betam.m
%  vrest=-69; y=4*exp(-(v-vrest)/18);
%  function y=alphah(v); % (5) alphah.m
%  vrest=-69; y=0.07*exp(-0.05*(v-vrest));
%  function y=betah(v); % (6) betah.m
%  vrest=-69; y=1/(exp(0.1*(30-(v-vrest)))+1);
%  function y=pulse(t); % (7) pulse.m
%  if t<.001
%  y=0;
%  elseif t<.002
%  y=-20;
%  else
%  y=0;
%  end
%  function y=rhsV(t,V,n,m,h); % (8) rhsV.m
% Ena=55; Ek=-82; El= -59; gkbar=24.34; gnabar=70.7;
% gl=0.3; cm=0.001;
% y=-(gnabar*m^3*h*(V-Ena)+gkbar*n^4*(V-Ek)+gl*(V-El)+pulse(t))/cm;
%  function y=rhsn(t,V,n,m,h); % (9) rhsn.m
%  y=1000*(alphan(V)*(1-n) - betan(V)*n);
%  function y=rhsm(t,V,n,m,h); % (10) rhsm.m
%  y=1000*(alpham(V)*(1-m) - betam(V)*m);
%  function y=rhsh(t,V,n,m,h); % (11) rhsh.m
%  y=1000*(alphah(V)*(1-h) - betah(V)*h);
%  function Yprime=hhRHS(t,Y);% (12) hhRHS.m
% Yprime=[rhsV(t,Y(1),Y(2),Y(3),Y(4));rhsn(t,Y(1),Y(2),Y(3),Y(4));rhsm(t,Y(1),Y(2),Y(3),Y(4));...
      rhsh(t,Y(1),Y(2),Y(3),Y(4))];
%
> vrest=-69;
> [t,sol]=ode45('hhRHS', [0 .02], [vrest; 0.315; 0.042; 0.608]);
> Vs=sol(:,1);
> plot(t,Vs)
```

*The Hodgkin–Huxley propagation equations predict impulse speed.*

*In vivo*, an axon is not clamped. Consequently, instead of the entire axon undergoing an action potential at the same time, an action potential is localized and propagates along the axon in time. Thus voltage is a function of position, $x$, along the axon as well as a function of time. Consider a small section of axon lying between $x$ and $x + \Delta x$. The basic equation states that the current in at $x$ minus the current out at $x + \Delta x$ and minus the membrane current must equal the charge buildup on that section of membrane, that is, must equal the capacitance current. Hence

$$i(x) - i(x + \Delta x) - (i_{\text{Na}} + i_{\text{K}} + i_{\ell})2\pi a \Delta x = C_{\text{m}} 2\pi a \Delta x \frac{dV}{dt}.$$

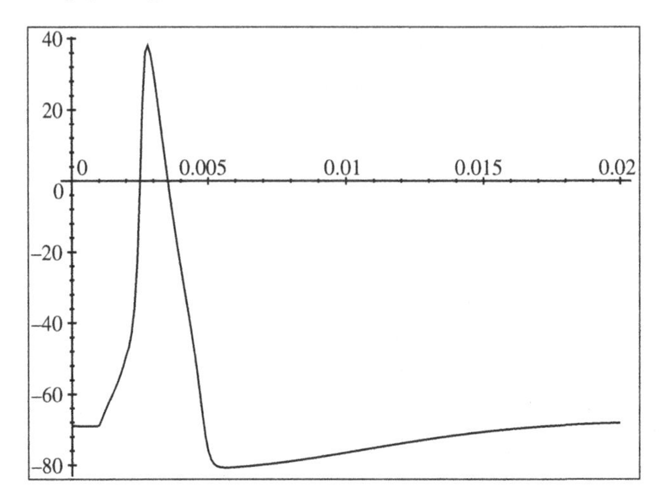


**Fig. 7.5.7.** A simulated action potential.

In this we have taken the radius of the axon to be $a$ and multiplied the per-square-centimeter quantities by the surface area of the section of axon in question. Divide by $\Delta x$, let $\Delta x \to 0$, and divide by $2\pi a$ to get

$$\frac{-1}{2\pi a}\frac{\partial i}{\partial x} - (i_{\mathrm{Na}} + i_{\mathrm{K}} + i_{\ell}) = C_{\mathrm{m}}\frac{dV}{dt}. \tag{7.5.19}$$

Next, suppose there is no membrane current. Then the voltage drop over the length of the axon is related to the current $i$ along the axon by Ohm's law, $V = iR$. In this $R$ is the total resistance of the axoplasm (not membrane resistance). But in fact, each section of axon of length $\Delta x$ contributes to the overall resistance in proportion to its length, namely, $\frac{R\Delta x}{L}$, where $L$ is the total length of the axon. Thus if position along the axon is denoted by $x$, then resistance as a function of $x$ increases linearly from 0 to $R$. In the meantime, voltage as a function of $x$ falls from $V$ to 0. In particular, Ohm's law as applied to the section $dx$ becomes

$$\frac{dV}{dx} = -i\frac{dR}{dx}, \tag{7.5.20}$$

where the negative sign indicates that $V$ is decreasing while $R$ is increasing.

In this $\frac{dR}{dx}$ is the resistance per unit length, which we take to depend only on the cross-sectional area of the axon. And this dependence is in inverse proportion; that is, if the area is halved, then the resistance is doubled. Hence, letting $\bar{R}$ denote the resistance per unit length per unit area, then

$$\frac{dR}{dx} = \frac{\bar{R}}{\pi a^2}, \tag{7.5.21}$$

where $a$ is the radius of the axon. Substituting (7.5.21) into (7.5.20) and solving for $i$ gives

$$i = -\frac{\pi a^2}{\bar{R}} \frac{dV}{dx}. \tag{7.5.22}$$

In this, the negative sign can be interpreted to mean that current moves down the voltage gradient.

Differentiate (7.5.22) with respect to $x$ and substitute (7.5.19) to get

$$\frac{a}{2\bar{R}} \frac{\partial^2 V}{\partial x^2} - (i_{\text{Na}} + i_{\text{K}} + i_{\ell}) = C_{\text{m}} \frac{dV}{dt}. \tag{7.5.23}$$

This equation, along with the equations for $n$, $m$, and $h$ corresponding to the membrane ion currents, (7.5.7), (7.5.12), and (7.5.13), constitute the Hodgkin–Huxley propagation equations. Hodgkin and Huxley numerically obtain from these equations the value 18.8 meters/second for the propagation velocity $c$ of an action potential. It is known to be 21 meters/second, and so the agreement is quite good.

We will not solve this partial differential equation system here, but instead we will show that the propagation velocity is proportional to the square root of the radius $a$ and inversely proportional to the axon resistance $\bar{R}$. Hodgkin and Huxley note that if the action potential propagates along the axon unchanged in shape, then its shape as a function of $x$ for fixed $t$ is equivalent to its shape as a function of $t$ for fixed $x$. This is formalized by the *wave equation* [11]

$$\frac{\partial^2 V}{\partial x^2} = \frac{1}{c^2} \frac{\partial^2 V}{\partial t^2}. \tag{7.5.24}$$

In this $c$ is the propagation velocity. Substituting the second derivative with respect to $x$ from (7.5.24) into (7.5.23) gives

$$\frac{a}{2\bar{R}c^2} \frac{d^2 V}{dt^2} - (i_{\text{Na}} + i_{\text{K}} + i_{\ell}) = C_{\text{m}} \frac{dV}{dt}. \tag{7.5.25}$$

In this equation, the only dependence on $a$ occurs in the first term. Since the other terms do not depend on $a$, the first term must be independent of $a$ as well. This can happen only if the coefficient is constant with respect to $a$,

$$\frac{a}{2\bar{R}c^2} = \text{constant}.$$

But then it follows that

$$c = (\text{constant})\sqrt{\frac{a}{\bar{R}}}.$$

Thus the propagation velocity is proportional to the square root of the axon radius and inversely proportional to the square root of axon resistance. The large size of the squid's axon equips it for fast responses, an important factor in eluding enemies.

## 7.6 Fitzhugh–Nagumo Two-Variable Action Potential System

*The Fitzhugh–Nagumo two-variable model behaves qualitatively like the Hodgkin–Huxley space-clamped system. But being simpler by two variables, action potentials and other properties of the Hodgkin–Huxley system may be visualized as phase-plane plots.*

*The Fitzhugh–Nagumo phase-plane analysis demonstrates all-or-nothing response.*

Fitzhugh [12] and Nagumo [13] proposed and analyzed the following system of two differential equations, which behaves qualitatively like the Hodgkin–Huxley space-clamped system:

$$\begin{aligned}\frac{dV}{dt} &= V - \frac{1}{3}V^3 - w,\\ \frac{dw}{dt} &= c(V + a - bw).\end{aligned} \tag{7.6.1}$$

In this $V$ plays the role of membrane potential, but $w$ is a general "refractory" variable not representing any specific Hodgkin–Huxley variable. The parameters $a$, $b$, and $c$ are the constants

$$a = 0.7, \qquad b = 0.8, \qquad c = 0.08.$$

The virtue of this system is in elucidating the regions of physiological behavior of membrane response.

The phase plane is the coordinate plane of the two dependent variables $V$ and $w$. A curve $V = V(t)$ and $w = w(t)$ in the phase plane corresponding to a solution of the differential equation system for given initial values $V_0 = V(0)$, $w_0 = w(0)$ is called a *trajectory*.

Two special curves in the phase plane are the *isoclines*. These are the curves for which either $\frac{dV}{dt}$ or $\frac{dw}{dt}$ is zero. The $w$ isocline

$$V + a - bw = 0$$

is a straight line with slope 1.25 and intercept 0.875. To the left of it, $\frac{dw}{dt} < 0$, and to the right, $\frac{dw}{dt} > 0$. The $V$ isocline

$$V - \frac{1}{3}V^3 - w = 0, \quad \text{or} \quad w = V\left(1 - \frac{1}{3}V^2\right),$$

is a cubic with roots 0 and $\pm\sqrt{3}$. Above it $\frac{dV}{dt} < 0$, and below it $\frac{dV}{dt} > 0$. The isoclines divide the phase plane into four regions, or quadrants. In the first, above the cubic and to the right of the line, $\frac{dV}{dt} < 0$ and $\frac{dw}{dt} > 0$. Hence a trajectory in this quadrant will tend toward decreasing $V$ and increasing $w$, upward and leftward. In quadrant 2, above the cubic and to the left of the line, trajectories tend downward and leftward. In quadrant 3, below the cubic and to the left of the line, trajectories tend downward and rightward. In quadrant 4, below the cubic and to the right of the line, the derivatives are $\frac{dV}{dt} > 0$ and $\frac{dw}{dt} > 0$, so trajectories tend upward and rightward. The isoclines and quadrants are shown in Figure 7.6.1.

The isoclines intersect at $V_e = -1.1994$ and $w_e = -0.6243$. At this point $\frac{dV}{dt} = 0$ and $\frac{dw}{dt} = 0$, and so a trajectory at that point does not move at all; it is a stationary point (see Section 2.4). For this particular system of differential equations, it can be shown that trajectories starting anywhere eventually lead to this stationary point. As

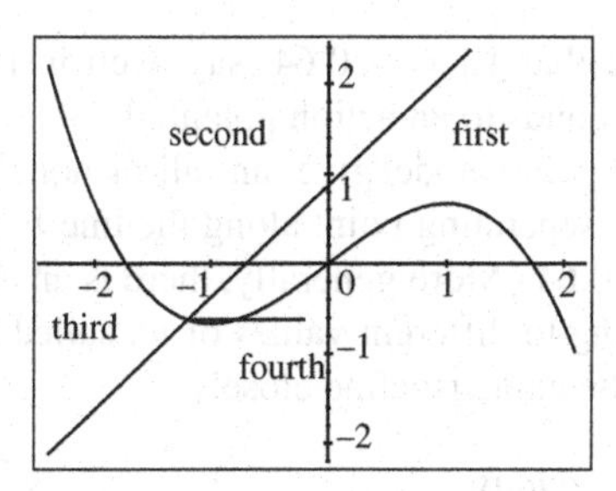


**Fig. 7.6.1.** Isoclines and "quadrants."

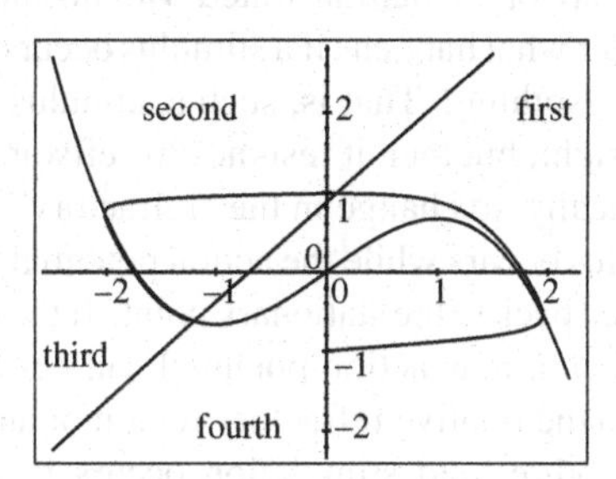


**Fig. 7.6.2.** Fitzhugh–Nagumo trajectory.

a result, it is known as a globally attracting stable stationary. It plays the role of the resting state in our description of an axon.

Consider the progress of a trajectory that is begun at the point $(V_0, w_0)$ located to the right of the ascending portion of the cubic isocline. Since this is in quadrant 4, the trajectory will tend rightward until it crosses the descending section of the same isocline. From there the trajectory will tend upward and leftward until it crosses the $w$ isocline. Proceeding from there leftward and downward, it next crosses the $V$ isocline again. Finally, it proceeds downward and rightward, ending up at the equilibrium point. See Figure 7.6.2.

Fix attention on the behavior of $V$ along this trajectory. It first increases in value until it reaches the descending branch of the cubic isocline. This will be its maximum value. Crossing this isocline, $V$ then decreases, eventually below the stationary point. Completing the trajectory, $V$ increases slowly until it is stationary. But this describes the behavior of the membrane potential, $V$, during an action potential.

Next, suppose a trajectory is begun immediately to the right of the stationary point. Then a very different thing happens. Assume that the starting point $(V_0, w_0)$ is taken so that $w_0 = w_e$ and $V_0 = V_e + \Delta V$. This starting point lies on the horizontal line through the stationary point as shown in Figure 7.6.1, which, in turn, intersects the cubic isocline at $-0.786$ and $1.98$ besides $-1.994$. Therefore, depending on the size of $\Delta V$, the starting point falls in either quadrant 1 or 4.

If it is in quadrant 1, then the response trajectory returns more or less directly to stationary. This is analogous to a subthreshold stimulation in an axon.

But if $\Delta V$ is so large that $V_0 = -0.64$, say, well inside quadrant 4, then the response trajectory corresponds to an action potential.

Thus this Fitzhugh–Nagumo model gives an "all-or-none" response to a stimulus, just as an axon does. The separating point along the line $w = -0.624$ lies between $V_0 = -0.65$ and $V_0 = -0.64$. More generally, there is an entire separating "all-or-none" curve, corresponding to different values of $w$, called a *separatrix*. It follows the ascending branch of the cubic isocline closely.

*Refractory and enhanced regions.*

Other features of axon behavior are demonstrated with this model as well. During an "action potential," consider what happens if a stimulus occurs while the trajectory lies above the cubic isocline. Nothing! That is, such a stimulus causes the trajectory to jump horizontally to the right, but then it resumes its leftward horizontal movement. In particular, there is basically no change in the "refractory" variable $w$.

Now suppose a stimulus occurs while the action potential trajectory is descending in the third quadrant headed back to the stationary point. If the stimulus is large enough to cross the separatrix, then a new action potential can be initiated. Consequently, this region corresponds to the relative refractory region of an axon's behavior.

Finally, suppose a subthreshold stimulation occurs from the stationary point. There is no action potential, but a second subthreshold stimulation might be sufficient to cross the all-or-none separatrix and initiate an action potential. Therefore the region between the stationary point and the separatrix corresponds to the enhanced state of an axon.

## Exercises/Experiments

The purpose of these experiments is to see the effect of varying some of the parameters in the Hodgkin–Huxley action potential model (use the code on pp. 220–221) and to gauge the sensitivity of these parameters.

In preparing a report on the following experiment, one should discuss the observations of the experiments. Also describe how an action potential changed with respect to the modification of each parameter and to what degree. In the sensitivity experiments, gauge the effect on the action potential of a 5% and a 10% change from the nominal value, both up and down, of the parameter. Please submit supporting graphs.

1. Run the code as is and note the maximum value of the response. Gradually lower pulse strength (the 20 in the pulse line) in several steps. At what strength is there no action potential?
2. Experiment with increasing the value of membrane capacitance $C_m$. Note the sensitivity of the response with respect to this parameter. What, physiologically, might cause the membrane capacitance to change in a living system?
3. Experiment with the leakage conductance $g_\ell$. This is principally the conductance of chloride ions. How sensitive is the response to this parameter?

4. Experiment with the sodium conductance $\bar{g}_{Na}$. This is the base level of conductance of sodium ions. How sensitive is the response to this parameter?

**Questions for Thought and Discussion**

1. Discuss the roles of voltage-gated channels and diffusion processes in the transmission of information across neuronal synapses.
2. Starting with the number 2, number the following events in the order in which they occur. ("Site A" is an arbitrary midaxonal location.)

| | |
|---|---|
| neuronal membrane is depolarized at site A by external stimulus | 1 |
| acetylcholine esterase breaks down neurotransmitter | ______ |
| $K^+$ channels open at site A | ______ |
| postsynaptic chemical-gated channels open | ______ |
| $Na^+$/K pump restores resting potential at site A | ______ |
| interior of neuron at site A is at positive potential with respect to exterior | ______ |

3. In what ways is the transmission of information by an action potential different from the transmission of electrical information by a copper wire?

## References and Suggested Further Reading


[1] NEURONAL BIOLOGY AND PHYSIOLOGY:
E. N. Marieb, *Human Anatomy and Physiology*, Benjamin/Cummings Publishing, Redwood City, CA, 1992.

[2] NEURONAL BIOLOGY:
P. H. Raven and G. B. Johnson, *Biology*, Mosby/Year Book, 3rd ed., St. Louis, 1992.

[3] NEURONAL BIOPHYSICS:
F. R. Hallett, P. A. Speight, and R. H. Stinson, *Introductory Biophysics*, Halstead/Wiley, New York, 1977.

[4] MOLECULAR BIOLOGY OF NEURONS:
B. Alberts, D. Bray, J. Lewis, M. Raff, K. Roberts, and J. D. Watson, *Molecular Biology of the Cell*, 3rd ed., Garland, New York, 1994.

[5] HODGKIN–HUXLEY EXPERIMENTS:
J. D. Murray, *Mathematical Biology*, Springer-Verlag, New York, 1989.

[6] HODGKIN–HUXLEY EXPERIMENTS:
A. L. Hodgkin and A. F. Huxley, Currents carried by sodium and potassium ions through the membrane of the giant axon of *Logio*, *J. Physiol.*, **116** (1952), 449–472.

[7] HODGKIN–HUXLEY EXPERIMENTS:
A. L. Hodgkin and A. F. Huxley, Components of membrane conductance in the giant axon of *Logio*, *J. Physiol.* **116** (1952), 473–496.

[8] HODGKIN–HUXLEY EXPERIMENTS:
A. L. Hodgkin and A. F. Huxley, The dual effect of membrane potential on sodium conductance in the giant axon of *Logio*, *J. Physiol.*, **116** (1952), 497–506.

[9] HODGKIN–HUXLEY EXPERIMENTS:
A. L. Hodgkin and A. F. Huxley, A quantitative description of membrane current and its application to conduction and excitation in nerve, *J. Physiol.*, **117** (1952), 500–544.

[10] HODGKIN–HUXLEY EXPERIMENTS:
A. L. Hodgkin, A. F. Huxley, and B. Katz, Measurement of current-voltage relations in the membrane of the giant axon of *Logio*, *J. Physiol.*, **116** (1952), 424–448.

[11] HODGKIN–HUXLEY EXPERIMENTS:
R. V. Churchill and J. W. Brown, *Fourier Series and Boundary Value Problems*, 4th ed., McGraw–Hill, New York, 1987.

[12] FITZHUGH–NAGUMO EQUATIONS:
R. Fitzhugh, Impulses and physiological states in theoretical models of nerve membrane, *Biophys. J.*, **1** (1961), 445–466.

[13] FITZHUGH–NAGUMO EQUATIONS:
J. S. Nagumo, S. Arimoto, and S. Yoshizawa, An active pulse transmission line simulating nerve axon, *Proc. Inst. Radio Engineers*, **20** (1962), 2061–2071.

# 8

# The Biochemistry of Cells

## Introduction

The purpose of this chapter is to present the structure of some of the molecules that make up a cell and to show how they are constructed under the supervision of hereditary elements of the cell. This will lead the way to a mathematical description of biological catalysis at the end of this chapter and is a necessary prelude to the discussion of the human immunodeficiency virus in Chapter 10. As a result, this chapter contains a lot of biological information.

We will see that biological molecules can be created outside of a cellular environment, but only very inefficiently. Inside a cell, however, the information for biomolecules is encoded in the genetic material called nucleic acid. Thus we will establish a direct relationship between the chemicals that constitute a cell and the cell's hereditary information.

The topical material of this chapter is organized along the lines of small to large. We begin by presenting a description of the atoms found in cells and then show how they are assembled into small organic molecules. Some of these small molecules can then be polymerized into large biochemical molecules, the biggest of which have molecular weights on the order of billions. These assembly processes are mediated by certain macromolecules which are themselves molecular polymers and whose own assembly has been mediated by similar molecular polymers. Thus we develop a key process in biology—self-replication.

## 8.1 Atoms and Bonds in Biochemistry

*Most of the atoms found in a cell are of common varieties: hydrogen, carbon, nitrogen and oxygen. They are, in fact, major components of air and dirt. What is it then that makes them so fundamental to life? To answer this question we must examine the ways that these atoms form bonds to one another—because it is through molecular organization that we will characterize living systems.*

*A living system is a highly organized array of atoms, attached to one another by chemical bonds. The bonds may be strong, requiring considerable energy for their*

R.W. Shonkwiler and J. Herod, *Mathematical Biology: An Introduction with Maple and Matlab*, Undergraduate Texts in Mathematics, DOI: 10.1007/978-0-387-70984-0_8,

*rearrangement. This leads to structures that are somewhat permanent and which can be changed only under special biochemical conditions. These bonds are said to be covalent, and they result from a process of "electron sharing." Carbon, nitrogen, and oxygen atoms can form a practically unlimited array, held together by covalent bonds.*

*Alternatively, some chemical bonds are weak, the heat energy available at room temperature being sufficient to break them. Because of their weakness, the structures they form are highly variable, leading to material movement and regional uniformity (among other things). The most important weak bond is called a hydrogen bond: it is the electrical attraction between a hydrogen nucleus on one molecule and an asymmetrically oriented electron on nitrogen or oxygen atoms of the same molecule or another one.*

*Organization is the key to living systems.*

In Section 3.3, we pointed out that the individual processes found in living systems are also found in nonbiological situations. We emphasized that the "signature" of life was the organization, or integration, of those processes into a unified system. We now extend that concept to physical organization at the atomic and molecular levels.

Calcium, phosphorus, potassium, sulfur, sodium, and chlorine account for about 3.9% of the atoms in our bodies.[1] Just four other elements make up the other 96%; they are hydrogen, carbon, nitrogen, and oxygen. These four elements most abundant in our bodies are also found in the air and earth around us—as $H_2O$, $CO_2$, $N_2$, $O_2$, and $H_2$. Thus if we want to explain why something has the special quality we call "life," it does not seem very fruitful to look for exotic ingredients; they aren't there. Where else might the explanation be?

An important clue can be found in experiments in which living systems are frozen to within a few degrees of 0°K, so that molecular motion is virtually halted. Upon reheating, these living systems resume life processes. The only properties conserved in such an experiment are static structural ones. We can conclude that a critical property of life lies in the special ways that the constituent atoms of living systems are organized into larger structures. We should therefore suspect that the atoms most commonly found in our bodies have special bonding properties, such that they can combine with one another in many ways. This is indeed the case: carbon, nitrogen, oxygen, and hydrogen are capable of a virtually infinite number of different molecular arrangements. In fact, it has been estimated that the number of ways that the atoms C, H, O, N, P, and S can be combined to make low-molecular-weight compounds ($MW < 500$) is in the billions [1]!

Of the large number of possible arrangements of C, N, O, and H, the forces of evolution have selected a small subset, perhaps a thousand or so, on which to base life. Members of this basic group have then been combined into a vast array of biomacromolecules. For example, the number of atoms in a typical biomacromolecule might range from a few dozen up to millions, but those with more than a few hundred atoms are always polymers of simpler subunits.

[1] About 15 more elements are present in trace amounts.

Living systems are assemblages of common atoms, each part of the system having a very specific organization at all size levels. In other words, all living things can be thought of as regions of great orderliness, or organization. Death is marked by the disruption of this organization—either suddenly, as in the case of an accident, or slowly, as in the case of degenerative disease. In any case, death is followed by decompositional processes that convert the body to gases, which are very disorganized.

Physicists use *entropy* as a measure of disorder; there is an important empirical rule, the *second law of thermodynamics*, which states that entropy in the universe increases in the course of every process. Living systems obey this rule, as they do all other natural chemical and physical principles. As an organism grows, it assembles atoms into an orderly, low-entropy arrangement; at the same time, the entropy of the organism's surroundings increases by even more, to make the net entropic change in the universe positive. This net increase is to be found in such effects as the motion of air molecules induced by the organism's body heat, in the gases it exhales, and in the natural waste products it creates.

Nature is full of good examples of the critical role played by organization in living systems. Consider that a bullfighter's sword can kill a 600-pound bull and that 0.01 micrograms of the neurotoxin *tetrodotoxin* from a puffer fish can kill a mouse. The catastrophic effects of the sword and the toxin seem out of proportion to their masses. In light of the discussion above, however, we now understand that their effects are not based on mass at all, but instead on the disruption of critically-organized structures, e.g., the nervous system [2].

*Covalent bonds are strong interactions involving electron sharing.*

A very strong attraction between two atoms results from a phenomenon called "*electron sharing*"; it is responsible for binding atoms into biochemical molecules. One electron from each of two atoms becomes somewhat localized on a line between the two nuclei. The two nuclei are electrostatically attracted to the electrons and therefore remain close to one another.

Figure 8.1.1 shows simple planetary models of two hydrogen atoms. (Later we will generalize our discussion to other atoms.) The radius of this orbit is about 0.05 nm, and so the nuclei are about 0.1 nm apart. At some time each of the electrons will find itself at a point immediately between the two nuclei. When this happens, each of the two nuclei will exert the same electrical attraction on the electron, meaning that the electron can no longer be associated with a particular nucleus. There being no reason to "choose" either the right or the left nucleus, the electron will spend more time directly in between the two than in any other location.[2] The two electrons in the center then act like a kind of glue, attracting the nuclei to themselves and thus

[2] The idea that an electron is more likely to be found in one region of space than in another is built into the quantum-mechanical formulation, which is outside the scope of this book. In the quantum-mechanical formulation, there are no orbits and the electron is represented as a probability cloud. The denser the cloud, the greater the probability of finding the electron there.

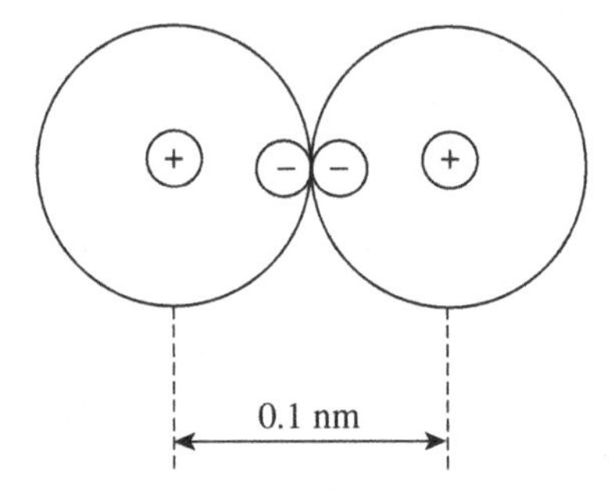


**Fig. 8.1.1.** A model of the hydrogen molecule. Planetary orbits are shown, but the electrons are equally attracted to both nuclei and therefore spend most of their time in the region directly between the two nuclei. This interaction is called a covalent bond.

toward each other. A stable molecule is thereby formed; the attraction between its constituent atoms is called a *covalent bond*.

A covalent bond always contains two electrons because of an unusual electronic property: An electron spins on its own axis. For quantum-mechanical reasons, an electron always pairs up with another electron having the opposite spin direction, leading to "*spin pairing*" in covalent bonds and in certain other situations. An atom or molecule with an odd number of electrons is called a *radical*; it is unstable, quickly pairing up with another radical via a covalent bond. For example, atomic hydrogen has a very transitory existence, quickly forming the diatomic hydrogen molecule $H_2$, in which the electrons' spins are paired. Thus electrons in stable chemicals appear in pairs. For a further discussion of this topic, see Yeargers [3].

Covalent bonds are very stable. In order to break one, i.e., to dissociate a biomolecule, it would require at least four electron volts of energy. For comparison, that much energy is contained by quanta in the ultraviolet region of the electromagnetic spectrum and exceeds that of the visible region of the solar spectrum. In passing, this helps us to understand why sunlight is carcinogenic—its ultraviolet component alters the chemistry of chemical components of our skin. If not for the fact that most of the sun's ultraviolet radiation is filtered out by the earth's atmosphere, life on earth would have to be chemically quite different from what it is.

Each kind of atom forms a fixed number of covalent bonds to its atomic neighbors; this number is called the *valence*. Table 8.1.1 gives the atomic numbers and valences of hydrogen, carbon, nitrogen, and oxygen.

**Table 8.1.1.**

| Atom | Symbol | Atomic number | Valence |
|---|---|---|---|
| Hydrogen | H | 1 | 1 |
| Carbon | C | 6 | 4 |
| Nitrogen | N | 7 | 3 |
| Oxygen | O | 8 | 2 |

Figure 8.1.2 shows the structures and names of several common organic molecules. The bond angles are shown; they may fluctuate by several degrees depending on the exact composition of the molecule. In each case the length of the bond is about 0.1 nm, again depending on the constituent atoms. Note also that double bonds are possible, but only if they are consistent with the valences given above.

You can see from Figure 8.1.2 that there are only two basic bonding schemes: If the molecule has only single bonds, the bond angles are 109°, and if there is a double bond the bond angles are 120°. Note that the former leads to a three-dimensional shape and the latter to a planar shape. This should become evident if you compare the structures of ethane and ethene.

*Hydrogen bonds are weak interactions.*

Figure 8.1.3 shows some more molecular models, containing oxygen and nitrogen. These molecules are electrically neutral: Unless ionized, they will not migrate toward either pole of a battery. Unlike hydrocarbons, however, their charges are not uniformly distributed. In fact, nitrogen and oxygen atoms in molecules have pairs of electrons (called lone pairs) that are arranged in a highly asymmetrical way about the nucleus. Figure 8.1.3 shows the asymmetrically oriented electrons of nitrogen and oxygen. There are three important points to be noted about these pictures: First, the reason that lone pair electrons are "paired" is that they have opposite spin directions from one another, as was described earlier. Second, the angles with which the lone pairs project outward are consistent with the 109° or 120° bond angles described earlier. Third, it must be emphasized that these molecules are electrically neutral—their charges are not uniformly distributed in space, but they total up to exactly zero for each complete molecule. The presence of lone pairs has important structural consequences to molecules that contain nitrogen or oxygen. Consider the water molecule shown in Figure 8.1.3. Two lone pairs extend toward the right and bottom of the picture, meaning that the right and lower ends of the molecule are negative. The entire molecule is neutral, and therefore the left and upper ends must be positive. We associate the negative charge with the lone pairs and the positive charge with the nuclei of the hydrogen atoms at the other end. Such a molecule is said to be *dipolar*.

Dipolar molecules can electrically attract one another, the negative end of one attracting the positive end of the other. In fact, a dipolar molecule might enter into several such interactions, called *hydrogen bonds* (H-bonds). Figure 8.1.4 shows the H-bonds in which a water molecule might participate. Note carefully that the ensemble of five water molecules is not planar.

Hydrogen bonds are not very strong, at least compared to covalent bonds. They can be broken by energies on the order of 0.1 eV, an energy that is thermally available at room temperature. There are two mitigating factors, however, that make H-bonds very important in spite of the ease with which they can be broken. The first is their sheer numbers. Nitrogen and oxygen are very common atoms in living systems, as mentioned earlier, and they can enter into H-bonding with neighboring, complementary H-bonding groups. While each H-bond is weak, there are so many of them that they can give considerable stability to systems in which they occur.

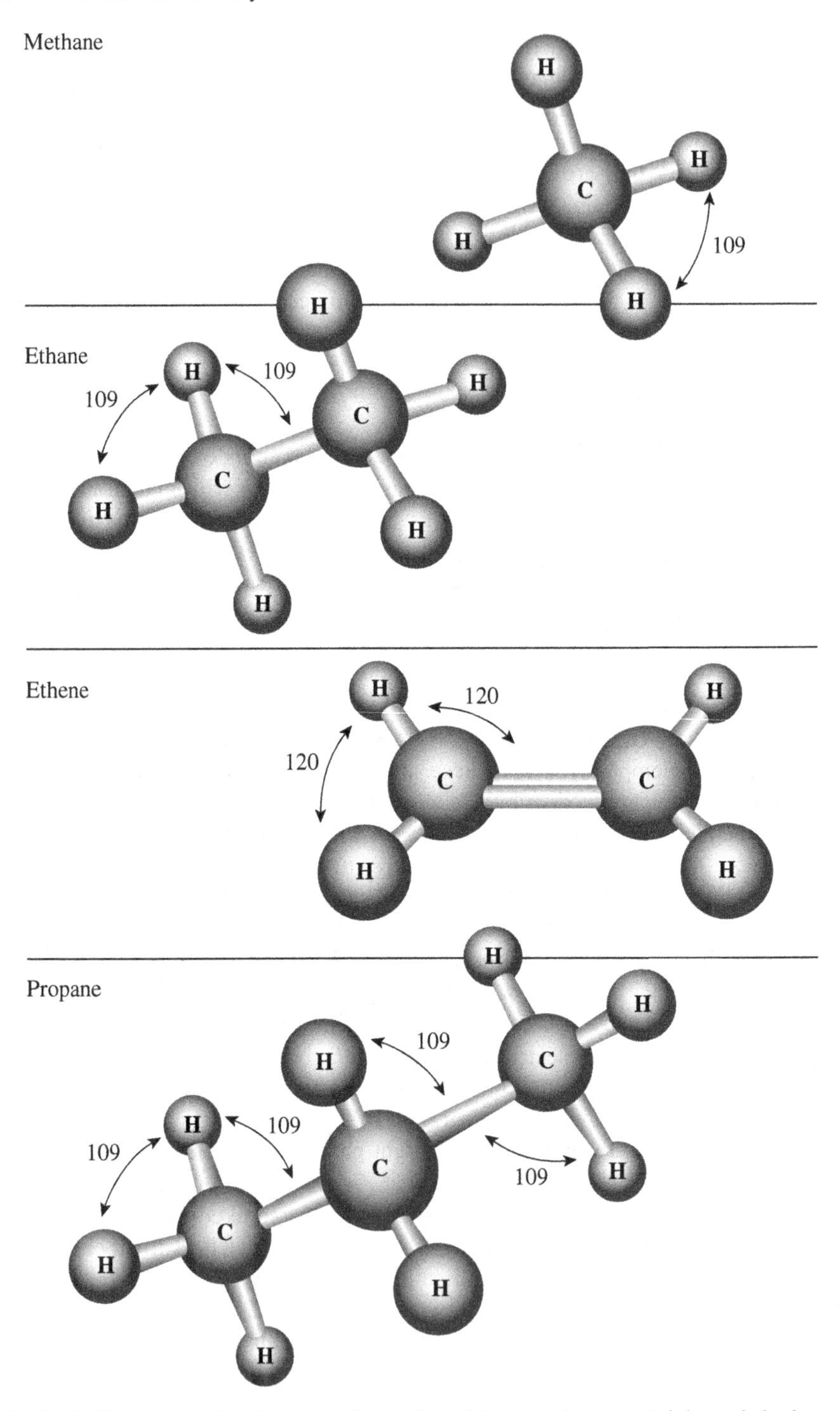


**Fig. 8.1.2.** Three-dimensional models of several small hydrocarbons (containing only hydrogen and carbon atoms). Bond angles are shown.

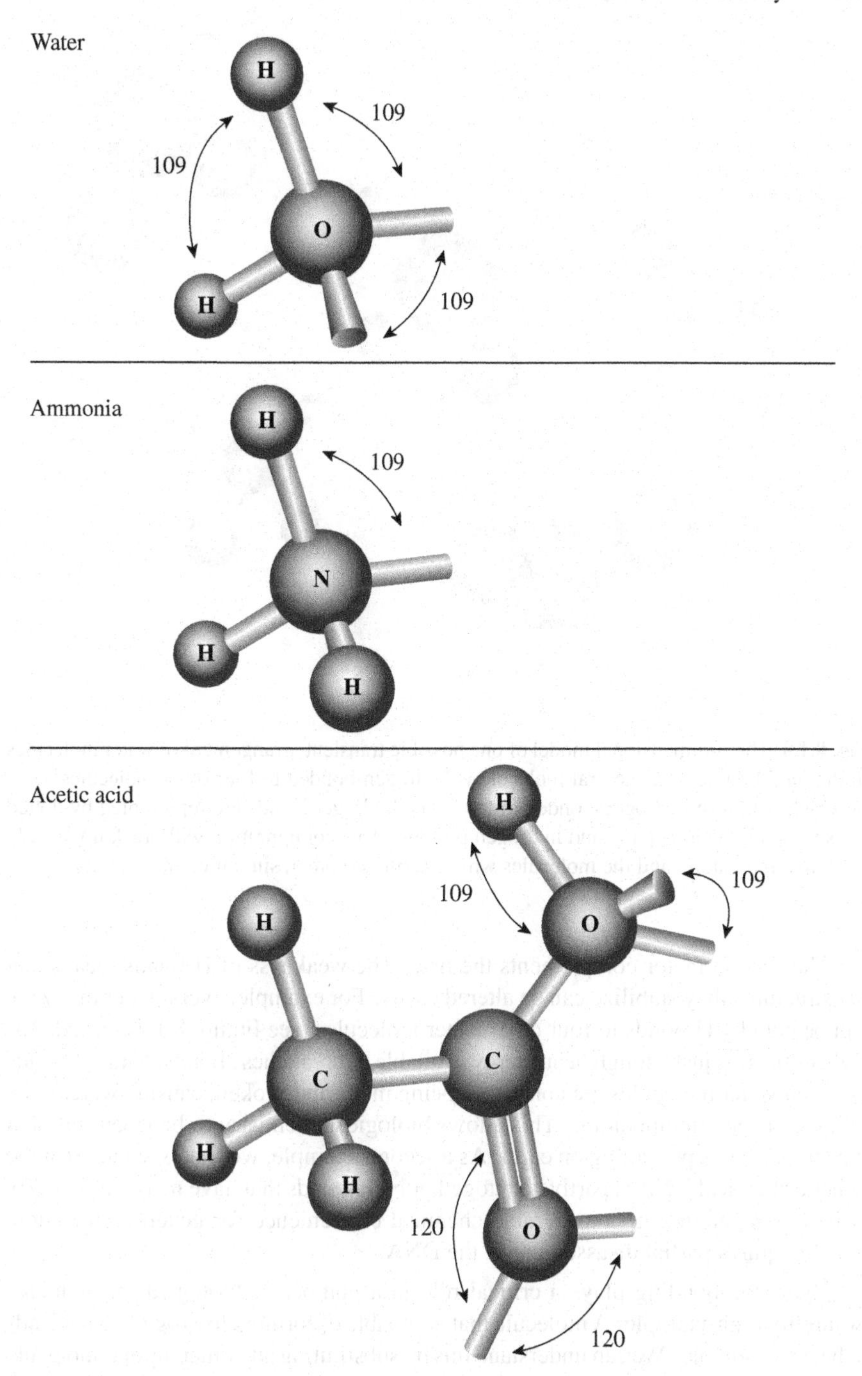


**Fig. 8.1.3.** Three-dimensional models of molecules containing oxygen and nitrogen. The stubs originating on the oxygen and the nitrogen atoms, but not connected to any other atom, represent lone pairs, or asymmetrically oriented electrons.

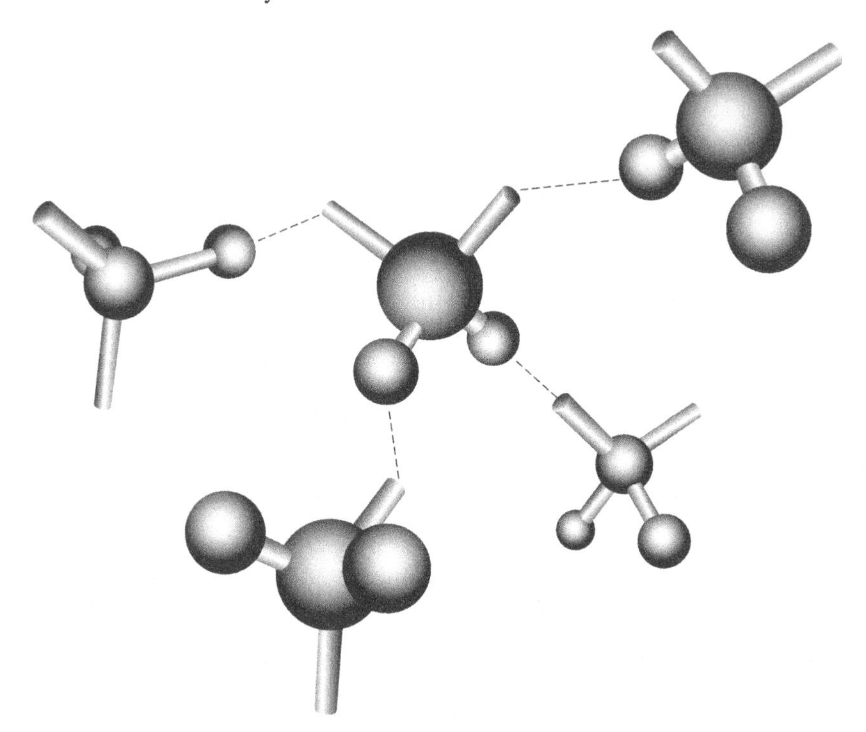

**Fig. 8.1.4.** Three-dimensional model of one possible transient arrangement of water molecules in the liquid phase. The central molecule is hydrogen-bonded to four other molecules, each of which is in turn hydrogen-bonded to four. The hydrogen bonds are represented by dotted lines between the lone pairs and hydrogen protons. This configuration will break up quickly at room temperature, and the molecules will re-form into other, similar configurations.

The second factor complements the first: The weakness of H-bonds means that the structures they stabilize can be altered easily. For example, every water molecule can be held by H-bonds to four other water molecules (see Figure 8.1.4). At 20–30° Celsius there is just enough heat energy available to break these bonds. Thus H-bonds between water molecules are constantly being made and broken, causing water to be a liquid at room temperature. This allows biological chemistry to be water-based at the temperatures prevailing on earth. As a second example, we shall see later that the genetic chemical DNA is partly held together by H-bonds that have marginal stability at body temperature, a considerable chemical convenience for genetic replication, which requires partial disassembly of the DNA.

Hydrogen-bonding plays a critical role in a number of biological phenomena. Solubility is an example: A molecule that is capable of forming hydrogen bonds tends to be water soluble. We can understand this by substituting any other dipolar molecule (containing one or more lone pairs) for the central water molecule of Figure 8.1.4. On the other hand, a molecule lacking lone pair electrons is not water soluble. Look at the propane molecule in Figure 8.1.2 and note that such *hydrocarbons* lack the ability to dissolve in water because they lack the necessary asymmetrical charges needed for H-bonding. We shall return to the topic of H-bonding when nucleic acids and heat storage are discussed later in this chapter. These topics, as well as other kinds of chemical bonding interactions, are discussed by Yeargers [3].

## 8.2 Biopolymers

*At the beginning of this chapter, it was pointed out that the attribute we call "life" is due to the organization, not the rarity, of constituent atoms. Figures* 8.1.2 *and* 8.1.3 *showed that sequences of carbon, oxygen, and nitrogen atoms, with their many bends and branches, can potentially combine to form elaborate three-dimensional macromolecules. What happens is that atoms combine to form molecular monomers having molecular weights on the order of a few hundred. In turn, these monomers are chained into linear or branched macromolecular polymers having molecular weights of up to a billion. The ability to create, organize, and maintain these giant molecules is what distinguishes living things from nonliving things.*

*Polysaccharides are polymers of sugars.*

A typical sugar is *glucose*, shown in Figure 8.2.1(a). The chemical characteristics that make glucose a sugar are the straight chain of carbons, the multiple −OH groups, and the double-bonded oxygen. Most of the other sugars we eat are converted to glucose, and the energy is then extracted via the conversion of glucose to carbon dioxide. This process is called *respiration*; it will be described below. A more common configuration for a sugar is exemplified by the ring configuration of glucose, shown in Figure 8.2.1(b).

The polymerization of two glucose molecules is a *condensation* reaction, shown in Figure 8.2.2. Its reverse is *hydrolysis*. We can extend the notion of sugar polymerization into long linear or branched chains, as shown by the arrows in Figure 8.2.2. The actual function of a polysaccharide, also called a *carbohydrate*, will depend on the sequence of component sugars, their orientations with respect to each other, and whether the chains are branched.

Polysaccharides serve numerous biological roles. For example, plants store excess glucose as *starch*, a polysaccharide found in seeds such as rice and wheat (flour is mostly starch). The structural matter of plants is mainly *cellulose*; it comprises most of what we call wood. When an animal accumulates too much glucose, it is polymerized into *glycogen* for storage in the muscles and liver. When we need glucose for energy, glycogen is hydrolyzed back to monomers. These and other functions of sugars will be discussed later in this and subsequent chapters.

*Lipids are polymers of fatty acids and glycerol.*

*Fatty acids*, exemplified in Figure 8.2.3, are distinguished from each other by their lengths and the positions of their double bonds. Note the organic acid group (-COOH) at one end. Fatty acids with double bonds are said to be *unsaturated*; *polyunsaturated*

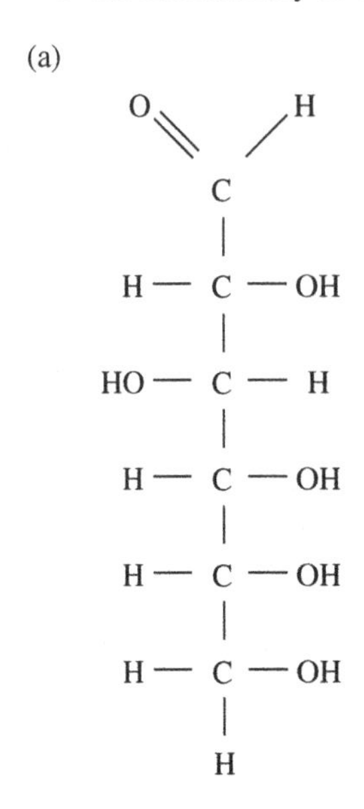


**Fig. 8.2.1.** (a) A model of the linear form of the glucose molecule. (b) A model of the ring form of the glucose molecule. The right-hand version, which omits many of the identifying symbols, is the more common representation.

fatty acids are common in plants, whereas saturated fatty acids, lacking double bonds, are common in animals. *Glycerol* and three fatty acids combine to form a *lipid*, or *fat*, or *triglyceride*, as pictured in Figure 8.2.4. The reverse reaction is again called hydrolysis.

Lipids are efficient at storing the energy of excess food that we eat; a gram of lipid yields about four times the calories of other foods, e.g., carbohydrates and proteins. Lipids are fundamental components of cell membranes: A common lipid of cell membranes is a phospholipid, pictured in Figure 8.2.5. You should now be able to put Figure 8.2.5 into the context of Figure 6.1.1. Note how the hydrocarbon regions of

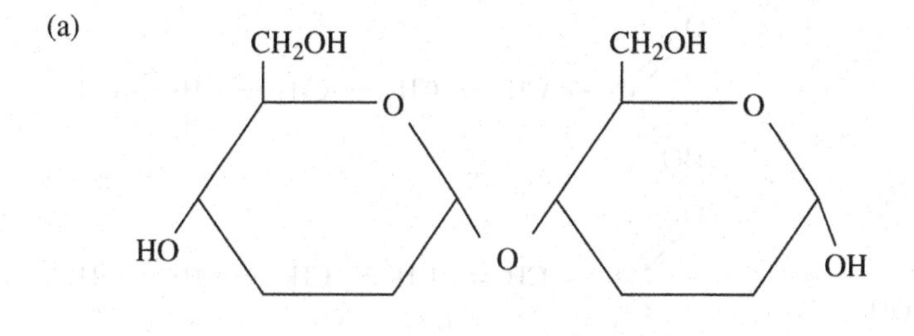




**Fig. 8.2.2.** (a) A model of a disaccharide, consisting of two glucose molecules. (b) A model showing the three possible directions that the polysaccharide of (a) could be extended. A large polysaccharide, with many such branches, would be very complex.

the phospholipid are in the interior of the membrane and how the hydrophilic oxygen groups (having lone pair electrons) are on the membrane's exterior, where they can hydrogen-bond to the surrounding water.

*Nucleic acids are polymers of nucleotides.*

Nucleic acids contain the information necessary for the control of a cell's chemistry. This information is encoded into the sequence of monomeric units of the nucleic acid, called *nucleotides*, and is expressed as chemical control through a series of processes called the *central dogma of genetics*—to be described below. When a cell reproduces asexually, its nucleic acids are simply duplicated and the resultant molecules are partitioned equally among the subsequent daughter cells, thus ensuring

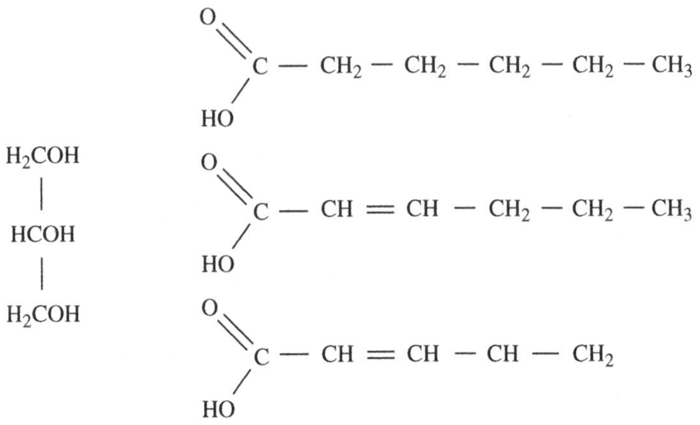

$H_2COH$ — $HCOH$ — $H_2COH$

$HO-C(=O)-CH_2-CH_2-CH_2-CH_2-CH_3$

$HO-C(=O)-CH=CH-CH_2-CH_2-CH_3$

$HO-C(=O)-CH=CH-CH-CH_2$

**Fig. 8.2.3.** A model of a glycerol molecule (left) and three arbitrary fatty acids (right).

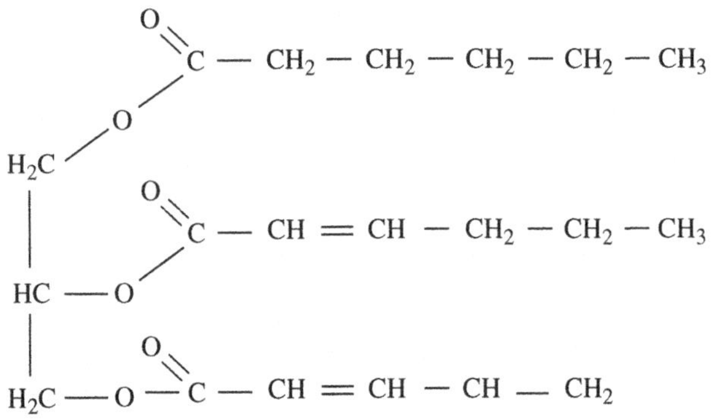

$H_2C-O-C(=O)-CH_2-CH_2-CH_2-CH_2-CH_3$

$HC-O-C(=O)-CH=CH-CH_2-CH_2-CH_3$

$H_2C-O-C(=O)-CH=CH-CH-CH_2$

**Fig. 8.2.4.** A model of a fat, or triglyceride. It consists of a glycerol and three fatty acids. Compare Figure 8.2.3.

that the daughter cells will have the same chemical processes as the original cell. In sexual reproduction, nucleic acids from two parents are combined in fertilization, resulting in an offspring whose chemistry is related by sometimes complex rules to that of its parents.

There are two kinds of nucleic acids: Their names are *deoxyribonucleic acid* (DNA) and *ribonucleic acid* (RNA). The monomer of a nucleic acid is a *nucleotide*, which is composed of three parts: a sugar, one or more phosphate groups, and a nitrogenous base. Figure 8.2.6 shows the components of a typical nucleotide.

DNA is a double helix. Figure 8.2.7 shows a model of the macromolecule, partially untwisted to reveal its underlying structure. Note that it is formed from two covalently linked linear polymers, which are wrapped around each other. The two single strands are H-bonded to one another, as shown by dotted lines in the figure. Figure 8.2.8 shows the details of the H-bonding between DNA nucleotides.

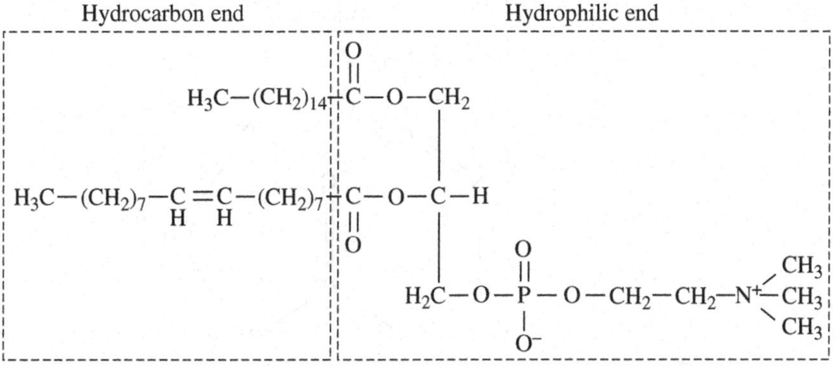

Hydrocarbon end | Hydrophilic end

$$H_3C-(CH_2)_{14}-\overset{O}{\overset{\|}{C}}-O-CH_2$$

$$H_3C-(CH_2)_7-\underset{H}{C}=\underset{H}{C}-(CH_2)_7-\underset{O}{\underset{\|}{C}}-O-C-H$$

$$H_2C-O-\overset{O}{\overset{\|}{P}}(O^-)-O-CH_2-CH_2-N^+(CH_3)_3$$

**Fig. 8.2.5.** A phospholipid, or phosphoglyceride, found in cell membranes. Note that it has a hydrophilic end that is attracted to water and a hydrocarbon (hydrophobic) end that is repelled by water. The hydrophilic end faces the aqueous outside world or the aqueous interior of the cell. The hydrophobic end of all such molecules is in the interior of the membrane, where there is no water. This picture should be compared to the schematic lipids shown in Figure 6.1.1: The circles on the phospholipids of Figure 6.1.1 correspond to the right-hand box of this figure, and the two straight lines of Figure 6.1.1 correspond to the two hydrocarbon chains in the left-hand box of this figure.

The DNA molecule is very long compared to its width. The double helix is $2.0 \times 10^{-9}$ m wide, but about $10^{-3}$ m long in a bacterium and up to 1 m long in a human. There are ten base pairs every $3.6 \times 10^{-7}$ m of length of double helix. Thus a 1-meter-long DNA molecule has about $3 \times 10^8$ base pairs. If any of the four nucleotides can appear at any position, there could exist $4^{3\times10^8}$ possible DNA molecules of length 1 m. Obviously, an incredible amount of information can be encoded into such a complex molecule. Note that DNA uses only a four-letter "alphabet" but can compensate for the small character set by writing very long "words."

There are some important structural details and functional consequences to be noted about Figures 8.2.7 and 8.2.8:

1. *Each of the two single-stranded polymers of a DNA molecule is a chain of covalently linked nucleotides.* All four possible nucleotides are shown, but there are no restrictions on their order in natural systems; any nucleotide may appear at any position on a given single strand. It is now possible experimentally to determine the sequences of long DNA chains; see Section 14.1.

2. *Once a particular nucleotide is specified at a particular position on one strand, the nucleotide opposite it on the other strand is completely determined.* Note that $A$ and $T$ are opposite one another, as are $C$ and $G$; no other base pairs are allowed in DNA. (From now on, we shall indicate the names of the nucleotides by their initials, i.e., $A$, $T$, $C$, and $G$.) There are very important physical and biological reasons for this *complementary* property. The physical reason can be seen by a close examination of the H-bonds between an $A$ and a $T$ or between a $C$ and a $G$ in Figure 8.2.8. Recall that an H-bond is formed between a lone pair of

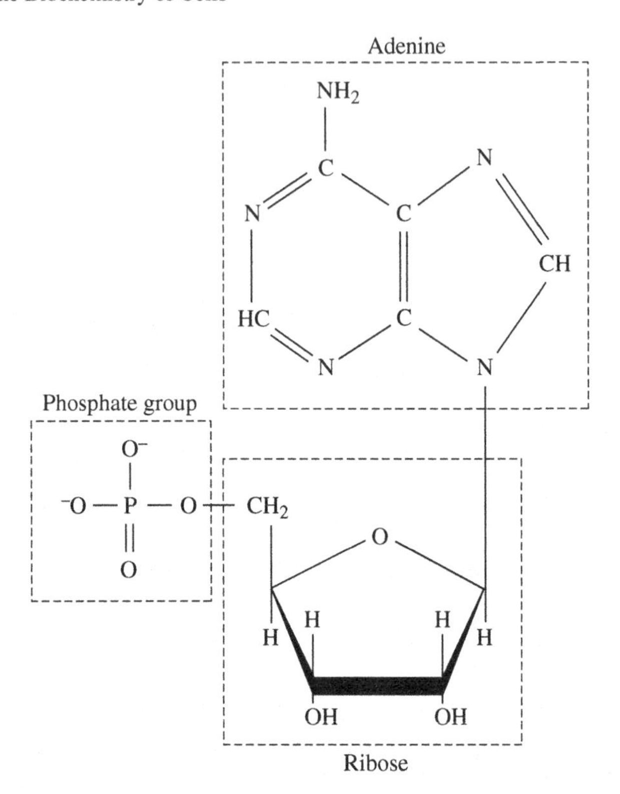


**Fig. 8.2.6.** A typical nucleotide, consisting of a nitrogenous base (adenine), a sugar (ribose), and a phosphate group. Other nucleotides can have other nitrogenous bases, a different sugar, and more phosphates.

electrons and a hydrogen nucleus, and note that two such bonds form between $A$ and $T$ and that three form between $C$ and $G$. There are no other ways to form two or more strong H-bonds between any of these nucleotides; thus the ways shown in Figure 8.2.7 are the only possibilities. For example, $A$ cannot effectively H-bond to $C$ or $G$. We should note that the property of complementary H-bonding requires that the two single strands have different nucleotide sequences, but that the sequence of one strand be utterly determined by the other.

3. *The helical configuration is a spontaneous consequence of H-bonding the two single strands together.* Helicity disappears if the H-bonds are disrupted. Recall from the discussion of the structure of water in Section 8.1 that H-bonds have marginal stability at room temperature. We should therefore expect that the two strands of helical DNA can be separated, i.e., the helix can be *denatured*, without expending much energy. In fact, DNA becomes denatured at around 45–55°C, only about 8 to 18 degrees above body temperature. Once thermal denaturation has occurred, however, the two strands can often spontaneously reassociate into their native double helical configuration if the temperature is then slowly reduced.

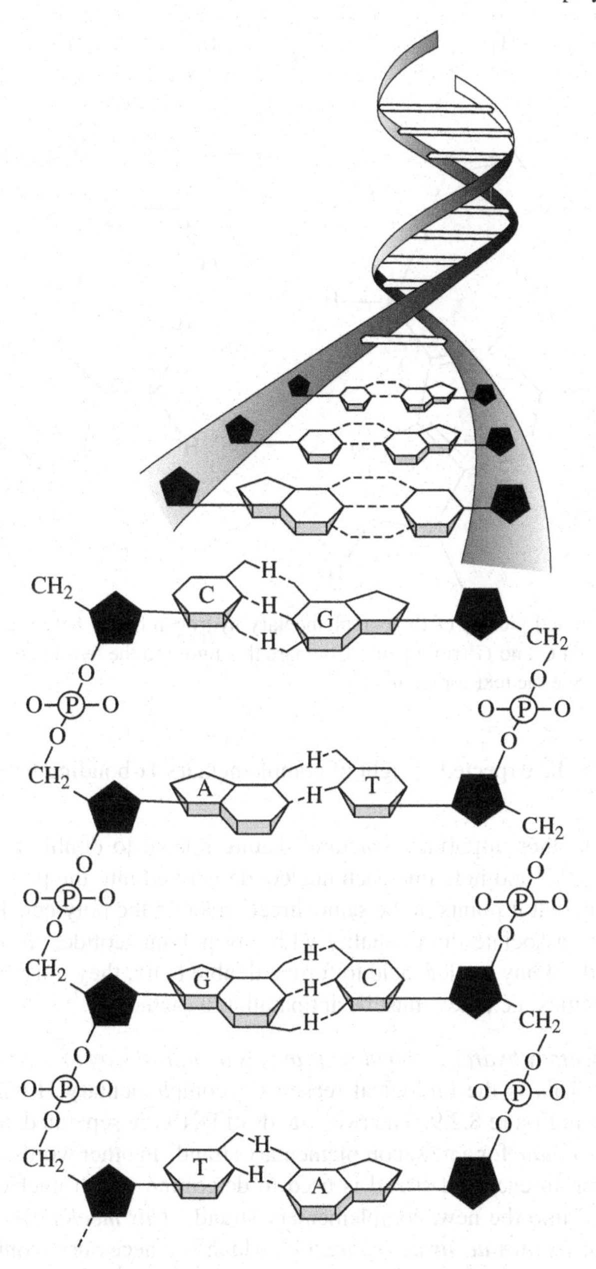


**Fig. 8.2.7.** A DNA molecule, showing the arrangement of the nucleotide components into two covalent polymers, each of which is hydrogen-bonded to the other. Note that *A* (adenine) and *T* (thymine) are hydrogen-bonded to each other, and *C* (cytosine) and *G* (guanine) are hydrogen-bonded to each other. The hydrogen bonds are indicated by the dashes. (Redrawn from C. Starr and R. Taggart, *Biology: The Unity and Diversity of Life*, 6th ed., Wadsworth, Belmont, CA, 1992. Used with permission.)

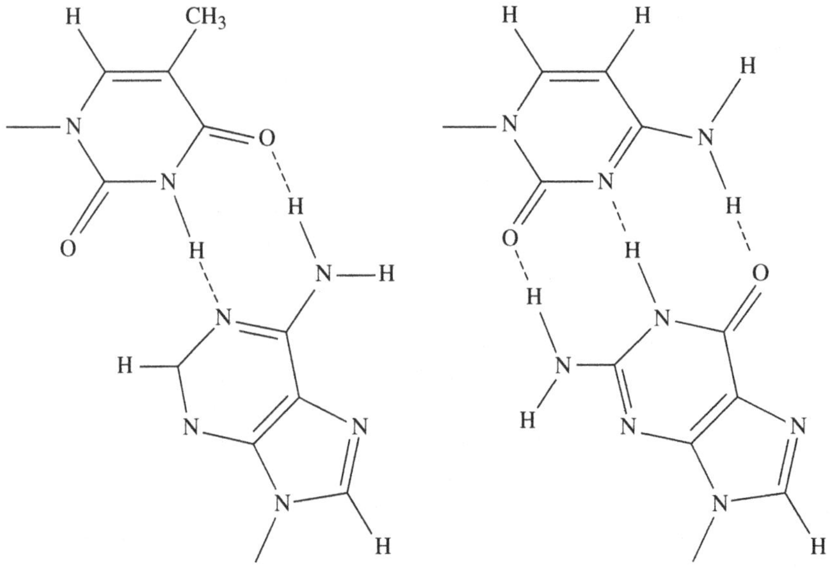

Fig. 8.2.8. A detailed picture of the complementary hydrogen bonds between *A* and *T* (left pair), and between *C* and *G* (right pair). Compare this figure to the hydrogen-bonded groups in Figure 8.2.7. See the text for details.

This should be expected in light of complementary H-bonding between the two strands.

There is another important structural feature related to double helicity: Look at Figure 8.2.7 and note that each nucleotide is fitted into the polynucleotide in such a way that it points in the same direction along the polymer. It is therefore possible to associate directionality with any polynucleotide. In order for the two strands of any nucleic acid to form a double helix, they must have opposite directionalities, i.e., they must be antiparallel to each other.[3]

4. *Complementary hydrogen-bonding provides a natural way to replicate DNA accurately.* This is the biological reason for complementary H-bonding and is illustrated in Figure 8.2.9. The two strands of DNA are separated, and each then acts as a *template* for a new, complementary strand. In other words, the sequence information in each old strand is used to determine which nucleotides should be inserted into the new, complementary strand. *This mechanism allows DNA to code for its own accurate replication*, which is a necessary requirement for a genetic chemical.

Occurring just prior to cell division, the process of DNA self-replication yields two double-stranded DNA molecules that are exact copies of the original. Then, during

[3] For example, look at the location of the methyl group (-$CH_2$-) between the phosphate group and the ribose group. Note how it is in a different position on the two strands.

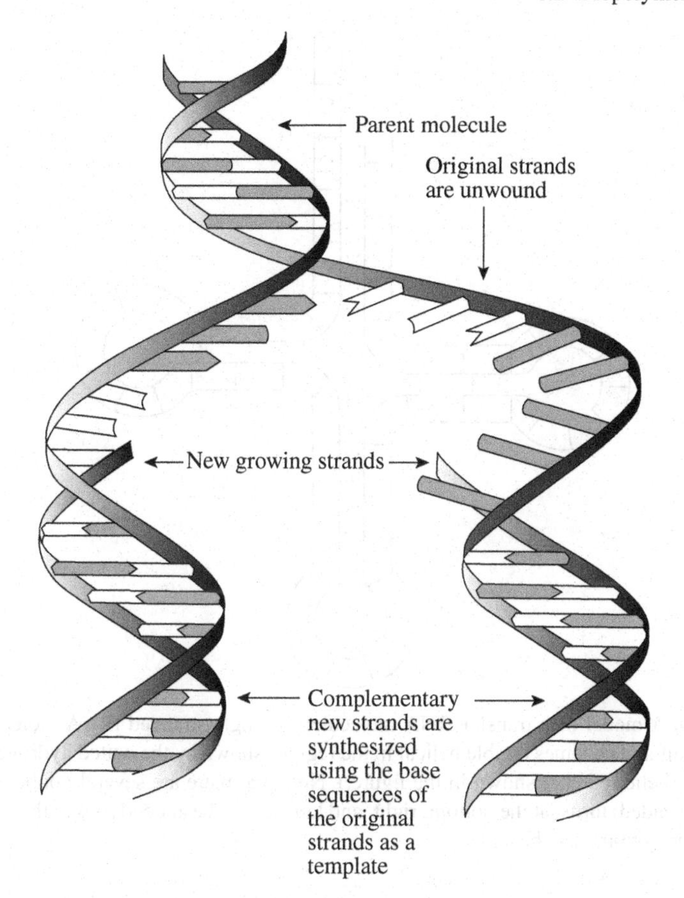


**Fig. 8.2.9.** A model of a replicating DNA molecule. The two strands of the parent double helix separate, and each one acts as a template for a new strand. Complementary hydrogen-bonding ensures that the two resulting double helices are exact copies of the original molecule. (Redrawn from J. Levine and K. Miller, *Biology*, 1st ed., D. C. Heath, Lexington, MA, 1991.)

cell division, each of the daughter cells gets one of the copies. The two daughter cells thus each end up with the same genetic material that the original cell had and should therefore also have the same life properties.

There are three classes of RNA molecules: The first is called *messenger RNA*, or mRNA. Each piece of mRNA averages about a thousand bases in length, but is quite variable. It is single stranded and nonhelical. The second kind of RNA is *transfer RNA*, or tRNA. There are several dozen distinguishable members of this class; they contain in the range of 75 to 95 bases, some of which are not the familiar *A*, *T*, *C*, and *G*. tRNA is single stranded but is double helical. This unexpected shape is due to the folding over of the tRNA molecule, as shown in Figure 8.2.10. The third kind of RNA is *ribosomal RNA*, or rRNA. This molecule accounts for most of a cell's RNA. It appears in several forms in cellular organelles associated with protein synthesis,

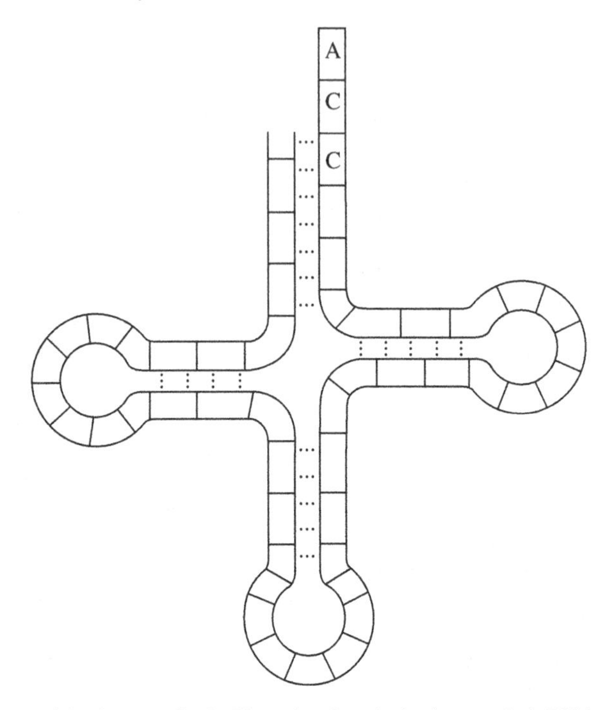


**Fig. 8.2.10.** A model of a transfer RNA molecule. A single-stranded tRNA molecule folds back on itself and becomes double helical in the regions shown by the dotted hydrogen bonds. (The actual helicity is not shown in the figure.) Note that there are several nonhelical (non-hydrogen-bonded) turns, at the bottom, right, and left sides. The anticodon is in the nonhelical region at the bottom; see the text.

and it has molecular weights ranging from around a hundred up to several thousand. The functions of the various RNAs will be discussed shortly.

*Proteins are polymers of amino acids.*

The monomer of a protein is an amino acid, a synonym for which is *residue*. A protein polymer is often called a *polypeptide*. While many amino acids can exist, only twenty are found in proteins. They share the general structure shown in Figure 8.2.11. The group labeled *R* can take on twenty different forms, thus accounting for all members of the group.[4] The right end (-COOH) is the *carboxyl* and the bottom (-$NH_2$) is the *amino* end.

Figure 8.2.12 shows how two amino acids are polymerized into a dipeptide (two residues). Note that the attachment takes place by combining the amino end of one residue with the carboxyl end of the other. The covalent bond created in this process is called a *peptide bond*, as shown in Figure 8.2.12.

An interesting feature of a dipeptide is that, like an individual amino acid, it has both a carboxyl end and an amino end. As a result, it is possible to add other residues

[4] We will ignore the fact that one of the amino acids is a slight exception.

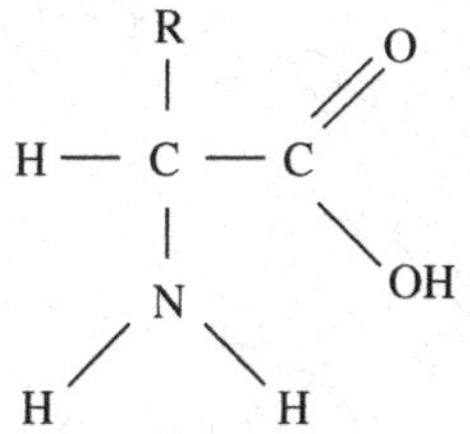

**Fig. 8.2.11.** A model of an amino acid, which is the monomer of a protein. The label $R$ stands for any one of twenty different groups. (The text mentions a slight exception.) Thus twenty different amino acids may be found in proteins.

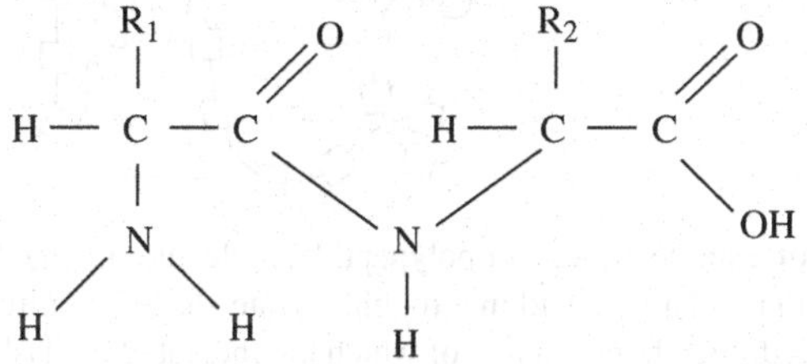

**Fig. 8.2.12.** A pair of amino acids bonded covalently into a dipeptide. The labels $R_1$ and $R_2$ can be any of the twenty groups mentioned in the caption of Figure 8.2.11. Thus there are 400 different dipeptides.

to the two ends of the dipeptide and thereby to extend the polymerization process as far as we like. It is quite common to find natural polypeptides of hundreds of residues and molecular weights over a hundred thousand. Figure 8.2.13 is an idealized picture of a polypeptide "backbone"; the individual amino acids are represented as boxes. Note that the polymer has a three-dimensional structure that includes helical regions and sheetlike regions, and that the whole three-dimensional shape is maintained by H-bonds and disulfide (-S-S-) bonds. The disulfide bonds are covalent, and the two amino acids that contribute the sulfur atoms are generally far from one another as measured along the polymer. They are brought into juxtaposition by the flexibility of the polymer and held there by the formation of the disulfide bond itself.

Our model of a protein is that of a long polymer of amino acids, connected by peptide bonds and folded into some kind of three-dimensional structure. At any location any of twenty different amino acids may appear. Thus there are $20^{100}$ possible polypeptides of 100 amino acids in length. Nowhere near this number have actual biological functions, but the incomprehensibly large number of possible amino acid sequences allows living systems to use proteins in diverse ways. Some of these ways will be described next.

*Some proteins are catalysts.*

There exists a very important class of proteins, called *enzymes*, whose function it is to to speed up the rate of biochemical reactions in cells (see [2] and [4]). In

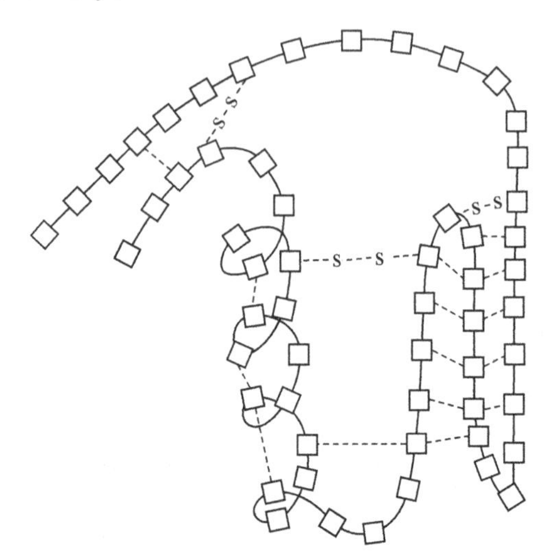


**Fig. 8.2.13.** A model of a single protein, or polypeptide, molecule. Each box corresponds to an amino acid. The resultant chain is held in a roughly ovate shape by sulfur-to-sulfur covalent bonds and by many hydrogen bonds, a few of which are indicated by dashed lines.

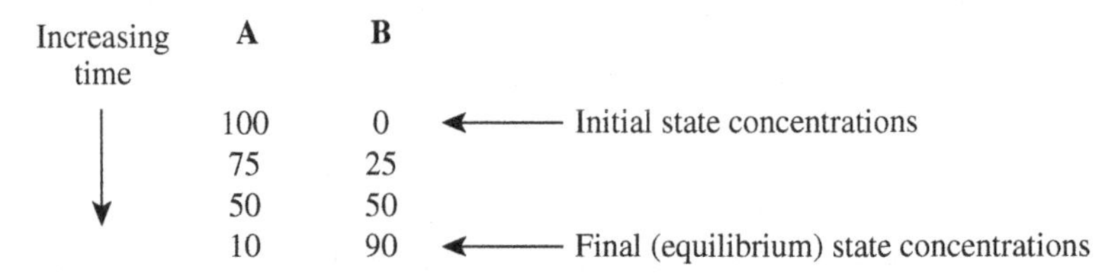

| Increasing time ↓ | A | B | |
|---|---|---|---|
| | 100 | 0 | ← Initial state concentrations |
| | 75 | 25 | |
| | 50 | 50 | |
| | 10 | 90 | ← Final (equilibrium) state concentrations |

**Fig. 8.2.14.** The progress of the reaction $A \leftrightarrow B$. The numbers give the amounts of the compounds $A$ and $B$ at various times. At the outset, there is no $B$, but as time passes, the amount of $B$ increases until $A$ and $B$ reach equilibrium at a ratio of $B : A = 9 : 1$.

order to understand this function we must understand what is meant by "reaction rate": Suppose there is a chemical reaction described by $A \longleftrightarrow B$, as shown in Figure 8.2.14. Let us suppose that initially there is lots of $A$ and no $B$. As time passes, some $A$ is converted to $B$, and also some $B$ back to $A$. Eventually, the relative amounts of $A$ and $B$ reach steady values, i.e., do not change with time. This final state is called an *equilibrium state*. The speed with which $A$ is converted to $B$ is the rate of the *reaction*. The observed rate evidently changes with time, starting out fast and reaching a net of zero at equilibrium, and therefore it is usually measured at the outset of the experiment, when there is lots of $A$ and no $B$.

There are several very important biological consequences of enzymatic catalysis. First, the essential effect of a catalyst is to speed up the rate of a reaction. A biochemical catalyst, i.e., an enzyme, can speed up the rate of a biochemical reaction by as much as $10^{13}$ times. This enormous potential increase has some very important consequences to cellular chemistry: First, catalyzed biochemical reactions are fast enough to sustain life, but uncatalyzed reactions are not. Second, if a reaction will not proceed at all in the absence of a catalyst, then no catalyst can ever make it proceed. After all, speeding up a rate of zero by $10^{13}$ still gives a rate of zero. Third, catalysts have no effect whatsoever on the relative concentrations of reactants and products at equilibrium, but they do affect the time the system takes to reach that equilibrium. Thus enzymes do not affect the underlying chemistry or net energetic requirements of the system in which they participate. Fourth, enzymes are very specific as to the reactions that they catalyze, their activity usually being limited to a single kind of reaction. This observation can be combined with the first one above (enzymatic increase in reaction rate) to yield an important conclusion: Whether a particular biochemical reaction goes at a high enough rate to sustain life depends entirely on the presence of specific enzyme molecules that can catalyze that particular reaction. Thus enzymes act like valves, facilitating only the reactions appropriate to a particular cell. No other reactions proceed fast enough to be significant and so they can be ignored.

The valvelike function of enzymes explains why a human and a dog can eat the same kind of food, drink the same kind of water, and breathe the same air, yet not look alike. The dog has certain enzymes that are different from those of the human (and, of course, some that are the same). Thus many biochemical reactions in a dog's cells proceed in a different direction from those in a human—in spite of there being the same initial reactants in both animals. Figure 8.2.15 shows how different metabolic paths can originate from the same starting point because of different enzyme complements.

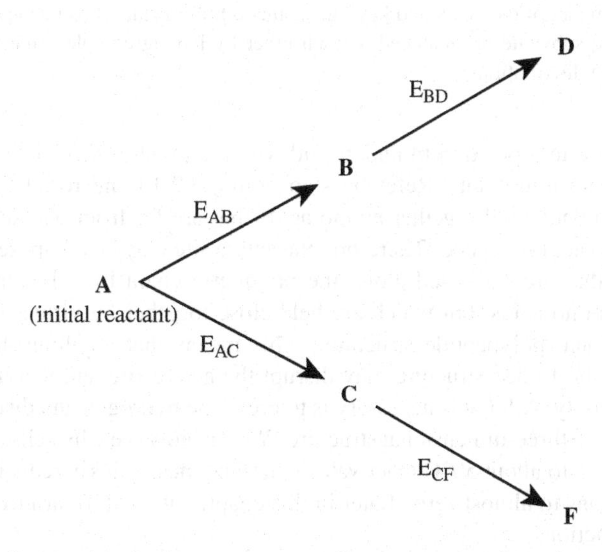


**Fig. 8.2.15.** A diagram showing how enzymes can direct sequences of reactions. *A* is the initial reactant, and the pair of enzymes $E_{AB}$ and $E_{BD}$ would catalyze the conversion of *A* to *D*. Alternatively, the enzymes $E_{AC}$ and $E_{CF}$ would catalyze the conversion of *A* to *F*. It is the enzymes, not the initial reactant, that determine what the end product will be. Of course, this does not mean that there will always exist an enzyme that can catalyze a particular reaction; rather, there will almost always exist an enzyme that can catalyze the particular reactions needed by a given cell.

The same reasoning explains why two people have different hair color, or numerous other differences.

The nature of the specificity of an enzyme for a single chemical reaction can be understood in terms of a "lock and key" mechanism: Suppose that we are again dealing with the reaction $A \longleftrightarrow B$, catalyzed by the enzyme $E_{AB}$. The catalytic event takes place on the surface of the enzyme at a specific location, called the *active site*, as shown in Figure 8.2.16. The compound $A$, or *substrate*, has a shape and electrical charge distribution that are complementary to the active site. This ensures that only the reaction $A \longleftrightarrow B$ will be catalyzed. Note that this reaction is reversible, and that the enzyme catalyzes in both directions.

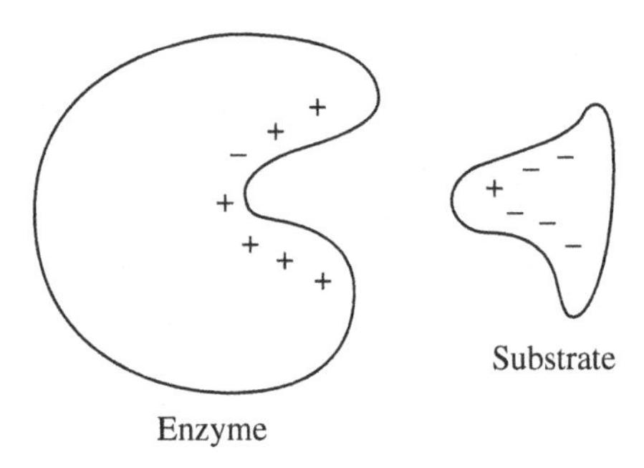


**Fig. 8.2.16.** A model of the "lock and key" mechanism for enzyme–substrate specificity. The enzyme and the substrate are matched to each other by having complementary shapes and electrical charge distributions.

Now we are in a position to understand why the three-dimensional structure of an enzyme is so important. Refer back to Figure 8.2.13 and recall that H-bonds and disulfide bonds hold together amino acids that are far from one another in the primary amino acid sequence. Therefore, the active site may be composed of several amino acids that are separated along the polymeric chain by a hundred or more intervening amino acids, but which are held close together by virtue of the folded three-dimensional polypeptide structure. This means that anything that disturbs, or denatures, the folded structure may disrupt the active site and therefore destroy enzymatic activity. All that is necessary is to break the hydrogen and disulfide bonds that maintain the three-dimensional structure. We can now see why cells are sensitive to heat: Heating to about 50°C inactivates their enzymes, quickly reducing the rates of their reactions to almost zero. Later in this chapter, we will return to the topic of enzymatic function.

*Noncatalytic proteins.*

The immense diversity of possible protein structures allows these macromolecules to be used for many biological purposes. Many of these have nothing to do with catalysis. We will divide these noncatalytic proteins into two somewhat arbitrary, but customary, categories and discuss them next.

*Category* 1: *Fibrous proteins*. These are called "fibrous" because they consist of large numbers of polypeptides arranged in parallel to yield long, stringlike arrays. Collagen, for example, is a fibrous protein found in skin and other organs. It consists of shorter protein molecules, each staggered one quarter length from the next one and thus linked into very long strings. Collagen acts as a binder, the long fibers helping to hold our bodies together.

Other examples of fibrous proteins are found in muscle tissue. Each muscle cell contains large numbers of fibrous proteins that are capable of sliding past one another and exerting force in the process. Our muscles can then move our skeletons and, therefore, our bodies. What we call "meat" is just muscle cut from an animal, and of course, it contains a lot of protein.

Another example of a fibrous protein is *keratin*, which appears in several forms in hair and nails, among other places. Some keratins form ropes of multiple strands, held together by disulfide bonds. Other keratins form sheetlike structures. One important form of keratin is silk, a threadlike exudation used in the wrapping of the cocoon of the silkworm *Bombyx mori*.

*Category* 2: *Globular proteins*. These proteins tend to be spherical or ovate and are often found dispersed, e.g., dissolved in solution. If aggregated, they do not form fibers. Enzymes are globular proteins, but we have already discussed them, and we will therefore restrict our discussion here to noncatalytic globular proteins.

As an example, the polypeptides *hormones* are typical noncatalytic globular proteins. They were introduced in Chapter 7. Hormones are biochemical communicators: They are manufactured in *endocrine glands* in one part of the body and are moved by the bloodstream to another part of the body, where they exert their effects on *target tissues*. At their target tissues hormones change the production and activity of enzymes and alter membrane permeability.

*Insulin*, a globular protein hormone, is produced by an organ called the pancreas and is released into the blood, to be carried throughout the body. The function of insulin is to regulate the metabolism of glucose in the body's cells. Lack of insulin has powerful metabolic consequences: The disorder *diabetes mellitus* is associated with the loss of insulin-producing cells of the pancreas, increases in the glucose levels of blood and urine, malaise, and even blindness.

Another class of noncatalytic globular proteins, introduced in Chapter 6, determines the selectivity of material transport by membranes. These proteins recognize and regulate the intercellular movements of specific compounds such as amino acids and various sugars and ions such as $Na^+$ and $Cl^-$. Called *transport proteins*, or *permeases*, they penetrate through membranes and have a sort of active site on one end to facilitate recognition of the material to be transported. They are, however, not catalysts in that the transported matter does not undergo a permanent chemical change as a result of its interaction with the transport protein.

Globular proteins are used to transport material in the body. One example, *hemoglobin*, which is discussed in Chapter 9, contains four polypeptide chains and four heme groups, the latter being organic groups with an iron atom. Hemoglobin is found in red blood cells, or *erythrocytes*. The principal use of hemoglobin is to carry oxygen from the lungs to the sites of oxygen-utilizing metabolism in the body.

Globular proteins are key molecules in our immune systems. A group of blood cells, called *lymphocytes*, are able to distinguish between "self" and "nonself" and therefore to recognize foreign material, such as pathogens, in our bodies. These foreign substances are often proteins but may be polysaccharides and nucleic acids; in any case, if they stimulate immune responses they are called *antigens* (Ag). Antigens stimulate lymphocytes to produce a class of globular proteins, called *antibodies* (Ab) or *immunoglobulins*, that can preferentially bind to Ag, leading to the inactivation of the Ag. The immune response will be discussed in some detail in Chapter 10.

Of particular importance to us in that chapter will be the globular proteins found in a covering, or *capsid*, of viruses. Viruses have very elementary structures, the simplest being a protein coat surrounding a core of genetic material. Viruses are so small that the amount of genetic material they can contain is very limited. Thus as an information-conserving mechanism, they use multiple copies of the same one or two polypeptides to build their protein coverings. Thus a typical virus may have an outer coat consisting of hundreds of copies of the same globular protein.

## 8.3 Molecular Information Transfer

*This section is a discussion of molecular genetics. The ability of DNA to guide its own self-replication was described in an earlier section. In this section, we will see how genetic information of DNA, coded into its polymeric base sequence, can be converted into base-sequence information of RNA. The base-sequence information of RNA can then be converted into amino acid–sequence information of proteins. The amino acid sequence of a protein determines its three-dimensional shape and therefore its function, i.e., participation in* $O_2$ *transport in erythrocytes, selection of material to cross a membrane, or catalysis of a specific biochemical reaction. The net process is contained in the following statement: DNA is the hereditary chemical because it provides an informational bridge between generations via self-replication, and it ultimately determines cellular chemistry. These processes are schematically condensed into the central dogma of genetics:*

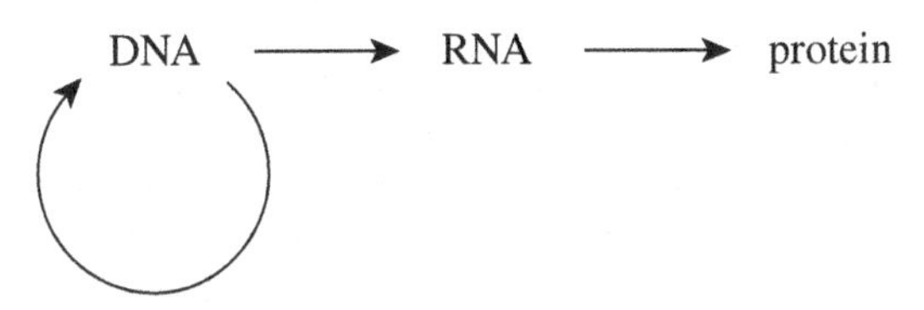


*It is very important to recognize that the arrows of the central dogma show the direction of information flow, not the direction of chemical reactions. Thus DNA passes its information on to RNA—the DNA is not chemically changed into RNA.*[5]

[5] We will modify this "dogma" somewhat in Chapter 10.

*Information flow from DNA to RNA is called transcription.*

Recall that enzymes determine which reactions in a cell effectively take place. For organisms other than certain viruses, DNA is the source of the information that determines which enzymes will be produced. In any case, there is an intermediary between DNA and proteins—it is RNA. This is expressed in the central dogma presented above (see [5] and [6]).

RNA production is shown schematically in Figure 8.3.1. The sequence of the single covalent strand of RNA nucleotides is determined by complementary H-bonding with one strand of a DNA molecule; in other words, the single, or coding, strand of DNA acts as a template for RNA production. Note the similarity between the use of a single-stranded DNA template for DNA production and the use of a single-stranded DNA template for RNA production. The differences are that RNA uses a different sugar and substitutes uracil in place of thymine.

The process of RNA production from DNA, called *transcription*, requires that the DNA molecule become denatured over a short portion of its length, as shown in Figure 8.3.1. This is a simple matter energetically because all that is required is to break a small number of H-bonds. The O-shaped denatured region moves along the DNA molecule, the double helix opening up at the leading edge of the "O" and closing at its trailing edge. RNA molecules, as mentioned earlier, are usually less than a thousand or so nucleotides long. Thus RNA replication normally begins at many sites in the interior of the DNA molecule, whose length may be on the order of millions of nucleotides.

*Information flow from RNA to enzymes is called translation.*

The process of protein production from RNA code brings together, one by one, all three kinds of RNA: ribosomal, messenger, and transfer. The three varieties are transcribed from the DNA of the cell and exported to sites away from the DNA. Here subcellular structures called *ribosomes* are constructed, in part using the rRNA. Ribosomes are the sites of protein synthesis, but the actual role of the rRNA is not well understood.

Several dozen different kinds of transfer RNA are transcribed from DNA. They all have a structure similar to that shown in Figures 8.2.10 and 8.3.2, which shows that tRNA is single stranded, but is helical by virtue of the folding of the polymer on itself. This requires that some regions on the strand have base sequences that are complementary to others, but in reverse linear order. (Recall from Figure 8.2.7 that a nucleic acid double helix requires that the two strands be antiparallel.) The various kinds of tRNA differ in their constituent bases and overall base sequences; the most important difference for us, however, is the base sequence in a region called the *anticodon*, at the bottom of the figure. The anticodon is actually a loop containing three bases that, because of the looping, are not H-bonded to any other bases in the tRNA molecule.

Let us consider the anticodon more closely. It contains three nucleotides that are not hydrogen-bonded to any other nucleotides. The number of such trinucleotides, generated at random, is $4^3 = 64$, so we might expect that there could be 64 different

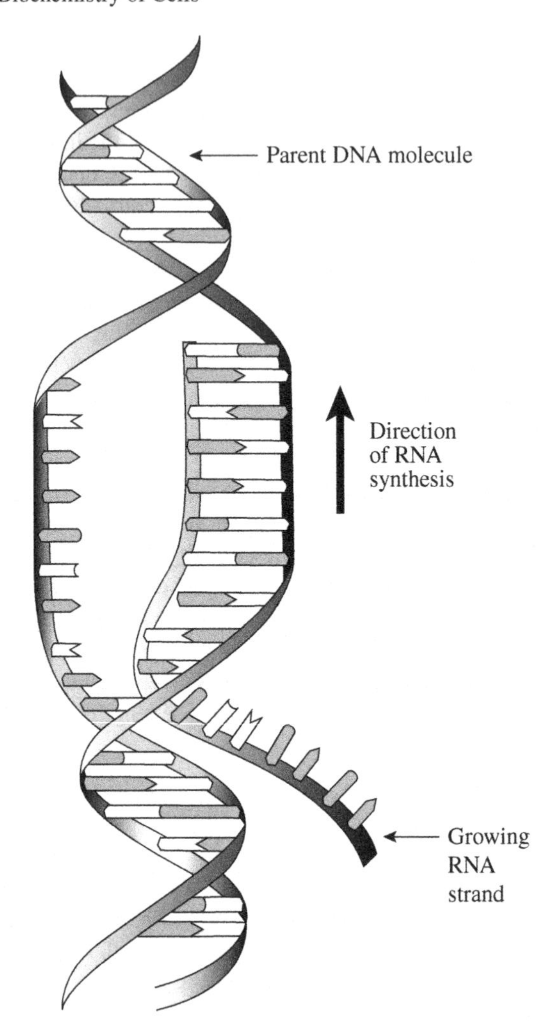


**Fig. 8.3.1.** A model showing the polymerization of RNA, using a DNA template. The DNA opens up to become temporarily single stranded over a short section of its length, and one of the two DNA strands then codes for the RNA. Complementary hydrogen-bonding between the DNA nucleotides and the RNA nucleotides ensures the correct RNA nucleotide sequence. (Redrawn from J. Levine and K. Miller, *Biology*, 1st ed., D. C. Heath, Lexington, MA, 1991. Used with permission.)

kinds of tRNA, if we considered only the anticodons. Actually, fewer than that seem to exist in nature, for reasons to be discussed shortly. The anticodon bases are not H-bonded to any other bases in the tRNA molecule but are arranged in such a three-dimensional configuration that they can H-bond to three bases on *another* RNA molecule.

All tRNA molecules have a short "pigtail" at one end that extends beyond the opposite end of the polymer. This pigtail always ends with the sequence $CCA$. An

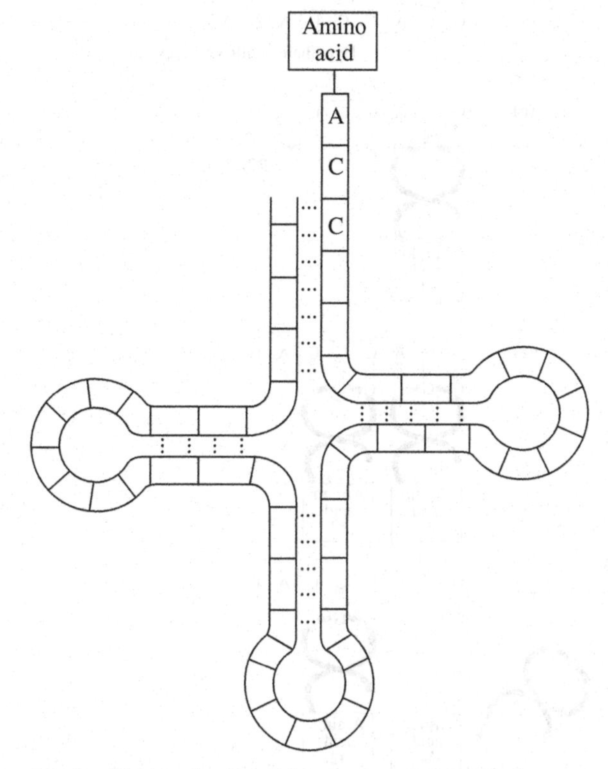


**Fig. 8.3.2.** A model of a tRNA molecule, with an amino acid attached to one end. An enzyme ensures that the tRNA molecule becomes covalently attached to its correct amino acid. The anticodon region is at the bottom. Compare this figure with Figure 8.2.10.

amino acid can be covalently attached to the terminal adenine, giving a tRNA amino acid molecule, as shown in Figure 8.3.2. A given type of tRNA, identified by its anticodon, can be attached to one, and only one, specific type of amino acid. No other pairings are possible. When we see such specificity in biochemistry, we should always suspect that enzymes are involved. In fact, there are enzymes whose catalytic function is to link up an amino acid with its correct tRNA. A tRNA molecule that is attached to its correct amino acid is said to be "charged."

Messenger RNA consists of strings of about 1000 or so nucleotides, but that is only an average figure—much mRNA is considerably longer or shorter. The reason for this variability is that each piece of mRNA is the transcription product of one or a few genes on DNA. Thus the actual length of a particular piece of mRNA corresponds to an integral number of DNA genes, and of course, that leads to a great deal of variability in length. After being exported from the DNA, the mRNA travels to a ribosome, to which it becomes reversibly attached.

The next part of this discussion is keyed to Figure 8.3.3:

(a) One end of a piece of mRNA is attached to a ribosome, the area of association covering at least six mRNA nucleotides.

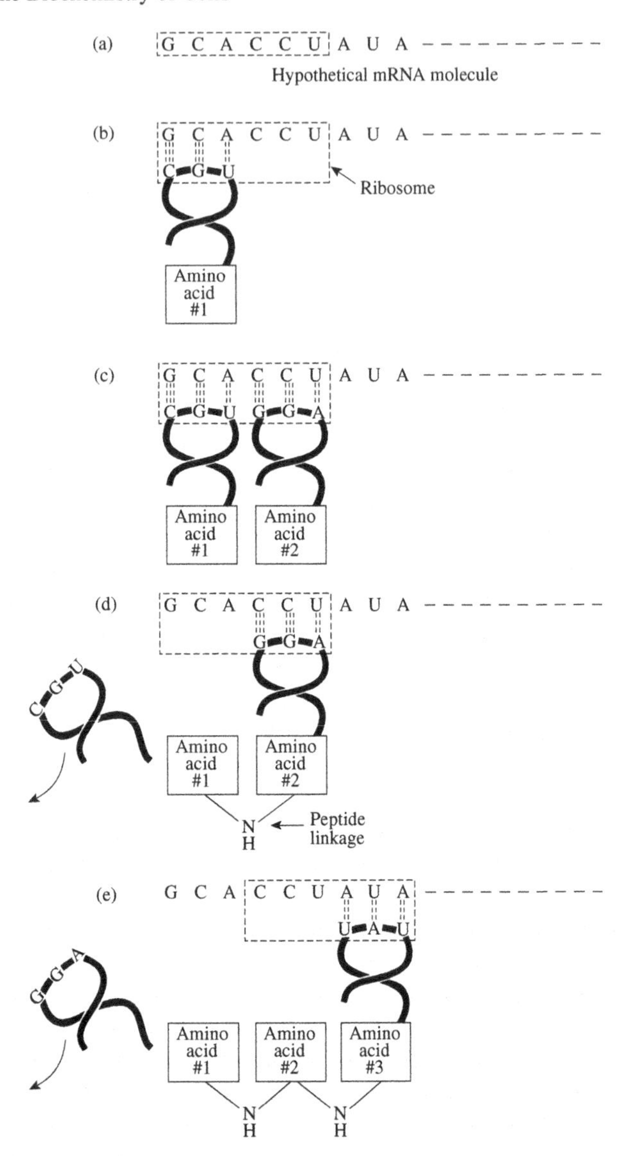


**Fig. 8.3.3.** The polymerization of a polypeptide, using DNA information and RNA intermediaries. (a) A ribosome attaches to the end of an mRNA molecule. (b) A molecule of tRNA, with its correct amino acid attached, hydrogen-bonds to the mRNA. The hydrogen-bonding is between the first three nucleotides of the mRNA (a codon) and the three tRNA nucleotides in a turn of tRNA (an anticodon). Each tRNA has several such turns, as depicted in Figures 8.2.10 and 8.3.2, but only one is the anticodon. (c) A second tRNA then hydrogen-bonds to the mRNA, thus lining up two amino acids. (d) The two amino acids are joined by a covalent bond, a process that releases the first tRNA. (e) The ribosome moves down the mRNA molecule by three nucleotides and a third tRNA then becomes attached to the mRNA. The process continues until an intact protein is formed. Note how the amino acid sequence is ultimately dictated by the nucleotide sequence of the DNA.

(b) A tRNA molecule, with its correct amino acid attached, forms complementary H-bonds between its anticodon and the first three nucleotides of the mRNA. The latter trinucleotide is called a *codon*. Note that codon–anticodon recognition mates up not only the correct anticodon with its correct codon, but in the process also matches up the correct amino acid with its codon.
(c) Next, a second charged tRNA hydrogen-bonds to the second mRNA codon.
(d) A peptide linkage forms between the two amino acids, detaching the first amino acid from its tRNA in the process.

Let us review what has happened so far: A sequence of DNA nucleotides comprising a small integral number of genes has been transcribed into a polymer of mRNA nucleotides. The sequence of the first six of these nucleotides has subsequently been translated into the sequence of two amino acids. There is a direct informational connection mapping the sequence of the original six DNA nucleotides into the sequence of the two amino acids. The correctness of this mapping is controlled by two physical factors: First, complementarity between DNA and mRNA and between mRNA and tRNA, and second, specific enzymatic attachment of tRNA to amino acids.

Now returning to Figure 8.3.3, the ribosome moves three nucleotides down the mRNA and a third charged tRNA attaches to the mRNA at the third codon:

(e) A third amino acid is then added to the growing polypeptide chain. The translation process continues and eventually a complete polypeptide chain is formed. The nucleotide sequence of the DNA has been converted into the primary structure of the polypeptide. Note how the conversion of nucleotide sequence to amino acid sequence was a transfer of information, not a chemical change of DNA to protein.

Figure 8.3.3 is really a pictorial representation of the central dogma. The overall process yields proteins, including enzymes of course. These enzymes determine what chemical reactions in the cell will proceed at a rate consistent with life. Two very important observations come out of this discussion: First, the chemistry of a cell is ultimately determined by the sequence of DNA nucleotides, and second, because of this, the replication and partitioning of DNA during cell division ensures that daughter cells will have the same chemistry as the parent cell. We can extend the latter conclusion: The union of a sperm and an egg in sexual reproduction combines genetic material from two parents into a novel combination of DNAs in a new organism, thus ensuring that the offspring has both chemical similarities to, and chemical differences from, each of the parents.

*A gene is enough nucleic acid to code for a polypeptide.*

The word "gene" is often loosely used to mean "a site of genetic information." A more exact definition from molecular biology is that a gene is a sequence of nucleotides that codes for a complete polypeptide. This definition, however, requires the elaboration of several points:

1. If a functioning protein's structure contains two separately created polypeptides, then by definition, two genes are involved.

2. As will be discussed below, some viruses eliminate DNA from their replicative cycle altogether. Their RNA is self-replicating. In those cases, their genes are made of RNA.
3. The nucleotide sequence for any one complete gene normally lies entirely on one strand of DNA, called the *coding strand.* Note that coding segments for a gene, called *exons*, are not generally contiguous. Noncoding stretches between coding segments are *introns*; see Section 14.6. Furthermore, not all genes need lie on the same one strand; transcription may jump from one strand to another between gene locations. There may even be overlapping genes on the same strand.

*The concept of coding.*

The aptly named *genetic code* can be presented in a chart showing the correspondence between RNA trinucleotides (codons) and the amino acids they specify, the *codon translation* table; see Table 8.3.1.

**Table 8.3.1.** Codon translation table.

| | *T* | *C* | *A* | *G* |
|---|---|---|---|---|
| *T* | *TTT* Phe(*F*) | *TCT* Ser(*S*) | *TAT* Tyr(*Y*) | *TGT* Cys(*C*) |
| | *TTC* Phe(*F*) | *TCC* Ser(*S*) | *TAC* Tyr(*Y*) | *TGC* Cys(*C*) |
| | *TTA* Leu(*L*) | *TCA* ” | *TAA* stop | *TGA* stop |
| | *TTG* ” | *TCG* ” | *TAG* stop | *TGG* Trp(*W*) |
| *C* | *CTT* Leu(*L*) | *CCT* Pro(*P*) | *CAT* His(*H*) | *CGT* Arg(*R*) |
| | *CTC* ” | *CCC* ” | *CAC* ” | *CGC* ” |
| | *CTA* | *CCA* ” | *CAA* Gln(*Q*) | *CGA* ” |
| | *CTG* ” | *CCG* ” | *CAG* ” | *CGG* ” |
| *A* | *ATT* Ile(*I*) | *ACT* Thr(*T*) | *AAT* Asn(*N*) | *AGT* Ser(*S*) |
| | *ATC* ” | *ACC* ” | *AAC* ” | *AGC* ” |
| | *ATA* ” | *ACA* ” | *AAA* Lys(*K*) | *AGA* Arg(*R*) |
| | *ATG* Met(*M*) | *ACG* ” | *AAG* ” | *AGG* ” |
| *G* | *GTT* Val(*V*) | *GCT* Ala(*A*) | *GAT* Asp(*D*) | *GGT* Gly(*G*) |
| | *GTC* ” | *GCC* ” | *GAC* ” | *GGC* ” |
| | *GTA* ” | *GCA* ” | *GAA* Glu(*E*) | *GGA* ” |
| | *GTG* ” | *GCG* ” | *GAG* ” | *GGG* ” |

Several interesting features emerge from considering such a table. There are 64 codons potentially available to specify 20 amino acids. It turns out, however, that there are only about half that many distinctive tRNA molecules, indicating that some tRNAs can bind to more than one codon. This redundancy is explained by the *wobble hypothesis*: Examination of tRNA structure shows that the nucleotide at one end of the anticodon has only a loose fit to the corresponding codon nucleotide—it wobbles. Thus H-bonding specificity is relaxed at this position and some tRNAs can bind to more than one codon.[6]

[6] Recall that polynucleotides have directionality; thus the two ends of a codon or anticodon are distinct. Only the one drawn at the right-hand end wobbles.

Not all possible codons specify an amino acid. Three of them are *termination codons*, or *stop codons*. They do not specify any amino acid; rather, they signal the ribosome to cease translation and to release the completed polypeptide. This is especially useful if one piece of mRNA codes for two adjacent genes: termination codons signal the translation machinery to release the first polypeptide before starting on the translation of the second one. Without the termination codons, the ribosome would continue to add the amino acids of the second polypeptide to the end of the first one, negating the biological functions of both.

*The nature of mutations.*

*Mutations* are changes in the nucleotide sequence of DNA. A base change in a codon would probably result in a new amino acid being coded at that point. For example, sickle-cell anemia results from a single incorrect amino acid being inserted into the protein fraction of hemoglobin. Suppose a nucleotide pair were deleted: Virtually every amino acid encoded thereafter (downstream) would be incorrect. Evidently, the severity of a *deletion error*, or an *insertion error*, for that matter, depends on how close to the start of transcription it occurs.

## 8.4 Enzymes and Their Function

*Two important concepts that have been presented in this chapter are the central dogma of genetics and the role of enzymes in facilitating specific chemical reactions in a cell. DNA, via RNA, codes for a specific set of cellular enzymes (among other proteins). Those enzymes can catalyze a specific set of chemical reactions and thereby determine the biological nature of the cell.*

*In this section, we will take a closer look at the way that enzymes work. Our approach will be a thermodynamic one, following the path of solar energy into biological systems, where it is used to create orderly arrangements of atoms and molecules in a cell. We will show how enzymes select from among the many possible configurations of these atoms and molecules to arrive at those that are peculiar to that type of cell.*

*The sun is the ultimate source of energy used by most biological systems.*

The sun is the ultimate source of energy available to drive biological processes. (We ignore the tiny amounts of energy available from geothermal sources.) Its contributions are twofold: First, solar energy can be captured by green plants and incorporated into chemical bonds, from which it can be then obtained by animals that eat the plants and each other. Second, solar energy heats the biosphere and thus drives biochemical reactions, virtually all of whose rates are temperature-dependent. Both of these considerations will be important in the discussion to follow.

*Entropy is a measure of disorder.*

A highly disordered configuration is said to have high *entropy*. The most disordered of two configurations is the one that can be formed in the most ways. To show how this definition conforms to our everyday experience, consider the possible outcomes of tossing three coins: HHH, HHT, HTH, THH, HTT, THT, TTH, TTT. There is only one way to get all heads, but there are six ways to get a mixture of heads and tails. Thus a mixture of heads and tails is the more disordered configuration. The condition of mixed heads and tails has high entropy (is a disorderly outcome), and the condition of all heads has low entropy (is an orderly outcome). Note that all eight specific configurations have the same probability ($\frac{1}{8}$), but that six of them contain at least one head and one tail.

Given that there generally are more disordered outcomes than there are ordered outcomes, we would expect that disorder would be more likely than order. This, of course, is exactly what we see in the case of the coins: Throw three coins and a mixture of heads and tails is the most common result, whereas all heads is a relatively uncommon result.

*The universe is proceeding spontaneously from lower to higher entropy.*

An empirical rule, the *second law of thermodynamics*, states that the entropy of the universe increases in every process. For instance, if a drop of ink is placed in a beaker of water, it will spontaneously spread throughout the water. There are few ways to put all the ink into one spot in the water and many ways to distribute it throughout the water, so we see that the entropy of the water/ink mixture increases. As other examples, consider what happens when the valve on a tank of compressed gas is opened or when a neatly arranged deck of cards is thrown up into the air. In each case, entropy increases.

The second law does not preclude a decrease in entropy in some local region. What it does require is that if entropy decreases in one place it must increase somewhere else by a greater absolute amount. There is no reason why the ink, once dispersed, cannot be reconcentrated. The point is that reconcentration will require some filtration or adsorption procedure that uses energy and generates heat. That heat will cause air molecules to move, and rapidly moving air molecules have more entropy (are more disordered) than slowly moving molecules. Likewise, the air can be pumped back into the tank and the cards can be picked up and resorted, both of which processes require work, which generates heat and therefore entropy.

Living systems are local regions of low entropy; their structures are highly organized, and even small perturbations in that organization can mean the difference between being alive and not being alive. From the earlier discussion, we can see that nothing in the second law forbids the low entropy of living systems, as long as the entropy of the universe increases appropriately during their formation.

*Entropy increases in a process until equilibrium is reached.*

Recall the examples of the previous section: The ink disperses in the water until it is uniformly distributed; the gas escapes the tank until the pressure is the same inside and outside of the tank; the cards flutter helter-skelter until they come to rest on a surface. In each case, the process of entropy-increase continues to some endpoint and then stops. That endpoint is called an *equilibrium state*.

Any equilibrium can be disrupted; more water can be added to the ink, the room containing the gas can be expanded, and the table bearing the cards can drop away. In each case, the system will then find a new equilibrium. Thus we can regard equilibria as temporary stopping places along the way to the maximal universal entropy predicted by the second law.

*Free energy is energy available to do useful work.*

Every organism needs energy for growing, moving, reproducing, and all the other activities we associate with being alive. Each of these activities requires organized structures. To maintain this organization, or low entropy, requires that the living system expend energy, much as energy was required to reconcentrate the ink or to re-sort the cards in the earlier examples.

*Free energy* is energy that can do useful work. In living systems, "useless" work is that which causes a volume change or which increases entropy. Whatever energy is left over is "free" energy. Living systems do not change their volume much, so entropy is the only significant thief of free energy in a cell. Therefore, free energy in a cell decreases when entropy increases. To a good approximation, we can assume that a living system begins with a certain amount of potential energy obtained from sunlight or food; some energy will then be lost to entropy production, and the remainder is free energy.[7]

To a physical chemist, the convenient thing about free energy is that it is a property of the system alone, thus excluding the surroundings. In contrast, the second law requires that one keep track of the entropy of the entire universe. As a result, it is usually easier to work with free energy than with entropy. We can summarize the relationship between the two quantities as they pertain to living systems by saying that entropy of the universe always increases during processes and that a system in equilibrium has maximized its entropy, whereas the free energy of a system decreases during processes and, at equilibrium, the system's free energy is minimized.

*Free energy flows, with losses, through biological systems.*

Thermonuclear reactions in the sun liberate energy, which is transmitted to the earth as radiation, which is absorbed by green plants. Some of the sun's radiation then heats the plant and its surroundings, and the rest is incorporated into glucose by *photosynthesis*. In photosynthesis, some of the free energy of the sun is used to create covalent bonds among parts of six carbon dioxide molecules, forming glucose, the six-carbon sugar, as shown in the following (unbalanced) reaction:[8]

---

[7] If you have studied physical chemistry, you will recognize this quantity specifically as Gibbs's free energy [4].

[8] The reason that water appears on both sides of the reaction equation is that the two water molecules are not the same: One is destroyed and the other is created in the reaction. The reaction shown is a summary of the many reactions that constitute photosynthesis.

$$CO_2 + H_2O \xrightarrow{\text{light energy}} \text{glucose} + H_2O + O_2$$

The plant, or an animal that eats the plant, then uses some of the free energy of the glucose to add a phosphate group to adenosine diphosphate (ADP) in the process called *respiration*:

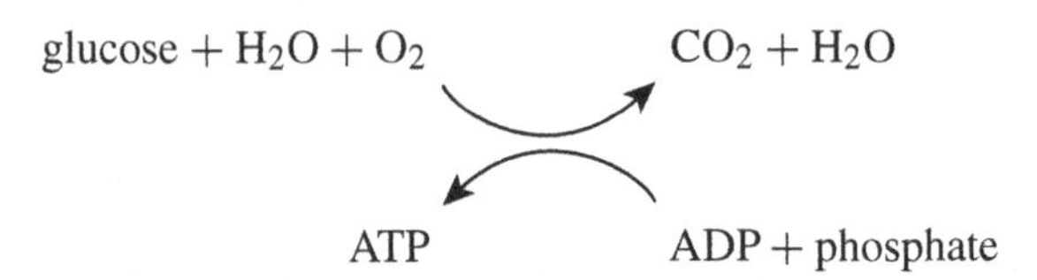


The resultant adenosine triphosphate now has some of the energy that originated in the sun. The ATP can then move around the cell by diffusion or convection and drive various life processes (moving, growing, repair, driving Na/K pumps, etc.):

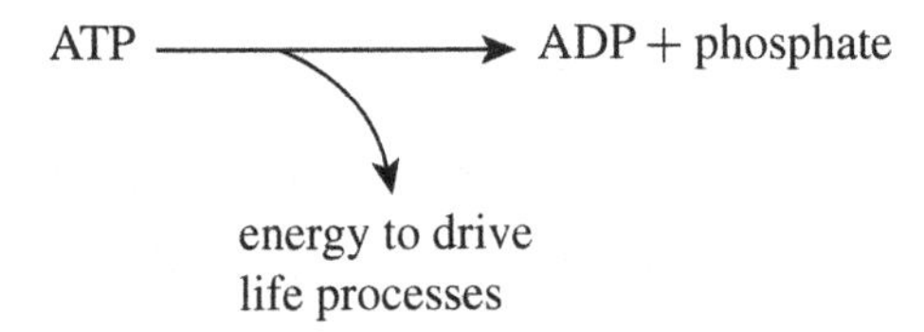


To recapitulate: Sunlight drives photosynthesis, in which carbon dioxide is combined to make glucose. The latter thus contains some of the energy that originated in the sun. In respiration, the plant or animal that eats the plant then converts some of the free energy in the glucose into free energy of ATP. Finally, at a site where it is needed, the ATP gives up its free energy to drive a biological process, e.g., contraction of a muscle.

At every step along the way from sun to, e.g., muscle movement, entropy is created and free energy is therefore lost. By the time an animal moves its muscle, only a small fraction of the original free energy the green plant got from the sun remains. If a subsequent carnivore should eat the herbivore, still more free energy would be lost. After the carnivore dies, decomposing organisms get the last of whatever free energy is available to living systems.

*The heat generated in biochemical reactions can help to drive other reactions.*

The earlier discussion pointed out that free energy, ultimately derived from the sun, is used to drive the processes we associate with being alive. As these processes occur, entropy is generated. Although the resultant heat energy will eventually be lost to the surroundings, it can be stored for a short while in the water of the cell and thus be used to maintain or increase the rates of cellular chemical reactions.

In order to understand how heat energy can promote chemical reactions, we need to digress a bit. If a process were able to occur spontaneously (increasing entropy; decreasing free energy), why would it not have already occurred? Water should spontaneously flow from a lake to the valley below, as shown in Figure 8.4.1(a). This has not happened because there is a dam in the way, but a siphon would take

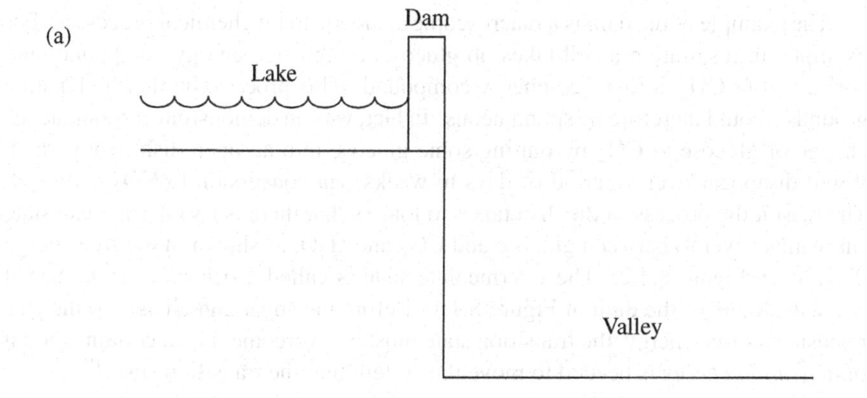


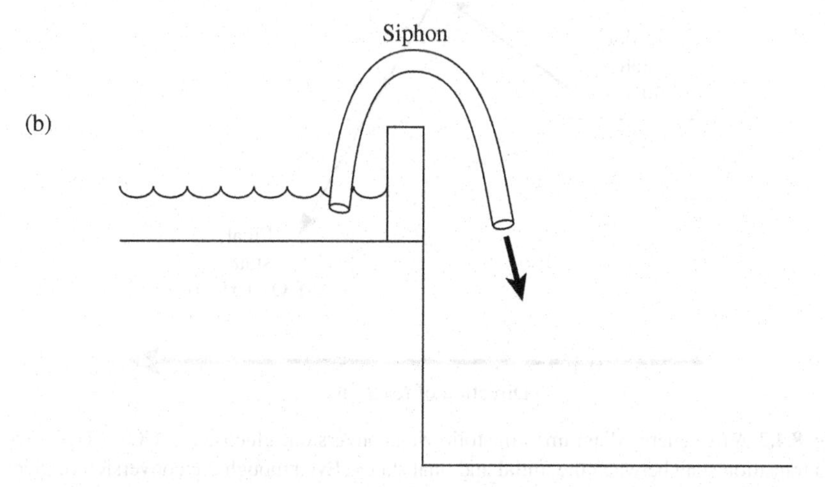


**Fig. 8.4.1.** (a) A lake holds back water above a valley; thus the water has a certain amount of potential energy with respect to the valley. (b) The water can get past the dam via a siphon, but the energy of the water with respect to the valley is not changed by the trip through the siphon. In other words, the energy yielded by the water in falling to the valley is independent of the path it takes. (We are assuming that friction is negligible.)

care of that without any net outlay of energy (Figure 8.4.1(b)). The latter point is critical: The water going up the siphon requires the same amount of energy that it gets back in going down the siphon.[9] From that point on, the water can fall to the valley, developing exactly as much kinetic energy as it would have if the dam had not existed in the first place.

[9] We are ignoring friction here.

The example of the dam is a macroscopic analogue to biochemical processes. For example, in respiration a cell takes up glucose, a high-free-energy compound, and converts it to $CO_2$, a low-free-energy compound. This process, on thermodynamic grounds, should therefore be spontaneous. In fact, we can demonstrate a spontaneous change of glucose to $CO_2$ by putting some glucose into an open dish, from which it will disappear over a period of days to weeks, via conversion to $CO_2$ and $H_2O$. The reason the process in the dish takes so long is that there is an intermediate state (in reality, several) between glucose and $CO_2$ and $H_2O$, as shown in the free energy diagram in Figure 8.4.2. The intermediate state is called a *transition state*, and it is the analogue of the dam in Figure 8.4.1. Before the sugar can change to the gas, releasing its free energy, the transition state must be overcome, i.e., a certain amount of *activation energy* is needed to move the system into the transition state.[10]

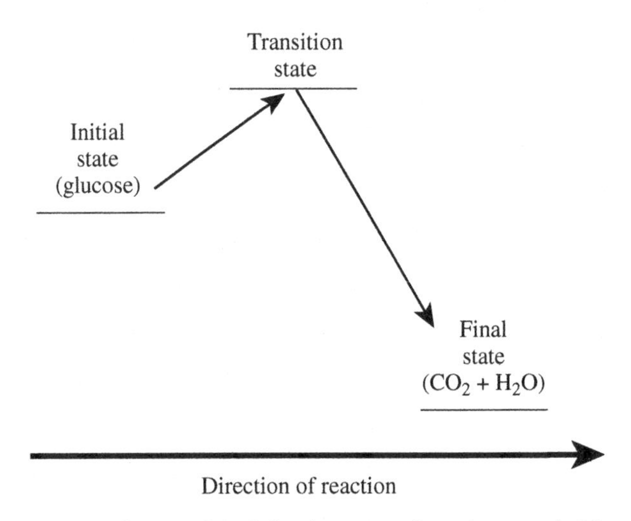


**Fig. 8.4.2.** A free energy diagram of the following conversion: glucose $\rightleftarrows$ $CO_2 + H_2O$. There is a transition state between the initial and final states. Even though the conversion of glucose to $CO_2$ and $H_2O$ is energetically downhill, it will not be a spontaneous conversion because of the transition state.

This energy is returned on the other side of the transition state, after which the chemical system behaves as if the transition state were not there. The examples of the dam and the glucose suggest a general conclusion: The net change in free energy between two states is independent of any intermediate states.

Transition states are the rule, not the exception, and the biochemical reactions of living systems are typical in that those that release free energy must first be activated.

[10] Figure 8.4.2 is, of course, only a model. The actual conversion of glucose to carbon dioxide in an open dish would involve numerous intermediate compounds, some of which would be real transition states and some of which would be more-or-less stable compounds. For instructive purposes, we represent the system as having a single transition state.

There are two sources of activation energy available to cells, however: First, most cells exist at 0–40°C, and second, heat energy is generated by the normal inefficiency of cellular processes.[11] This heat energy is stored in H-bond vibrations in the water of the cell, at least until it is finally lost to the external environment. While this heat energy is in the cell it is available to push systems into their transition states, thus promoting chemical reactions. After serving its activation function, the heat energy is returned unchanged.

The preceding discussion explains how heat serves a vital cellular function in providing activation energy to drive cellular biochemical reactions. This, however, does not close the subject, because activation energy is tied in with another observation: The glucose in a dish changes to $CO_2$ and $H_2O$ over a period of months, and the same change can occur in a cell in seconds or less. Yet the temperatures in the dish and in the cell are the same, say 37°C. The difference is that the reactions in the cell are catalyzed by enzymes.

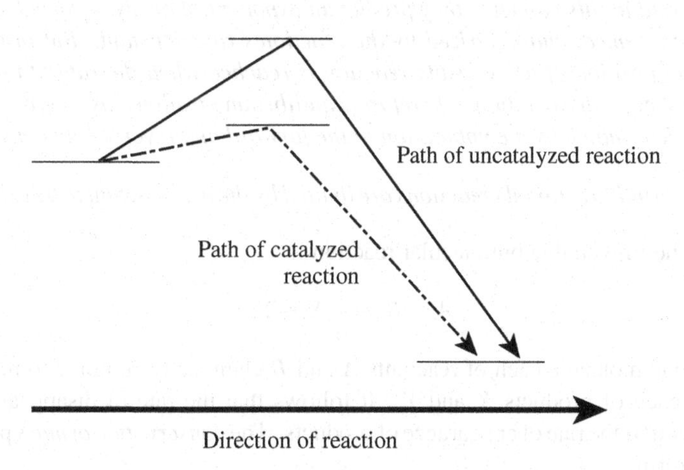


**Fig. 8.4.3.** The effect of enzymatic catalysis on the height of a transition state. The enzyme lowers the energy of the transition state, but as in Figure 8.4.1, the overall change in energy is independent of the path. Lowering the transition state does, however, permit the reaction to proceed spontaneously in the presence of a little thermal energy.

In brief, the catalytic function of an enzyme is to reduce the energy of the transition state and thereby to lessen the amount of heat energy needed by the system to meet the activation energy requirement. In this manner, the enzyme speeds up the rate at which the reaction proceeds from the initial state (100% reactant) toward the final, equilibrium state (almost 100% product). Figure 8.4.3 is a free-energy diagram for a biochemical system in its catalyzed and uncatalyzed conditions. The enzyme catalyst lowers the activation energy and makes it much easier for the initial state to be converted into the transition state and thus into the final state. The dependence of reaction rate on activation energy is exponential; thus a small change in activation

[11] Direct sunlight is also used by many "cold-blooded" animals to heat up their bodies.

energy can make a very big difference in reaction rate. For comparison, enzymatic catalysis potentially can speed up the rates of reactions by as much as $10^{13}$ times.

How much energy is actually available? At 30°C the average amount of heat energy available is about 0.025 eV per molecule, but the energy is unevenly distributed, and some substrate molecules will have more and some will have less. Those that have more will often have enough to get to the transition states made accessible by enzymatic catalysis.

## 8.5 Rates of Chemical Reactions

*Stoichiometric rules are not sufficient to determine the equilibrium position of a reversible chemical reaction; but adding reaction rate principles makes the calculation possible. Primarily, rate equations were designed to foretell the speed of specific reactions, and in this capacity, they predict an exponential decaying speed, as reactants are consumed, characterized by the reaction's rate constant. But in fact, the equilibrium position of a reversible reaction is reached when the rate of formation equals the rate of dissociation. Therefore, equilibrium positions, as well as reaction rates, are determined by a combination of the forward and reverse rate constants.*

*Irreversible (unidirectional) reactions are limited by the first reactant to be exhausted.*

Consider the irreversible bimolecular reaction

$$A + B \longrightarrow X + Y, \tag{8.5.1}$$

in which one molecule each of reactants $A$ and $B$ chemically combine to make one molecule each of products $X$ and $Y$. It follows that the rate of disappearance of reactants equals the rate of appearance of products. The *conservation of mass* principle takes the form

$$\frac{dX}{dt} = \frac{dY}{dt} = -\frac{dA}{dt} = -\frac{dB}{dt}.$$

If $M_0$ denotes the initial number of molecules of species $M$, by integrating each member of this chain of equalities from time zero to time $t$, we get

$$X(t) - X_0 = Y(t) - Y_0 = -A(t) + A_0 = -B(t) + B_0. \tag{8.5.2}$$

Equation (8.5.2) gives the amount of each species in terms of the others, so if any one of them is known, then they all are. But in order to know the amount of any one of them, we must know how fast the reaction occurs. This is answered by the *law of mass action* (due to Lotka): The rate at which two or more chemical species simultaneously combine is proportional to the product of their concentrations. Letting $[M]$ denote the concentration of species $M$, the mass action principle states that the rate at which product is formed is equal to

$$k[A][B], \tag{8.5.3}$$

where the constant of proportionality $k$ is characteristic of the reaction.

So far, our considerations have been completely general, but now we must make some assumptions about where the reaction is occurring. We suppose this to occur in a closed reaction vessel, such as a beaker with a fixed amount of water. In this case, concentration of a given species is the number of its molecules divided by the (fixed) volume of the medium.[12] There is the possibility that one or more of the products, $X$ or $Y$, is insoluble and precipitates out of solution. This is one of the main reasons that a bimolecular reaction may be irreversible. In what follows, we use the notation $m(t)$ to denote the concentration of species $M$. In case some species, say $X$, precipitates out of solution, we can still interpret $x(t)$ to mean the number of its molecules divided by the volume of the medium, but it will no longer be able to participate in the reaction. In this way, we can calculate the amount of $X$ that is produced. While a product may precipitate out without disturbing the (forward) reaction, the reactants must remain dissolved.

Combining the mass action principle with (8.5.2), we get

$$\frac{dx}{dt} = kab = k(a_0 + x_0 - x)(b_0 + x_0 - x) \tag{8.5.4}$$

with initial value $x(0) = x_0$. The stationary points of (8.5.4) are given by setting the right-hand side to zero and solving to get (see Section 2.4)

$$x = a_0 + x_0 \quad \text{or} \quad x = b_0 + x_0. \tag{8.5.5}$$

The first of these says that the amount of $X$ will be its original amount plus an amount equal to the original amount of $A$. In other words, $A$ will be exhausted. The second equation says the reaction stops when $B$ is exhausted.

Suppose, just for argument, that $a_0 < b_0$; then also $a_0 + x_0 < b_0 + x_0$. While $x(t) < a_0 + x_0$, the right-hand side of (8.5.4) is positive; therefore, the derivative is positive and so $x$ increases. This continues until $x$ asymptotically reaches $a_0 + x_0$, whereupon the reaction stops. The progression of the reaction as a function of time is found by solving (8.5.4), which is variables separable:

$$\frac{dx}{(a_0 + x_0 - x)(b_0 + x_0 - x)} = kdt.$$

Note the similarity of this equation to the Lotka–Volterra system of Section 4.4. The left-hand side can be written as the sum of simpler fractions:

$$\frac{1}{(a_0 + x_0 - x)(b_0 + x_0 - x)} = \frac{1}{b_0 - a_0}\frac{1}{a_0 + x_0 - x} - \frac{1}{b_0 - a_0}\frac{1}{b_0 + x_0 - x}.$$

Thus (8.5.4) may be rewritten as

$$\left[\frac{1}{a_0 + x_0 - x} - \frac{1}{b_0 + x_0 - x}\right] dx = (b_0 - a_0)kdt.$$

[12] By contrast, for an open reaction vessel, such as the heart or a chemostat, the concentrations are determined by that of the inflowing reactants.

Integrating gives the solution

$$-\ln(a_0 + x_0 - x) + \ln(b_0 + x_0 - x) = (b_0 - a_0)kt + q,$$
$$\ln\left(\frac{b_0 + x_0 - x}{a_0 + x_0 - x}\right) = (b_0 - a_0)kt + q,$$

where $q$ is the constant of integration. Now this may be solved for in terms of $x$,

$$x = \frac{(a_0 + x_0)Qe^{(b_0-a_0)kt} - (b_0 + x_0)}{Qe^{(b_0-a_0)kt} - 1}, \tag{8.5.6}$$

where $Q = e^q$ is a constant. This equation is graphed in Figure 8.5.1. The procedure described above can be performed by the computer using code such as the following:

```
MAPLE
> k:=1; a0:=2; b0:=3; x0:=1/2;
> dsolve({diff(x(t),t)=k*(a0+x0-x(t))*(b0+x0-x(t)),x(0)=x0},{x(t)});
> simplify(%);
> x:=unapply(rhs(%),t);
> plot([t,x(t),t=0..4],t=-1..3,tickmarks=[3,3],labels=['t','x(t)']);

MATLAB
% make up an m-file, chemRate.m, with
%  function xprime=chemRate(t,x);
%  k=1; a0=2; b0=3; x0=0.5; xprime = k*(a0+x0-x)*(b0+x0-x);
> x0=0.5; [t,x]=ode23('chemRate',[0 4],x0);
> plot(t,x)
```

The result is

$$x(t) = \frac{1}{2}\frac{-15 + 14e^{-t}}{-3 + 2e^{-t}}.$$

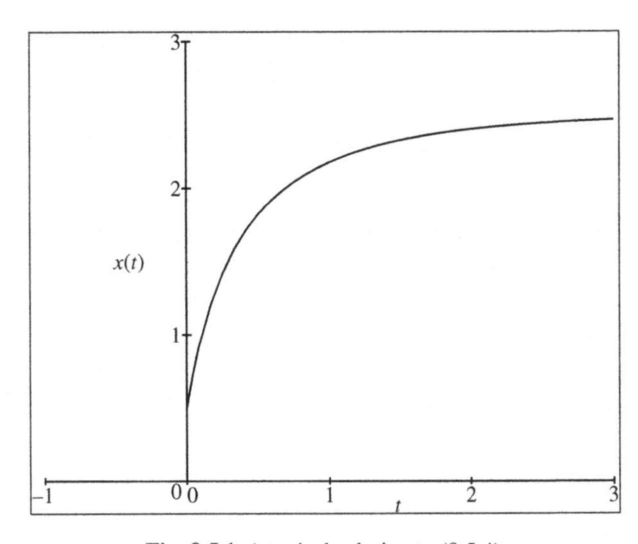


**Fig. 8.5.1.** A typical solution to (8.5.4).

**Example.** Suppose 2 moles of silver nitrate ($AgNO_3$) are mixed with 3 moles of hydrochloric acid (HCl). A white precipitate, silver chloride, is formed and the reaction tends to completion:

$$AgNO_3 + HCl \longrightarrow AgCl \downarrow + HNO_3.$$

From above, asymptotically, the reaction stops when the 2 moles of silver nitrate have reacted, leaving 2 moles of silver chloride precipitate and 1 mole of hydrochloric acid unreacted.

*Kinetics for reversible reactions work the same way.*

Now assume that reaction (8.5.1) is reversible,

$$A + B \rightleftarrows X + Y, \tag{8.5.7}$$

with the reverse reaction also being bimolecular. This time there is a backward rate constant, $k_{-1}$, as well as a forward one, $k_1$. From the mass action principle applied to the reverse reaction, we have

$$\text{rate of conversion of } X + Y = k_{-1}[X][Y].$$

Under normal circumstances, the forward and backward reactions take place independently of each other, and consequently the net rate of change of any species, say $X$, is just the sum of the effects of each reaction separately. It follows that the net rate of change in $X$ is given by

$$\begin{aligned} \frac{dx}{dt} &= (\text{conversion rate of } A + B) - (\text{conversion rate of X+Y}) \\ &= k_1[A][B] - k_{-1}[X][Y] \\ &= k_1(a_0 + x_0 - x)(b_0 + x_0 - x) - k_{-1}x(y_0 - x_0 + x), \end{aligned} \tag{8.5.8}$$

where (8.5.2) has been used in the last line. Circumstances under which the forward and backward reactions are not independent include precipitation of one of the species, as we have seen above. Another occurs when one of the reactions is highly exothermic. In that case, conditions of the reaction radically change, such as the temperature.

The analysis of (8.5.8) goes very much like that of (8.5.4). The stationary points are given as the solutions of the $\frac{dx}{dt} = 0$ equation

$$\begin{aligned} 0 &= k_1(a_0 + x_0 - x)(b_0 + x_0 - x) - k_{-1}x(y_0 - x_0 + x) \\ &= (k_1 - k_{-1})x^2 - (k_1(a_0 + b_0 + 2x_0) + k_{-1}(y_0 - x_0))x + k_1(a_0 + x_0)(b_0 + x_0). \end{aligned} \tag{8.5.9}$$

As one can see, if $k_1 \neq k_{-1}$, this is a quadratic equation and therefore has two roots, say $x = \alpha$ and $x = \beta$, which may be found using the quadratic formula, $\frac{1}{2a}(-b \pm \sqrt{b^2 - 4ac})$. The right-hand side of (8.5.8) thus factors into the linear factors

$$\frac{dx}{dt} = (k_1 - k_{-1})(x - \alpha)(x - \beta). \tag{8.5.10}$$

Again, just as above, this variable-separable differential equation is easily solved, but the nature of the solution depends on whether the roots are real or complex, equal or distinct. To decide about that, we must examine the discriminant of the quadratic formula, $b^2 - 4ac$. By direct substitution of the coefficients from (8.5.9) into the discriminant and simplifying, we get

$$b^2 - 4ac = k_1^2(a_0 - b_0)^2 + 2k_1k_{-1}(a_0 + b_0 + 2x_0)(y_0 - x_0) + k_{-1}^2(y_0 - x_0)^2. \tag{8.5.11}$$

The first and last terms are squares and so are positive (or zero). We see that if $y_0 \geq x_0$, then the discriminant is always positive or zero and the two roots are real. Since $X$ was an arbitrary choice, we can always arrange that $y_0 \geq x_0$, and so we assume that this is so.

Unless the initial concentrations are equal, $a_0 = b_0$ and $y_0 = x_0$, the roots will be distinct. We assume without loss of generality that

$$\alpha < \beta. \tag{8.5.12}$$

Then in a similar way to the derivation of (8.5.6), the solution of (8.5.10) is

$$\ln\left(\frac{x - \beta}{x - \alpha}\right) = (\beta - \alpha)(k_1 - k_{-1})t + q,$$

where $q$ is the constant of integration. This may be solved in terms of $x$,

$$x = \frac{\beta - Qe^{rt}}{1 - Qe^{rt}}, \tag{8.5.13}$$

where $Q$ is a constant and

$$r = (\beta - \alpha)(k_1 - k_{-1}).$$

If the discriminant is zero, then $\beta = \alpha$, and in that case the solution is

$$\frac{-1}{x - \alpha} = (k_1 - k_{-1})t + q,$$

or

$$x = \alpha - \frac{1}{(k_1 - k_{-1})t + q},$$

where $q$ is again the constant of integration.

### Exercises/Experiments

1. Suppose that $A + B \to C$, that the initial concentrations of $A$, $B$, and $C$ are $\frac{1}{2}$, $\frac{1}{3}$, and 0, respectively, and that the rate constant is $k$.

(a) Show that this leads to the differential equation in $z(t) = [C(t)]$ given by

$$z' = k\left(\frac{1}{2} - z\right)\left(\frac{1}{3} - z\right), \quad z(0) = 0.$$

(b) Solve this equation.
(c) Show that the corresponding equation for $x(t) = [A(t)]$ is

$$x' = kx\left(\frac{1}{6} - x\right), \quad x(0) = \frac{1}{2}.$$

(d) Solve this equation. Show by adding the solutions $x$ and $z$ that the sum is constant.
(e) At what time is 90% of the steady-state concentration of $C$ achieved?
(f) Suppose that $k$ is increased 10%. Now rework question (e).

2. Suppose that $A + B \leftrightarrow C + D$ is a reversible reaction, that the initial concentrations of $A$ and $B$ are $\frac{4}{10}$ and $\frac{5}{10}$, respectively, and that the initial concentrations of $C$ and $D$ are 0. Take $k_1 = 10$ and $k_{-1} = \frac{5}{2}$.
(a) Show that this leads to the differential equation

$$y' = 10(0.4 - y)(0.5 - y) - \frac{5y^2}{2}, \quad y(0) = 0.$$

(b) What is the equilibrium level of $[C]$. Draw two graphs: one where $k_{-1} = \frac{5}{2}$ and one where $k_{-1} = \frac{5}{4}$.

## 8.6 Enzyme Kinetics

*Enzymes serve to catalyze reactions in living systems, enabling complex chemical transformations to occur at moderate temperatures, many times faster than their uncatalyzed counterparts. Proteins, serving as the catalysts, are first used and then regenerated in a multistep process. Overall, the simplest enzyme-catalyzed reactions transform the enzyme's specific substrate into product, possibly with the release of a by-product. Referred to as enzyme saturation, these reactions are typically rate limited by the amount of enzyme itself. The degree to which saturation occurs relative to substrate concentration is quantified by the Michaelis–Menten constant of the enzyme–substrate pair.*

*Enzyme-catalyzed reactions are normally rate limited by enzyme saturation.*

The importance of enzyme-catalyzed reactions along with a general description of the biochemical principles of enzyme catalysis was given in Section 8.4. Here we will consider an enzyme, $E$, that acts on a single substrate, $S$, and converts it to an alternative form that is regarded as the product $P$. The enzyme performs this function by temporarily forming an enzyme–substrate complex, $C$, which then decomposes into product plus enzyme:

$$\begin{aligned} S + E &\rightleftarrows C \\ C &\longrightarrow P + E. \end{aligned} \tag{8.6.1}$$

The regenerated enzyme is then available to repeat the process.[13] Here we will work through the mathematics of enzyme kinetics. The general principles of chemical kinetics discussed in the previous section apply to enzyme kinetics as well. However, due to the typically small amount of enzyme compared to substrate, the conversion rate of substrate to product is limited when the enzyme becomes *saturated* with substrate as enzyme–substrate complex.

As in the previous section, we let $m$ denote the concentration of species $M$. The forward and reverse rate constants for the first reaction will be denoted by $k_1$ and $k_{-1}$, respectively, while the rate constant for the second will be taken as $k_2$. The rate equations corresponding to the reactions (8.6.1) are[14]

$$\begin{aligned} \frac{dc}{dt} &= k_1 es - k_{-1}c - k_2 c, \\ \frac{ds}{dt} &= -k_1 es + k_{-1}c, \\ \frac{de}{dt} &= -k_1 es + k_{-1}c + k_2 c, \\ \frac{dp}{dt} &= k_2 c. \end{aligned} \tag{8.6.2}$$

Note that complex $C$ is both formed and decomposed by the first reaction and decomposed by the second. Similarly, enzyme $E$ is decomposed and formed by the first reaction and formed by the second. The first three equations are independent of the formation of product $P$, and so for the present, we can ignore the last equation. As before, we denote by subscript 0 the initial concentrations of the various reactants. In particular, $e_0$ is the initial, and therefore total, amount of enzyme, since it is neither created nor destroyed in the process.

By adding the first and third equations of system (8.6.2), we get

$$\frac{dc}{dt} + \frac{de}{dt} = 0.$$

Integrating this and using the initial condition that $c_0 = 0$, we get

$$e = e_0 - c. \tag{8.6.3}$$

---

[13] Compare this scheme to Figure 8.4.3; $S + E$ constitutes the initial state, $C$ is the transition state, and $P + E$ is the final state.

[14] The units of $k_1$ are different from those of $k_{-1}$ and $k_2$, since the former is a bimolecular constant, while the latter are unimolecular.

We may use this to eliminate $e$ from system (8.6.2) and get the following reduced system:

$$\frac{dc}{dt} = k_1 s(e_0 - c) - (k_{-1} + k_2)c,$$
$$\frac{ds}{dt} = -k_1 s(e_0 - c) + k_{-1}c. \tag{8.6.4}$$

In Figure 8.6.1, we show some solutions of this system of differential equations.

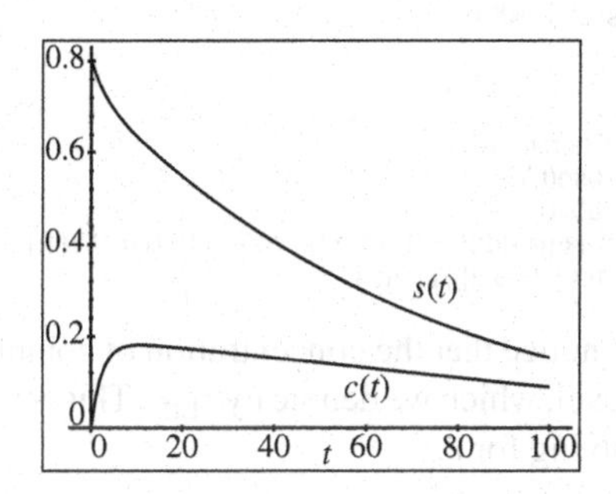


**Fig. 8.6.1.** Solutions to (8.6.2).

For the purpose of drawing the figure, we take the constants to be the following:

```
MAPLE
> k1:=1/10; km1:=1/10; k2:=1/10; e0:=4/10; (km1+k2)/k1;
```

The equations are nonlinear and cannot be solved in closed form. Consequently, we use numerical methods to draw these graphs. It should be observed that the level of $S$, graphed as $s(t)$, drops continuously toward zero. Also, the intermediate substrate $C$, graphed as $c(t)$, starts at zero, rises to a positive level, and gradually settles back to zero. In the exercises, we establish that this behavior is to be expected.

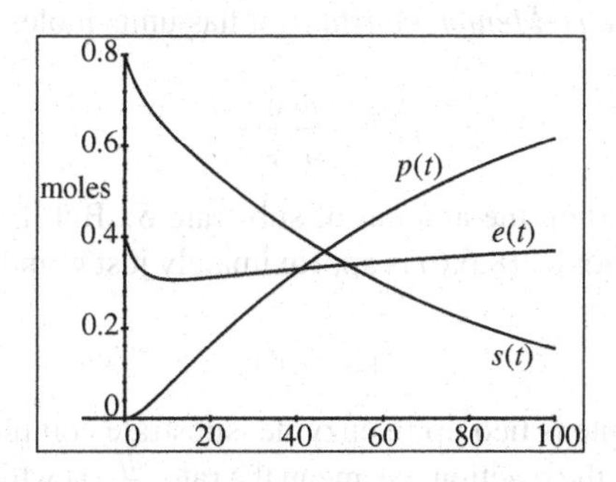


**Fig. 8.6.2.** Solutions to (8.6.1).

Also in the exercises we provide techniques to draw what may be a more interesting graph: Figure 8.6.2. In particular, we draw graphs of $s(t)$, $p(t)$, and $e(t)$. The first two of these are, in fact, the most interesting, since they demonstrate how much of $S$ is left and how much of $P$ has been formed. The addition of a graph for $e(t)$ illustrates that during the intermediate phase, some of the enzyme is tied up in the enzyme–substrate complex, but as the reaction approaches equilibrium, the value of $e(t)$ returns to its original value.

MAPLE

```
> with(plots): with(DEtools): # recall the parameters assigned on the previous page
> enz:=diff(c(t),t)=k1*s(t)*(e0-c(t))-(km1+k2)*c(t),diff(s(t),t)=-k1*s(t)*(e0-c(t))+km1*c(t);
> sol:=dsolve({enz,c(0)=0,s(0)=8/10},{c(t),s(t)}, type=numeric, output=listprocedure);
> csol:=subs(sol,c(t)); ssol:=subs(sol,s(t));
> J:=plot(csol,0..100): K:=plot(ssol,0..100):
> display({J,K});
```

MATLAB

```
% make up an m-file, enzymeRate.m, with
% function Yprime=enzymeRate(t,Y);
% k1=0.1; km1=0.1; k2=0.1; e0=0.4;
% Yprime=[k1*Y(2)*(e0-Y(1))-(km1+k2)*Y(1);-k1*Y(2)*(e0-Y(1))+km1*Y(1)];
> [t,Y]=ode23('enzymeRate',[0 100],[0; 0.8]); plot(t,Y)
```

From Figure (8.6.1), notice that the concentration of complex rises to a relatively invariant (“effective”) level, which we denote by $c_{\text{Eff}}$. This is found by setting $\frac{dc}{dt} = 0$ in system (8.6.4) and solving for $c$,

$$0 = k_1 s(e_0 - c) - (k_{-1} + k_2)c,$$

or

$$s(e_0 - c) = \frac{k_{-1} + k_2}{k_1} c.$$

The combination $k_M$ of rate constants

$$k_M = \frac{k_{-1} + k_2}{k_1} \tag{8.6.5}$$

is known as the *Michaelis–Menten constant*; it has units moles per liter. Solving for $c$ above, we get

$$c = \frac{s e_0}{k_M + s}, \tag{8.6.6}$$

which is seen to depend on the amount of substrate $S$. But if $s$ is much larger than $k_M$, then the denominator of (8.6.6) is approximately just $s$, and we find the invariant level of complex to be

$$c_{\text{Eff}} \approx e_0. \tag{8.6.7}$$

Thus most of the enzyme is tied up in enzyme–substrate complex.

By the *velocity* $v$ of the reaction, we mean the rate, $\frac{dp}{dt}$, at which product is formed. From system (8.6.2), this is equal to $k_2 c$. When the concentration of substrate is large, we may use $c_{\text{Eff}}$ as the concentration of complex and derive the maximum reaction velocity,

$$v_{\max} = k_2 e_0. \tag{8.6.8}$$

```
MAPLE
> fcn:=s->vmax*s/(kM+s); vmax:=10: kM:=15:
> crv:=plot([x,fcn(x),x=0..150],x=-20..160,y=-2..12,tickmarks=[0,0]):
> asy:=plot(10,0..150,tickmarks=[0,0]):
> midline:=plot(5,0..15.3,tickmarks=[0,0]):
> vertline:=plot([15.3,y,y=0..5],tickmarks=[0,0]):
> a:=0: A:=0: b:=13: B:=13*vmax/kM:
> slope:=x->A*(x-b)/(a-b)+B*(x-a)/(b-a):
> slopeline:=plot(slope,a..b):
> with(plots):
> display({crv,asy,midline,vertline,slopeline});
```

```
MATLAB
> vmax=10; kM=15;
> s=0:.1:150;
> v=vmax.*s./(kM+s);
> plot(s,v)
> hold on
> asy=vmax*ones(size(s));
> plot(s,asy)
> x=[0 13]; y=[0 13*vmax/kM];
> plot(x,y)
> x=[0 15.3]; y=[vmax/2 vmax/2];
> plot(x,y)
> x=[15.3 15.3];
y=[0 vmax/2];
> plot(x,y)
```

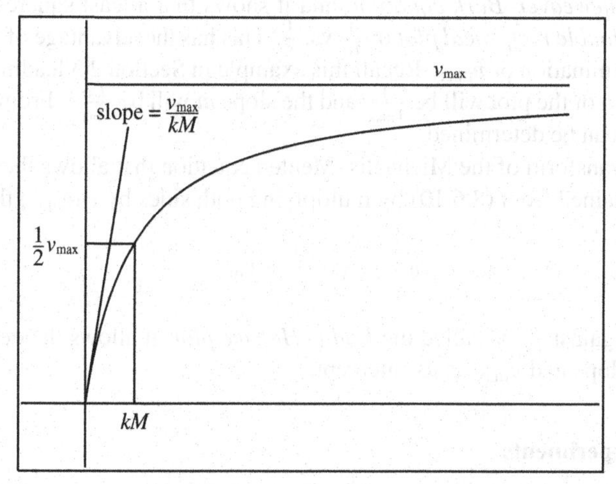


**Fig. 8.6.3.** Michaelis–Menten plot.

Likewise, from (8.6.6) and (8.6.8), the initial reaction velocity, $v_0$, is given by

$$v_0 = \left.\frac{dp}{dt}\right|_{t=0} = k_2\frac{se_0}{k_M + s} = \frac{v_{\max}s}{k_M + s}. \tag{8.6.9}$$

This is the *Michaelis–Menten equation*, the rate equation for a one-substrate, enzyme-catalyzed reaction. Its graph is shown in Figure 8.6.3.

The value of $k_M$ for an enzyme can be experimentally found from a plot of initial velocity vs. initial substrate concentration at fixed enzyme concentrations. This graph has the form of a rectangular hyperbola because at low substrate concentrations, $v_0$ is nearly proportional to substrate concentration $[S]$. On the other hand, at high substrate concentrations the reaction rate approaches $v_{\max}$ asymptotically because at these concentrations, the reaction is essentially independent of substrate concentration. By experimentally measuring the initial reaction rate for various substrate concentrations, we can make a sketch of the graph. Working from the graph, the substrate level that gives $\frac{1}{2}v_{\max}$ initial velocity is the value of $k_M$, seen as follows: From (8.6.9) with $v_0 = \frac{v_{\max}}{2}$,

$$\frac{1}{2}v_{\max} = \frac{v_{\max}s}{k_M + s},$$

and solving for $k_M$ gives

$$k_M = s.$$

Thus we interpret $k_M$ as the substrate concentration at which the reaction rate is half-maximal. By inverting the Michaelis–Menten equation (8.6.9), we get

$$\frac{1}{v_0} = \frac{k_M + s}{v_{\max}s} = \frac{k_M}{v_{\max}}\frac{1}{s} + \frac{1}{v_{\max}}. \tag{8.6.10}$$

This is the *Lineweaver–Burk equation*, and it shows that a least squares fit may be made to this *double reciprocal plot* of $\frac{1}{v_0}$ vs. $\frac{1}{s}$. This has the advantage of allowing an accurate determination of $v_{\max}$. Recall this example in Section 2.3 leading to (2.3.1). The intercept $b$ of the plot will be $\frac{1}{v_{\max}}$ and the slope $m$ will be $\frac{k_M}{v_{\max}}$. From these, both $v_{\max}$ and $k_M$ can be determined.

Another transform of the Michaelis–Menten equation that allows the use of least squares is obtained from (8.6.10) by multiplying both sides by $v_0 v_{\max}$; this yields

$$v_0 = -k_M\frac{v_0}{s} + v_{\max}. \tag{8.6.11}$$

A plot of $v_0$ against $\frac{v_0}{[S]}$ is called the *Eadie–Hofstee plot*; it allows the determination of $k_M$ as its slope and $v_{\max}$ as its intercept.

## Exercises/Experiments

**1.** Our intuition for the long-range forecast for (8.6.1) is that some of the reactants that move from $S$ to $C$ move on to $P$. But the assumption is that the second reaction is only one-way, so that the products will never move back toward $S$.[15] This suggests that $S$ will be depleted. We conjecture that $s_\infty = 0$ and $c_\infty = 0$. We confirm this with the notions of stability that we studied in Section 2.5.

(a) Find all the stationary solutions by observing that setting $\frac{ds}{dt} = 0$ and $\frac{dc}{dt} = 0$ leads to the equations

$$k_1 s(e_0 - c) - (k_{-1} + k_2)c = 0,$$

[15] In the context of a free-energy diagram (Figures 8.4.2 and 8.4.3), the one-way nature of the process $C \to P$ is due to a lack of sufficient free energy in the environment to cause the reaction $P \to C$.

$$-k_1 s(e_0 - c) + k_{-1}c = 0.$$

While it is clear that $s = 0$ and $c = 0$ is a solution, establish that this is the only solution for the equations as follows:

```
MAPLE
> restart;
> solve(k1*s*(e0-c)-(km1+k2)*c=0,c);
> subs(c=%, -k1*s*(e0-c)+km1*c); normal(%);
> numer(%)/denom(%%)=0;

MATLAB
% MATLAB cannot symbolically solve the system but we can proceed this way: add the two
% equations and notice that the first terms cancel and the second terms nearly cancel, leaving
% -k2*c = 0. This shows that c=0. With c=0 in either equation it is easy to see that s=0 too. Now
% we find the Jacobian numerically.
% Make an m-file, enzyme96.m, with
%  k1=1; k2=2; km1=1.5; e0=5;
%  csPrime=[k1*s.*(e0-c)-(km1+k2)*c; -k1*s.*(e0-c)+km1*c];
%
% The Jacobian = the matrix whose first column is the derivative of the component functions
% with respect to c and the second column is with respect to s. Take derivatives at c=s=0.
> J1=(enzyme96(0+eps,0)-enzyme96(0,0))/eps; J2=(enzyme96(0,0+eps)-enzyme96(0,0))/eps;
> J=[J1 J2]; % Jacobian at (0,0)
> eig(J) % both values negative real, so (0,0) stable
```

Substitute this into the second equation and set the resulting equation equal to zero. Argue that $s$ must be zero and $c$ must be zero.

(b) Establish that $s = c = 0$ is an attracting stationary point, by finding the linearization about this one and only stationary point. (Recall Section 4.4.)

```
MAPLE
> with(LinearAlgebra): with(VectorCalculus):
> Jacobian([k1*s*(e0-c)-(km1+k2)*c,-k1*s*(e0-c)+km1*c],[c,s]);
> subs({c=0,s=0},%);
> Eigenvalues(%);
> expand((km1+k2+k1*e0)^2-4*k2*k1*e0);
```

(c) Verify that the eigenvalues of the linearization are

$$-\frac{1}{2}\left((k_{-1} + k_2 + k_1 e_0) \pm \sqrt{(k_{-1} + k_2 + k_1 e_0)\hat{\ }2 - 4k_2 k_1 e_0}\right)$$

and that both these are negative. Argue that this implies that $\{0, 0\}$ is an attracting stationary point for $\{c(t), s(t)\}$.

**2.** Draw the graph of Figure (8.6.3). Here is the syntax that does the job:

```
MAPLE
> restart;
> k1:=1/10: k2:=1/10: km1:=1/10: s0:=8/10: e0:=4/10:
> sol:=dsolve({diff(s(t),t)=-k1*e(t)*s(t)+km1*c(t),diff(c(t),t)=k1*e(t)*s(t)-(km1+k2)*c(t),
     diff(p(t),t)=k2*c(t),diff(e(t),t)=-k1*e(t)*s(t)+(km1+k2)*c(t),
     s(0)=s0, c(0)=0, p(0)=0, e(0)=e0},{s(t),c(t),p(t),e(t)},numeric,output=listprocedure);
> es:=subs(sol,e(t)); ps:=subs(sol,p(t)); ss:=subs(sol,s(t));
> plot({es,ps,ss},0..100,color=[red,green,black]);

MATLAB
% For problem 2 and 3
% Make an m-file, exer962.m, with
%  function Yprime=exer962(t,Y); % Y(1)=c, Y(2)=s, Y(3)=e, Y(4)=p
%  k1=0.1; k2=0.1; km1=0.1; % problem 2
%  k1=1; k2=0.1; km1=0.025; % problem 3
% Yprime=[k1*Y(3).*Y(2)-(km1+k2)*Y(1);-k1*Y(3).*Y(2)+km1*Y(1);
%       -k1*Y(3).*Y(2)+(km1+k2)*Y(1); k2*Y(1)];
```

```
> s0=0.8; e0=0.4; [t,Y]=ode23('exer962',[0 100],[0;s0; e0; 0]);
> plot(t,Y)
```

**3.** Draw the graph of the solution $c(t)$ in system (8.6.2) with constants chosen so that $k_M \approx 1$ and $S = 10$. The point to observe is that $c(t) \approx e_0$ for large values of $t$.

```
MAPLE
> restart;
> k1:=1: k2:=1/10: km1:=1/40: s0:=10: e0:=4/10: (km1+k2)/k1: s0/(%+s0);
> sol:=dsolve({diff(s(t),t)=-k1*e(t)*s(t)+km1*c(t),diff(c(t),t)=k1*e(t)*s(t)-(km1+k2)*c(t),
        diff(p(t),t)=k2*c(t),diff(e(t),t)=-k1*e(t)*s(t)+(km1 + k2)*c(t),
        s(0)=s0,c(0)=0,p(0)=0,e(0)=e0},{s(t),c(t),p(t),e(t)},numeric,output=listprocedure);
> cs:=subs(sol,c(t));
> plot(cs,0..150);
```

```
MATLAB
% continued from previous problem and rerun for the second part change k1 and km1 above to k1=1;
% km1=0.025;
> s0=10; e0=0.4;
> [t,Y]=ode23('exer962',[0 100],[0;s0; e0; 0]);
> plot(t,Y(:,1)) % graph of c
```

**4.** Suppose that $A + B \to C$, that the initial concentrations of $A$, $B$, and $C$ are 2, 3, and 0, respectively, and that the rate constant is $k$.

(a) The concentration of $C$ is sampled at $t = \frac{3}{2}$ and is found to be $\frac{3}{5}$. What is an approximation for $k$?

(b) Instead of determining the concentration of $C$ at just $t = \frac{3}{2}$, the concentration of $C$ is found at five times (see Table 8.6.1).

**Table 8.6.1.**

| Time | Concentration |
|---|---|
| $\frac{1}{2}$ | 0.2 |
| 1.0 | 0.4 |
| $\frac{3}{2}$ | 0.6 |
| 2.0 | 0.8 |
| $\frac{5}{2}$ | 1.0 |

Estimate $k$. Plot your data and the model that your $k$ predicts on the same graph.

**5.** We have stated in this chapter that the addition of an enzyme to a reaction could potentially speed the reaction by a factor of $10^{13}$. This problem gives a glimpse of the significance of even a relatively small increase in the reaction rate. Suppose that we have a reaction

$$A \leftrightarrow B \to C.$$

Suppose also that $k_{-1} = k_2 = 1$, that the initial concentration of $A$ is $a_0 = 1$, and the initial concentrations of $B$ and $C$ are zero.

(a) Show that the differential equations model for this system is

$$\frac{da}{dt} = -k_1 a(t) + k_{-1} b(t),$$

$$\frac{db}{dt} = k_1 a(t) - (k_{-1} + k_2) b(t),$$
$$\frac{dc}{dt} = k_2 b(t).$$

(b) Find $a(t)$, $b(t)$, and $c(t)$ for $k_1 = 1$ and for $k_1 = 10$. Plot the graphs for the three concentrations in both situations.

MAPLE
```
> with(LinearAlgebra):
> k1:=1; km1:=1; k2:=1;
> A:=Matrix([[-k1,km1,0],[k1,-km1-k2,0],[0,k2,0]]);
> u:=evalm(MatrixExponential(A,t) &* [1,0,0]):
> a:=unapply(u[1],t); b:=unapply(u[2],t); c:=unapply(u[3],t):
> plot({a(t),b(t),c(t)},t=0..7,color=[red,blue,black]);
> fsolve(c(t)=.8,t,1..7);
```

MATLAB
```
% contents of m-file exer964.m:
%  function Yprime=exer964(t,Y); km1=1; k2=1; k1=1;
%  Yprime=[-k1*Y(1)+k1*Y(2); k1*Y(1)-(km1+k2)*Y(2); k2*Y(2)];
%
> a0=1; b0=0; c0=0;
> [t1,Y1]=ode23('exer964',[0 7],[a0; b0; c0]);
> plot(t1,Y1);
% now change k1 above to k1=10; note the y-axis scale
> [t10,Y10]=ode23('exer964',[0 7],[a0; b0; c0]);
> plot(t10,Y10)
% compare product c directly; be sure to check the y-axis scale
> plot(t1,Y1(:,3),'b')
> hold on
> plot(t10,Y10(:,3),'r')
```

(c) Take $k_1 = 1, 10, 20, 30, 40,$ and $50$. Find $T_k$ such that $c(T_k) = 0.8$ for each of these $k$s. Plot the graph of the pairs $\{k, T_k\}$. Find an analytic fit for these points.

## Questions for Thought and Discussion

**1.** Draw the structural formulas ("stick models") for
   (a) butane, the four-carbon hydrocarbon, having all carbons in a row, and no double bonds;
   (b) isopropanol, having three carbons, an -OH group on the middle carbon, and no double bonds;
   (c) propene, with three carbons and one double bond.

**2.** Relate this set of reactions to a free-energy level diagram: $A+E \leftrightarrow B \rightarrow C+E$, where $E$ is an enzyme. What effect does $E$ have on the energy levels?

**3.** Assume the reaction $A \leftrightarrow C \leftrightarrow B$, where the intermediate state $C$ has a *lower* free energy than $A$ or $B$. Knowing what you do about the behavior of free energy, what would happen to the free energy difference between $A$ and $B$ if the free energy of $C$ were changed?

**4.** A mechanical analogue of the situation in Question 3 is a wagon starting at $A$, rolling downhill to $C$, and then rolling uphill to $B$. There is a frictional force on the wagon wheels. What do you think will be the effect of varying the depth of $C$?

**5.** Describe the chemical differences between RNA and DNA. What are their biological (functional) differences?

**6.** Outline the process of information flow from DNA to the control of cellular chemistry.

**7.** Name six kinds of proteins and describe their functions.

## References and Suggested Further Reading

[1] THERMODYNAMICS:
H. J. Morowitz, *Energy Flow in Biology*, Academic Press, New York, 1968.

[2] BIOCHEMICAL STRUCTURE:
L. Stryer, *Biochemistry*, 2nd ed., W. H. Freeman, 1981.

[3] BIOCHEMICAL STRUCTURE AND THERMODYNAMICS:
E. K. Yeargers, *Basic Biophysics for Biology*, CRC Press, Boca Raton, FL, 1992.

[4] THERMODYNAMICS AND ENZYME FUNCTION:
P. W. Atkins, *Physical Chemistry*, W. H. Freeman, 3rd ed., New York, 1986.

[5] CHEMICAL GENETICS:
D. T. Suzuki, A. J. F. Griffiths, J. H. Miller, and R. C. Lewontin, *An Introduction to Genetic Analysis*, W. H. Freeman, 3rd ed., New York, 1986.

[6] CHEMICAL GENETICS:
J. D. Watson, N. W. Hopkins, J. W. Roberts, J. A. Steitz, and A. M. Weiner, *Molecular Biology of the Gene*, 4th ed., Benjamin/Cummings, Menlo Park, CA, 1987.

# Part II

## Systems and Diseases

# 9

# The Biological Disposition of Drugs and Inorganic Toxins

## Introduction

This chapter is a discussion of how some foreign substances get into the body, how they become distributed, what their effects are, and how they are eliminated from the body. Lead is the exemplar in the biological discussion, but the biological concepts can be applied to many other substances. The mathematical discussion focuses on lead poisoning and on pharmaceuticals.

Lead can be eaten, inhaled, or absorbed through the skin; it is then distributed to other tissues by the blood. Some of it is then removed from the body by excretion and defecation. Any lead that is retained in the body can have unpleasant biological consequences—anemia and mental retardation, for instance. These processes can be understood only at the levels of organ systems and of the tissues that make up those organs. Thus this chapter includes discussions of the lungs, the digestive tract, the skin, blood, the circulatory system, bones, and the kidneys, all of which are involved in the effects of lead on humans.

## 9.1 The Biological Importance of Lead

*No biological requirement for lead has ever been demonstrated. Rather, there is much experimental evidence that it is toxic. Lead is sparsely distributed in nature, but mining activities to support the manufacture of batteries, leaded gasoline, and other products have concentrated it.*

*Trace metals play an important role in nutrition.*

A number of metallic elements play crucial roles in our nutrition, usually in small amounts. Sodium is necessary to nerve conduction, iron is an essential part of hemoglobin, and magnesium is a component of chlorophyll. In many cases, traces of metals are required for the correct functioning of biomolecules; examples are copper, manganese, magnesium, zinc, and iron. On the other hand, no human metabolic

R.W. Shonkwiler and J. Herod, *Mathematical Biology: An Introduction with Maple and Matlab*, Undergraduate Texts in Mathematics, DOI: 10.1007/978-0-387-70984-0_9,

need for lead has ever been demonstrated, whereas the toxic effects of lead are well documented.

*Most lead enters the biosphere through human-made sources.*

Lead is found in the earth's crust in several kinds of ore. Its natural concentration in any one place is almost always quite low and is of little concern to biology. Commercial uses for lead, however, have produced locally high concentrations of the metal in air, water, and soil (see [1] and [2]).

About 40% of the refined lead in the Western world is used in the manufacture of batteries, and another 10% finds its way into antiknock compounds in gasoline. Cable sheathing and pipes account for about another 15%. Still more is used in paints, glassware, ceramics, and plastics. Fortunately, the use of lead in interior paints and gasoline has lately been approaching zero in the USA and some other countries.

*A biological compartment model shows the paths of lead into, around, and out of an organism.*

It often happens that a particular material, introduced into one part of an organism, quickly reaches a common concentration in several other parts of the organism. The various parts in which this equilibration occurs constitute a single *biological compartment*. Note that the parts of a compartment can be organ systems, organs, or parts of an organ, and that they need not be near each other if a suitable distribution system is available (see [3]).

Figure 9.1.1 illustrates these concepts in a biological compartment model for lead. One compartment is the blood, in which flow and turbulence cause rapid mixing. The blood carries the lead to "soft tissues," which we take to mean all tissues that are neither bone nor blood. All these soft tissues behave similarly toward lead and take it up to about the same degree, so they can be considered to constitute a second compartment.[1] A third compartment is bone, in which lead has a very long half-life. The fourth compartment is implied: It is the environment, from which the lead originates and to which it must ultimately return, either through living biological processes or at the death of the organism. The arrows connecting the compartments show the direction of lead movement between the various compartments.

## 9.2 Early Embryogenesis and Organ Formation

*In discussions of the uptake and metabolism of toxins and drugs, the relevant compartments are almost always organs and groups of organs. This section presents some information about how organs develop in embryos. We show that organs and*

[1] This is a somewhat rough approximation: The aorta, the main artery out of the heart, is a soft tissue that seems not to equilibrate lead quickly with other soft tissues. Further, the behavior of a few other soft tissues toward lead seems to depend on the national origin and age of the cadavers used in the data collection. (See Schroeder and Tipton [4].)

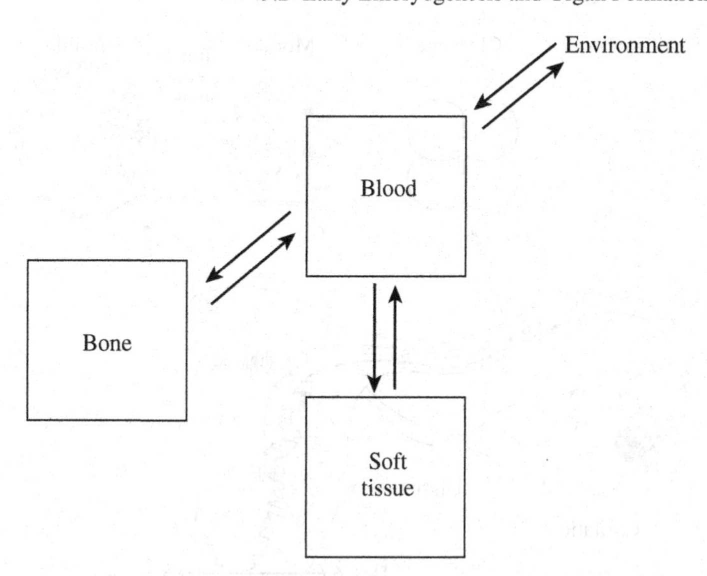


**Fig. 9.1.1.** A compartment model of the three biological tissue types among which lead becomes distributed, and the relationship of the compartments to the environmental compartment.

*their interdependence on one another develop from the very start. This will allow us to treat organs as a group of interacting compartments for mathematical purposes. Thus the stage will be set for our discussion of lead poisoning.*

*Specific organs have evolved in multicellular organisms to perform very specific biological tasks. This functional specialization has a benefit and a cost. The benefit is that a given organ is usually very efficient at performing its assigned biological tasks. The cost is that the organ generally can do little else and must therefore rely on other organs to support it in other functions. As examples, the heart is a reliable blood pump and the kidneys efficiently remove nitrogenous wastes from blood. However, the heart is dependent on the kidneys to remove blood wastes that would be detrimental to the heart, and the kidneys are dependent on the heart to pump blood at a high enough pressure to make the kidneys work.*

*Early divisions of fertilized eggs result in an unchanged total cell mass.*

A fertilized egg, or *zygote*, starts to divide by mitosis soon after fertilization. In the case of humans, this means that division starts while the zygote is still in the oviduct. Early divisions merely result in twice the number of cells, without any increase in total cell mass, and are thus called *cleavage divisions* (see [5]).

A one-celled zygote contains the full complement of genes available to the organism.[2] As mitosis proceeds, many of the new cells *differentiate*, or take up more specialized functions, a process that coincides with the inactivation of unneeded

[2] In Chapter 10, we will discuss an exception to this statement: Certain viruses can insert genes into cells.

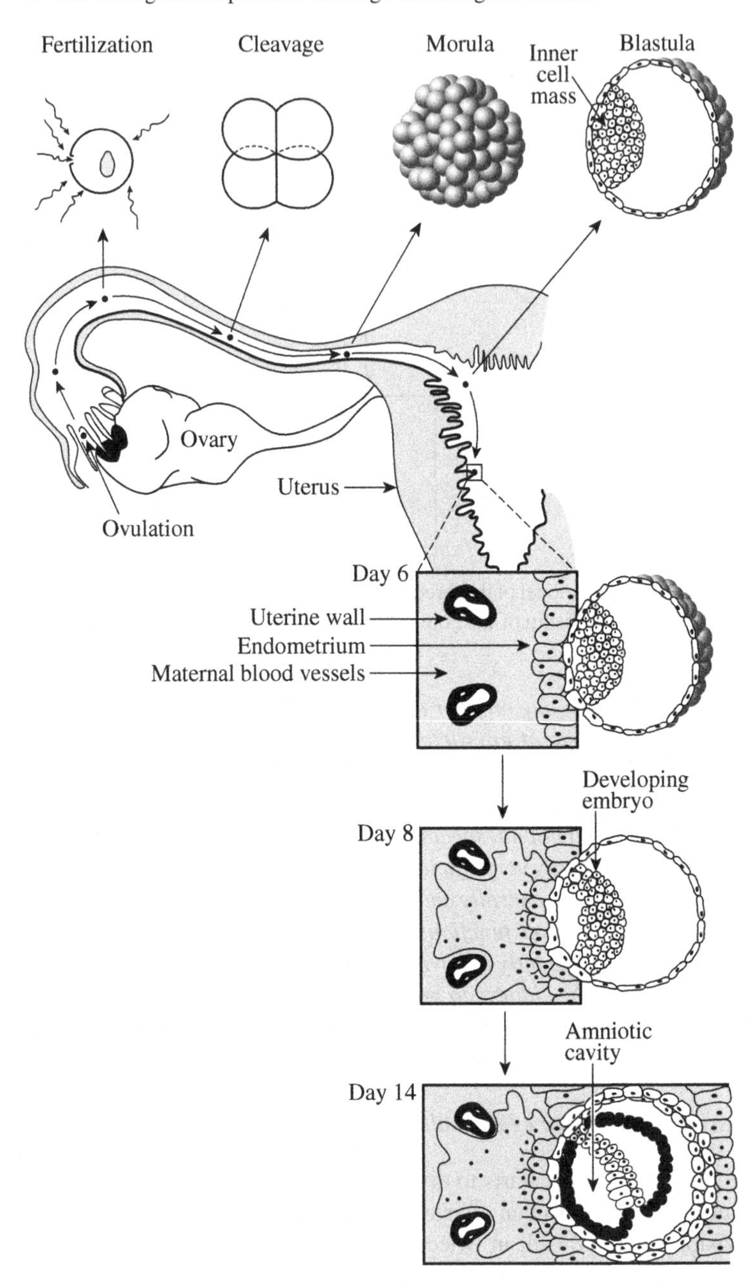


**Fig. 9.2.1.** The stages of development of a mammalian fetus, arranged with respect to where they occur in the female reproductive system. A single cell becomes a mass of cells (called a morula) and then forms a hollow ball (a blastula), with the inner cell mass at one side. (Redrawn from W. Purves, G. Orians, and C. Heller, *Life: The Science of Biology*, 4th ed., Sinauer Associates, Sunderland, MA, 1995. Used with permission.)

genes. The genes that remain active in a particular cell are those that determine the functional nature of that cell. Thus cells of the heart and of the liver require different sets of active genes (although the two sets certainly overlap).[3] The gene inactivation process starts early: Even before the heart organ itself becomes obvious, there are heart precursor cells that individually contract. Soon, however, a functioning heart and blood vessels develop in concert with the fetus's need for a continuing blood supply.

*Higher organisms have a three-layer body plan.*

Embryos of each species have their own unique behavior, but there are several events in embryogeny that are shared among most multicellular species: Early cleavage generates a solid mass of cells, which then hollows out. Next, cell proliferation and movement create three basic tissue layers. Finally, the various organ systems develop out of these basic tissue layers. We will follow a human zygote through these steps.

Fertilization of the human egg occurs in the upper third of the oviduct, after which the zygote moves toward the *uterus*, a powerful muscular organ (see Figure 9.2.1). The initial cleavage divisions take place in the oviduct, and result in a solid ball of cells that resembles a mulberry; its name, *morula*, is taken from the Latin word for that fruit. The morula hollows out to form the structure shown in Figure 9.2.1. It is called a *blastocyst* and consists of an outer sphere of cells, with an inner cell mass at one end. When the blastocyst reaches the uterus, it embeds into the uterine wall, a process called *implantation*, at about day 8.

Figures 9.2.2 and 9.2.3 illustrate embryonic development after implantation. To start, the embryo is little more than a disk of cells, but an elongated structure called the *primitive streak* soon forms along what will be the head-to-foot axis. Cells on the outer margins of the primitive streak migrate toward it, downward through it, and into the interior of the disk of cells. (Figure 9.2.2 is a perspective view and Figure 9.2.3 is a cross-section through the disk.) This migration, called *gastrulation*, establishes three *germ layers* of tissue from which all subsequent organs will develop. These germ layers are the *endoderm*, the *mesoderm*, and the *ectoderm*. Next, the tips of the embryo fold down and around, a process that establishes the tubular nature of the embryo (Figure 9.2.3(c)). The endoderm will form much of the digestive tract and lungs, the ectoderm will form the skin and nervous system, and the mesoderm, lying in between the other two layers, will form many of the internal organs. Figure 9.2.3(c) is a cross-section of the fetus, which projects outward from the plane of the diagram.

*The placenta is the interface between mother and fetus.*

Mammalian fetuses are suspended in a watery *amniotic fluid* in a membranous sac

[3] Inactivation of unneeded genes by a cell is a common event. For example, different genes are active in different parts of a single division cycle. Even the inactivation of embryonic genes during early development is not irreversible: Crabs can grow new claws, plant stems can be induced to grow roots, and the genes for cell division are reactivated in cancer cells.

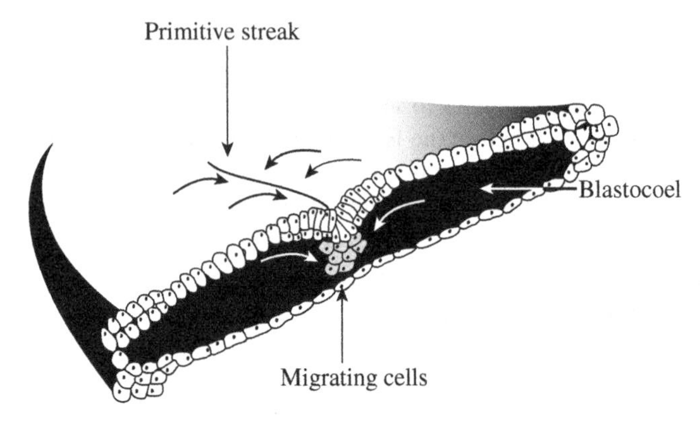


**Fig. 9.2.2.** A perspective view of gastrulation. Cells move from the sides, into the primitive streak, and downward into the embryo. (Redrawn from W. Purves, G. Orians, and C. Heller, *Life: The Science of Biology*, 4th ed., Sinauer Associates, Sunderland, MA, 1995. Used with permission.)

called the *amnion*; this structure cushions the fetus against mechanical injury.[4] As pointed out in Chapter 6, the blood of the mother and her fetus do not mix, a fact that necessitates a special structure for the exchange of maternal and fetal materials: A flat, platelike organ called the *placenta* develops between mother and child at the point of implantation; recall the discussion in Section 6.3. All materials exchanged between mother and child cross at the placenta, one side of which is composed of fetal tissues and the other side of which is composed of maternal uterine tissue. The two sides of the placenta have interdigitating projections into each other to increase their area of contact, thus facilitating material exchange. Lead is among the many substances that cross the placenta; thus a mother can poison her fetus by passing lead to it.

Shortly after it leaves the fetal heart, the fetus's blood is shunted outward into a vessel in the *umbilical cord* and into the placenta. At the placenta the fetal blood takes $O_2$ and nutrients from the mother's blood and gives up $CO_2$ and wastes. The fetus's blood, after having been enriched in $O_2$ and nutrients and cleansed of $CO_2$ and wastes, returns to the fetal body through the umbilical cord.

*The evolutionary development of the coelom facilitated the evolution of large animals.*

During embryogenesis in the higher animals, a cavity forms in the center of the mesoderm (Figure 9.2.3). This cavity is the *coelom*, and it has played a major role in the evolution of large animals (larger than $\approx$ 1 cm). By definition, a body cavity is called a coelom only if it is completely surrounded by mesoderm, the latter being identified by its creation during gastrulation and its role in forming specific internal organs, e.g., bones, muscles, heart, sex organs, and kidneys.

[4] An amnion and amniotic fluid are also found in the eggs of egg-laying mammals and reptiles and in bird eggs.

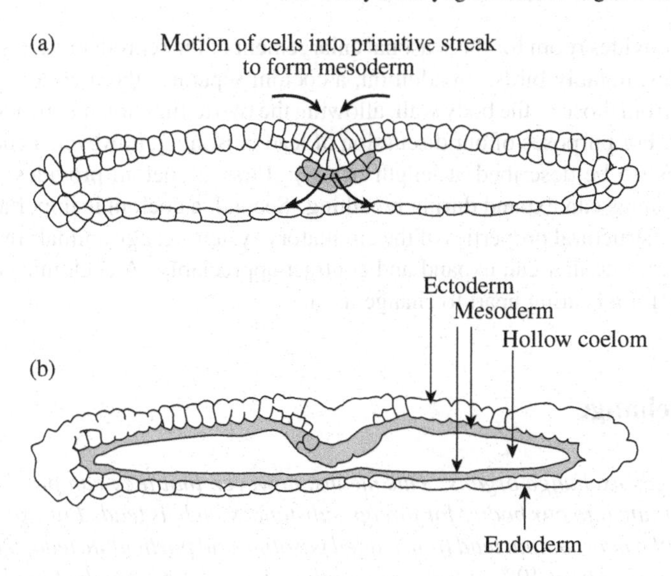


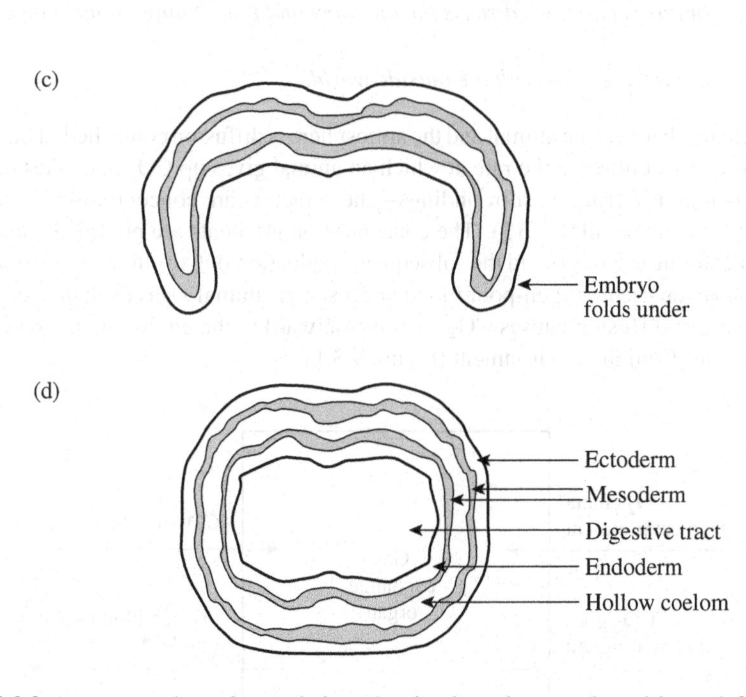


**Fig. 9.2.3.** A cutaway view of gastrulation, showing how the mesodermal layer defines the coelom. All the structures shown project out of the plane of the paper. (a) The cells move toward the primitive streak, then down into the interior of the zygote. (b) These cells form the mesodermal lining of the coelom. (c) This process is followed by a curling of the embryo to form the digestive tract. (d) This is a cross-section of the elongated fetus. It forms a tube, with the digestive tract running down the center.

A coelom provides room for the seasonal enlargement of the reproductive systems of some animals, notably birds. In addition, a coelom separates the muscles of the digestive tract from those of the body wall, allowing the two to function independently of one another. For purposes of our discussion of lead poisoning, however, a coelom plays two roles (to be described at length below): First, a coelom provides room into which the lungs can expand during breathing. Second, a coelom is important in determining the structural properties of the circulatory system: Large animals require a powerful heart, one that can expand and contract appreciably. A coelomic cavity provides space for a beating heart to change its size.

## 9.3 Gas Exchange

*The lungs are gas exchange organs. This means, however, that they can provide an efficient entry route into our bodies for foreign substances such as lead. For example, the air we breathe may contain lead from leaded gasoline and particulate lead, mainly from lead smelters. About 40% of the lead we inhale is absorbed into the blood from the lungs. In this section, we discuss the anatomy and functioning of our lungs.*

*Animals can exchange gas with the outside world.*

Gas exchange between an animal and the atmosphere is diffusion controlled. Thus two physical factors influence the rate at which an animal gives up $CO_2$ and takes up $O_2$ across membranes from its surroundings—these factors are concentration gradients and surface area; recall (6.2.13). The concentration gradients are provided naturally by the metabolic use of $O_2$ and the subsequent production of $CO_2$ in respiring tissues. Carbon dioxide–rich/oxygen-poor blood arrives at an animal's gas exchange organs, where passive diffusion causes $CO_2$ to move outward to the environment and $O_2$ to move inward from the environment (Figure 9.3.1).

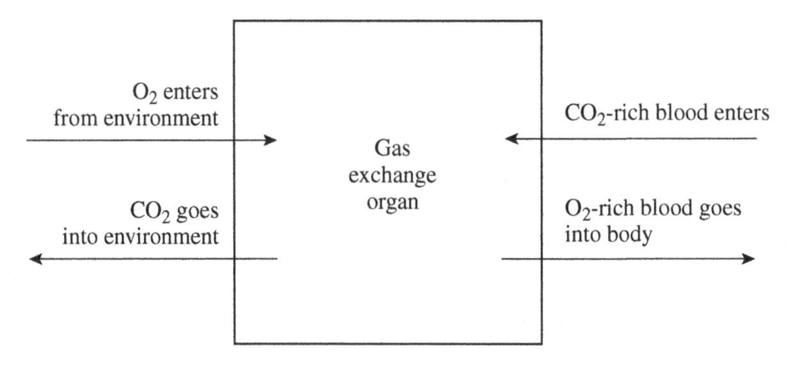


**Fig. 9.3.1.** A gas exchange organ. Blood with an excess of $CO_2$ and a deficiency of $O_2$ arrives at the gas exchange organ. Concentration gradients drive $CO_2$ from the blood into the atmosphere and $O_2$ from the atmosphere into the blood.

Several organs other than lungs can serve as sites of gas exchange: In insects, tiny *tracheal tubes* carry air directly to and from target tissues. The skin of some animals, notably amphibians, is a gas exchange organ. Many aquatic animals have *gills*, which are feathery structures that contain networks of blood vessels over which water can flow. The water brings in $O_2$ and carries away $CO_2$. We are most concerned, however, with humans, which, along with all other mammals and birds, exchange gases in *lungs* (see [6] and [7]).

Lungs themselves have no muscles, and are therefore incapable of pumping air in and out. Instead, air is pushed in and out of the lungs indirectly: Inhalation occurs when a set of *intercostal* muscles moves the ribs so as to increase the front-to-back size of the chest cavity (i.e., the coelom). At the same time, a dome-shaped sheet of muscle (the *diaphragm*) at the base of the chest cavity contracts and moves downward (Figure 9.3.2). The two actions expand the volume of the chest cavity and thereby draw air into the lungs. On the other hand, exhalation occurs when those intercostal muscles and the diaphragm relax.[5]

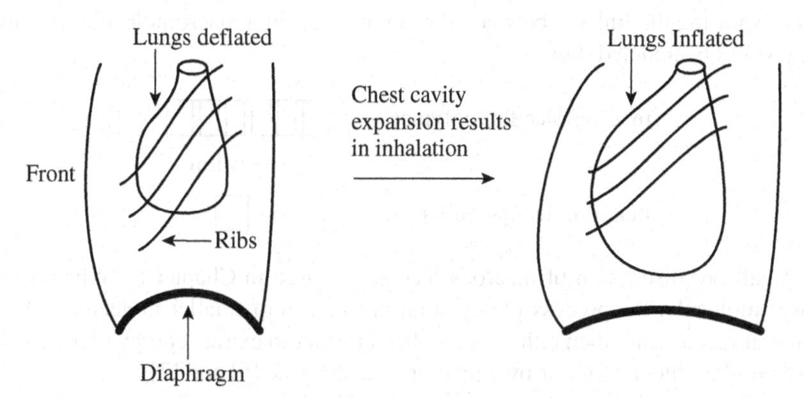


**Fig. 9.3.2.** The process of inhalation. The lungs have no muscles of their own. Their expansion and contraction are driven by the expansion and contraction of the chest cavity. Exhalation can also occur if the person simply relaxes, allowing the chest cavity to become smaller.

Air enters the lungs through tubes that branch out profusely, and which lead to small sacs called *alveoli*. It is the large number of alveoli that gives the lungs their extensive surface exposure to the outside world—about 100 $m^2$ of area. The alveoli are lined with tiny blood vessels that exchange $CO_2$ and $O_2$ with the atmosphere, giving up $CO_2$ to the environment and then carrying $O_2$ to tissues.

[5] Forcible exhalation of air does not result from the reverse action of the muscles mentioned for inhalation. Muscles can exert force only in contraction, and thus a second set of intercostal muscles and certain abdominal muscles act to push air forcibly from the lungs by moving the front of the ribs *downward* and decreasing the volume of the chest cavity.

## 9.4 The Digestive System

*This section is a discussion of the function of our digestive tract, a system of organs uniquely able to take nutrient materials, as well as toxins like lead, from our environment and route them into our blood.*

*The digestive system is an important path for lead intake. Rain washes atmospheric lead into municipal water supplies and people drink it. Lead in the pipes and solder of home water distribution systems leaches into drinking water.*[6] *Wine may have a high lead content. Plants absorb lead through their roots or bear it on their surfaces; the plants may then be consumed by humans. Organ meats, particularly kidneys, may contain high lead concentrations. Children may eat lead-based paint from old furniture. The fraction of ingested lead that is absorbed by the digestive tract is usually about* 10–15% *but may approach* 45% *during fasting (perhaps because it does not compete with food for absorption).*

*Digestion is the splitting of biopolymers into smaller pieces.*

The word "digestion" has a very restricted meaning in physiology: It is a particular way of splitting the linkage between the components in a macromolecular polymer. The process is modeled thus:

macromolecular polymer: □□□□□□

↓ digestion

macromolecular subunits: $6 \cdot$ □

Recall our discussion of macromolecular structure in Chapter 8. When we eat a large molecule, the process of digestion breaks it into smaller molecules. These smaller molecules are then either metabolized further to extract energy or are used to make building blocks for our own macromolecules (see [6] and [7]).

*The mammalian digestive tract is a series of specialized organs.*

At a fairly early stage in the evolution of animals, different parts of the digestive tract assumed different roles. In particular, the various organs of the digestive system reflect the animal's lifestyle. Figure 9.4.1 is a diagram of a mammalian digestive system; we will use it for the ensuing discussion.

(a) *Dentition.* Teeth are used for cutting, piercing, and grinding. Extreme development of the piercing teeth is typical of predators; cutting and grinding teeth are prominent in herbivores. Human dentition is more characteristic of herbivores than of carnivores.

---

[6] Lead is poorly soluble in water unless oxygen is present, a condition unfortunately met in most drinking water.

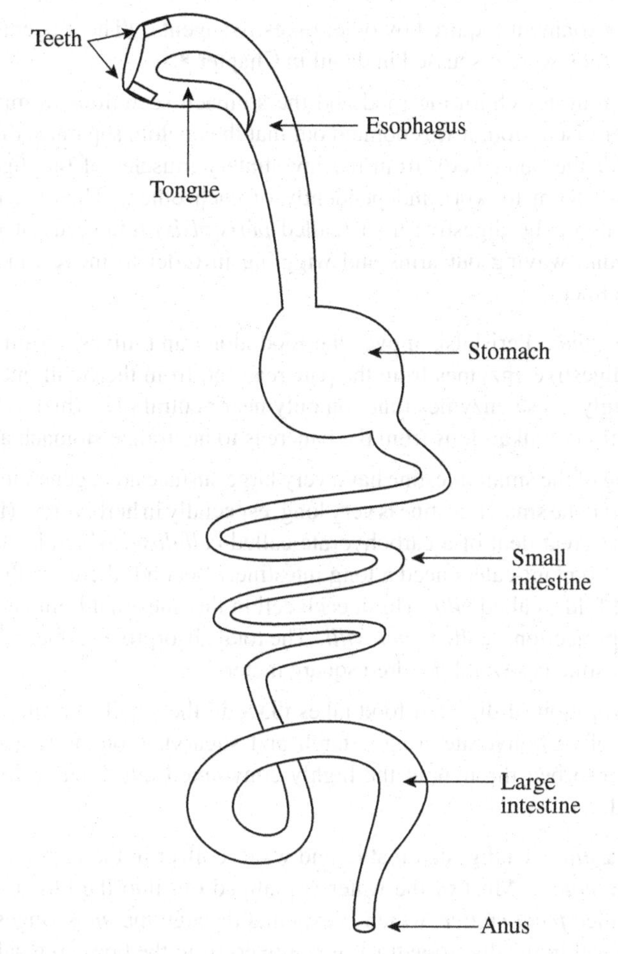


**Fig. 9.4.1.** The digestive tract is a convoluted tube, with different portions performing different specialized functions. See text for details.

(b) *Tongue*. This organ pushes food back into the esophagus, which leads to the stomach. The act of swallowing pushes a small flap over the opening to the windpipe, or *trachea*, which minimizes the possibility of choking on food.

(c) *Esophagus*. This is the tube leading from the back of the mouth cavity to the stomach.

(d) *Stomach*. This is an organ of mixing and, to a lesser extent, digestion; except for small molecules such as water and ethanol, very little absorption takes place from the stomach into the blood. The presence of food in the stomach triggers the process of digestion: Glands in the wall of the stomach generate hydrochloric acid, which contributes directly to the chemical breakup of food and creates the acidic environment required by other digestive agents. (These agents are called *enzymes*; they were discussed in detail in Chapter 8.)

Stomach muscles churn the food and the stomach secretions to mix them. In our earlier discussion, it was pointed out that the coelom separates the voluntary muscles of the outer body from the involuntary muscles of the digestive tract and allows them to work independently of each other. Thus the involuntary movements of the digestive tract (called *peristalsis*) relieve us of the need to walk around waving our arms and wiggling in order to move food along our digestive tract.

(e) *Small intestine*. Peristalsis moves the food along and mixes it with a large variety of digestive enzymes from the pancreas and from the small intestine itself. Surprisingly, these enzymes function only near neutral pH, which is attained by the secretion of alkali ions from the pancreas to neutralize stomach acid.

The lining of the small intestine has a very large surface area, generated in several ways. First, the small intestine is very long, especially in herbivores. (Plant matter contains a great deal of a carbohydrate called *cellulose*, which is very difficult to digest, so plant-eaters need a long intestine.) Second, there are the numerous intestinal folds, called *villi*. Third, each cell in the intestinal lining has hundreds of small projections, called *microvilli*. The total absorptive surface of the human small intestine is several hundred square meters!

Most absorption of digested food takes place in the small intestine. Molecular subunits of carbohydrates (e.g., starch and sugars), proteins (e.g., meat) and fats are absorbed throughout the highly convoluted intestinal surface and into the blood.

(f) *Large intestine*. Undigested matter and water collect in the large intestine, also called the *colon*. Most of the water is pumped out into the blood, leaving the waste, called *fecal matter*, which is expelled through the *anus*. Ingested lead, if not absorbed in the digestive tract, is removed from the body in fecal matter.

## 9.5 The Skin

*The third pathway by which chemicals can enter our bodies is through the skin. We will examine the unique properties of our skin and the conditions under which lead can cross it.*

*Skin is a sensor of, and a waterproof barrier to, the outside world.*

Skin is uniquely constructed to be the interface to our environment. Working from the inside to the outside of our skin, there is first a layer rich in small blood vessels, or *capillaries*. Capillaries bring nutrients and oxygen to the skin to support the needs of the many nearby nerve endings and other specialized cells that detect stimuli such as pain, pressure, and heat.

The next skin layer, also requiring materials brought by the blood, is a group of rapidly dividing, pancake-shaped cells. As these cells divide, they push toward the outside and die. The final, outer skin layer, called the *stratum corneum*, consists of these dead cells. Thus we are surrounded by a layer of dead cells, of which the membranes are a principal remnant. It is this layer that we need to examine more closely in our discussion of lead poisoning.

Cellular membranes were described in Chapter 6; we can review that discussion by pointing out that the principal structural components of cell membranes are closely related to fats. Thus cell membranes are waterproof, a property that makes good sense. After all, most biological chemistry is water-based, and therefore we need to protect our interior aqueous environment from our exterior environment, which is also mostly aqueous.

There are, however, chemicals that can penetrate the cell membranes of the stratum corneum and other cells and thereby get into our bloodstreams. Because membranes contain a lot of fatlike molecules, we should not be surprised that some fat-soluble compounds may move across membranes via a temporary state in which the substances become transiently dissolved in the membrane. Absorption of compounds across the skin is said to be *percutaneous*.[7] Examples of such compounds are tetramethyl lead and tetraethyl lead, "antiknock" compounds found in leaded gasoline. It was common in days past to see people hand-washing machine parts in leaded gasoline, an activity virtually guaranteed to cause lead absorption.

## 9.6 The Circulatory System

*Our circulatory system partitions chemicals throughout the body. It picks them up at the lungs, the digestive tract, and the skin and distributes them to other body tissues. In the case of lead poisoning, the circulatory system plays another key role: One of the most important toxic effects of lead is to interfere with the synthesis of the oxygen-carrying pigment hemoglobin, found in red blood cells. We will describe the circulatory system and show how its anatomy promotes the rapid distribution of materials to all other body tissues.*

*The discussion of oxygen transfer across the placenta in Section* 6.3 *was a short introduction to some of the material in this section.*

*Circulatory systems move a variety of materials around an animal's body.*

Living organisms are open thermodynamic systems, which means that they are constantly exchanging energy and matter with their surroundings. The exchange of materials between the cells of a multicellular animal and the animal's environment

[7] The compound dimethylsulfoxide (DMSO) rapidly penetrates the skin and is sold as a remedy for certain bone joint disorders. It is possible to taste DMSO by sticking one's finger into it: The compound crosses the skin of the finger, gets into the bloodstream, and goes to the tongue, where it generates the sensation of taste—said variously to be like garlic or oysters.

is mediated by a circulatory system. This system picks up $O_2$ at the lungs, gills, tracheal tubes, or skin, and delivers it to metabolizing cells. The $CO_2$ that results from the metabolic production of energy is returned to the gas exchange organ and thus to the organism's surroundings. The circulatory system also picks up nutrients at the digestive tract and delivers them to the body's cells; there it picks up wastes, which it takes to the kidneys or related organs for excretion. Further, the blood carries minerals, proteins, and chemical communicators, or *hormones*, from one part of the body to the other. Hormones regulate such activities as growth, digestion, mineral balances, and metabolic rate (see [6] and [7]).

*Open circulatory systems are convenient, but inefficient.*

The blood of most small invertebrate animals spends most of its time circulating leisurely around the animal's internal organs. An example is shown in Figure 9.6.1. Note that the blood is not necessarily confined to vessels at all; rather, it merely bathes the internal organs. This kind of circulatory system is said to be *open*.

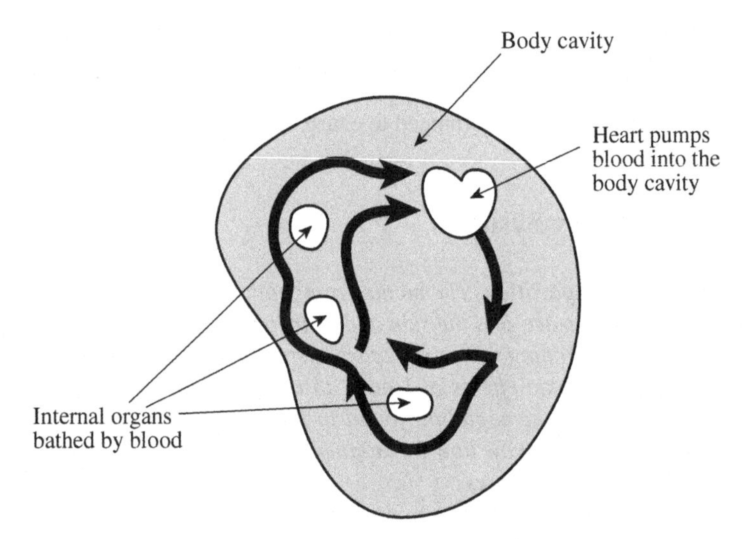


**Fig. 9.6.1.** An open circulatory system. The heart pumps blood into an open body cavity. Organs in the body cavity are bathed by the blood, exchanging gases, nutrients, and wastes with it. The blood eventually returns to the heart.

The correct functioning of an open circulatory system relies on a very important physical property: The materials carried by the blood, and listed in the previous section, can diffuse effectively over distances of no more than about one millimeter (see Section 6.2 and Dusenbery [8]). Thus an upper limit is set for the size of internal organs; no part of any organ can be more than about 1 mm from the blood. Of course, this also sets a limit on the size of the entire organism; animals with open circulation seldom exceed a few centimeters in size. Typical examples are snails and houseflies; atypical examples are giant clams and lobsters, but being aquatic, they have the advantage of being bathed in water, which provides for easy waste and $CO_2$ removal. Further, these large animals have low specific metabolic rates, which minimizes their $O_2$ and nutrient needs, as well as their production of wastes.

Open circulatory systems are structurally simple, but the size restriction they place on organisms is a major shortcoming. There are plenty of ecological niches into which large animals could fit, especially on land, but to do so has required the evolution of a different kind of circulatory system. We examine that next.

*Closed circulatory systems are efficient, but require a powerful heart.*

An open circulatory system can be likened to a large fan at the front of a classroom. Air gets moved from the fan toward the rear of the room and it eventually returns to the back of the fan. Most people in the room feel a small breeze from time to time. On the other hand, a *closed circulatory system* can be likened to an enclosed air pump, with a network of conduits that take air directly to each person individually and return the air directly to the pump.

As pointed out above, the problem with open circulation is that most materials can diffuse distances of no more than about one millimeter during biologically realistic times, thus limiting the size of the organism. Closed circulatory systems remove this restriction by taking blood, via tiny vessels, directly to the immediate vicinity of all metabolizing cells. This blood distribution is independent of the size of the organism, how far the cells are from the heart, and how deep the cells are inside an organ.

A vertebrate closed circulatory system contains a heart that pumps blood into a thick-walled artery, called the *aorta*. The latter then progressively branches into smaller arteries and then into capillaries, whose walls are only one cell thick, across which material exchange between blood and other tissues must take place (Figure 9.6.2). Capillaries have such narrow lumens that blood cells must be bent in order to get through. Thus capillaries have the large surface-to-volume ratio necessary for their material exchange function. Eventually, capillaries join together in groups to form small veins that combine into a few large veins and return blood to the heart.

A closed circulatory system is very efficient because it delivers blood right to the doorstep of metabolizing tissues; no cell is very far from a capillary. An important problem is built into closed circulatory systems, however: The frictional resistance to blood flow in the capillaries is much greater that that in arteries. This is true in spite of the fact that the total cross-sectional area of the artery leading from the heart (the aorta) is much less than the total cross-sectional area of all the body's capillaries.

The reason for this apparent contradiction is well known: Blood is viscous and thus adheres to vessel walls as it passes. Only in the center of the vessel lumen is this friction reduced. The difficulty is that capillaries have a very small lumen and a lot of wall area (to increase their material-exchange properties). Therefore, the frictional resistance of capillaries to blood flow is very high. A general rule is that the friction encountered by a viscous material passing through a tube is inversely proportional to the fourth power of the radius of the tube (see (6.2.17) and Vogel [9]). Thus if the

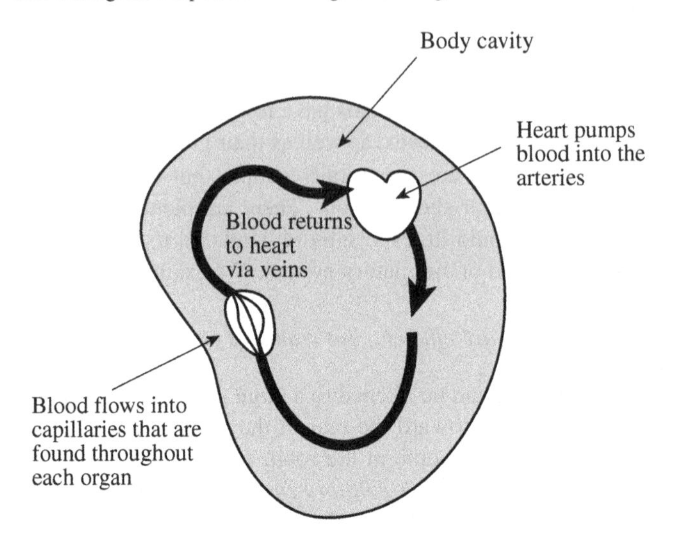


**Fig. 9.6.2.** A closed circulatory system. The heart pumps blood into a system of arteries, which deliver the blood directly to capillaries in organs. The blood in these capillaries exchanges gases, nutrients, and wastes with tissues in their immediate vicinity. The blood returns to the heart via a system of veins.

radius is halved, the frictional force goes up by 16 times. The end result is that a closed circulatory system requires a very powerful heart to force blood through the many tiny capillaries, in spite of the large total cross-sectional area of the latter.

*The cellular fraction of blood serves a variety of functions.*

Whole blood has both a liquid fraction, called *plasma*, and a cellular fraction. Plasma is mostly water, but it also contains many inorganic ions, hormones, biochemicals such as sugars, fats, amino acids, and proteins. The levels of all these plasma solutes are critical and thus are maintained at relatively constant levels.

Blood cells are usually classified on the basis of their appearance. All of them originate from common precursor cells in the bone marrow and become specialized later.

(a) *Red blood cells*, or *erythrocytes*, will be of special importance to us in our later discussion of lead poisoning. Mammalian erythrocytes lose their nuclei during formation; the remaining cytoplasm contains large quantities of a biochemical called *hemoglobin*, which has a high affinity for $O_2$. We introduced the biological role of hemoglobin in Section 6.3; we now elaborate on it.

Erythrocytes pick up $O_2$ at the lungs:

*Reaction* 1:[8]

$$\underset{\text{(hemoglobin)}}{\text{Hb}} + O_2 \longrightarrow \underset{\text{(oxyhemoglobin)}}{O_2\text{-Hb}}$$

[8] Recall from Chapter 6 that one hemoglobin molecule can bind up to four $O_2$ molecules.

The erythrocytes are then carried to the sites of metabolism, where the oxygen is needed, and Reaction 1 is reversed to free the oxygen for use in metabolic processes:

*Reaction* 2:

$$Hb + O_2 \longleftarrow O_2\text{-}Hb$$

Interestingly, the affinity of hemoglobin for $CO_2$ is fairly low; therefore, most $CO_2$ is carried from the sites of respiration back to the lungs in the form of carbonic acid or the bicarbonate ion, dissolved in the water of the plasma:

*Reaction* 3:

$$CO_2 + H_2O \longrightarrow \underset{\text{(carbonic acid)}}{H_2CO_3} \longrightarrow H^+ + \underset{\text{(bicarbonate ion)}}{HCO_3^-}$$

At the lungs, the $CO_2$ is reconstituted from the bicarbonate ion and carbonic acid, and then exhaled:

*Reaction* 4:

$$CO_2 + H_2O \longleftarrow H_2CO_3 \longleftarrow H^+ + HCO_3^-$$

Reactions 1–4 are related: Reaction 4 takes place at the lungs and raises the blood pH by removing carbonic acid. The rate of Reaction 1 is pH-dependent; conveniently, it goes faster at higher pH. Thus the removal of $CO_2$ at the lungs promotes the attachment of hemoglobin to $O_2$. The opposite occurs in metabolizing tissues: The production of $CO_2$ lowers the blood pH there via Reaction 3, and the lower pH promotes Reaction 2, the release of oxygen from oxyhemoglobin.[9]

While the affinity of hemoglobin for $CO_2$ is low, its affinity for carbon monoxide (CO) is very high. As a result, the presence of even small amounts of CO can prevent the attachment of $O_2$ to hemoglobin, accounting for the lethal effect of CO.

The best-studied toxic effect of lead is its role in causing *anemia*, a reduction in erythrocyte concentration. This in turn reduces the oxygen-carrying ability of the blood. The anemia is evidently the result of two processes: First, lead interferes with hemoglobin synthesis, and second, lead causes the lifetimes of mature erythrocytes to be reduced from the usual four months.

(b) *Platelets* are blood cells involved in clotting; they are actually cell fragments that lack nuclei. Platelets collect at the site of an injury and disintegrate. This releases platelet proteins that generate a cascade of reactions, finally resulting in the formation of a clot consisting of a plasma protein called *fibrin*.

(c) *Leukocytes* are nucleated and are often called white blood cells; they are used to fight off infections. One class of leukocytes, the lymphocytes, will be discussed in Chapter 10 in the context of HIV infections. A second group, the granulocytes, functions in certain general responses to infections and allergens.

---

[9] At least *some* $CO_2$ does bind to hemoglobin. It has the useful effect that it decreases the oxygen affinity of the hemoglobin (in the vicinity of metabolizing tissues).

*There are four overlapping functional paths in our circulation: systemic, pulmonary, lymphatic, and fetal.*

Blood cells and plasma can take several routes around the human body, the differences between the routes being both anatomical and functional.

Look at Figure 9.6.3. The mammalian heart has four chambers: The two upper ones are atria and the two lower ones are the muscular *ventricles*. Blood, rich in $CO_2$ from metabolizing tissues, enters the heart at the right atrium and goes to the right ventricle. The powerful ventricle pushes the blood to capillaries in the lungs; these capillaries surround the many small air sacs (alveoli), which are filled with air when we inhale. Here $CO_2$ is moved from the plasma into the alveoli for exhalation, and the erythrocytes pick up $O_2$ from the alveoli; the chemistry for these two processes was outlined in Reactions 1 and 4 in the last section. The blood then returns to the heart at the left atrium. The path of the blood just described—from the heart to the lungs and back to the heart—is called the *pulmonary circulation*.

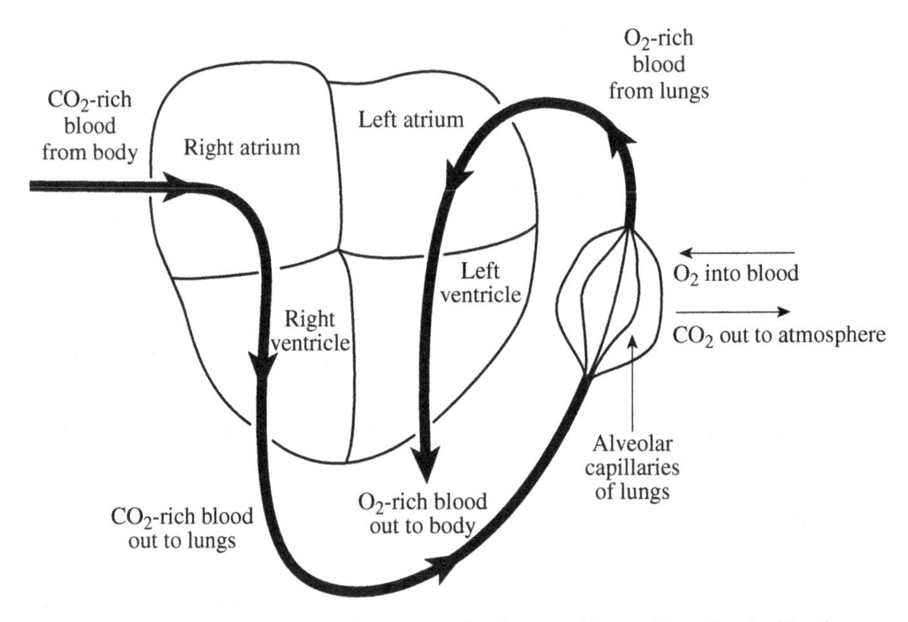


**Fig. 9.6.3.** The flow of blood through a mammalian heart and lungs. Note that the blood enters and leaves the heart twice.

> *Pulmonary circulation in outline form*:
> right ventricle ⟶ pulmonary artery ⟶ lung capillaries ⟶ pulmonary vein ⟶ left atrium

After moving from the left atrium to the left ventricle, blood exits the heart and goes to respiring tissues all over the body via arteries and then capillaries. At the capillaries, $O_2$ is given up to the respiring cells and $CO_2$ is taken from them into the plasma.[10] These reactions were described earlier as Reactions 2 and 3. The blood now returns to the heart by way of veins. The path of blood from the heart to respiring cells and back to the heart is called *systemic circulation.*

> *Systemic circulation in outline form*:
> left ventricle ⟶ aorta ⟶ other arteries ⟶ capillaries of respiring tissues ⟶ veins ⟶ right atrium

Note the importance of the powerful ventricles—they must overcome the high frictional resistance that blood meets in the narrow lumens of the capillaries.

Figure 9.6.4 shows a connected group, or *bed*, of capillaries. The lumen of the artery narrows down as the blood approaches the capillaries, and as a result, the frictional resistance increases dramatically. For much of the blood, the effect is almost like hitting a dead end. Consequently, the hydrostatic pressure in the blood at the front (upstream) end of the capillary bed rises sharply. The high hydrostatic pressure pushes some of the liquid fraction of the plasma through gaps in the vessel walls (Figure 9.6.5). The only part of the plasma that cannot be pushed out is the plasma protein fraction, because these molecules are too big. Thus this *interstitial fluid* forced out of capillaries lacks the large plasma proteins.

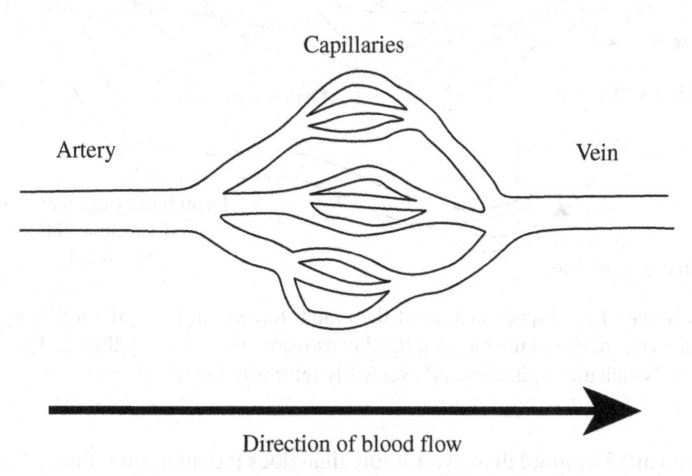


**Fig. 9.6.4.** A capillary bed. As blood moves from arteries to capillaries, the overall cross-section of the circulatory system increases, but overall frictional resistance increases also.

At the far (downstream) end of the capillary bed, the capillaries join together, friction decreases, and the hydrostatic pressure also decreases. The blood plasma at this point contains everything that the interstitial fluid contains, plus plasma proteins.

[10] The phrase "respiring cells" here means cells that are breaking down sugar to get energy and that are therefore giving off carbon dioxide.

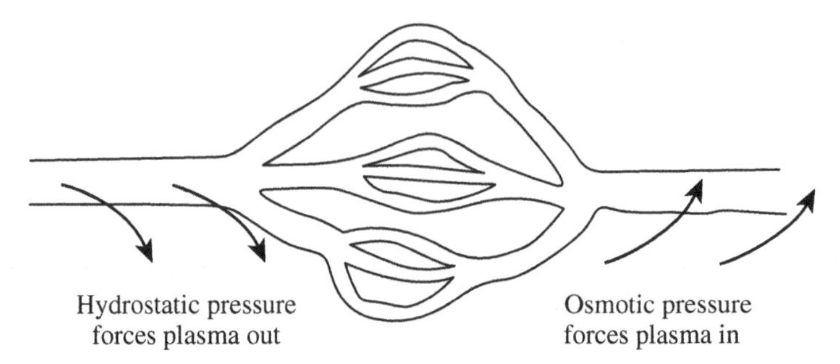


**Fig. 9.6.5.** The increase in frictional resistance to blood flow imposed by capillaries generates a high hydrostatic pressure upstream from the capillaries. This hydrostatic pressure forces some of the liquid fraction of the blood out of the upstream end of the capillaries and into the surrounding tissues. On the downstream end of the capillaries, the blood, now with a high concentration of dissolved substances, draws some of the liquid from the surrounding tissues back into the circulatory system by osmotic pressure.

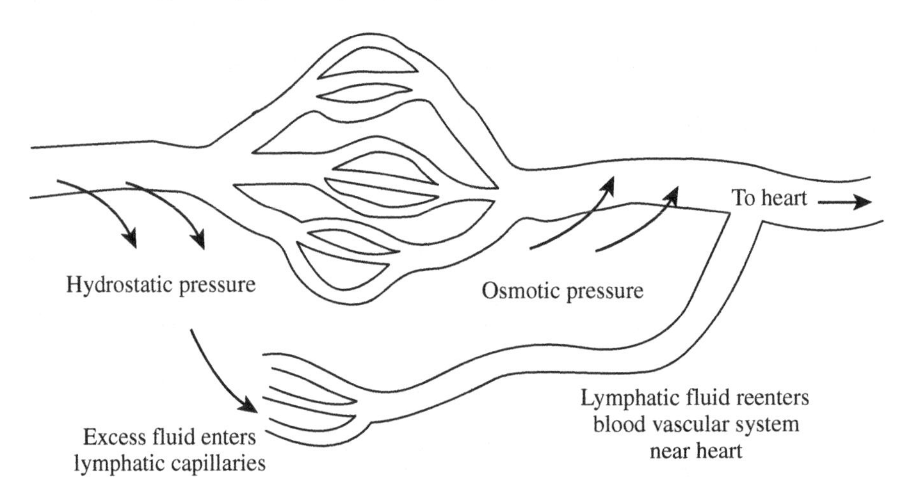


**Fig. 9.6.6.** Some of the liquid fraction of the blood, forced out of capillaries by hydrostatic pressure, does not return to the blood at the downstream end of the capillaries. This excess is picked up by lymphatic capillaries and eventually returns to the bloodstream near the heart.

Thus the plasma has more dissolved solute than does the interstitial fluid. As a result, most of the interstitial fluid is osmotically drawn back into the vessels to dilute the plasma proteins. Not all the interstitial fluid makes it back to the blood, however; there is a positive pressure differential of several millimeters of mercury between the hydrostatic pressure on the upstream end of the capillary bed and the osmotic pressure on the downstream end. This would cause a buildup of fluid in the tissues if it were not for the fact that there is another path of circulation to collect the excess fluid.

The extra fluid is collected in a set of capillaries of the *lymphatic circulation* and brought by lymphatic veins to the upper body, where they empty into blood veins near the heart (see Figure 9.6.6). During the journey to the heart, the flow of the *lymphatic*

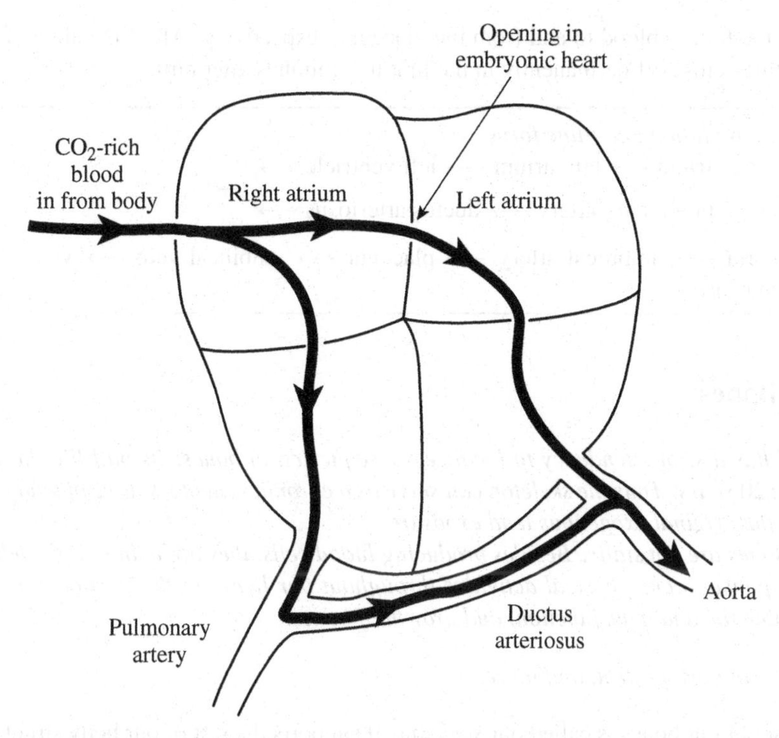


**Fig. 9.6.7.** A diagram of fetal circulation. An opening between the two atria, and a vessel called the *ductus arteriosus*, shunt fetal blood away from the lungs.

*fluid*, as the interstitial fluid is called at this point, is pushed by contractions of the nearby skeletal (voluntary) muscles. This movement should be bidirectional, but lymphatic veins have valves that allow flow only toward the heart. Along the way the lymphatic veins pass through *lymph nodes*, packets of lymphatic tissue that filter out pathogens.

In addition to the interstitial fluid, some white blood cells can also move between the blood circulation and the lymphatic circulation. They do so by squeezing between the endothelial cells that constitute the walls of blood capillaries.

> *Lymphatic circulation in outline form*:
> heart ⟶ arteries ⟶ systemic capillaries ⟶ intercellular spaces ⟶
> lymphatic capillaries ⟶ lymphatic veins ⟶ blood veins ⟶ heart

The fourth path of the circulatory system is found in fetuses and is called the *fetal circulation*. A fetus does not breathe, so its pulmonary circulation is minimized via two shunts: There is an opening in the fetal heart (the foramen ovule), between the right and left atria, that directs blood away from the pulmonary circulation (Figure 9.6.7). Secondly, a special vessel, called the *ductus arteriosus*, carries blood from the pulmonary artery directly to the aorta. Finally, the umbilical artery and vein carry the fetus's blood to and from the placenta, respectively. All of the above fetal structures close off permanently in the first few minutes after birth.

*Fetal circulation in outline form*:

right atrium ⟶ left atrium ⟶ left ventricle ⟶
↘ pulmonary artery ⟶ ductus arteriosus ⟶ (both join) ⟶

aorta ⟶ umbilical artery ⟶ placenta ⟶ umbilical vein ⟶ veins to heart

## 9.7 Bones

*Lead has a strong tendency to localize, or sequester, in bones; its half-life there is about 20 years. Thus the skeleton can serve as a chronic systemic source of lead long after the original exogenous lead exposure.*

*Bones are not static. Besides producing blood cells, they are being "remodeled" in response to our physical activities throughout our lives. In this section, we will examine the anatomy, function, and growth of bones.*

*Bones support, protect, and move.*

The set of our bones is called our *skeleton*. It supports the rest of our body structure. All large terrestrial animals require an internal skeleton because the nonskeletal tissue would collapse under its own weight and a hard external skeleton (like an insect's) would weigh too much. Our skeleton surrounds our internal organs, protecting them from mechanical injury. In the cases in which organs are not protected by bone, e.g., eyeballs and testicles, reflexive reactions and very low pain thresholds are necessary for protection. With the aid of our voluntary muscles, we can use our skeleton to project effects at a distance—walking, reaching, and hugging, for instance.

*Bone marrow is the source of blood cells.*

In the earlier section on the circulatory system a number of blood cells were described. All of these cells originate from a single variety of cell in the core, or *marrow*, of bones. These cells are called *stem cells*, and they divide rapidly, generating large numbers of daughter cells. The subsequent fate of a daughter cell depends on the conditions of its maturation environment. Some lymphocytes, for instance, mature in the thymus gland, just behind the breastbone of children. Red blood cells mature in the marrow, synthesize hemoglobin, and then (in mammals) lose their nuclei. We will have more to say about blood cell origins in Chapter 10.

*Bony tissue is replaced throughout life.*

There are special cells, called *osteoblasts*, in bone that secrete a protein about which the compound *hydroxyapatite* (mainly calcium phosphate) crystallizes; thus hard bone is formed. When the muscles to which a bone is attached become stronger, the stresses cause the bone to thicken to adapt to the new need. This is initiated by another group of cells, called *osteoclasts*, which secrete chemicals to dissolve hard bone tissue. Osteoblasts then reconstitute the bone in a new, thicker form.[11] It has been estimated that our bones are replaced up to ten times in our lives.

Virtually all bone growth after adolescence affects the thickness of a bone. Bone growth before that time may be in thickness, but may also be in the length of the bone, accounting for the dramatic rate of change in a child's height.

Lead tends to be deposited in the region of bones near their ends, as revealed by X-ray pictures. The lead is then very slowly released over a period of many years. This long half-life means that lead tends to accumulate in bones. Indeed, about 95% of a person's total-body lead can be found in the bones [2].

## 9.8 The Kidneys

*Our kidneys provide a mechanism for ridding the body of water-soluble substances. They are very selective, however: They can maintain a constant chemical composition in our bodies by removing materials in excess and retaining materials in short supply. Thus we should correctly expect that the kidneys would help to excrete lead. The problem is that much lead becomes sequestered in bone and is therefore not solubilized in blood where the kidneys can get at it. Nevertheless, if absorbed lead is to be removed from the body, kidney excretion will be the major route out.*

*The kidneys remove nitrogenous wastes and help regulate the concentration of materials in the blood.*

When we eat protein, e.g., the muscle from a cow's leg, the process of digestion breaks the protein down into its component compounds, called *amino acids*. In our bodies, amino acids have two possible fates: First, they can be incorporated into our own proteins. We can synthesize most, but not all, of the amino acids we need from related precursors we get in our diet. Second, ingested amino acids can be broken down to extract some of their energy. In this section, we will be concerned with one aspect of the latter of these two fates—the removal of a certain chemical group (-$NH_2$) from an amino acid prior to extraction of the amino acid's energy.

Unless it is to be used in the synthesis of other nitrogenous compounds in our bodies, the amino group (-$NH_2$) of ingested amino acids must be removed from our body as waste. The problem is that the amino group quickly forms *ammonia* ($NH_3$), which is toxic. Aquatic animals can get rid of ammonia by releasing it directly into the surrounding water, in which the ammonia is highly soluble. Many terrestrial animals, humans included, convert the ammonia to urea ($H_2NCOH_2$), which is moderately

[11] It should now be evident how archaeologists can tell so much about an animal by examining a few bones. The pattern of bumps and thickenings on a bone constitute a graphic history of the animal's life.

water-soluble and much less toxic than ammonia.[12] The urea is then removed from our bodies by our kidneys in urine.

At the kidneys, blood pressure forces some of the liquid fraction of blood into small kidney tubules (see Figure 9.8.1). This liquid contains water, ions, many essential biological molecules, and, of course, urea. As the liquid moves along these convoluted tubes, most of the water and most of the desirable dissolved substances are resorbed back into the blood. The aqueous liquid left behind in the kidneys is *urine*, which contains a high concentration of urea. Urine also contains other dissolved substances that the blood has in excess of normal needs. Thus the kidneys serve the *homeostatic function* of maintaining the concentration of dissolved substances in the blood and other tissues at normal levels.[13]

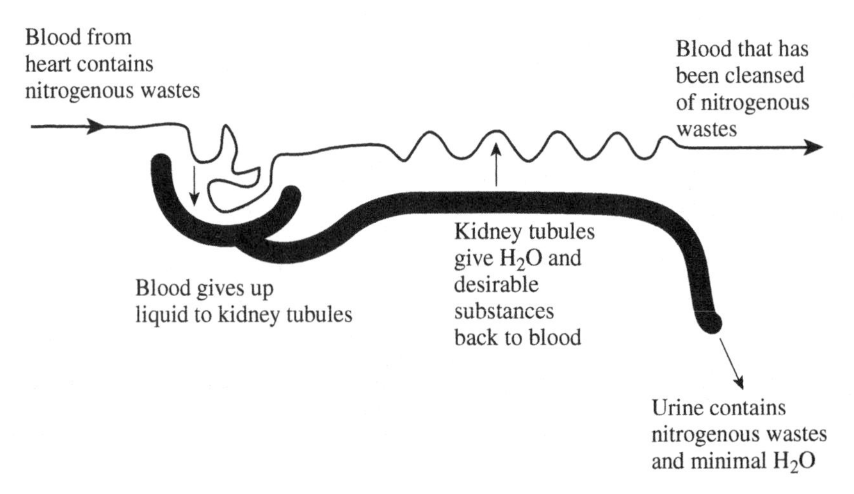


**Fig. 9.8.1.** A simplified model of the mammalian kidney. Blood pressure at the kidneys forces some of the liquid fraction of the blood into kidney tubules. This liquid fraction contains wastes, but also contains useful solutes, such as sugars. Later, the kidney takes back the useful substances and most of the water, leaving urine with the concentrated wastes (to be voided).

Lead is absorbed into bones about 100 times faster than it is released. Thus an important therapeutic approach to lead poisoning is to try to prevent the lead from becoming sequestered in bone, from which it would be slowly released back into the blood over a period of decades. The trick is to make the lead very soluble shortly after exposure so the kidneys can excrete it efficiently. There is a class of chemical compounds, called *metal chelates*, that react rapidly with lead and many

[12] Egg-laying animals go one step further and convert the urea to uric acid, which is water-insoluble. Thus uric acid can accumulate inside an avian or reptilian egg and not harm the fetus prior to hatching.

[13] Patients with *diabetes mellitus* produce insufficient quantities of the protein *insulin*. This leads to an excess of the sugar glucose in their blood. The kidneys remove some of the excess *glucose* in urine—thus providing a means of medical diagnosis.

other metals to form very soluble compounds that the kidneys can excrete. *Penicillamine*, for example, can be administered orally, and the compound it forms with lead, penicillamine-lead, is highly soluble in water. The structure of penicillamine-lead is shown here:[14]

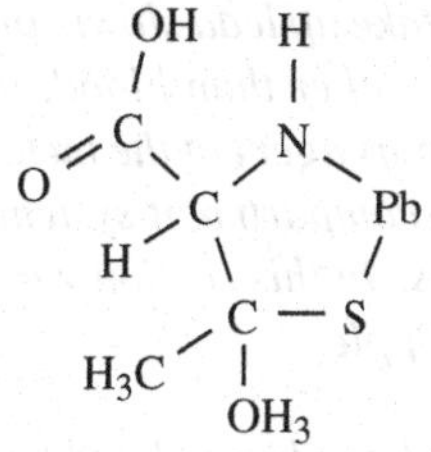

```
         OH    H
         /     |
     O//C\   /N\
          C       Pb
        H/  \     /
             C — S
        H3C/ |
             OH3
```

In summary, kidneys first remove most dissolved materials, both useful and waste, from the blood and later put the useful materials back into the blood. We could easily imagine a simpler system, in which only a small amount of water, all the urea, and any other material in excess is removed directly, without benefit of resorption, but that is not how our kidneys work. The forces of evolution do not necessarily yield the simplest system, but instead yield a modification of a preexisting system. This frequently generates a more complicated system, but if the new system provides a selective advantage, it can become fixed in the population.

Lead has been shown to have a direct pathological effect on kidney function. Acute exposure in children leads to malfunctioning of the resorption process, resulting in a high concentration of glucose and other desirable compounds in the urine. Chronic lead exposure eventually results in general kidney failure.

## 9.9 Clinical Effects of Lead

*Lead poisoning is indicated by a variety of clinical symptoms, including gastrointestinal and mental disorders.*

*Lead poisoning causes a wide variety of general symptoms.*

We conclude with a short description of symptoms of lead poisoning that a physician might see in a patient. In adults, there are gastrointestinal disorders such as vomiting and pain. In children, there are central nervous system disorders, e.g., drowsiness, irritability, and speech disturbances, as well as gastrointestinal symptoms. An interesting symptom in some cases is a blue line along the gums, formed when lead reacts with sulfur, the accumulation of the latter being associated with poor dental hygiene.

The effect of lead on IQ is an interesting one. As mentioned above, lead-induced neurophysiological disorders are especially noted in children. This is not unexpected because lead seems to affect the velocity of nerve impulse conduction. Evidence suggests that lead poisoning can reduce a child's IQ by about five points.

[14] Chelating agents are not without risks: A chelating agent that picks up lead may also pick up other divalent metals, e.g., calcium. Loss of blood calcium can lead to uncontrollable muscle tremors and even death.

## 9.10 A Mathematical Model for Lead in Mammals

*While lead interacts differently with the various tissues of the body, as a first approximation we need only distinguish three tissue types: bone, blood, and the other soft tissue of the body. Bone tends to take up lead slowly but retain it for very long periods of time, in contrast to soft tissue, other than blood, in which the turnover of lead is much quicker. Blood is the transport agent of the metal. The disposition of lead in the body can be followed as a three-compartment system by tracking its movement into and out of these three tissue types. In this section we analyze such a model proposed by Rabinowitz, Wetherill, and Kopple.*

*The uptake and movement of lead can be modeled by the use of compartments.*

The activity of lead in the body depends on the tissue in which it is located (recall the end of Section 9.1). To construct a mathematical model for the movement of lead, at least three distinct tissue types must be recognized: bone, blood, and soft tissue (other than blood).[15] These will be our mathematical compartments. Lead enters the system by ingestion and through the skin and lungs. These intake paths usher the substance to the blood. From the blood, lead is taken up by bone and by soft tissue. This uptake is reversible: Lead is also released by these organic reservoirs back to the blood. However, especially for bone, lead's half-life in that tissue is very long. Lead can be shed from the body via the kidneys from the blood and to a lesser extent, through hair. Thus blood is the main conduit through which lead moves among our compartments.

To begin the model, let compartment 1 be the totality of the victim's blood, compartment 2 the soft tissue, and compartment 3 the skeletal system. We must also treat the environment as another compartment to account for lead intake and elimination; we designate it as compartment 0. Let $x_i$, $i = 1, \dots, 3$, denote the amount of lead in compartment $i$ and let $a_{ij}$, $i = 0, \dots, 3$, $j = 1, \dots, 3$, denote the rate of movement of lead *to* compartment $i$ *from* compartment $j$. The product $a_{ij}x_j$ is the rate at which the amount of lead increases in compartment $i$ due to lead in compartment $j$. There is no reason that $a_{ij}$ should equal $a_{ji}$, and as noted above, the rate of movement from blood to bone is very different from the reverse rate. The units of the $a_{ij}$s are per day.

Because we will not keep track of the amount of lead in the environment, this is an *open compartment* model. Instead, we account for environmental intake by including a separate term, $I_L(t)$, applied to compartment 1, the blood. From the discussion above, some of the rates are zero; namely, $a_{03} = a_{23} = a_{32} = 0$, signifying no direct elimination to the environment from bone and no direct exchange between bone and soft tissue. In addition, all rates $a_{i0}$ are 0, since there is no $x_0$ term. Finally, there is no need for terms of the form $a_{ii}$, since a compartment is our finest unit of resolution.

With these preparations, we may now present the model that derives from the simple fact that the rate of change of lead in a compartment is equal to the difference

[15] As in the discussion ending Section 9.1, throughout this section, by soft tissue, we mean soft tissue other than blood.

between the rate of lead entering and the rate leaving:

$$\begin{aligned}
\frac{dx_1}{dt} &= -(a_{01} + a_{21} + a_{31})x_1 + a_{12}x_2 + a_{13}x_3 + I_L(t), \\
\frac{dx_2}{dt} &= a_{21}x_1 - (a_{02} + a_{12})x_2, \\
\frac{dx_3}{dt} &= a_{31}x_1 - a_{13}x_3.
\end{aligned} \tag{9.10.1}$$

In words, the first equation, for example, says that lead leaves the blood for the environment, soft tissue, and bone at a rate in proportion to the amount in the blood; lead enters the blood from the soft tissue and bone in proportion to their respective amounts; and lead enters the blood from the environment according to $I_L(t)$. The algebraic sum of these effects is the rate of change of lead in the blood. In line with our discussion of Section 2.4, this system can be written in matrix form as

$$\mathbf{X}' = A\mathbf{X} + \mathbf{f}. \tag{9.10.2}$$

Here $\mathbf{X}$ is the vector of $x$s, $\mathbf{f}$ is the vector

$$\mathbf{f} = \begin{bmatrix} I_L(t) \\ 0 \\ 0 \end{bmatrix},$$

and $A$ is the $3 \times 3$ matrix

$$A = \begin{bmatrix} -(a_{01} + a_{21} + a_{31}) & a_{12} & a_{13} \\ a_{21} & -(a_{02} + a_{12}) & 0 \\ a_{31} & 0 & -a_{13} \end{bmatrix}.$$

From (2.4.12), the solution is

$$\mathbf{X} = e^{At}\mathbf{X}_0 + e^{At}\int_0^t e^{-As}\mathbf{f}(s)ds. \tag{9.10.3}$$

We will now suppose that the intake of lead, $I_L(t)$, is constant; this is a reasonable assumption if the environmental load remains constant. Then $\mathbf{f}$ is also constant, and we may carry out the integration on the right-hand side of (9.10.3). In keeping with the result that $-a^{-1}e^{-at}$ is the integral of the ordinary exponential function $e^{-at}$, we get

$$\begin{aligned}
e^{At}\int_0^t e^{-As}\mathbf{f}(s)ds &= e^{At}[-A^{-1}e^{-As}]\Big|_0^t \mathbf{f} \\
&= -e^{At}[e^{-At} - I]A^{-1}\mathbf{f} \\
&= -[I - e^{At}]A^{-1}\mathbf{f}.
\end{aligned}$$

Substitution of this result into (9.10.3) gives

$$\mathbf{X} = e^{At}\mathbf{X}_0 - [I - e^{At}]A^{-1}\mathbf{f} = e^{At}[\mathbf{X}_0 + A^{-1}\mathbf{f}] - A^{-1}\mathbf{f}. \tag{9.10.4}$$

This is the solution of system (9.10.1), provided $A^{-1}$ exists. We can obtain solutions provided that the exponential $e^{At}$ is computable.[16]

*The long-term predictions of the model.*

Recall from the discussion of Section 2.6 that the long-term behavior of the solution is predicted by knowledge of the eigenvalues of the matrix $A$. But it is easily seen that this is a compartment matrix (cf. Section 2.6). The diagonal terms are all negative, the first column sum is $-a_{01}$, the second column sum is $-a_{02}$, and the third column sum is 0. Therefore, by the Gershgorin circle theorem, the eigenvalues of $A$ have negative or zero real parts. In the case that they are all strictly negative, then the exponential $e^{At}$ tends to the zero matrix as $t \to \infty$. As a result, the long-term fate of the lead in the body is given by the term $A^{-1}\mathbf{f}$,

$$\mathbf{X} \to -A^{-1}\mathbf{f} \quad \text{as } t \to \infty. \tag{9.10.5}$$

*A study on human subjects.*

Rabinowitz, Wetherill, and Kopple studied the lead intake and excretion of a healthy volunteer living in an area of heavy smog. Their work is reported in [3] and extended by Batschelet, Brand, and Steiner in [11]. (See also [12].) The data from this study were used to estimate the rate constants for the compartment model (9.10.1). Lead is measured in micrograms and time in days. For example, the rate 49.3 is given below as the ingestion rate of lead in micrograms per day, and the other coefficients are as given in Table 9.10.1.

**Table 9.10.1.** Lead exchange rates.

| **coefficients** | $a_{01}$ | $a_{12}$ | $a_{13}$ | $a_{21}$ | $a_{02}$ | $a_{31}$ | $I_L$ |
|---|---|---|---|---|---|---|---|
| **value** | 0.0211 | 0.0124 | 0.000035 | 0.0111 | 0.0162 | 0.0039 | 49.3 |

This model has significant biological implications. The output of the computation of the exponential of this matrix does not seem to merit printing. More important for the purposes of understanding this model is the graph of the solutions. These graphs are shown in Figure 9.10.1. This figure shows graphs of the total lead in compartments 1, 2, and 3 over a period of 365 days. The horizontal axis is days and the vertical axis is in units of micrograms of lead.

Our calculation of the solution for (9.10.1) follows the procedure of (9.10.4) exactly. As we will see, the eigenvalues for this matrix are negative. Further, since the trend of the solution is independent of the starting condition—recall (9.10.5)—we take the initial value to be

[16] See [10] for a delightful discussion of the problems involved.

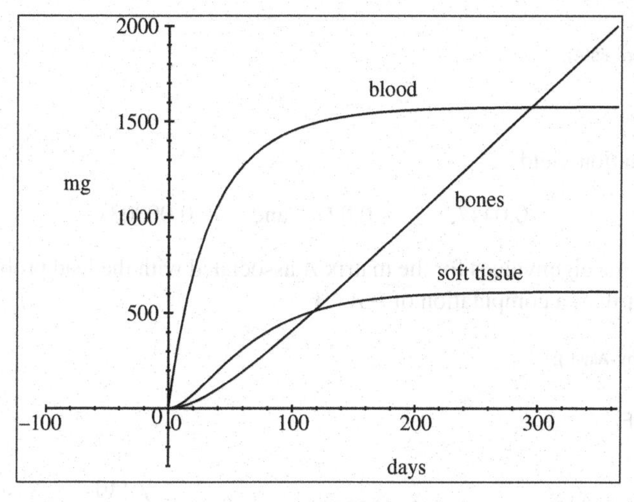


**Fig. 9.10.1.** Solutions for (9.10.1).

$$X_0 = \begin{pmatrix} 0 \\ 0 \\ 0 \end{pmatrix}.$$

The solution is computed and graphed with

```
MAPLE
> with(LinearAlgebra):
> A:=Matrix(3,3,[-0.0361, 0.0124, 0.000035, 0.0111, -0.0286, 0.0, 0.0039, 0.0, -0.000035]);
> etA:=MatrixExponential(A,t);
> AinvF:=evalm(MatrixInverse(A)&*vector([493/10,0,0]));
> u:=evalm(-AinvF+etA&*AinvF);
> plot({u[1](t),u[2](t),u[3](t)},t=0..365);
MATLAB
> A=[-0.0361 0.0124 0.000035; 0.0111 -0.0286 0.0; 0.0039 0 -0.000035];
> f=[49.3; 0; 0]; % column vector
> t=0:2:366; % in steps of 2
  % make an m-file, At.m, with function Y=At(A,t), Y=t*A;
> AinvF=inv(A)*f;
  % make up the sequence of solution vectors
> u=[]; s=size(t);
> for i=1:s(2)
  u=[u,-AinvF+expm(At(A,t(i)))*AinvF];
  end
> plot(t,u)
```

One observation is that the level of lead in the blood and soft tissue approaches the steady state quickly. Lead achieves a steady state in the blood of about 1800 units and about 700 in the soft tissue. The level of lead in the bones continues to rise after a period of one year. In this model, the bones continue to absorb lead because of the constant rate of input. On the other hand, the bones release lead slowly and steadily. As we have already seen in the discussion in Section 9.9, a high level of lead in the bones has implications for severe and chronic biological problems.

The levels of lead in the steady state is the next subject for discussion. For this lead model, the long-term behavior of solutions can be computed as follows:

MAPLE
```
> lambda:=Eigenvals(A):
```

MATLAB
```
> [evect,eval]=eig(A)
```

This computation yields

$$-0.0447, \qquad -0.02, \quad \text{and} \quad -0.00003.$$

We find that the eigenvalues for the matrix $A$ associated with the lead problem are all negative. Further, a computation of $-A^{-1}\mathbf{f}$

MAPLE
```
> leadlim:=evalm(-AinvF);
```

MATLAB
```
> leadlim=-AinvF
```

yields

$$-A^{-1}\mathbf{f} = (1800, 698, 200582), \quad \text{where } \mathbf{f} = \left(\frac{439}{10}, 0, 0\right).$$

Hence this model predicts the long-range forecast summarized as follows: The levels of lead in the blood will rise to about 1800 micrograms, the level of lead in the other soft tissues will rise to about 698 micrograms, and the level of lead in the bones will rise to about 200582 micrograms.

It should be recognized that the coefficients for our absorption of lead are not constants. By way of data for long-range forecast, Ewers and Schlipkoter point out that after age 20, the lead content of most soft tissue does not show age-related changes [1]. The lead content of bones and teeth increases throughout life, since lead becomes localized and accumulates in human calcified tissues (bone and teeth). This accumulation begins with fetal development and continues to approximately the age of 60 years. At least for adults, various studies show that approximately 95% of the total body lead is lodged in the bones.

### Exercises/Experiments

1. This exercise is an investigation of the lead model. The exercise is broken into a set of questions that can be answered by modification of the syntax in this chapter.
   (a) What is the long-range forecast for lead in each of the compartments using model (9.10.1)?
   (b) Approximately what is the lead level achieved in each of the compartments in "1 year"?
   (c) Ewers and Schlipkoter state that 95% of the total body lead of human adults is lodged in bones [1]. Schroeder and Tipton state that "Bones contain 91% of the total body lead" [4]. What percentage of the total body lead does this model place in the bones?
   (d) Redo (a) and (b) by doubling or halving the ingestion rate. What is the revised long-range forecast for each of the compartments? Approximately what is the revised lead level achieved in each of the compartments in "1 year"?

**2.** All the remaining questions are concerned with a person that has lived in a lead-contaminated environment so long that a level of 2500 micrograms of lead has accumulated in the bones. You may continue to assume that in the contaminated environment 49.3 micrograms per day are absorbed. In their 1968 paper, Schroeder and Tipton stated that an average of 17 micrograms of lead are retained per day. We take this absorption rate to be that in a "clean" environment.

(a) What level of lead do you expect in the tissue and bones for a person living in a contaminated environment long enough that 2500 micrograms of lead has accumulated in the bones?

(b) Suppose that this person described in the previous question is moved to a relatively lead-free environment. What is the approximate level of lead in the bones, tissue, and blood at the end of one year after living in this new environment?

(c) We have seen that there are drugs that alleviate the effects of lead in the bones by increasing the rate of removal from the bones. What should that rate be to cut in half the amount of lead in the bones at the end of one year in the cleaner environment?

(d) Suppose that the person takes the drug you have designed but is not moved to the cleaner environment. What are the levels of lead in the bones, tissue, and blood after one year of taking the drug while living in the contaminated environment?

(e) Ewers and Shlipkoter give the half-life of lead in blood as 19 days, in soft tissue as 21 days, and in bones as 10 to 20 years [1]. What is the half-life as assumed in our model?

(f) According to this model, what percentage of the lead ingested into the body is returned to the environment during the 100th day in the initial situation?

## 9.11 Pharmacokinetics

*The routes for dispersion of drugs through the body follow the same pattern as those of lead. The previous section followed lead through the body. The model of this section examines how the body handles the ingestion of a decongestant. We keep track of this drug in two compartments of the body: the gastrointestinal tract and the circulatory system. The mathematical importance of this model is that the limit for the system is a periodic function.*

*A two-compartment pharmacokinetic model is used to construct a drug utilization scenario.*

Among all the means for the delivery of therapeutic drugs to the bloodstream, oral ingestion/gastrointestinal absorption is by far the most popular. In this section, we study this delivery mechanism, following closely the work of Spitznagel [13]. The working hypothesis of the study is the following series of events. The drug is taken orally on a periodic basis resulting in a *pulse* of dosage delivered to the gastrointestinal (GI) tract. From there, the drug moves into the bloodstream, without delay, at a rate proportional to its concentration in the GI tract and independent of its concentration in the blood. Finally, the drug is metabolized and cleared from the blood at a rate proportional to its concentration there.

Evidently, the model should have two compartments: Let $x(t)$ denote the concentration of drug in the GI tract and $y(t)$ its concentration in the blood. In addition, we need the drug intake regimen; denote by $D(t)$ the drug dosage, as seen by the GI tract, as a function of time $t$.[17] The governing equations are

$$\begin{aligned}\frac{dx}{dt} &= -ax + D,\\ \frac{dy}{dt} &= ax - by.\end{aligned} \tag{9.11.1}$$

Since the equations in (9.11.1) constitute a linear system with forcing function $D(t)$, its solution, in matrix form, is given by (2.4.12), which we repeat here:

$$\mathbf{Y} = e^{Mt}\mathbf{Y}_0 + e^{Mt}\int_0^t e^{-Ms}\mathbf{P}(s)ds. \tag{9.11.2}$$

In this equation, $\mathbf{Y}$ and $\mathbf{P}$ are the vectors

$$\mathbf{Y} = \begin{bmatrix} x \\ y \end{bmatrix} \quad \text{and} \quad \mathbf{P} = \begin{bmatrix} D(s) \\ 0 \end{bmatrix},$$

and $M$ is the coefficient matrix

$$\begin{bmatrix} -a & 0 \\ a & -b \end{bmatrix}.$$

Note that as a compartment model, the diagonal terms of this matrix are negative and the column sums are negative or zero. Consequently, the first term of the solution, $e^{Mt}\mathbf{Y}_0$, is transient, that is, it tends to 0 with time. Therefore, asymptotically the solution tends to the second term,

$$\mathbf{Y} \to e^{Mt}\int_0^t e^{-Ms}\begin{bmatrix} D(s) \\ 0 \end{bmatrix} ds, \tag{9.11.3}$$

which is periodically driven by $D$.

*Periodic solutions predict serum concentration cycles.*

In conjunction with specific absorption and metabolism rates for a given drug, system (9.11.1) and its solution, (9.11.2), may be used to predict cycles of drug concentration in the blood. Fortunately, the required data are available for a variety of drugs, such as PPA and CPM, as reported and defined in [13]. As mentioned above, the exact

[17] With the use of time-release capsules, a drug may not be immediately available to the GI tract even though the medication has been ingested.

shape of the dosage profile, $D(t)$, depends on how the producer, the pharmaceutical company, has buffered the product. We assume that the drug is taken every six hours (four times a day) and dissolves within about $\frac{1}{2}$ hour, providing a unit-pulse dosage with height 2 and pulse width $\frac{1}{2}$ on the interval $[0, 6]$; see Figure 9.11.1. The rate parameters $a$ and $b$ are typically given as half-lives; cf. (3.5.8). For PPA, we use a $\frac{1}{2}$-hour half-life in the GI tract, so $a = 2\ln(2)$, and a 5-hour half-life in the blood, $b = \frac{\ln(2)}{5}$.

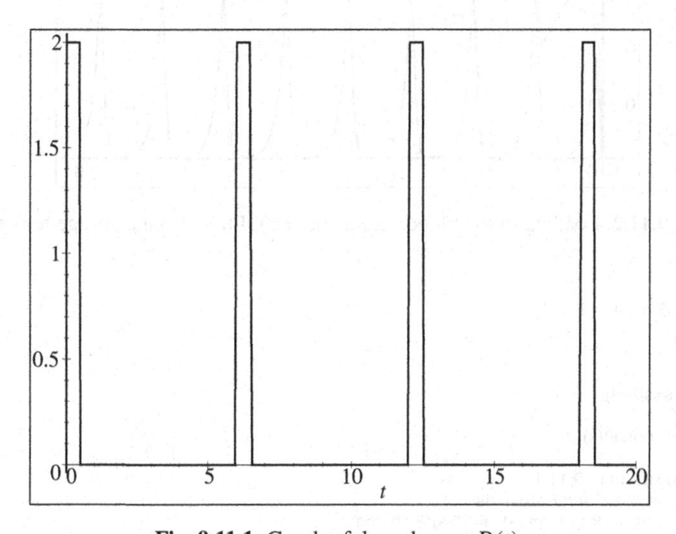


**Fig. 9.11.1.** Graph of drug dosage $D(t)$.

For a numerical solution, we take zero initial conditions, $x(0) = y(0) = 0$, that is, initially no drug is present in the GI tract or circulatory system, and use a Runge–Kutta method to obtain Figure 9.11.2. The behavior of $x(t)$ is the lower graph and predicts an oscillating increase and decrease of concentration in the GI tract. On the other hand, the concentration in the circulatory system, $y(t)$, is predicted to be an oscillation superimposed on a gradually increasing level of concentration.

```
MAPLE
> restart;
> a:=ln(2)*2; b:=ln(2)/5;
> Dose1:=t->sum((signum(t-n*6)-signum(t-(n*6+1/2))),n=0..10);
> plot(Dose1(t),t=0..20);
> with(plots): with(DEtools):
> J:=DEplot({diff(x(t),t)=Dose1(t)-a*x(t),diff(y(t),t)=a*x(t)-b*y(t)},{x(t),y(t)},t=0..50,{[0,0,0]},stepsize=0.5,
        scene=[t,x],linecolor=RED):
> K:=DEplot({diff(x(t),t)=Dose1(t)-a*x(t),diff(y(t),t)=a*x(t)-b*y(t)},{x(t),y(t)},t=0..50,{[0,0,0]},stepsize=0.5,
        scene=[t,y],linecolor=BLACK):
> plots[display]({J,K});
```

```
MATLAB
% make up an m-file dose.m with
% function H=dose(t)
% H=0;
% for n=0:50
```

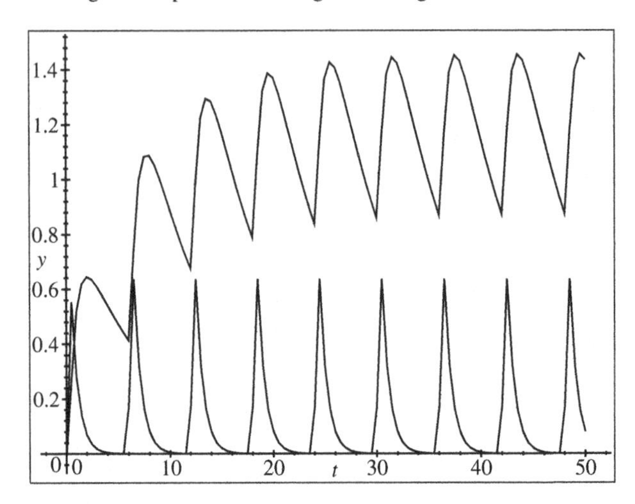


**Fig. 9.11.2.** Loading of the bloodstream and the GI tract from a dosage regime.

```
  %  if (t >= 6*n)&(t < 6*n+.5)
  %     H=1;
  %  end
  % end
> t=0:.05:50; s=size(t);
> for k=1:s(2)
    Dvect(k)=2*dose(t(k));
  end
> plot(t,Dvect) % Figure 9.11.1
  % make up an m-file drugRate.m as
  % function Yprime=drugRate(t,x); a=2*log(2); b=log(2)/5;
  % Yprime=[-a*x(1)+2*dose(t); a*x(1)-b*x(2)];
> [t,Y] = ode23('drugRate',[0 50],[0;0]);
> plot(t,Y) % Figure 9.11.2
```

In Figure 9.11.3, we show the phase-plane plot, $x$ vs. $y$, of this solution. It shows that, asymptotically, the solution tends to a periodic (but nonsinusoidal) oscillation; this is called a *limit cycle*.

```
MAPLE
> phaseportrait([diff(x(t),t)=Dose1(t)-a*x(t), diff(y(t),t)=a*x(t)-b*y(t)],[x(t),y(t)],t=0..50,{[0,0,0]},stepsize=0.5);

MATLAB
> plot(Y(:,1),Y(:,2))
```

From the figure, the high concentration level in the blood is about 1.8, while the low is about 1.1 on the limit cycle. In designing a drug, it is desirable to keep the concentration as uniform as possible and to come up to the limit cycle as quickly as possible. Toward that end, the parameter $a$ can be more easily adjusted, for example by a "time release" mechanism. The parameter $b$ tends to be characteristic of the drug itself.

The asymptotic periodic solution may be found from (9.11.3) or by the following manual scheme.

We can solve for $x(t)$ explicitly. There are two parts: one part for $0 < t < \frac{1}{2}$, where the dosage function has value 2, and the other part for $\frac{1}{2} < t < 6$, where the

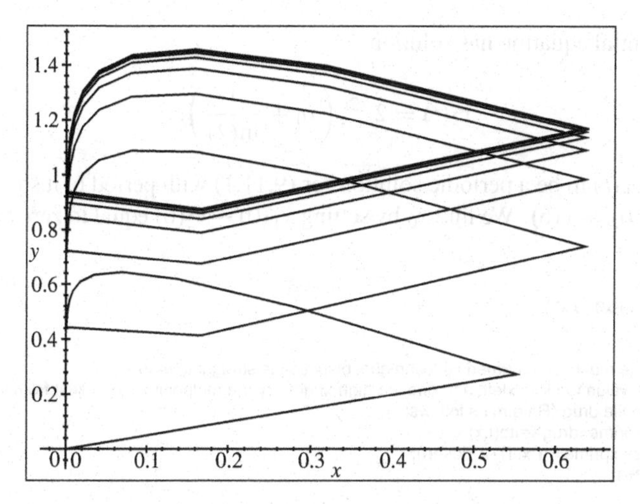


**Fig. 9.11.3.** $\{x(t), y(t)\}$ with limit cycle.

dosage function has value 0. We call the first part $x_1(t)$ and find the solution here. In this case, the input from $D(t)$ is 2 and the initial value for the periodic solution is yet unknown. Call the initial value $x_0$. We have

$$x(t) = -ax(t) + 2$$

with $x(0) = x_0$.

```
MAPLE (symbolic)
> dsolve({diff(x(t),t)+a*x(t)=2,x(0)=x0},x(t));
> x1:=unapply(op(2,%),t);
> simplify(x1(t));
```

Noting that $2^{2t} = 4^{-t}$, we see that this equation has solution

$$x_1(t) = \frac{1}{\ln(2)} + \left(x_0 - \frac{1}{\ln(2)}\right) 2^{-2t}.$$

Follow this part of the solution from $t = 0$ to $t = \frac{1}{2}$. Next, we compute the solution $x(t)$ for (9.11.1) with $\frac{1}{2} < t < 6$. In this case, the input from $D(t) = 0$ and the initial value for the continuation of the periodic solution starts where the first part left off. Thus

$$x(t) = -ax(t)$$

with

$$x\left(\frac{1}{2}\right) = x_1\left(\frac{1}{2}\right).$$

```
MAPLE (symbolic)
> dsolve({diff(x(t),t)+a*x(t)=0,x(1/2) = x1(1/2)},x(t));
> x2:=unapply(op(2,%),t);
> simplify(x2(t));
```

This differential equation has solution

$$x_2(t) = 2^{-2t}\left(x_0 + \frac{1}{\ln(2)}\right).$$

In order for $x(t)$ to be a periodic solution for (9.11.1) with period 6, it should be true that $x_0 = x(0) = x(6)$. We find $x_0$ by setting $x_1(0) - x_2(6)$ equal to zero and solving for $x_0$:

```
MAPLE
> x0:=solve(x2(6)-x0=0,x0);
```

```
MATLAB
% In MATLAB we must use a numerical technique; bisection is straightforward
% start with a value too low, xleft, and one too high xright, try the midpoint xmid, adjust from there
% make an m-file drugXRate.m as follows:
%  function xprime=drugXrate(t,x)
%  a=2*log(2); xprime=-a*x(1)+2*dose(t);
> xleft=0; xright=1;
> for k=1:16
>   xmid=(xleft+xright)/2;
  % solve ode on 0 to 0.5
>   [t,x]=ode23('drugXrate',[0 .5],xmid); s=size(x); x05=x(s(1)); % save the ending value of x
  % solve ode on 0.5 to 6 with that ending value as starting value
>   [t,x]=ode23('drugXrate',[.5 6],x05); s=size(x); x6=x(s(1)); % save the final ending value
  % we want x6 to equal xmid; adjust if too big or too small
>   if x6 > xmid % end value bigger than start
>     xleft=xmid; % raise the start value
>   else
>     xright=xmid; % lower the start value
>   end
> end
> x0periodic=xmid
```

The solution is

$$x_0 = \frac{1}{\ln(2)\cdot 4095}.$$

It remains to find the periodic solution $y$ for the second equation. Equation (9.11.2) can be rewritten, now that we have a formula for $x(t)$:

$$y(t) + by(t) = ax(t).$$

The function $y_1(t)$ will be the solution for $0 < t < \frac{1}{2}$ and $y_2(t)$ is the solution for $\frac{1}{2} < t < 6$.

Now continue the solution for $\frac{1}{2} < t < 6$ and for $y(\frac{1}{2}) = y_1(\frac{1}{2})$:

```
MAPLE
> dsolve({diff(y(t),t)=a*x1(t)-b*y(t), y(0) = y0},y(t));
> simplify(rhs(%)); y1:=unapply(%,t);
> dsolve({diff(y(t),t)=a*x2(t)-b*y(t), y(1/2)=y1(1/2)},y(t));
> y2:=unapply(op(2,%),t);
  # require that y2(6)=y1(0)
> solve(y2(6)=y0,y0); y0:=evalf(%);
```

```
MATLAB
% now find y0periodic
> yleft=0; yright=1;
> for k=1:16
>   ymid=(yleft+yright)/2; [t,Y]=ode23('drugRate',[0.5],[x0periodic;ymid]);
>   y=Y(:,2); s=size(y); y05=y(s(1));
```

```
> [t,Y]=ode23('drugRate',[.5 6],[x05; y05]); y=Y(:,2); s=size(y); y6=y(s(1));
>   if y6 > ymid
>     yleft=ymid;
>   else
>     yright=ymid;
>   end
> end
> y0periodic=ymid
```

The result is that $y_0$ is about 0.8864. Figure 9.11.4 shows one period of $x(t)$ and $y(t)$ superimposed on the same graph, and Figure 9.11.5 is the parametric plot $(x(t), y(t))$. These are produced by the following computer codes. For Figure 9.11.4:

MAPLE
```
> plot({[t,x1(t),t=0..1/2],[t,x2(t),t=1/2..6],[t,y1(t),t=0..1/2],[t,y2(t),t=1/2..6]},color=BLACK);
```

MATLAB
```
> [t,Y]=ode23('drugRate',[0 6],[x0periodic; y0periodic]);
> plot(t,Y)
```

And for Figure 9.11.5:

MAPLE
```
> plot({[x1(t),y1(t),t=0..1/2],[x2(t),y2(t),t=1/2..6]},view=[0..1,0..2],color=BLACK);
```

MATLAB
```
> plot(t,Y); plot(Y(:,1),Y(:,2))
```

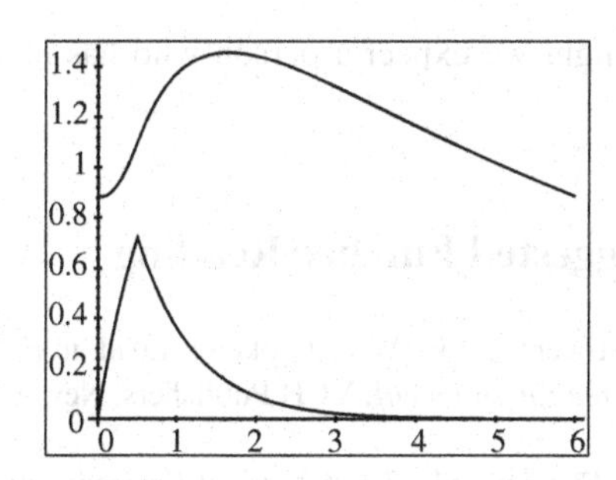


**Fig. 9.11.4.** Superimposed plots of $x(t)$ (lower) and $y(t)$ (upper).

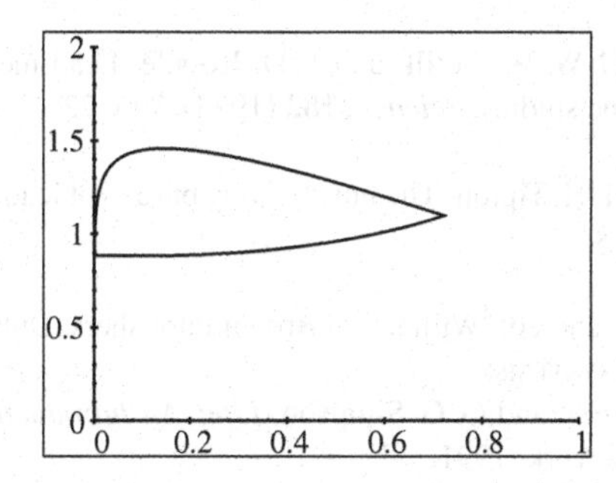


**Fig. 9.11.5.** Parametric plot $(x(t), y(t))$.

This model continues to raise many questions. Because $y(t)$ represents the level of the drug in the circulatory system, we note that the level should be large enough for the drug to be effective, and but not so large as to cause side effects. The job of the pharmaceutical company is to determine the appropriate level that $y$ should have, and to adjust $a$ and $b$ to maintain that level.

**Exercises/Experiments**

1. A part of the interest in [13] was to contrast the behaviors of PPA and CPM. CPM has a half-life of one hour in the GI tract and 30 hours in the blood system. This contrast can be made by modifying the code suggested in Section 9.11.

2. With the model of this section, suppose that $a$ and $b$ are as specified for the two drugs. What could be an initial intravenous injection level, $x_0$, with subsequent equal dose levels as specified and taken orally, so that the periodic steady state is achieved in the first six hours and maintained thereafter?

**Questions for Thought and Discussion**

1. For what anatomical and molecular reasons do we expect the lead in leaded gasoline to move percutaneously into our bodies?

2. Describe the path of ingested lead from mouth to bone marrow.

3. For what reasons might we expect a person who has ingested lead to become anemic?

## References and Suggested Further Reading

[1] LEAD POISONING: U. Ewers and H.-W. Schlipkoter, Lead, in E. Merian, ed., *Metals and Their Compounds in the Environment*, VCH Publishers, New York, 1991, 971–1014.

[2] LEAD POISONING:
J. Lerihan and W. W. Fletcher, eds. *The Chemical Environment*, Academic Press, 1977, New York, 28–92.

[3] LEAD POISONING:
M. B. Rabinowitz, G. W. Wetherill, and J. D. Kopple, Lead metabolism in the normal human: Stable isotope studies, *Science*, **182** (1973), 725–727.

[4] LEAD POISONING:
H. A. Schroeder and I. H. Tipton, The human body burden of lead, *Arch. Environ. Health*, **17**-6 (1968), 965–978.

[5] EMBRYOLOGY:
L. Johnson, *Biology*, 2nd ed., William C. Brown Publishers, Dubuque, IA, 1987.

[6] BIOLOGICAL ORGAN SYSTEMS:
W. S. Beck, K. F. Liem, and G. G. Simpson, *Life: An Introduction to Biology*, 3rd ed., Harper–Collins, New York, 1991.

[7] BIOLOGICAL ORGAN SYSTEMS:
E. P. Solomon, L. R. Berg, D. W. Martin, and C. Villee, *Biology*, 3rd ed., Saunders College Publishing, Fort Worth, TX, 1993.

[8] DIFFUSION DISTANCES IN CELLS:
D. Dusenbery, *Sensory Ecology*, W. H. Freeman, San Francisco, 1992, Chapter 4.

[9] RESISTANCE TO FLUID FLOW:
S. Vogel, *Life in Moving Fluids: The Physical Biology of Flow*, Princeton University Press, Princeton, NJ, 1989, 165–169.

[10] EXPONENTIAL OF A MATRIX:
C. B. Moler and C. F. Van Loan, Nineteen dubious ways to compute the exponential of a matrix, *SIAM Rev.*, **20**-4 (1978), 801–836.

[11] SMOG:
E. Batschelet, L. Brand, and A. Steiner, On the kinetics of lead in the human body, *J. Math. Biol.*, **8** (1979), 15–23.

[12] COMPUTER MODELING:
R. L. Borrelli, C. S. Coleman, and W. E. Boyce, *Differential Equations Laboratory Workbook*, Wiley, New York, 1992.

[13] PHARMACOKINETICS:
E. Spitznagel, *Two-Compartment Pharmacokinetic Models C-ODE-E*, Harvey Mudd College, Claremont, CA, 1992.

# 10

# A Biomathematical Approach to HIV and AIDS

## Introduction

Acquired immunodeficiency syndrome (AIDS) is medically devastating to its victims, and wreaks financial and emotional havoc on everyone, infected or not. The purpose of this chapter is to model and understand the behavior of the causative agent of AIDS—the human immunodeficiency virus (HIV). This will necessitate discussions of viral replication and immunology. By the end of this chapter, the student should have a firm understanding of the way that HIV functions and be able to apply that understanding to a mathematical treatment of HIV infection and epidemiology.

Viruses are very small biological structures whose reproduction requires a host cell. In the course of viral infection the host cell is changed or even killed. The host cells of HIV are specific and very unique: They are cells of our immune system. This is of monumental importance to the biological and medical aspects of HIV infection and its aftermath. HIV infects several kinds of cells, but perhaps its most devastating cellular effect is that it kills helper T-lymphocytes. Helper T-lymphocytes play a key role in the process of gaining immunity to specific pathogens; in fact, if one's helper T-lymphocytes are destroyed, the entire specific immune response fails. Note the irony: HIV kills the very cells that are required by our bodies to defend us from pathogens, including HIV itself! The infected person then contracts a variety of (often rare) diseases to which uninfected persons are resistant, and that person is said to have AIDS.

## 10.1 Viruses

*Viruses are small reproductive forms with powerful effects. A virus may have only four to six genes, but those genes enable it to take over the synthetic machinery of a normally functioning cell, turning it into a small biological factory producing thousands of new viruses. Some viruses add another ability: They can insert their nucleic acid into that of the host cell, thus remaining hidden for many host cell generations prior to viral reproduction.*

R.W. Shonkwiler and J. Herod, *Mathematical Biology: An Introduction with Maple and Matlab*, Undergraduate Texts in Mathematics, DOI: 10.1007/978-0-387-70984-0_10,

*HIV is an especially versatile virus. It not only inserts its genetic information into its host's chromosomes, but it then causes the host to produce new HIV. Thus the host cells, which are immune system components, produce a steady stream of HIV particles. Eventually, this process kills the host cells and the patient becomes incapable of generating critical immune responses.*

*A virus is a kind of parasite.*

Each kind of virus has its own special anabolic ("building up") needs, which, because of its genetic simplicity, the virus may be unable to satisfy. The host cell then must provide whatever the virus itself cannot. This requires a kind of biological matching between virus and host cell analogous to that between, say, an animal parasite and its host. Host specificity is well developed in viruses: As examples, the rabies virus infects cells of our central nervous system, cold viruses affect cells of our respiratory tract, and the feline leukemia virus affects certain blood cells of cats (see [1]).

*The basic structure of a virus is a protein coat around a nucleic acid core.*

Simple viruses may have only four to six genes, but most viruses have many more than that. In the most general case the viral nucleic acid, either DNA or RNA, is surrounded by a protein coat, called a *capsid* (see Figure 10.1.1). In addition, many viruses have outer layers, or *envelopes*, which may contain carbohydrates, lipids, and proteins. Finally, inside the virus there may be several kinds of enzymes along with the nucleic acid.

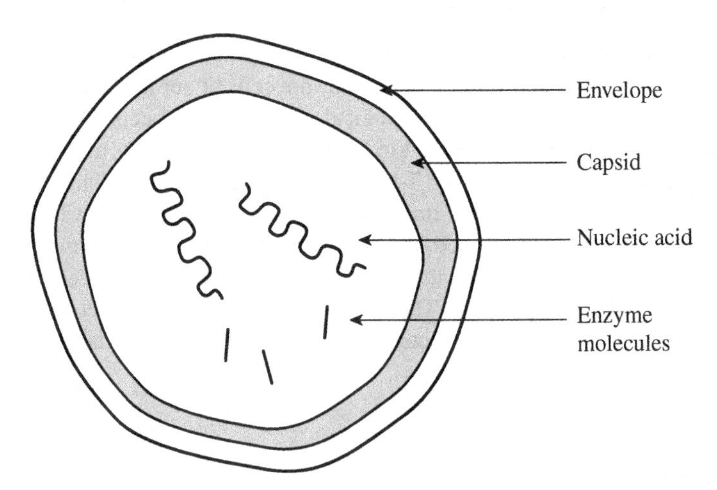


**Fig. 10.1.1.** A generalized drawing of a virus. In a given real case the envelope and/or enzyme molecules may be absent and the nucleic acid may be DNA or RNA.

A virus cannot reproduce outside a host cell, which must provide viral building materials and energy. All the virus provides is instructions via its nucleic acids and, occasionally, some enzymes. As a result, viruses are not regarded as living things.

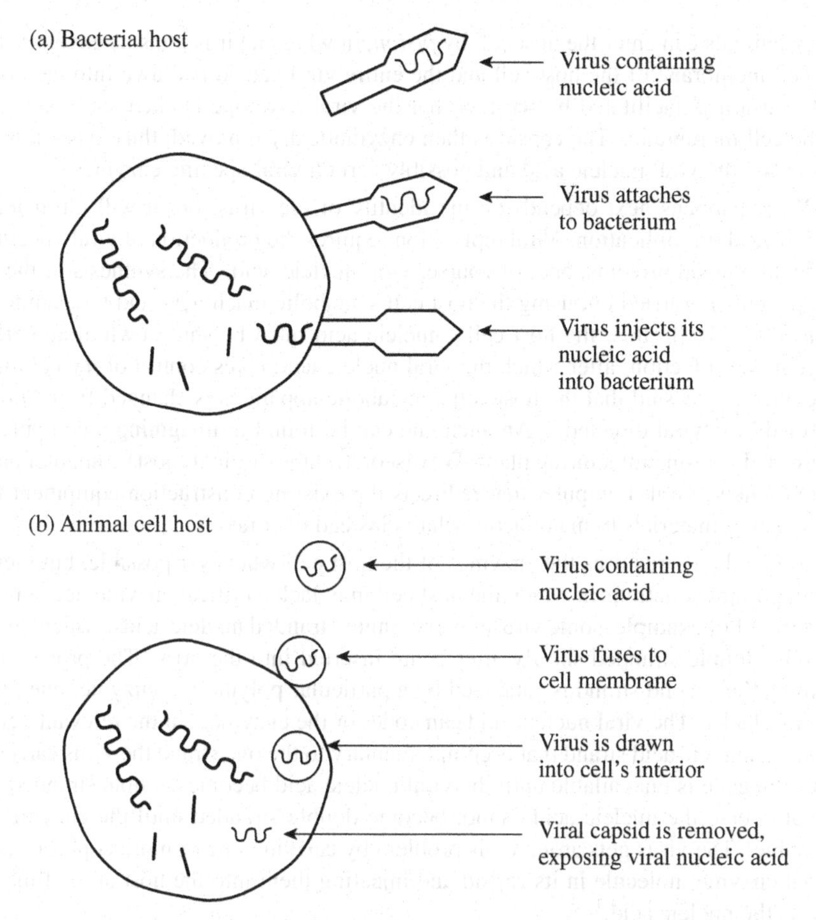
(a) Bacterial host
Virus containing nucleic acid
Virus attaches to bacterium
Virus injects its nucleic acid into bacterium
(b) Animal cell host
Virus containing nucleic acid
Virus fuses to cell membrane
Virus is drawn into cell's interior
Viral capsid is removed, exposing viral nucleic acid

**Fig. 10.1.2.** Some models of viral infection. (a) A virus whose host is a bacterium recognizes some molecular feature of the correct host, attaches to it, and injects its nucleic acid into it. (A virus whose host is a bacterium is called a *bacteriophage*.) (b) A virus whose host is an animal cell recognizes some molecular feature of the correct host cell and is then drawn into the host cell, where the capsid is removed.

*Viral nucleic acid enters the host cell and redirects the host cell's metabolic apparatus to make new viruses.*

A virus attaches to its specific host's outer covering, host–virus specificity being ensured by host-to-viral molecular recognition. The molecules involved are proteins or *glycoproteins*, a sugar–protein combination. At this point the viral nucleic acid enters the host cell, the precise means of entry depending on the nature of the virus (see Figure 10.1.2). For instance, viruses called *bacteriophages* infect bacteria. Bacteriophages have no envelope and seem to inject their nucleic acid into the bacterium, leaving the viral protein capsid outside. Alternatively, nucleic acids from viruses that infect animals can enter the host cell by *fusion*, in which a virus joins its envelope to the cell membrane of the host cell and the entire viral capsid is drawn into the host cell. Fusion is facilitated by the fact that the viral envelope is chemically similar to the cell membrane. The capsid is then enzymatically removed, thus exposing its contents—the viral nucleic acid and possibly certain viral-specific enzymes.

What happens next depends on the identity of the virus, but it will ultimately lead to viral multiplication. Viral replication requires the production of viral-specific enzymes, capsid proteins, and, of course, viral nucleic acid. The synthesis of these components is carried out using the host cell's anabolic machinery and biochemical molecules. To do this, the host cell's nucleic acid must be shut down at an early stage in the infection, after which the viral nucleic acid takes control of the cellular machinery. It is said that the host cell's metabolic apparatus is changed from "host directed" to "viral directed." An analogue can be found in imagining a computer-controlled sofa-manufacturing plant. We disconnect the original (host) computer and install a new (viral) computer that redirects the existing construction equipment to use existing materials to manufacture chairs instead of sofas.

Typically a virus uses the enzymes of the host cell whenever possible, but there are important situations in which the host cell may lack a critical enzyme needed by the virus. For example, some viruses carry single-stranded nucleic acids, which must become double stranded shortly after being inserted into the host. The process of forming the second strand is catalyzed by a particular polymerase enzyme, one that the host lacks. The viral nucleic acid can code for the enzyme, but the relevant gene is on the nucleic acid strand that is complementary to the one strand the virus carries. Thus the gene is unavailable until the viral nucleic acid becomes double stranded—but of course the nucleic acid cannot become double stranded until the enzyme is available! The virus gets around this problem by carrying one or more copies of the actual enzyme molecule in its capsid and injecting them into the host at the time it injects the nucleic acid.[1]

As the virus's various component parts are constructed, they are assembled into new, intact viruses. The nucleic acid is encapsulated inside the protein capsid, perhaps accompanied by some critical viral enzymes. The assembly of the capsid is spontaneous, like the growth of a crystal. The newly assembled viruses then escape from the host cell and can start the infection process anew.

*Many RNA viruses do not use DNA in any part of their life cycle.*

The central dogma was presented in Chapter 8 to show the path of genetic information flow:

[1] Recall from Chapter 8 that in a given segment of DNA, only one of the two DNA strands actually codes for RNA. That strand is called the *coding strand*. In the example given above, the coding strand would be the strand formed after infection. Thus its genes would not be available until after the nucleic acid became double stranded.

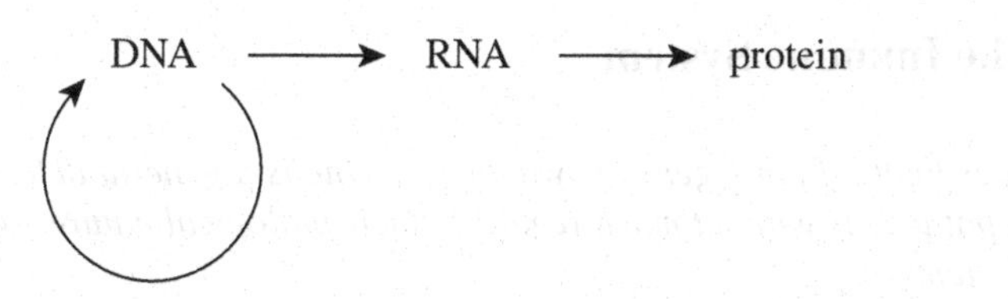


Note that because RNA is complementary to DNA, it should be possible to skip the DNA part of the scheme. All that is necessary to justify this assertion is to demonstrate that RNA can code for its own self-replication. While this does not seem to happen in cellular systems, it is well known in viruses: Viral RNA replicates just as DNA does, using complementary base-pairing. After replication, the RNA is packaged into new viruses.[2]

Our revised statement of the central dogma, accounting for RNA self-replication, now looks like this:

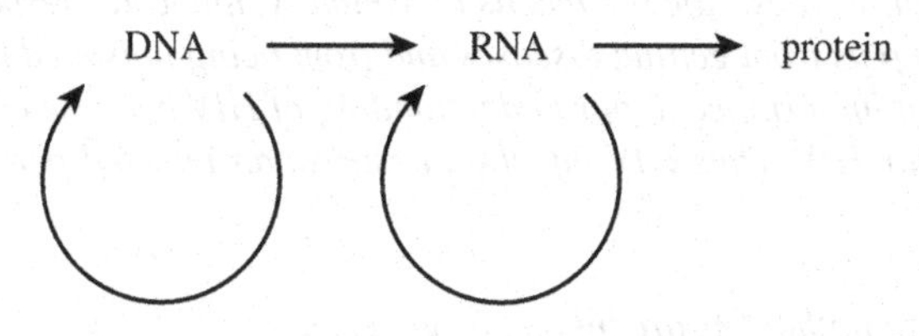


*There are several variations in the host-cell-escape mechanism for viruses.*

Some viruses merely replicate their nucleic acid, translate out the necessary proteins, encapsulate, and then burst out of the host cell an hour or two after infection. This bursting process kills the host cell and is called *lysis*; the virus is said to be *lytic*.

Other viruses, said to be *lysogenic*, delay the lytic part of their reproductive process. For example, the DNA of some DNA viruses is inserted into the host cell body and then into the host's DNA. Thus when the host's DNA is replicated at cell division, so is the viral DNA. The inserted viral DNA is called a *provirus*, and it can remain inserted in the host DNA for many cell generations. Sooner or later, depending on the lysogenic virus, host, and culture conditions, the provirus begins to replicate its nucleic acid and produces RNA, which then produces viral proteins. New viruses are then assembled and lyse the host to get out.

There is an alternative to lysis in the escape process: When the viruses exit the host cell, they may instead *bud off* from the host cell, in a process that is the reverse of fusion. In the process, they take a piece of the cell membrane for an envelope, but do not kill the host cell. Cells that release viruses by budding can therefore act as virtually unending sources of new viruses. This, in fact, is the behavior of certain blood cells infected with HIV.

---

[2] There are single-stranded and double-stranded RNA viruses, just as there are single- and double-stranded DNA viruses. HIV is a single-stranded RNA virus—its conversion to double-stranded form will be described in Section 10.3.

## 10.2 The Immune System

*Our bodies fight off pathogens by two means. One is a general defense system that removes pathogens without much regard to their biological nature; stomach acid is such a system.*

*Of more concern to us in our considerations of HIV is a second, specific response to pathogens (and other foreign substances); this response is tailored to each infective agent. Specialized blood cells called lymphocytes have the ability to recognize specific molecular parts of pathogens and to mount a chemical response to those fragments. Initially, we have at most only a few lymphocytes that can recognize each such fragment, but upon contact with the fragment, the lymphocyte will start to divide extensively to provide a clone of cells. Thus there results a large clone of identical lymphocytes, all of which are chemically "tuned" to destroy the pathogen.*

*In this section, we describe the means by which lymphocytes respond to foreign substances to keep us from getting diseases and from being poisoned by toxins. This subject is of great importance to our understanding of HIV because certain lymphocytes are hosts for HIV. Thus HIV infection destroys an infected person's ability to resist pathogens.*

*Some responses to pathogens are innate, or general.*

We possess several general mechanisms by which we can combat pathogens. These mechanisms have a common property: They are essentially nondiscriminatory. Each one works against a whole class of pathogens and does not need to be adapted for specific members of that class. For example, tears and egg white contain an enzyme that lyses the cell walls of certain kinds of bacteria. Stomach acid kills many pathogens that we eat. Damaged tissue attracts blood- clotting agents and dilates capillaries to allow more blood to approach the wound. Finally, there are blood cells that can simply engulf pathogens; these cells are *granulocytes* and *macrophages*.

The problem with the innate response is that it cannot adapt to new circumstances, whereas many pathogens are capable of rapid genetic change. Thus many pathogens have evolved ways to circumvent the innate response. For such pathogens, we need an immune response that can change in parallel with the pathogen (see [1] and [2]).

*Blood cells originate in bone marrow and are later modified for different functions.*

Humans have bony skeletons, as do dogs, robins, snakes, and trout, but sharks and eels have cartilaginous skeletons. In the core, or *marrow*, of our bones is the blood-forming tissue, where all of our blood cells start out as *stem cells*. Repeated division of stem cells results in several paths of cellular specialization, or *cell lines*, as shown in Figure 10.2.1. Each cell line leads to one of the various kinds of mature blood cells described in Section 9.6. One cell line becomes red blood cells. Another line generates cells involved in blood clotting. Still other lines have the ability to engulf and digest pathogens. Finally, there is a cell line that generates cells capable of specifically adapted defenses to pathogenic agents. They are called *lymphocytes*.

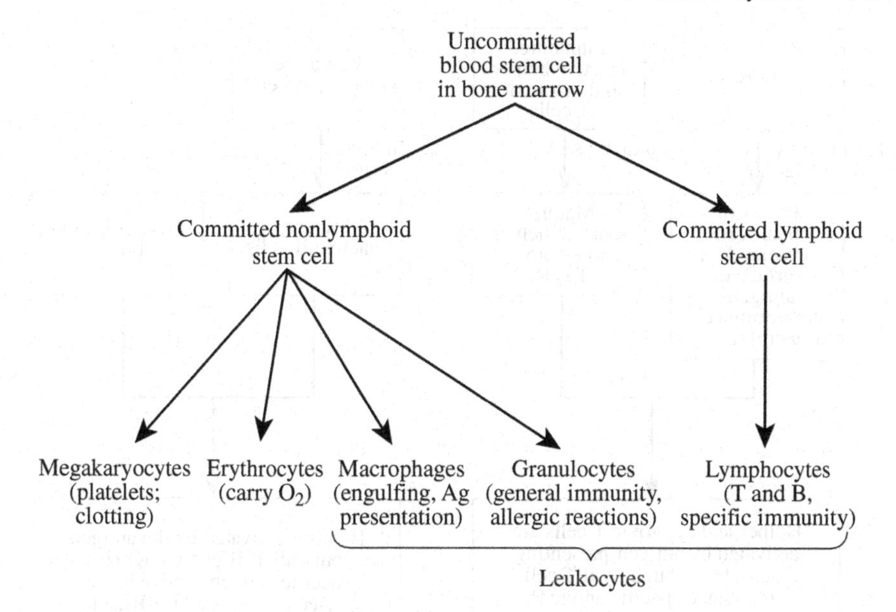


**Fig. 10.2.1.** A flow chart showing the development of mammalian blood cells from their generalized state to their final, differentiated state.

*Some immune responses are adaptive, or specific, to the pathogen.*

Our immune system is capable of reactions specifically tailored to each foreign substance, or *antigen*; in other words, each and every antigen elicits a unique response. At first glance, we might think that the finite amount of information that a cell can contain would place a ceiling on the number of specific responses possible. We will see that the restriction is not important because the specific immune system works against as many as $10^{12}$ distinct antigens![3]

Certain cell lines, derived from bone marrow stem cells, mature in our lymphatic system to become *lymphocytes*. For example, T-lymphocytes, or *T-cells*, mature in the thymus gland, which is found prominently under the breastbone of fetuses, infants, and children. B-lymphocytes, or *B-cells*, mature in bone marrow. These two kinds of lymphocytes are responsible for the adaptive immune responses, but they play different and complementary roles.

*T-cells are responsible for the cell-mediated immune response.*

We will be especially interested in two groups of T-cells: *helper T-cells* and *cytotoxic T-cells* (see Figure 10.2.2). After they mature in the thymus of neonatal and prenatal animals, these T-cells are *inactive*. On their outer surfaces, inactive T-cells have recognition proteins that can bind to antigens (via hydrogen bonds and other interac-

[3] The size of this number, even its order of magnitude, is subject to some debate. In any case, it is *very* big.

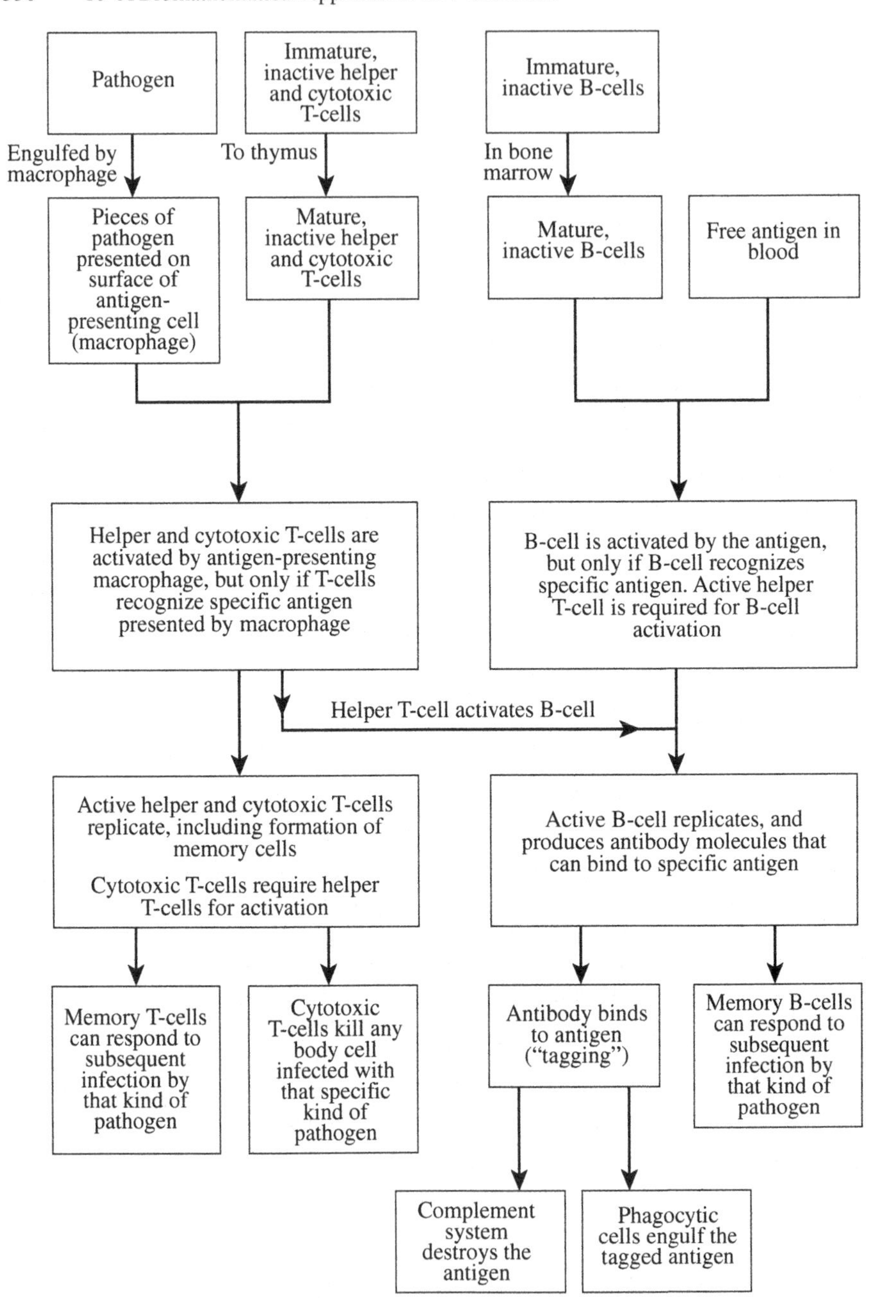


**Fig. 10.2.2.** A flow chart showing the events and interactions surrounding the specific immune response. The cell-mediated response begins at the top left and the humoral response begins at the top center. The two responses interact at the center of the page. The details are described in the text.

tions). This binding cannot take place, however, unless some preliminary steps have already occurred: First, one or more *antigen-presenting cells*, or *macrophages*, must ingest the pathogen. Second, the antigen-presenting macrophages must then break off various molecular pieces of the pathogen and move them to their own surface, i.e., *present* the various antigenic fragments (called *epitopes*) to inactive T-cells. This presentation activates the T-cells and causes them to divide repeatedly into clones, each of which consists of identical, active helper T-cell or cytotoxic T-cells. In fact, there should result a clone of active helper and cytotoxic T-cells for each of the various epitopes that the antigen-presenting cells display, one clone originating from each activated T-cell.[4] An important point: The active helper T-cells are required in the activation of the cytotoxic T-cells. The active cytotoxic T-cells then approach and kill cells infected with the pathogen, thus killing the pathogen at the same time. The cytotoxic T-cell recognizes the infected cells because the infected cells, like macrophages, present epitopes on their surfaces. The T-cell response is often called *cell-mediated* immunity because the effect requires the direct and continued involvement of intact T-cells.

The concept of an adaptive response, or *immunological specificity*, is associated with the recognition of an infected antigen-presenting cell by a helper T-cell or cytotoxic T-cell. An inactive T-cell will be activated only if its specific receptors recognize the specific antigenic fragment being presented to it. Evidence suggests that the surface receptors of each individual inactive T-cell are unique, numerous, and of a single kind. Because there are upward of a trillion or so different inactive T-cells in our bodies, the presented parts of virtually every pathogen should be recognized by at least a few of the T-cells.

*B-cells are responsible for the humoral immune response.*

Like T-cells, B-cells are inactive at the time they mature and have recognition proteins on their surfaces. As with helper T-cells, these surface receptors vary from cell to cell and can recognize antigens. However, while helper T-cells require that the antigen appear on an antigen-presenting cell, B-cells can recognize an antigen that is free in the liquid fraction of the blood. When an inactive B-cell recognizes and binds to the antigen to which its surface proteins are complementary, the B-cell is then activated, and it subsequently divides many times to form a clone of identical active B-cells, sometimes called *plasma cells*. Active B-cells then secrete large quantities of a single kind of protein molecule, called an *antibody*, into the blood. These antibodies are able to bind to the antigen, an act that "labels" the antigen for destruction by either of two mechanisms: A set of chemical reactions, collectively called *complement*, can kill certain antibody-tagged bacteria, or tagged antigens can attract macrophages, which devour the antigen. The B-cell response is often called the *humoral* immune response, meaning "liquid-based."

[4] When antigen-presenting cells cut up a pathogen, many different antigenically active epitopes may result. Potentially, each epitope can activate a different T-cell upon presentation. Thus a single infecting bacterium could activate many different T-cell clones.

The concept of specificity for B-cell activation arises in a way similar to that for T-cells, namely in the recognition of the antigen by B-cell surface receptors. Evidently, all or most of our approximately one trillion inactive B-cells have different surface receptors. The recognition by a B-cell of the exact antigen for which that particular B-cell's surface is "primed" is an absolute requirement for the activation of that B-cell. Fortunately, most pathogens, bacteria, and viruses, for example, have many separate and distinct antigenic regions; thus they can trigger the activation of many different B-cells.

*Intercellular interactions play key roles in adaptive immune responses.*

The specificity of both T- and B-cell interactions with pathogens cannot be overemphasized; no adaptive immune response can be generated until receptors on these lymphocytes recognize the one specific antigen to which they can bind.

Note how T- and B-cells provide interlocking coverage: The cytotoxic T-cells detect the presence of intracellular pathogens (by the epitopes that infected cells present), and B-cells can detect extracellular pathogens. We would therefore expect T-cells to be effective against already-infected cells and B-cells to be effective against toxins, such as snake venom, and free pathogens, such as bacteria, in the blood.

Our discussion so far has emphasized the individual roles of T- and B-cells. In fact, correct functioning of the adaptive immune system requires that these two kinds of cells interact with each other. It was pointed out earlier that the activation of cytotoxic T-cells requires that they interact with active helper T-cells. In fact, helper T-cells are also needed to activate B-cells, as shown in Figure 10.2.2. Note the pivotal role of active helper T-cells: They exercise control over cell-mediated immunity *and* humoral immunity as well, which covers the entire adaptive immune system.

*Lymphocytes diversify their receptor proteins as the cell matures.*

At first glance, the central dogma of genetics would seem to suggest that the information for the unique surface protein receptor of each inactive lymphocyte should originate in a different gene. In other words, every inactive lymphocyte would merely express a different surface receptor gene. In each person, there seem to be about $10^{12}$ unique inactive lymphocytes, and therefore there would have to be the same number of unique genes! Actually, independent estimates of the *total* number of genes in a human cell indicate that there are only about 30,000; see Section 14.3.

The many variant forms of lymphocyte surface receptor proteins originate as the cell matures, and are the result of the random scrambling of genetic material—which leads to a wide variety of amino acid sequences without requiring the participation of a lot of genetic material. As an example, Figure 10.2.3 shows a length of hypothetical DNA that we will use to demonstrate the protein diversification process. We imagine the DNA to consist of two contiguous polynucleotide strings, or classes, labeled A and B. Each class has sections 1 through 4. The protein to be coded by the DNA will contain two contiguous polypeptide strings, one coded by a single section of A and one coded by a single section of B. Thus there are 16 different proteins that could result. To generate a particular protein, the unneeded sections of genetic material

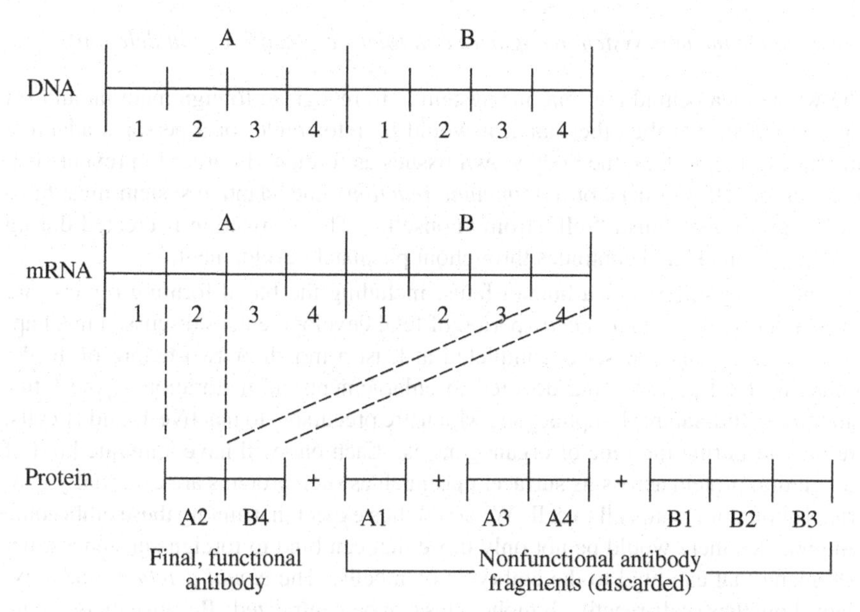


**Fig. 10.2.3.** A simplified picture of the creation of a specific antibody by a single lymphocyte. The final antibody molecule is coded from one section each of DNA regions A and B. Because the two sections are picked at random there are 16 possible outcomes. This figure shows how many possible antibodies could be coded from a limited amount of DNA. In a real situation, there would be many sections in many DNA regions, and the final, functional antibody would contain contributions coded by several regions.

will be enzymatically snipped out, either at the DNA stage or the mRNA stage. The protein that ultimately results in the figure is derived from DNA sections A2 and B4. The selection of A2 and B4 was random; any of the other 15 combinations was equally likely.

In a real situation, namely, the DNA coding for one of the proteins in B-cell antibodies, there are about 240 sections distributed among four classes. Of these, perhaps seven sections are actually expressed in a given cell, meaning that there are thousands of combinations of seven sections that were not expressed in that cell. These other combinations will be expressed in other B-cells, thereby generating a large number of lymphocytes with different surface proteins.

There are still other ways that lymphocytes generate diverse recognition proteins. For example, B-cells form antibodies by combining two completely separate polypeptides, each of which uses the random "pick and choose" method described in the previous two paragraphs. Further, when maturing, the nucleic acid segments that code for lymphocyte surface recognition proteins mutate very rapidly, more so than do other genes. All of this leads to the great variability in recognition proteins that is so crucial to the functioning of the adaptive immune system, and it does so while requiring a minimum amount of DNA for its coding.

*The adaptive immune system recognizes and tolerates "self" (clonal deletion).*

The whole idea behind the immune system is to recognize foreign material and rid the body of it. On the other hand, it would be intolerable for a person's adaptive immune system to treat the body's own tissues as foreign. In order to prevent such rejection of self-products, or *autoimmune reactions*, the adaptive system must have some way to distinguish "self" from "nonself." This distinction is created during fetal development and continues throughout postnatal development.

The organ systems of a human fetus, including the blood-forming organs, are formed during the *organogenetic* period of fetal development, as discussed in Chapter 9. Most organogenesis is completed at least a month or two before birth, the remaining fetal period being devoted to enlargement and maturation of the fetus. Embryonic (immature) lymphocytes, which are precursors to inactive T- and B-cells, are present during the time of organogenesis. Each one will have a unique kind of recognition protein across its surface, inasmuch as such proteins are essentially generated at random from cell to cell. We could thus expect that among these embryonic lymphocytes there would be not only those that can bind to foreign substances, but also many that can bind to the embryo's own cells. The *clonal deletion* model explains how these self-reactive lymphocytes can be neutralized: Because there are no pathogens in a fetus, the only cells capable of binding to lymphocytes would be the fetus's own cells. Therefore, embryonic B- or T-cells that bind to *any* cell in the fetus are killed or inactivated. Only self-reacting embryonic lymphocytes should be deleted by this mechanism. This reduces the possibility of maturation of a lymphocyte that could subsequently generate an autoimmune response.

There is good evidence for clonal deletion: Mouse embryos can be injected early *in utero* with a virus or other material that would normally be antigenic in a postnatal mouse. After birth, the treated mouse is unable to respond immunologically to subsequent injections of the virus. The mouse has developed an *acquired immunological tolerance* to the antigen. What has happened at the cellular level is that any embryonic mouse lymphocytes that reacted with the prenatally injected virus were killed or inactivated by clonal deletion—the virus was treated as "self." Thus there can be no mature progeny of these lymphocytes after birth to react to the second exposure to the virus.

*There is another mechanism for killing self-reacting lymphocytes.*

Clonal deletion reduces the possibility of an autoimmune response, but does not eliminate it. Recall that clonal deletion requires that self-products meet up with embryonic lymphocytes; mature lymphocytes will not do. The fact is that some embryonic lymphocytes slip through the clonal deletion net by not meeting the self-products that would have inactivated them. In addition, lymphocytes seem to mutate frequently, a process that postnatally may give them receptors that can react with self-products. Finally, the thymus gland, while much reduced in adults, continues to produce a limited number of new T-cells throughout life. These new cells, with receptors generated at random, may be capable of reacting with self-products.

There is a mechanism for getting rid of mature T-cells that can react with their own body's cells: Recall that a T-cell is activated when an infected antigen-presenting cell presents it with a piece of antigen. In fact, this activation has another requirement: The antigen-presenting cell must also display a *second* receptor, one that is found only on *infected* antigen presenters. If a mature T-cell should bind to an uninfected antigen presenter, one lacking the second receptor, the T-cell itself is inactivated (because that binding is a sign that the T-cell receptors are complementary to uninfected self-products). On the other hand, if a mature T-cell binds to an infected antigen presenter, the infection being signaled by the second receptor, that binding is acceptable, and the normal activation of the T-cell ensues.

*Inactive lymphocytes are selected for activation by contact with an antigen (clonal selection).*

The clonal deletion system described above results in the inactivation or killing of immature T- and B-cells if they react with any antigen. This process provides the individual with a set of lymphocytes that can react only with nonself products. These surviving T- and B-cells then remain in our blood and lymphatic circulatory systems in an inactive state until they come into contact with the antigens to which their surface receptors are complementary. This will be either as free, extracellular antigens for B-cells or on an antigen-presenting cell in the case of T-cells.

Once the proper contact is made, the lymphocyte is activated and begins to divide rapidly to form a clone of identical cells. But what if the correct antigen never appears? The answer is an odd one—namely, the lymphocyte is never activated and remains in the blood and lymphatic systems all of our life. What this means is that only a tiny fraction of our lymphocytes ever become activated in our lifetimes; the rest just go around and around our circulation or remain fixed in lymph nodes. This process of activating only those few lymphocytes whose activity is needed, via contact with their appropriate antigens, is called *clonal selection*.

The notion of clonal selection suggests an immense amount of wasted effort on the part of the immune system. For example, each of us has one or more lymphocytes capable of initiating the rejection of a skin transplant from the seventieth president of the United States (in about a century), and others that would react against a cold virus that people contracted in Borneo in 1370 AD. None of us will ever need those capabilities, but we have them nevertheless. It might seem that a simpler mechanism would have been the generation of a single generic kind of lymphocyte and then its adaptation to each individual kind of antigen. This process is called the *instructive mechanism*, but it is not what happens.

*The immune system has a memory.*

Most people get mumps or measles only one time. If there are no secondary complications these diseases last about a week, which is the time it takes for the activation of T- and B-cells by a pathogen and the subsequent destruction of the pathogen. Surely these people are exposed to mumps and measles many times in their lives, but they seem to be unaffected by the subsequent exposures. The reason for this is well understood: First, they have antibodies from the initial exposure, and second, among the results of T- and B-cell activation are "memory" cells, whose surface recognition proteins are complementary to the antigenic parts of the activating pathogen (refer back to Figure 10.2.2). These memory cells remain in our blood and lymphatic systems for the rest of our lives, and if we are infected by the same pathogen again, they mount a response just like the original one, but much more intensely and in a much shorter time. The combination of preexisting antibodies from the initial exposure and the intense, rapid secondary response by memory cells usually results in our being unaware of the second exposure.

Why, then, do we get so many colds if the first cold we get as babies generates memory cells? The answer lies in two facts: The adaptive immune response is very specific, and the cold virus mutates rapidly. The memory cells are as specific for antigen as were their original inactive lymphocyte precursors. They will recognize only the proteins of the virus that caused the original cold; once having gotten a cold from that particular strain of cold virus, we won't be successfully infected by it again. The problem is that cold viruses mutate rapidly, and one effect of mutation is that viral-coat proteins (the antigens) change their amino acid sequences. Once that happens, the memory cells and antibodies from a previous infection don't recognize the new, mutated strain of the virus and therefore can't respond to it. The immune response must start all over, and we get a cold that requires a week of recovery (again). If it is possible to say anything nice about mumps, chicken pox, and such diseases, it is that their causative agents do not mutate rapidly and we therefore get the diseases only once, if at all. We shall see in the next section that rapid mutation characterizes HIV, allowing the virus to stay one step ahead of the specific immune system's defenses.

*Vaccinations protect us by fooling the adaptive immune system.*

The idea behind immunization is to generate the immune response without generating the disease. Thus the trick is to inactivate or kill the pathogen without damaging its antigenic properties. Exposure to this inactive pathogen then triggers the immune responses described earlier, including the generation of memory cells. During a subsequent exposure, the live, active pathogen binds to any preexisting antibody *and* activates memory cells; thus the pathogen is inactivated before disease symptoms can develop. As an example, vaccination against polio consists in swallowing live-but-inactivated polio virus. We then generate memory cells that will recognize active polio viruses if we should be exposed to them at a later date.

Exposure to some pathogenic substances and organisms is so rare that vaccination of the general population against them would be a waste of time and money. Poisonous snake venom is a case in point: The active agent in snake venom is a destructive enzyme distinctive to each kind of snake genus or species, but fortunately almost no one ever gets bitten by a snake. Snake venom is strongly antigenic, as we would expect a protein to be, but the symptoms of snake bite appear so rapidly that the victim could die long before the appropriate lymphocytes could be activated. Unless the snakebite victim already has preexisting antibodies or memory T-cells against the venom, say, from a previous survivable exposure to the venom, he or she could be in a lot of trouble. The way around this problem is to get another animal, like a horse, to generate the antibodies by giving it a mild dose of the venom. The antivenom antibodies are then extracted from the horse's blood and stored in a hospital refrigerator until a snakebite victim arrives. The antibodies are then injected directly into the bitten area, to tag the antigenic venom, leading to its removal.

A snakebite victim probably won't take the time to identify the species of the offending reptile, and each snake genus or species can have an immunologically distinctive venom. To allow for this, hospitals routinely store mixtures of antibodies against the venoms of all the area's poisonous snakes. The mixture is injected at the bite site, where only the correct antibody will react with the venom—the other antibodies do nothing and eventually disappear without effect.[5] This kind of immunization is said to be passive, and it has a very important function in prenatal and neonatal babies, who get passive immunity via interplacental transfer of antibodies and from the antibodies in their mother's milk. This protects the babies until their own immune systems can take over.

## 10.3 HIV and AIDS

*The human immunodeficiency virus defeats the immune system by infecting, and eventually killing, helper T-cells. As a result, neither the humoral nor the cell-mediated specific immune responses can function, leaving the patient open to opportunistic diseases.*

*As is true of all viruses, HIV is very fussy about the host cell it chooses. The problem is that its chosen hosts are immune system cells, the very same cells that are required to fend it off in the first place. Initially, the victim's immune system responds to HIV infection by producing the expected antibodies, but the virus stays ahead of the immune system by mutating rapidly. By a variety of mechanisms, some poorly understood, the virus eventually wears down the immune system by killing helper T-cells, which are required for the activation of killer T-cells and B-cells. As symptoms of a low T-cell count become manifested, the patient is said to have AIDS.*

*In this section, we will describe the reproduction of HIV as a prelude to a mathematical treatment of the behavior of HIV and the epidemiology of AIDS.*

*The human immunodeficiency virus (HIV) infects T-cells and macrophages, among others.*

The outer coat of HIV is a two-layer lipid membrane, very similar to the outer membrane of a cell (see Figure 10.3.1). Projecting from the membrane are sugar–protein projections, called gp120. These gp120 projections recognize and attach to a protein called CD4, which is found on the surfaces of helper T-cells, macrophages, and monocytes (the latter are macrophage precursors). The binding of gp120 and CD4

[5] Note that the unneeded antibodies do not provide a "memory" because there is no activation of lymphocytes—hence no memory cells.

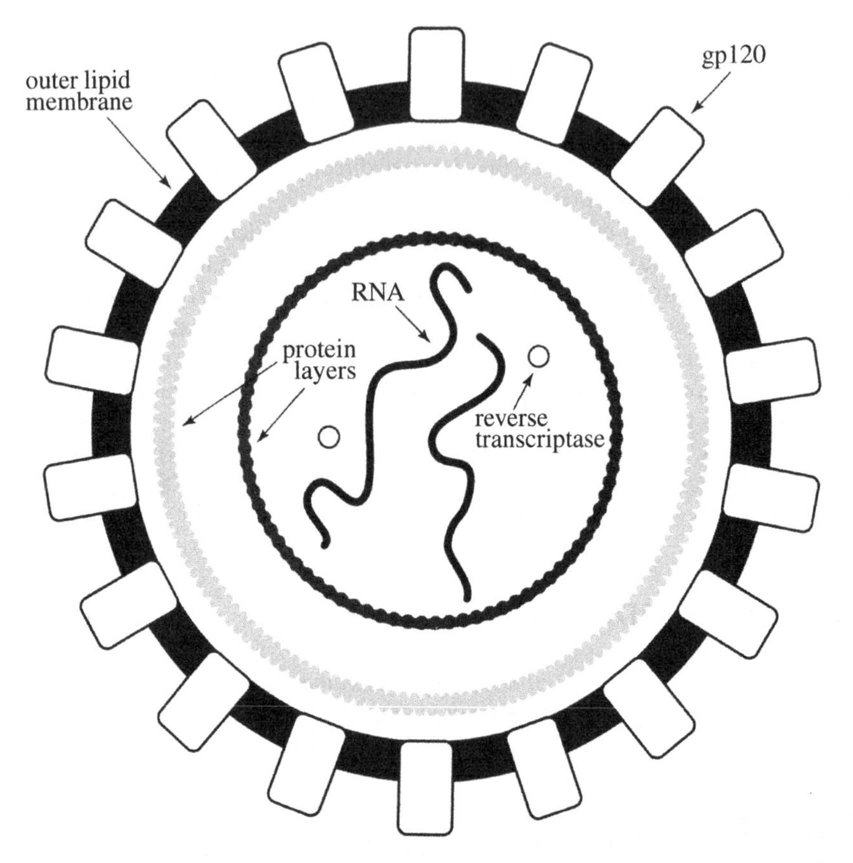


**Fig. 10.3.1.** A model of the human immunodeficiency virus (HIV). The outer membrane of the HIV is derived from the outer membrane of the host cell. Thus an antibody against that part of the HIV would also act against the host cell. Note that the HIV carries copies of the reverse transcriptase enzyme.

leads to the fusion of the viral membrane and the cell membrane. Then the viral capsid is brought into the blood cell (see [3] and [4]).

*HIV is a retrovirus.*

The HIV capsid contains two identical single strands of RNA (no DNA). The capsid is brought into the host cell by fusion between the viral envelope and the cell membrane, as described in Section 10.1. The capsid is then enzymatically removed. The HIV RNA information is then converted into DNA information, a step that is indicated by the straight left-pointing arrow in the following central dogma flow diagram:[6]

[6] This is our final alteration to "dogma."

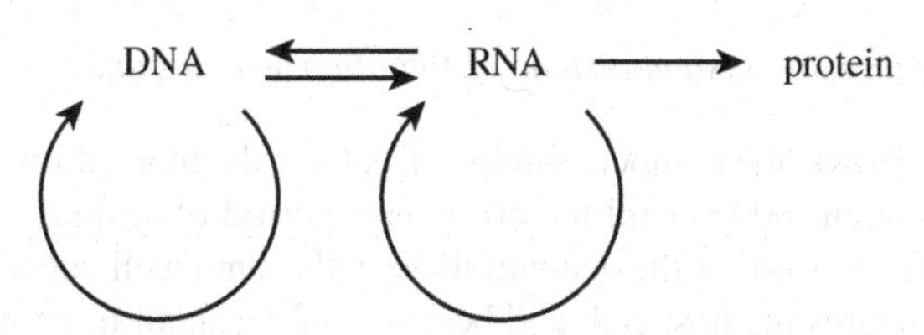


The conversion of RNA information into DNA information involves several steps and is called *reverse transcription*: First, the single-stranded HIV RNA acts as a template for the creation of a strand of DNA. This process entails complementary H-bonding between RNA nucleotides and DNA nucleotides, and it yields a hybrid RNA-DNA molecule. The RNA is then removed and the single-strand of DNA acts as a template for the creation of a second, complementary, strand of DNA. Thus a double helix of DNA is manufactured, and it carries the HIV genetic information.

The chemical process of covalently polymerizing DNA nucleotides and depolymerizing RNA nucleotides, like most cellular reactions involving covalent bonds, requires enzymatic catalysis to be effective. The enzyme that catalyzes reverse transcription is called *reverse transcriptase*. Reverse transcriptase is found inside HIV, in close association with the viral RNA, and it enters the host cell right along with the RNA, ready for use. Once HIV DNA is formed it is then spliced into the host cell's own DNA; in other words, it is a provirus.

In a general sense, a provirus becomes an integral part of the host cell's genetic material; for instance, proviruses are replicated right along with the host cell's genome at cell division. It should therefore not be surprising that the physiology and morphology of the host cell changes as a result of the incorporated provirus. For example, one important consequence of HIV infection is that gp120 projections appear on the lymphocyte's surface.

Once in the form of a provirus, HIV starts to direct the host cell's anabolic machinery to form new HIV. As the assembled viruses exit the host cell by budding, they pick up a part of the cell's outer lipid bilayer membrane, along with some of the gp120 placed there by the provirus. The newly formed virus is now ready to infect a new cell.

The budding process does not necessarily kill the host cell. In fact, infected macrophages seem to generate unending quantities of HIV. T-cells do eventually die in an infected person, but as explained below, it is not clear that they die from direct infection by the virus.

The flow of information from RNA to DNA was omitted when the central dogma was first proposed because at the time, no one believed that information flow in that direction was possible. As a consequence, subsequent evidence that it existed was ignored for some years—until it became overwhelming. The process of RNA-to-DNA informational flow is still called "reverse transcription," the key enzyme is called "reverse transcriptase," and viruses in which reverse transcription is important are still called "retroviruses," as though something were running backward. Of course, there is nothing actually "backward" about such processes; they are perfectly normal in their natural context.

*HIV destroys the immune system instead of the other way around.*

As Figure 10.3.2 shows, the number of helper T-cells in the blood drops from a normal concentration of about 800 per ml to zero over a period of several years following HIV infection. The reason for the death of these cells is not well understood, because budding usually spares the host cell, and besides, only a small fraction of the T-cells in the body ever actually become infected by the HIV in the first place. Nevertheless, all the body's helper T-cells eventually die. Several mechanisms have been suggested for this apparent contradiction: Among them, the initial contact between HIV and a lymphocyte is through the gp120 of the HIV and CD4 of the T-cell. After a T-cell is infected, gp120 projections appear on its own surface, and they could cause that infected cell to attach to the CD4 receptors of other, *uninfected* T-cells. In this way, one infected lymphocyte could attach to many uninfected ones and disable them all. In fact, it has been observed that if cells are artificially given CD4 and gp120 groups, they clump together into large multinuclear cells (called *syncitia*).

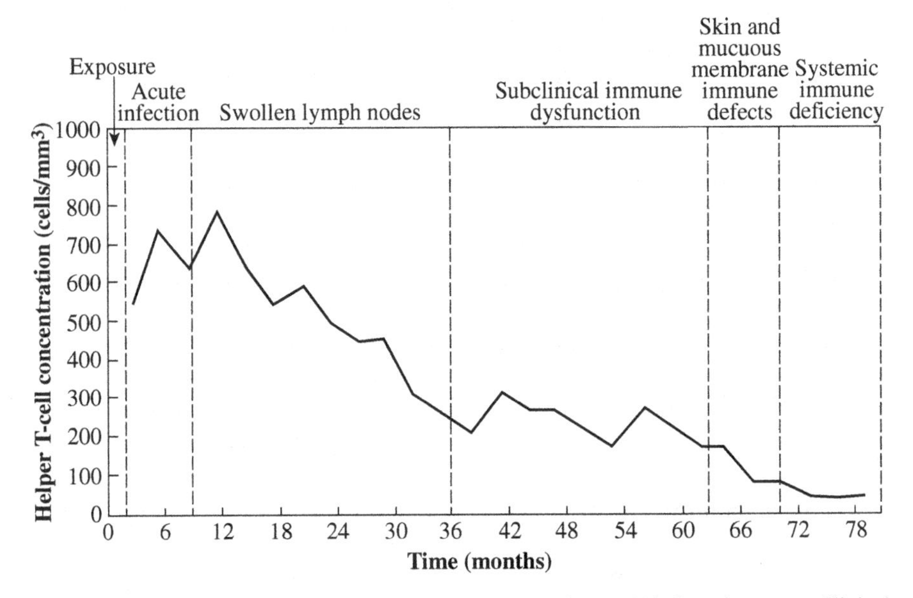


**Fig. 10.3.2.** A graph of the helper T lymphocyte count of an HIV-infected person. Clinical symptoms are indicated along the top of the figure. Note the correlation between the decrease in T-cell count and the appearance of the clinical symptoms. (Redrawn from R. Redfield and D. Burke, HIV infection: The classical picture," *Sci. Amer.*, **259**-4 (1988), 90–98; copyright ©1988 by Scientific American, Inc. All rights reserved.)

A second possible way that helper T-cells might be killed is suggested by the observation that the infected person's lymph nodes atrophy. The loss of those parts of the lymphatic system may lead to the death of the T-cells.

Third, a normal function of helper T-cells is to stimulate killer T-cells to kill viral-infected cells. It may be that healthy helper T-cells instruct killer T-cells to kill infected helper T-cells. Eventually, this normal process could destroy many of the body's T-cells as they become infected, although, as noted earlier, only a small fraction of helper T-cells ever actually become infected.

Fourth, it has been demonstrated that if an inactive HIV-infected lymphocyte is activated by antigen, it yields greatly reduced numbers of memory cells. In fact, it seems that the activation process itself facilitates the reproduction of HIV by providing some needed stimulus for the proper functioning of reverse transcriptase.

*HIV infection generates a strong initial immune response.*

It is shown in Figure 10.3.3 that the immune system initially reacts vigorously to HIV infection, producing antibodies as it should.[7]

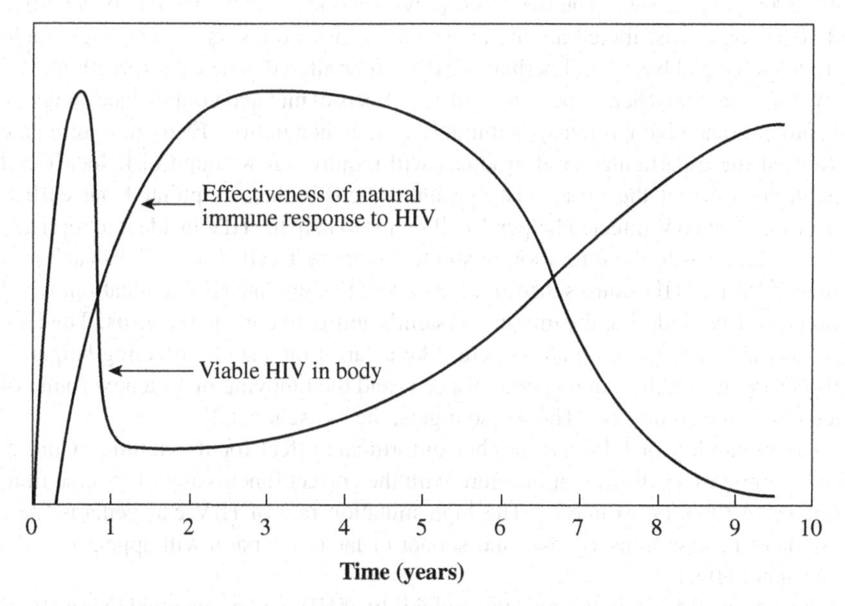


**Fig. 10.3.3.** A graph of immune response and viral appearance vs. time for an HIV-infected person. The initial infection generates a powerful immune response. That response, however, is later overwhelmed by the virus, which kills the helper T-lymphocytes that are required by the humoral and cell-mediated immune responses. (Redrawn from R. Redfield and D. Burke, HIV infection: The classical picture, *Sci. Amer.*, **259**-4 (1988), 90–98; copyright ©1988 by Scientific American, Inc. All rights reserved.)

[7] The presence of antibodies against HIV is the basis for the diagnosis of HIV infection. Note that it takes several months to get a measurable response.

Nonetheless, the circulating helper T-cell count soon begins an irreversible decrease toward zero, as discussed above. As helper T-cells die off, the ability of the adaptive immune system to combat any pathogen, HIV or other, also vanishes.

In Section 10.4, we will describe a mathematical model for the interaction between helper T-cells and HIV.

*The high mutability of HIV demands continued response from the adaptive immune system.*

Mutations occur commonly in HIV RNA, and the reason is reasonably well understood: Reverse transcriptase lacks a "*proofreading*" capacity. This proofreading ability is found in enzymes that catalyze the polymerization of DNA from a DNA template in the "forward" direction of the central dogma. Thus the occasional mismatching of H-bonds between nucleotides, say the pairing of A opposite G, gets corrected. On the other hand, reverse transcriptase, which catalyzes DNA formation from an RNA template, seems not to be able to correct base-pairing errors, and this leads to high error rates in base placement—as much as one mismatched base out of every 2000 polymerized. The two RNA polynucleotides of HIV have between 9000 and 10000 bases distributed among about nine genes, so this error rate might yield up to five changed bases, and perhaps three or four altered genes, per infection.

We are concerned here especially with the effects of mutated viral surface antigens, e.g., proteins and glycoproteins, on immune system recognition. Every new antigenic version of these particular viral products will require a new round of helper T-cell activation to defeat the virus. The problem there is that, as pointed out earlier, activation of an HIV-infected helper T-cell seems to help the HIV inside it to replicate and, further, leads to the formation of stunted memory T-cell clones. Thus each new antigenic form of HIV causes the immune system to stimulate HIV replication, while simultaneously hindering the immune system's ability to combat the virus. The HIV stays just ahead of the immune system, like a carrot on a stick, affecting helper T-cells before the T-cells can respond properly, and then moving on to a new round of infections. One could say, "The worse it gets, the worse it gets!"

The mutability of HIV has another unfortunate effect for its victims. Current therapy emphasizes drugs that interfere with the correct functioning of reverse transcriptase; AZT is an example. The high mutation rate of HIV can generate new versions of reverse transcriptase, and sooner or later, a version will appear that the drug cannot affect.

In Section 10.5, we will model the mutability of HIV and its eventual overwhelming of the immune system.

*HIV infection leads to acquired immunodeficiency syndrome* (*AIDS*).

A person infected with HIV is said to be "HIV positive." Such people may be asymptomatic for a considerable time following infection, or the symptoms may be mild and transient; the patient is, however, infectious. Eventually, the loss of helper T-cells will leave the person open to infections, often of a rare kind (see Figure 10.3.2). As examples, pneumonia caused by a protozoan called *Pneumocystis carinii* and a cancer of blood vessels, called Kaposi's sarcoma, are extremely rare in the general population, yet they frequently are found in HIV-positive people. Everyone is exposed to the pathogens that cause these diseases, but people do not get the disease if their immune systems are working properly. When HIV-positive persons exhibit unusual diseases as a result of low helper T-cell counts, they are said to have AIDS.

## 10.4 An HIV Infection Model

*A model for HIV infection involves four components: normal T-cells, latently infected T-cells, infected T-cells actively replicating new virus, and the virus itself. Any proposed model should incorporate the salient behavior of these components and respect biological constraints. In this section, we present such a model and show that it has a stationary solution. This model was developed and explored by Perelson, Kirschner, and coworkers.*

*T-cell production attempts to maintain a constant T-cell serum concentration.*

In this section, we will be presenting a model for T-cell infection by HIV, as described in Section 10.2 (see [5, 6, 7, 8]). This model tracks four components, three types of T-cells and the virus itself, and therefore requires a four-equation system for its description. As a preliminary step toward understanding the full system of equations, we present first a simplified version, namely, the equation for T-cells in the absence of infection. In forming a mathematical model of T-cell population dynamics based on the discussion of Section 10.2, we must incorporate the following assumptions:

- Some immunocompetent T-cells are produced by the lymphatic system; over relatively short periods of time, their production rate is constant and independent of the number of T-cells present. Over longer periods of time their production rate adjusts to help maintain a constant T-cell concentration, even in adulthood. Denote this *supply rate* by $s$.
- T-cells are produced through clonal selection if an appropriate antigen is present, but the total number of T-cells does not increase unboundedly. Model this using a logistic term, $rT(1 - \frac{T}{T_{\max}})$, with per capita growth rate $r$ (cf. Section 4.3).
- T-cells have a finite natural lifetime after which they are removed from circulation. Model this using a death rate term, $\mu T$, with a fixed per capita death rate $\mu$.

Altogether, the differential equation model is

$$\frac{dT}{dt} = s + rT\left(1 - \frac{T}{T_{\max}}\right) - \mu T. \tag{10.4.1}$$

In this, $T$ is the T-cell population in cells per cubic millimeter.

We want the model to have the property that solutions, $T(t)$, that start in the interval $[0, T_{\max}]$ stay there. This will happen if the derivative $\frac{dT}{dt}$ is positive when $T = 0$ and negative when $T = T_{\max}$. From (10.4.1),

$$\left.\frac{dT}{dt}\right|_{T=0} = s,$$

and since $s$ is positive, the first requirement is fulfilled. Next, substituting $T = T_{\max}$ into (10.4.1), we get the condition that must be satisfied for the second requirement,

$$\left.\frac{dT}{dt}\right|_{T=T_{\max}} = s - \mu T_{\max} < 0,$$

or, rearranged,

$$\mu T_{\max} > s. \tag{10.4.2}$$

The biological implication of this statement is that when the number of T-cells has reached the maximum value $T_{\max}$, then there are more cells dying than are being produced by the lymphatic system.

Turning to the stationary solutions of system (10.4.1), we find them in the usual way, by setting the right-hand side to zero and solving for $T$:

$$-\frac{r}{T_{\max}}T^2 + (r - \mu)T + s = 0.$$

The roots of this quadratic equation are

$$T = \frac{T_{\max}}{2r}\left((r - \mu) \pm \sqrt{(r - \mu)^2 + 4s\frac{r}{T_{\max}}}\right). \tag{10.4.3}$$

Since the product $\frac{4sr}{T_{\max}}$ is positive, the square root term exceeds $|r - \mu|$,

$$\sqrt{(r - \mu)^2 + \frac{4sr}{T_{\max}}} > |r - \mu|,$$

and therefore one of the roots of the quadratic equation is positive, while the other is negative. Only the positive root is biologically important, and we denote it by $T_0$, as the "zero virus" stationary point (see below). We now show that $T_0$ must lie between 0 and $T_{\max}$. As already noted, the right-hand side of (10.4.1) is positive when $T = 0$ and negative when $T = T_{\max}$. Therefore, it must have a root between 0 and $T_{\max}$, and this is our positive root $T_0$ calculated from (10.4.3) by choosing the + sign. We will refer to the difference $p = r - \mu$ as the T-cell *proliferation rate*; in terms of it, the globally attracting stationary solution is given by

$$T_0 = \frac{T_{\max}}{2r}\left(p + \sqrt{p^2 + 4s\frac{r}{T_{\max}}}\right). \tag{10.4.4}$$

This root $T_0$ is the only (biologically consistent) stationary solution of (10.4.1).

Now consider two biological situations.

**Table 10.4.1.** Parameters for Situation 1.

| Parameter | Description | Value |
|---|---|---|
| $s$ | T-cell from precursor supply rate | 10/mm$^3$/day |
| $r$ | normal T-cell growth rate | 0.03/day |
| $T_{\max}$ | maximum T-cell population | 1500/mm$^3$ |
| $\mu$ | T-cell death rate | 0.02/day |

**Situation 1: Supply rate solution.** In the absence of an infection, or at least an environmental antigen, the clonal production rate $r$ can be small, smaller than the natural death rate $\mu$, resulting in a negative proliferation rate $p$. In this case, the supply rate $s$ must be high in order to maintain a fixed T-cell concentration of about 1000 per cubic millimeter. The data in [6] confirm this.

With these data, calculate the stationary value of $T_0$ using (10.4.3) as follows:

```
MAPLE
> f:=T->s+r*T*(1-T/Tmax)- mu *T;
> s:= 10; r:=.03; mu:=.02; Tmax:=1500;
> fzero:=solve(f(T) = 0,T);
> T0:=max(fzero[1],fzero[2]);

MATLAB
> s=10; r=.03; mu=.02; Tmax=1500;
> p=[-r/Tmax (r-mu) s];
> T0=max(roots(p))
```

Next calculate and display trajectories from various starting points:

```
MAPLE
> deq:={diff(T(t),t)=f(T(t))};
> with(DEtools):
> inits:={[0,0],[0,T0/4],[0,T0/2],[0,(T0+Tmax)/2],[0,Tmax]};
> phaseportrait(deq,T(t),0..25,inits,stepsize=1,arrows=NONE);

MATLAB
% make up an m-file, hiv1.m, with
% function Tprime=hiv1(t,T); s=10; r=.03; mu=.02, Tmax=1500;
%  Tprime=s+r*T*(1-T/Tmax)-mu*T;
> [t,T]=ode23('hiv1',[0 50],0);
> plot(t,T); hold on
> [t,T]=ode23('hiv1',[0 50],T0/4); plot(t,T)
> [t,T]=ode23('hiv1',[0 50],T0/2); plot(t,T)
> [t,T]=ode23('hiv1',[0 50],(T0+Tmax)/2); plot(t,T)
> [t,T]=ode23('hiv1',[0 50],Tmax); plot(t,T)
```

**Situation 2: Clonal production solution.** An alternative scenario is that adult thymic atrophy has occurred, or a thymectomy has been performed. As a hypothetical and limiting situation, take $s$ to equal zero and ask how $r$ must change to maintain a comparable $T_0$. Use the parameters in Table 10.4.2.

```
MAPLE
> s:= 0; r:=.06; mu:=.02; Tmax:=1500;
> fzero:=solve(f(T)=0,T);
> T0:=max(fzero[1],fzero[2]);
> deq:={diff(T(t),t)=f(T(t))};
> inits:={[0,0],[0,T0/4],[0,T0/2],[0,(T0+Tmax)/2],[0,Tmax]};
> phaseportrait(deq,T(t),t=0..25,inits,stepsize=1,arrows=NONE);

MATLAB
> s=0; r=.06; mu=.02; Tmax=1500;
```

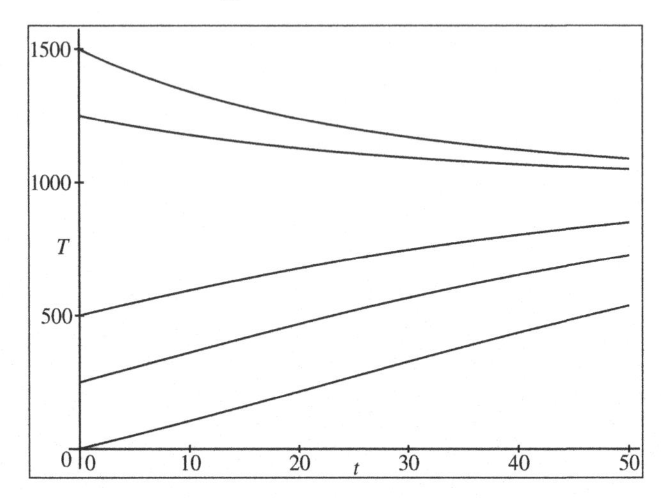


**Fig. 10.4.1.** Time vs. number of T-cells per cubic millimeter.

**Table 10.4.2.** Parameters for Situation 2.

| Parameter | Description | Value |
|---|---|---|
| $s$ | T-cell from precursor supply rate | $0/\text{mm}^3/\text{day}$ |
| $r$ | normal T-cell growth rate | 0.06/day |
| $T_{\max}$ | maximum T-cell population | $1500/\text{mm}^3$ |
| $\mu$ | T-cell death rate | 0.02/day |

```
> p=[-r/Tmax (r-mu) s];
> T0=max(roots(p))
  % make an m-file, hiv2.m, same as before execpt s=0; r=.06; mu=.02; Tmax=1500;
> hold off
> [t,T]=ode23('hiv2',[0 50],0);
> plot(t,T); hold on
> [t,T]=ode23('hiv2',[0 50],T0/4); plot(t,T)
> [t,T]=ode23('hiv2',[0 50],T0/2); plot(t,T)
> [t,T]=ode23('hiv2',[0 50],(T0+Tmax)/2); plot(t,T)
> [t,T]=ode23('hiv2',[0 50],Tmax); plot(t,T)
```

As above, $T_0$ is again about 1000 T-cells per cubic millimeter. Trajectories in this second situation are plotted in Figure 10.4.2; contrast the convergence rate to the stationary solution under this clonal T-cell production situation with the supply rate convergence of Situation 1.

**Remark.** Contrasting these situations shows that upon adult thymic atrophy or thymectomy, the response of the T-cell population is much slower. This suggests that one would find differences in the dynamics of T-cell depletion due to an HIV infection in people of different ages. Clearly, there is a need for $r$, the T-cell growth rate, to be large in compensation when the supply rate, $s$, is small. How can one influence one's value of $r$? The answer should be an inspiration for continuing biological and medical research.

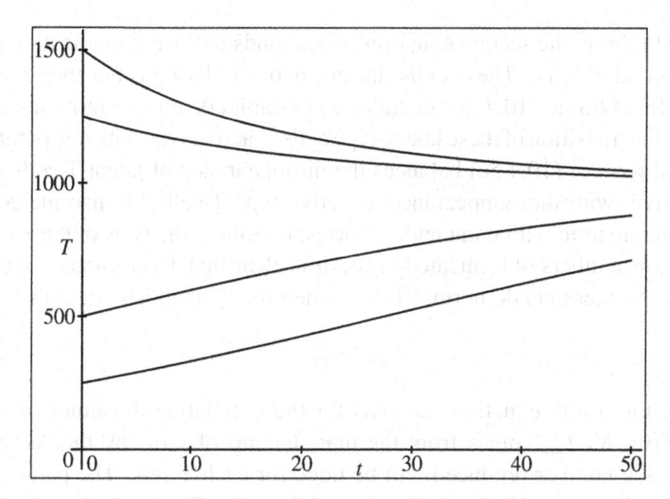


**Fig. 10.4.2.** Time vs. T-cell count with a reduced thymus function.

*A four-equation system is used to model T-cell–HIV interaction.*

To incorporate an HIV infection into the above model, we follow the approach taken by Perelson, Kirschner, and DeBoer [6] and differentiate three kinds of T-cells: Besides the normal variety, whose number is denoted by $T$ as before, there are T-cells infected with provirus, but not producing free virus. Designate the number of these *latently* infected T-cells by $T_L$. In addition, there are T-cells that are infected with virus and are *actively* producing new virus. Designate the number of these by $T_A$. The interaction between virus, denoted by $V$, and T-cells is reminiscent of a predator–prey relationship; a mass action term is used to quantify the interaction (see Section 4.4). However, only the active type T-cells produce virus, while only the normal T-cells can be infected.

We now present the model and follow with a discussion of its four equations separately:

$$\frac{dT}{dt} = s + rT\left(1 - \frac{T + T_L + T_A}{T_{\max}}\right) - \mu T - k_1 V T, \tag{10.4.5a}$$

$$\frac{dT_L}{dt} = k_1 V T - \mu T_L - k_2 T_L, \tag{10.4.5b}$$

$$\frac{dT_A}{dt} = k_2 T_L - \beta T_A, \tag{10.4.5c}$$

$$\frac{dV}{dt} = N\beta T_A - k_1 V T - \alpha V. \tag{10.4.5d}$$

The first equation is a modification of (10.4.1) with the inclusion of an infection term having mass action parameter $k_1$. When normal T-cells become infected, they immediately become reclassified as the latent type. In addition, note that the sum of all three types of T-cells counts toward the T-cell limit, $T_{\max}$.

The first term in the second equation corresponds to the reclassification of newly infected normal T-cells. These cells disappear from (10.4.5a) but then reappear in (10.4.5b). In addition, (10.4.5b) includes a per capita death rate term and a term to account for the transition of these latent-type cells to active type with rate parameter $k_2$.

The first term of (10.4.5c) balances the disappearance of latent T-cells upon becoming active, with their appearance as active-type T-cells. It also includes a per capita death rate term with parameter $\beta$ corresponding to the lysis of these cells after releasing vast numbers of replicated virus. It is clear that T-cells active in this sense perish much sooner than do normal T-cells, therefore $\beta$ is much larger than $\mu$:

$$\beta \gg \mu. \tag{10.4.6}$$

Finally, the fourth equation accounts for the population dynamics of the virus. The first term, $N\beta T_A$, comes from the manufacture of virus by the "active"-type T-cells, but the number produced will be huge for each T-cell. The parameter $N$, a large value, adjusts for this many-from-one difference. The second term reflects the fact that as a virus invades a T-cell, it drops out of the pool of free virus particles. The last term, with per capita rate parameter $\alpha$, corresponds to loss of virus through the body's defense mechanisms.

**Remark.** Note that in the absence of virus, i.e., $V = 0$, then both $T_L$ and $T_A$ are 0 as well, and setting these values into system (10.4.4), we see that this new model agrees with the old one, (10.4.1).

*The T-cell–HIV model respects biological constraints.*

We want to see that the model is constructed well enough that no population goes negative or goes unbounded. To do this, we first establish that the derivatives $\frac{dT}{dt}$, $\frac{dT_L}{dt}$, $\frac{dT_A}{dt}$, and $\frac{dV}{dt}$ are positive whenever $T$, $T_L$, $T_A$, or $V = 0$, respectively. This means that each population will increase, not decrease, at low population sizes.

But from (10.4.5a), if $T = 0$, then

$$\frac{dT}{dt} = s > 0,$$

and if $T_L = 0$, then (10.4.5b) gives

$$\frac{dT_L}{dt} = k_1 V T > 0;$$

likewise, if $T_A = 0$, then from (10.4.5c),

$$\frac{dT_A}{dt} = k_2 T_L > 0,$$

and finally (10.4.5d) becomes, when $V = 0$,

$$\frac{dV}{dt} = N\beta T_A > 0.$$

We have assumed that all the parameters are positive, and so these derivatives are also positive as shown.

Following Perelson, Kirschner, and DeBoer [6], we next show that the total T-cell population as described by this model remains bounded. This total, $T_\Sigma$ is defined to be the sum $T_\Sigma = T + T_L + T_A$, and it satisfies the differential equation obtained by summing the right-hand side of the first three equations in system (10.4.5),

$$\frac{dT_\Sigma}{dt} = s + rT\left(1 - \frac{T_\Sigma}{T_{\max}}\right) - \mu T - \mu T_L - \beta T_A. \tag{10.4.7}$$

Now suppose $T_\Sigma = T_{\max}$; then from (10.4.7),

$$\frac{dT_\Sigma}{dt} = s - \mu T - \mu T_L - \beta T_A + \mu T_A - \mu T_A,$$

and combining the second, third, and last terms as $-\mu T_{\max}$ gives

$$\frac{dT_\Sigma}{dt} = s - \mu T_{\max} - (\beta - \mu)T_A < s - \mu T_{\max},$$

where (10.4.6) has been used to obtain the inequality. Recalling condition (10.4.2), we find that

$$\frac{dT_\Sigma}{dt} < 0 \quad \text{if } T_\Sigma = T_{\max},$$

proving that $T_\Sigma$ cannot increase beyond $T_{\max}$.

In summary, system (10.4.5) has been shown to be consistent with the biological constraints that solutions remain positive and bounded.

*The T-cell infected stationary solution is stable.*

To find the stationary points of the T-cell–HIV model, that is, (10.4.5), we must set the derivatives to zero and solve the resulting (nonlinear) algebraic system, four unknowns and four equations. Solving the third equation, namely, $0 = k_2 T_L - \beta T_A$, for $T_A$ gives $T_A = (\frac{k_2}{\beta})T_L$, which may in turn be substituted for in all its other occurrences. This reduces the problem to three unknowns and three equations. Continuing in this way, we arrive at a polynomial in, say, $T$ whose roots contain the stationary points. We will not carry out this approach here. Instead, we will solve this system numerically, below, using derived parameter values. However, in [6] it is shown symbolically that at the uninfected stationary point $T_0$, (10.4.4) is stable (see Section 2.4) if and only if the parameter $N$ satisfies

$$N < \frac{(k_2 + \mu)(\alpha + k_1 T_0)}{k_2 k_1 T_0}.$$

By defining the combination of parameters on the right-hand side as $N_{\text{crit}}$, we may write this as

$$N < N_{\text{crit}}, \quad \text{where } N_{\text{crit}} = \frac{(k_2 + \mu)(\alpha + k_1 T_0)}{k_2 k_1 T_0}. \tag{10.4.8}$$

In Table 10.4.3, we give values of the parameters of system (10.4.5) as determined in [6].

This model reflects the clinical picture as presented in Greene [9].

**Table 10.4.3.** Parameters of the HIV infection model.

| Parameter | Description | Value |
|---|---|---|
| $s$ | T-cell from precursor supply rate | 10/mm$^3$/day |
| $r$ | normal T-cell growth rate | 0.03/day |
| $T_{\max}$ | maximum T-cell population | 1500/mm$^3$ |
| $\mu$ | normal/latently infected T-cell death rate | 0.02/day |
| $\beta$ | actively infected T-cell death rate | 0.24/day |
| $\alpha$ | free virus death rate | 2.4/day |
| $k_1$ | T-cell infection rate by free virus | $2.4 \times 10^{-5}$ mm$^3$/day |
| $k_2$ | latent-to-active T-cell conversion rate | $3 \times 10^{-3}$/day |
| $N$ | virus produced by an active T-cell | taken as 1400 here |

## Exercises/Experiments

1. In the uninfected situations, for both $s = 0$ and $s = 10$, derive the numerical solution $T$ for $f(T) = 0$. Which of the roots for this equation is in the interval $[0, T_{\max}]$?
2. In the virus-free situation, give a biological interpretation for $r$. Suppose that $r$ is increased to $r_n$ so that
$$\frac{r_n - r}{r} = 0.10.$$
That is, $r$ is increased by 10%. What is the percentage of increase of the steady state of $T$ cells corresponding to a 10% increase in $r$?
3. With the parameters as stated for the infected situation, what is the numerical value for each of these: $T_{\max}$, the uninfected steady state of $T$ cells, the infected steady state of $T$ cells, and $N_{\text{crit}}$. Is $N_{\text{crit}}$ more or less than the $N$ used in these parameters? What are the implications of this last answer?
4. Sketch a graph of how $T$, $T_L$, $T_A$, and $V$ evolve during the first year and move toward equilibrium. Continue the graph for two more years. Here is syntax that will accomplish this integration of the equations:

MAPLE
```
> deq:=diff(T(t),t)=-mu*T(t)+r*T(t)*(1-(T(t)+TL(t)+TA(t))/Tmax)-k1*V(t)*T(t),
    diff(TL(t),t)=k1*V(t)*T(t)-mu*TL(t)-k2*TL(t),
    diff(TA(t),t)= k2*TL(t)-b*TA(t),
    diff(V(t),t)=N*b*TA(t)-k1*V(t)*T(t)-a*V(t);
> s:=10; r:=0.03; Tmax:=1500; mu:=0.02; N:=1400;
> b:=.24; a:=2.4; k1:=0.000024; k2:=0.003; N:=1400;
> init:=T(0)=1000,TL(0)=0, TA(0)=0,V(0)=0.001;
> Digits:=16;
> sol:=dsolve({deq,init},{T(t),TL(t),TA(t),V(t)},numeric,output=listprocedure);
> Tsol:=subs(sol,T(t));
> TAsol:=subs(sol,TA(t));
> TLsol:=subs(sol,TL(t));
> Vsol:=subs(sol,V(t));
> plot('Tsol(t)','t'=0..900);
> plot('TLsol(t)','t'=0..600);
> plot('TAsol(t)','t'=0..365);
> plot('Vsol(t)','t'=0..365);
```

MATLAB
```
% contents of m-file exer104.m:
```

```
% function Yprime=exer104(t,Y)
% % Y(1)=T, Y(2)=TL, Y(3)=TA, Y(4)=V
% s=10; r=0.03; Tmax=1700; mu=0.02; b=.24;
% a=2.4; k1=0.000024; k2=0.003; N=1400;
% Yprime=[s-mu*Y(1)+r*Y(1).*(1-(Y(1)+Y(2)+Y(3))/Tmax)-k1*Y(4).*Y(1);...
%           k1*Y(4).*Y(1)-mu*Y(2)-k2*Y(2); k2*Y(2)-b*Y(3); N*b*Y(3)-k1*Y(4).*Y(1)-a*Y(4)];
%
> [t,Y] = ode23('exer104',[0 365],[1000; 0; 0; 0.001]);
> plot(t,Y)
% try out to about 3 1/2 years
> [t,Y] = ode23('exer104',[0 1200], [1000; 0; 0; 0.001]);
> plot(t,Y)
```

## 10.5 A Model for a Mutating Virus

*The model of the previous section illustrated the interaction of HIV with T-cells. It did not account for mutations of HIV. The following is a model for evolving mutations of an HIV infection and an immune system response. This model is based on one introduced into the literature by Nowak, May, and Anderson.*

*Any model of an HIV infection should reflect the high mutability of the virus.*

In Section 10.3, we discussed the high degree of mutability characteristic of the HIV virus, which results in a large number of viral quasi-species. However, the human immune system seems able to mount an effective response against only a limited number of these mutations. Furthermore, the activation of a latently infected helper T-cell appears to stimulate viral reproduction, with the result that every time a new mutant activates a T-cell, vigorous viral population growth ensues. The immune system's T-cell population evidently can endure this cycle only a limited number of times. The objective of this section is to modify the T-cell–HIV model to reflect these facts in the model. In this, we follow Nowak, May, and Anderson [10]; see also Nowak and McMichael [4].

**Key assumptions:**

1. The immune response to a viral infection is to create subpopulations of immune cells that are specific to a particular viral strain and that direct immunological attack against that strain alone. The response is directed against the highly variable parts of the virus.
2. The immunological response to the virus is also characterized by a response that is specific to the virus but that acts against all strains. In other words, it acts against parts of the virus conserved under mutations.
3. Each mutant of the initial viral infection can cause the death of all immune system cells whether those cells are directed at variable or conserved regions.

In this modified model, we keep track of three sets of populations. Let $\{v_1, v_2, \ldots, v_n\}$ designate the various subpopulations of viral mutants of the initial HIV infection. Let $\{x_1, x_2, \ldots, x_n\}$ designate the populations of specific lymphocytes created in response to these viral mutations. And let $z$ designate the immune response that can destroy all variations of the infective agent. The variable $n$, for the number of viral mutations that arise, is a parameter of the model. We also include a parameter, the *diversity threshold* $N_{\text{div}}$, representing the number of mutations that can be accommodated before the immune system collapses.

The equation for each HIV variant, $v_i$, consists of a term, with parameter $a$, for its natural population growth rate; a term, with parameter $b$, for the general immune response; and a term, with parameter $c$, for the specific immune response to that variant,

$$\frac{dv_i}{dt} = v_i(a - bz - cx_i), \quad i = 1, \ldots, n. \tag{10.5.1}$$

The equation for each specific immune response population $x_i$ consists of a term, with parameter $g$, that increases the population in proportion to the amount of its target virus present, and a term, with parameter $k$, corresponding to the destruction of these lymphocytes by any and all viral strains,

$$\frac{dx_i}{dt} = gv_i - kx_i(v_1 + v_2 + \cdots + v_n), \quad i = 1, \ldots, n. \tag{10.5.2}$$

Finally, the equation for the general immune response population $z$ embodies a term, with parameter $h$, for its increase in proportion to the sum total of virus present but also a mass action term for its annihilation upon encounter with any and all virus,

$$\frac{dz}{dt} = (h - kz)(v_1 + v_2 + \cdots + v_n). \tag{10.5.3}$$

In order to compute with the model later on, we enter these differential equations into the computer system now. Take the value of $n$ to be 6, as will be explained shortly. Although the code is lengthy, it is all familiar and very repetitious:

```
MAPLE
> mutatingSystem:=diff(v1(t),t)=(a-b*z(t)-c*x1(t))*v1(t),
        diff(v2(t),t)=(a-b*z(t)-c*x2(t))*v2(t),diff(v3(t),t)=(a-b*z(t)-c*x3(t))*v3(t),
        diff(v4(t),t)=(a-b*z(t)-c*x4(t))*v4(t),diff(v5(t),t)=(a-b*z(t)-c*x5(t))*v5(t),
        diff(v6(t),t)=(a-b*z(t)-c*x6(t))*v6(t),
        diff(x1(t),t)=g*v1(t)-k*x1(t)*(v1(t)+v2(t)+v3(t)+v4(t)+v5(t)+v6(t)),
        diff(x2(t),t)=g*v2(t)-k*x2(t)*(v1(t)+v2(t)+v3(t)+v4(t)+v5(t)+v6(t)),
        diff(x3(t),t)=g*v3(t)-k*x3(t)*(v1(t)+v2(t)+v3(t)+v4(t)+v5(t)+v6(t)),
        diff(x4(t),t)=g*v4(t)-k*x4(t)*(v1(t)+v2(t)+v3(t)+v4(t)+v5(t)+v6(t)),
        diff(x5(t),t)=g*v5(t)-k*x5(t)*(v1(t)+v2(t)+v3(t)+v4(t)+v5(t)+v6(t)),
        diff(x6(t),t)=g*v6(t)-k*x6(t)*(v1(t)+v2(t)+v3(t)+v4(t)+v5(t)+v6(t)),
        diff(z(t),t)=(h-k*z(t))*(v1(t)+v2(t)+v3(t)+v4(t)+v5(t)+v6(t));
```

```
MATLAB
% make up an m-file, hivMVRate.m, with
% function Yprime=hivMVRate(t,Y);
% % Y(1)=v1,..., Y(6)=v6, Y(7)=x1,...,Y(12)=x6, Y(13)=z
% a=5; b=4; c=5; g=1; h=1; k=1;
% Yprime=[(a - b*Y(13) - c*Y(7))*Y(1); (a - b*Y(13) - c*Y(8))*Y(2);
%        (a - b*Y(13) - c*Y(9))*Y(3); (a - b*Y(13) - c*Y(10))*Y(4);
%        (a - b*Y(13) - c*Y(11))*Y(5); (a - b*Y(13) - c*Y(12))*Y(6);
%        g*Y(1) - k*(Y(1)+Y(2)+Y(3)+Y(4)+Y(5)+Y(6))*Y(7);
%        g*Y(2) - k*(Y(1)+Y(2)+Y(3)+Y(4)+Y(5)+Y(6))*Y(8);
%        g*Y(3) - k*(Y(1)+Y(2)+Y(3)+Y(4)+Y(5)+Y(6))*Y(9);
%        g*Y(4) - k*(Y(1)+Y(2)+Y(3)+Y(4)+Y(5)+Y(6))*Y(10);
%        g*Y(5) - k*(Y(1)+Y(2)+Y(3)+Y(4)+Y(5)+Y(6))*Y(11);
%        g*Y(6) - k*(Y(1)+Y(2)+Y(3)+Y(4)+Y(5)+Y(6))*Y(12);
%        (h-k*Y(13))*(Y(1)+Y(2)+Y(3)+Y(4)+Y(5)+Y(6))];
%
```

*The fate of the immune response depends on a critical combination of parameters.*

Again drawing on [10], we list several results that can be derived from this modified model. The model adopts one of two asymptotic behaviors depending on a combination of parameters, denoted by $N_{\mathrm{div}}$, defined by

$$N_{\mathrm{div}} = \frac{cg}{ak - bh}, \quad \text{where } \frac{a}{b} > \frac{h}{k}. \tag{10.5.4}$$

If the number $n$ of viral variants remains below or equal to $N_{\mathrm{div}}$, then the virus population eventually decreases and becomes subclinical. On the other hand, if $n > N_{\mathrm{div}}$, then the virus population eventually grows unchecked.

Note that $N_{\mathrm{div}}$ depends on the immune response to the variable and conserved regions of the virus in different ways. If specific lymphocytes rapidly respond (a large $g$) and are very effective (a large $c$), then $N_{\mathrm{div}}$ will be large in proportion to each, meaning a large number of mutations will have to occur before the virus gains the upper hand. By contrast, the general immune response parameters, $h$ and $b$, appear as a combination in the denominator. Their effect is in direct opposition to the comparable viral parameters $a$ and $k$.

Naturally, the size of $N_{\mathrm{div}}$ is of considerable interest. Assuming that the denominator of (10.5.4) is positive, $ak > bh$, we make three observations; their proofs may be found in [10].

**Observation 1.** *The immune responses, the $x_i$s and $z$, in total have only a limited response to the HIV infection. That is, letting $X = x_1 + x_2 + \cdots + x_n$ be the sum of the specific immunological responses, then*

$$\begin{aligned} \lim_{t \to \infty} X(t) &= \frac{g}{k}, \\ \lim_{t \to \infty} z(t) &= \frac{h}{k}, \end{aligned} \tag{10.5.5}$$

*where the parameters $g$, $h$, and $k$ are as defined as in* (10.5.2)–(10.5.4). *The implication is that even though the virus continues to mutate, the immune system cannot mount an increasingly larger response.*

The next observation addresses the possibility that after some time, all the immune subspecies populations are decreasing.

**Observation 2.** *If all mutant subspecies populations $v_i$ are decreasing after some time $\tau$, then the number of mutants will remain less than $N_{\mathrm{div}}$ and the infection will be controlled. That is, if there is a time $\tau$ such that all derivatives $v_i'(t) < 0$ are negative for $t > \tau$, $i = 1, \ldots, n$, then the number of mutations $n$ will not exceed $N_{\mathrm{div}}$.*

In the next observation, we see that if the number of variations increases to some level determined by the parameters, then the viral population grows without bound.

**Observation 3.** *If the number of mutations exceeds $N_{\mathrm{div}}$, then at least one subspecies increases without bound. In fact, in this case, the sum $V(t) \equiv v_1 + v_2 + \cdots + v_n$ increases faster than a constant times $e^{at}$ for some positive number $a$.*

**Observation 4.** *If $ak < bh$, the immune system will eventually control the infection.*

*Numerical studies illustrate the observations graphically.*

In what follows, we give parameters with which computations may be made to visualize the results discussed here. These parameters do not represent biological reality; likely the real parameters are not known. The ones used illustrate the features of the model. In [10], the authors choose $a = c = 5$, $b = 4.5$, and $g = h = k = 1$. This choice yields the diversity threshold as 10 ($N_{\text{div}} = 10$). To keep our computation manageable, we choose the same values except $b = 4$:

MAPLE
```
> a:=5: b:=4: c:=5: g:=1: h:=1: k:=1: Ndiv:= c*g/(a*k-b*h);
```

MATLAB
```
> a=5; b=4; c=5; g=1; h=1; k=1; Ndiv=c*g/(a*k-b*h)
```

Hence for this set of parameters $N_{\text{div}} = 5$.

Suppose that first there is an initial infection and the virus runs its course without mutation. We can achieve this in our model, and see the outcome, by taking some initial infection, $v_1(0) = \frac{5}{100}$, for the original virus but zero level of infection initially for all the mutants.

MAPLE
```
> initialVals:=v1(0)=5/100,v2(0)=0,v3(0)=0,v4(0)=0,v5(0)=0,v6(0)=0,
            x1(0)=0,x2(0)=0,x3(0)=0, x4(0)=0,x5(0)=0,x6(0)=0,z(0)=0;
> sol1:=dsolve({mutatingSystem, initialVals},{v1(t),v2(t),v3(t),v4(t),v5(t),v6(t),
            x1(t),x2(t),x3(t),x4(t),x5(t),x6(t),z(t)},numeric,output=listprocedure);
> v1sol1:=subs(sol1,v1(t)); x1sol1:=subs(sol1,x1(t)); zsol1:=subs(sol1,z(t));
> plot(['t','v1sol1(t)','t'=0..10]);
```

MATLAB
```
> hold on
> init1=[0.05;0;0;0;0;0;0;0;0;0;0;0;0];
> [t1,Y]=ode23('hivMVRate',[0 .5],init1);
> X=Y(:,7:12); % sum the x's
> S1=X(:,1)+X(:,2)+X(:,3)+X(:,4)+X(:,5)+X(:,6);
> z1=Y(:,13); % retain the z's
> V=Y(:,1:6); plot(t1,V)
```

The result, shown in Figure 10.5.1, is a plot of the number of (unmutated) viral particles against time. One sees that the infection flares up but is quickly controlled by the immune system.

Now we explore what happens when there is one mutation of the original virus, effectively $n = 2$ in this case, where $n$ counts the number of genetically distinct viruses. Following our technique above, only the original virus and one mutant will be given a nonzero initial value. Further, to incorporate a delay in the onset of mutation, we take $v_2(t) = 0$ for $0 \le t < T_2$ and $v_2(T_2) = \frac{1}{100}$, where $T_2 = \frac{1}{2}$ is the time of the first mutation. From the run above, we know the size of the population of original virus at this time, $v_1(T_2)$.

MAPLE
```
> initialVals:=v1(1/2)=v1sol1(1/2), v2(1/2)=1/100, v3(1/2)=0, v4(1/2)=0, v5(1/2)=0, v6(1/2)=0,
            x1(1/2)=x1sol1(1/2), x2(1/2)=0, x3(1/2)=0, x4(1/2)=0, x5(1/2)=0, x6(1/2)=0, z(1/2)=zsol1(1/2);
> sol2:=dsolve({mutatingSystem, initialVals}, {v1(t),v2(t),v3(t),v4(t),v5(t),v6(t),
            x1(t),x2(t),x3(t),x4(t),x5(t),x6(t),z(t)}, numeric,output=listprocedure);
> v1sol2:=subs(sol2,v1(t)); v2sol2:=subs(sol2,v2(t));
> x1sol2:=subs(sol2,x1(t)); x2sol2:=subs(sol2,x2(t)); zsol2:=subs(sol2,z(t));
> ###
> initiaVals:=v1(1)=v1sol2(1),v2(1)=v2sol2(1), v3(1)=1/100, v4(1)=0,v5(1)=0,v6(1)=0,
```

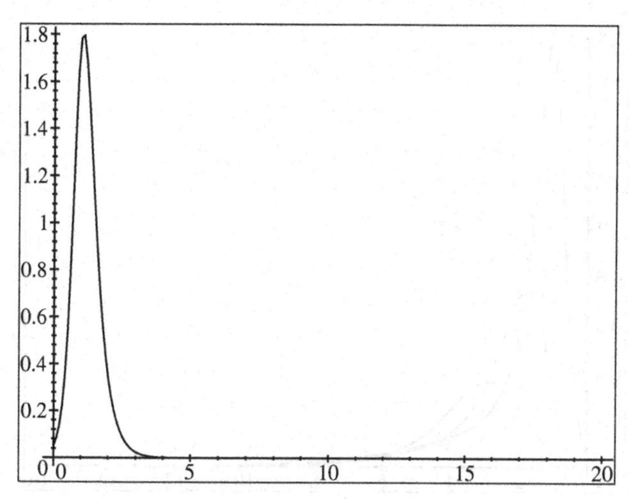

**Fig. 10.5.1.** A viral infection with no mutation.

```
          x1(1)=x1sol2(1),x2(1)=x2sol2(1),x3(1)=0, x4(1)=0,x5(1)=0, x6(1)=0, z(1)=zsol2(1);
> sol3:=dsolve({mutatingSystem, initialVals}, {v1(t), v2(t),v3(t),v4(t),v5(t),v6(t),
          x1(t),x2(t),x3(t),x4(t),x5(t),x6(t),z(t)}, numeric,output=listprocedure);
> v1sol3:=subs(sol3,v1(t)); v2sol3:=subs(sol3,v2(t)); v3sol3:=subs(sol3,v3(t));
> x1sol3:=subs(sol3,x1(t)); x2sol2:=subs(sol3,x2(t));
> x3sol3:=subs(sol3,x3(t)); zsol3:=subs(sol3.z(t));
> plot({['t','v1sol1(t)','t'=0..1/2], ['t','v1sol2(t)', 't'=1/2..1], ['t','v1sol3(t)','t'=1..10],
     ['t','v2sol2(t)','t'=1/2..1], ['t','v2sol3(t)', 't'=1..10], ['t','v3sol3(t)','t'=1..10]}, color=black);
```

MATLAB

```
> s=size(t1);
  % the ending values = last row = Y(s(1),:), add 1/100 to its 2nd component,
  % that becomes start values for next period
> init2=Y(s(1),:); init2(2)= init2(2)+0.01;
> [t2,Y]=ode23('hivMVRate',[.5 1],init2);
> X=Y(:,7:12); S2=X(:,1)+X(:,2)+X(:,3)+X(:,4)+X(:,5)+X(:,6);
> z2=Y(:,13); V=Y(:,1:6); plot(t2,V)
> s=size(t2); % add 2nd mutant
> init3=Y(s(1),:); init3(3)= init3(3)+0.01;
> [t3,Y]=ode23('hivMVRate',[1 1.5],init3);
> X=Y(:,7:12); S3=X(:,1)+X(:,2)+X(:,3)+X(:,4)+X(:,5)+X(:,6);
> z3=Y(:,13); V=Y(:,1:6); plot(t3,V)
```

We see the result in Figure 10.5.2. Each new mutant strain engenders its own flare-up, but soon the immune system gains control.

So far, the number of mutations has been less than $N_{\text{div}}$, but we now jump ahead and allow six mutations to occur:

MAPLE

```
> restart; with(plots):
> a:=5: b:=4: c:=5: g:=1: h:=1: k:=1:
> Ndiv:=c*g/(a*k-b*h);
> eq:=seq(diff(v[n](t),t)=(a-b*z(t)-c*x[n](t))*v[n](t), n=1..7),
     seq(diff(x[n](t),t)=g*v[n](t)-k*x[n](t)*(sum(v[p](t), p=1..7)),n=1..7),
        diff(z(t),t)=(h-k*z(t))*sum(v[q](t),q=1..7):
> NG:=6; #number of mutations to generate initial conditions for the infections
> init[1]:=v[1](0)=5/100,seq(v[n](0)=0,n=2..7),seq(x[n](0)=0,n=1..7),z(0)=0:
> for p from 1 to NG do
> sol[p]:=dsolve({eq,init[p]},numeric,output=listprocedure):
```

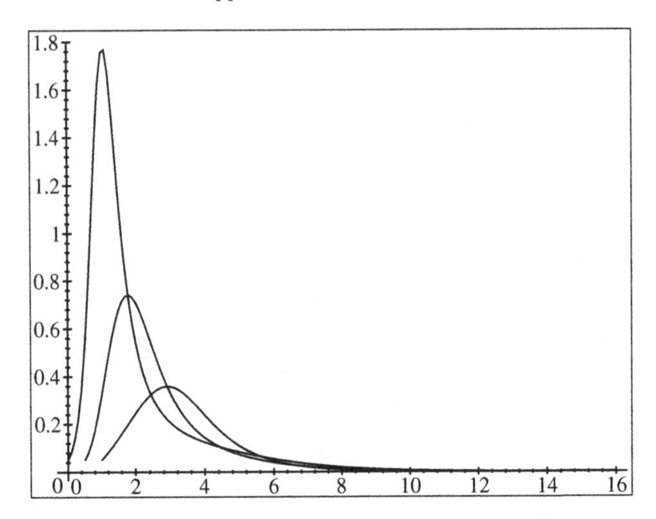


**Fig. 10.5.2.** Infection with two mutations.

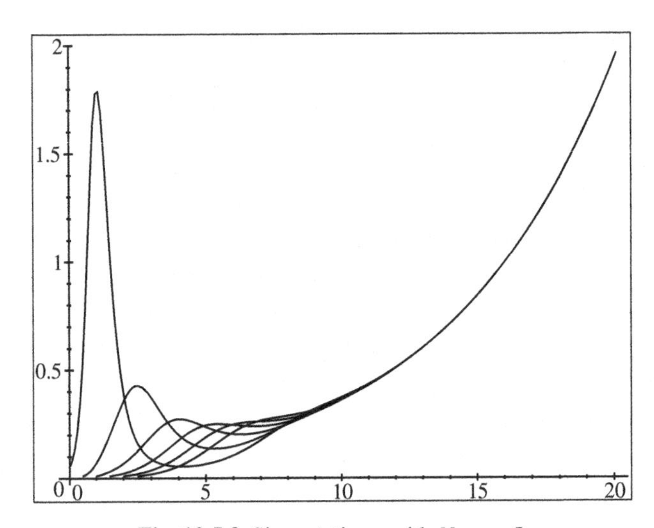


**Fig. 10.5.3.** Six mutations, with $N_{\text{div}} = 5$.

```
> for n from 1 to 7 do
    vs[p,n]:=subs(sol[p],v[n](t)):
    xs[p,n]:=subs(sol[p],x[n](t)):
  od:
> zs[p]:=subs(sol[p],z(t)):
> vs[p,p+1](p/2):=1/100;
> init[p+1]:=seq(v[m](p/2)=vs[p,m](p/2),m=1..7),seq(x[m](p/2)=xs[p,m](p/2),m=1..7),z(p/2)=zs[p](p/2):
> od:
> for m from 1 to NG do
  J[m]:=plot([seq([t,vs[n,m](t),t=(n-1)/2..n/2],n=1..NG),[t,vs[NG,m](t),t=NG/2..15]],color=BLACK):
  od:
> display([seq(J[i],i=1..NG)]);
```

MATLAB
```
> s=size(t3);init4=Y(s(1),:);
> init4(4)= init4(4)+0.01;
> [t4,Y]=ode23('hivMVRate',[1.5 2.0],init4);
> X=Y(:,7:12);
> S4=X(:,1)+X(:,2)+X(:,3)+X(:,4)+X(:,5)+X(:,6);
> z4=Y(:,13); V=Y(:,1:6); plot(t4,V)
  %%% end 4th step
> s=size(t4); init5=Y(s(1),:);
> init5(5)= init5(5)+0.01;
> [t5,Y]=ode23('hivMVRate',[2.0 2.5],init5);
> X=Y(:,7:12);
> S5=X(:,1)+X(:,2)+X(:,3)+X(:,4)+X(:,5)+X(:,6);
> z5=Y(:,13); V=Y(:,1:6); plot(t5,V)
  %%% end 5th step
> s=size(t5);init6=Y(s(1),:);
> init6(6)= init6(6)+0.01;
> [t6,Y]=ode23('hivMVRate',[2.5 20],init6);
> X=Y(:,7:12);
> S6=X(:,1)+X(:,2)+X(:,3)+X(:,4)+X(:,5)+X(:,6);
> z6=Y(:,13); V=Y(:,1:6); plot(t6,V)
  %%% end 6th step
> hold off
> t=[t1;t2;t3;t4;t5;t6]; S=[S1;S2;S3;S4;S5;S6];
> z=[z1;z2;z3;z4;z5;z6]; plot(t,S); plot(t,z)
```

One sees that the result is unexpected. At first things go as before: After an initial flare-up, the immune system begins to gain control and viral population tends downward. But then something happens: The immune system is overwhelmed.

Observation 2 predicts that since more mutations have occurred than $N_{\text{div}}$, the population will grow without bound. The graphs show this.

We now verify Observation 3 for these parameters. First, note that from (10.5.2), the sum of the $x_i$s satisfies the differential equation

$$X' = V * (g - kX),$$

where $X = x_1 + x_2 + \cdots + x_6$ and $V = v_1 + v_2 + \cdots + v_6$. We could compute the solution of this equation and expect that

$$\lim_{t \to \infty} (x_1(t) + x_2(t) + \cdots + x_6(t)) = \frac{g}{k}.$$

Or, using the computations already done, add the $x_i$s. The level of the $x_i$s are kept in this syntax in the eighth through thirteenth positions. The "time" variable is kept in the first position of the output. We add these $x_i$s in each output.

MAPLE
```
> plot({['t','x1sol1(t)','t'=0..1/2], ['t','x1sol2(t)+x2sol2(t)','t'=1/2..1],
     ['t','x1sol3(t)+x2sol3(t)+x3sol3(t)','t'=1..3/2],['t','x1sol4(t)+x2sol4(t)+x3sol4(t)+x4sol4(t)','t'=3/2..2],
     ['t','x1sol5(t)+x2sol5(t)+x3sol5(t)+x4sol5(t)+x5sol5(t)','t'=2..5/2],
     ['t','x1sol6(t)+x2sol6(t)+x3sol6(t)+x4sol6(t)+x5sol6(t)+x6sol6(t)','t'=5/2..10]},color=BLACK);
```

MATLAB
```
% These plots done via code above.
```

The plot of the response $z$, no matter where you start, should have asymptotic limit $\frac{h}{k}$ and looks essentially the same as that for $(x_1 + x_2 + \cdots + x_6)$. Here is one way to plot the values of $z$. What you should see is that the $z$ reaches a maximum:

MAPLE
```
> plot(['t','zsol1(t)','t'=0..1/2],['t','zsol2(t)','t'=1/2..1],['t','zsol3(t)','t'=1..3/2],['t','zsol4(t)','t'=3/2..2],
>      ['t','zsol5(t)','t'=2..5/2],['t','zsol6(t)','t'=5/2..10],color=BLACK);
```

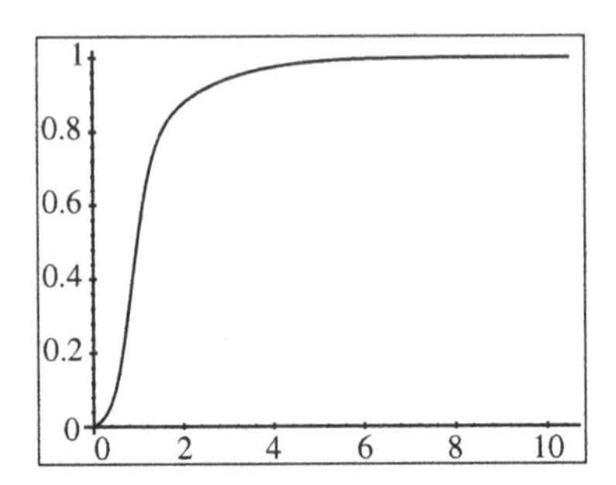


**Fig. 10.5.4.** Graph of $x_1 + x_2 + \cdots + x_n$.

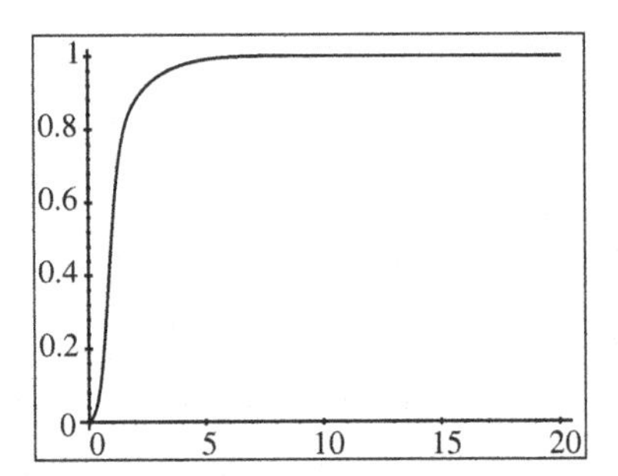


**Fig. 10.5.5.** Graph of $z(t)$ from (10.5.3).

*Viral suppression is possible with some parameters.*

It was stated in Observations 1 and 4 that there are two ways to achieve viral suppression. These are experiments that should be run. One could choose parameters such that $ak < bh$; then the immune system will eventually control the infection. No change need be made in the syntax, only $a = 4$ and $b = 5$. Other parameters could remain the same.

The simple models as presented in these two sections give a good first understanding of the progress from infection to remission to AIDS. Such an understanding provokes further study.

## 10.6 Predicting the Onset of AIDS

*Most diseases have a latency or incubation period between the time of infection and the onset of symptoms; AIDS is no exception. The latency period for AIDS varies greatly from individual to individual, and so far, its profile has not been accurately determined. However, assuming a given form of the incubation profile, we show that the onset of symptoms occurs, statistically, as the time of infection convolved with this profile.*

*AIDS cases can be statistically predicted by a convolution integral.*

In this chapter, we have discussed the epidemiology of the HIV infection and subsequent appearance of AIDS. For most diseases, there is a period of time between infection by the causative agent and the onset of symptoms. This is referred to as the *incubation period*; an affliction is asymptomatic during this time. Research is showing that the nature of this transition for HIV is a complicated matter. Along with trying to learn the mechanism of this process, considerable work is being devoted in an attempt to prolong the period between HIV infection and the appearance of AIDS. This period varies greatly among different individuals and appears to involve, besides general health, particular characteristics of the individual's immune system. See [5, 6, 7, 8] for further details.

The incubation period can be modeled as a probability density function $p(t)$ (see Section 2.8), meaning that the probability that AIDS onset occurs in a $\Delta t$ time interval containing $t$ is

$$p(t) \cdot \Delta t.$$

To discover the incubation density, records are made, when possible, of the time between contraction of HIV and the appearance of AIDS. See Bacchetti [12] for several comments by other researchers, and for a comprehensive bibliography. At the present this probability density is not known, but some candidates are shown in Figure 10.6.1(a)–(d).

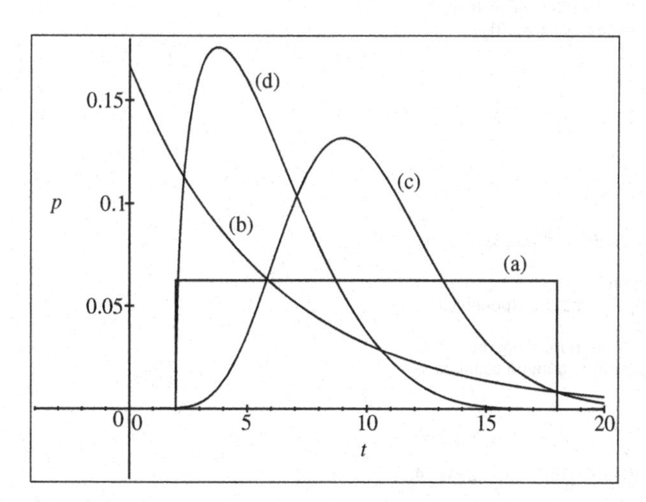


**Fig. 10.6.1.** Some HIV incubation probability densities. Graphs of (a) uniform distribution, (b) $\frac{e^{-t/6}}{6}$, (c) $t^9 e^{-t}$ normalized, (d) $\sqrt{\frac{t-2}{16}}(1 - \frac{t-2}{16})^4$ normalized.

Figure 10.6.1(a) is a uniform distribution over the period of 2 to 18 years. This distribution has no preferred single incubation moment, but incubation is guaranteed to occur no sooner than two years after infection and no later than 18 (18.0) years afterward. It is unlikely that this is the operating distribution, but we include it for comparison purposes.

Figure 10.6.1(b) is an exponential distribution. This distribution pertains to many "arrival time" processes in biology such as times for cell division (which occurs upon the "arrival" of cell maturation). As can be seen, incubation is likely to occur right away and diminishes with time. The incubation period can be infinitely long. (A fraction of those infected with HIV have, so far, remained asymptomatic "indefinitely.")

Figure 10.6.1(c) is a gamma distribution incorporating both a preferred incubation "window" and the possibility of an indefinitely long incubation period.

Figure 10.6.1(d) is a beta distribution. It allows for a preferred window, but as with the uniform distribution, incubation must occur between given times. Their graphs are illustrated in Figure 10.6.1.

The functions we have used to draw Figure 10.6.1 are $p_1$, $p_2$, $p_3$, and $p_4$ as defined in the following:

```
MAPLE
> c1:=int(t^9*exp(-t),t=0..infinity);
> c2:=evalf(Int(sqrt((t-2)/16)*(1-(t-2)/16)^4,t=2..20));
> p1:=t->1/16*(Heaviside(t-2)-Heaviside(t-18));
> p2:=t->exp(-t/6)/6;
> p3:=t->t^9*exp(-t)/c1;
> p4:=t->sqrt((t-2)/16)*(1-(t-2)/16)^4/c2;
> plot({p1(t),p2(t),p3(t),p4(t),t=0..20});
```

```
MATLAB
% make an m-file, incubationProfile.m:
% function y=incubationProfile(t);
% if t<2
%   y=0;
% elseif t<18
%   y=1/16;
% else
%   y=0;
% end
> t=linspace(0,20,100)
> for k=1:100
>   p1(k)=incubationProfile(t(k)); % uniform
> end
> plot(t,p1); hold on
> p2=exp(-t/6)/6; plot(t,p2) % exponential
> p3=t.^9.*exp(-t);
> c1=trapz(t,p3); % for normalization
> p3=p3/c1; plot(t,p3) % gamma distribution
> for k=1:100
>   if t(k)<2
>     p4(k)=0;
>   elseif t(k)<20
>     p4(k)=sqrt((t(k)-2)/16)*(1-(t(k)-2)/16)^4;
>   else
>     p4(k)=0;
>   end
> end
> c2=trapz(t,p4);
> p4=p4/c2; plot(t,p4) % the beta distribution
```

To derive a mathematical relationship for the appearance of AIDS cases, we will assume that the probability distribution for the incubation period can be treated as a deterministic rate. Let $h(t)$ denote the HIV infection density, that is,

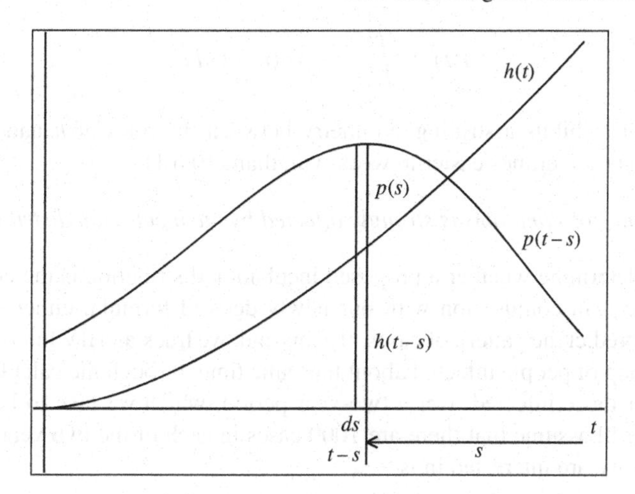


**Fig. 10.6.2.** $a(t) \approx \sum_s h(t-s) \cdot ds \cdot p(s)$.

$$\text{the number of new HIV infections during } [t, t+\Delta t) = h(t) \cdot \Delta t,$$

and let $a(t)$ denote the AIDS density; thus

$$\text{the number of new AIDS cases during } [t, t+\Delta t) = a(t) \cdot \Delta t.$$

We wish to determine $a(t)$ from $h(t)$. How many AIDS cases occur now, on day $t$ due to infections $s$ days ago? See Figure 10.6.2. The number of newly infected persons during the interval from $t-(s+ds)$ to $t-s$ is $h(t-s) \cdot ds$ and the fraction of them to become symptomatic $s$ days later is $p(s)$. Hence the contribution to $a(t)$ here is

$$h(t-s) \cdot ds \cdot p(s).$$

Since $a(t)$ is the sum of such contributions over all previous times, we get

$$a(t) = \int_0^\infty h(t-s)p(s)ds. \tag{10.6.1}$$

This sort of integral is known as a *convolution*; such integrals occur widely in science and engineering.

Convolution integrals have an alternative form under change of variable. Let $u = t-s$; then $s = t-u$ and $ds = -du$. Since $u = t$ when $s = 0$ and $u = -\infty$ when $s = \infty$, the integral of (10.6.1) becomes

$$a(t) = -\int_t^{-\infty} h(u)p(t-u)du.$$

Absorbing the minus sign into a reversal of the limits of integration and replacing the dummy variable of integration $u$ by $s$ gives

$$a(t) = \int_{-\infty}^{t} h(s)p(t-s)ds. \tag{10.6.2}$$

This equation exhibits a striking symmetry between the roles of $h$ and $p$. Equation (10.6.2) is sometimes easier to work with than (10.6.1).

*The occurrence of symptoms is strongly affected by the incubation distribution.*

In order to determine whether a proposed incubation distribution is the correct one, we must use it in conjunction with our newly derived formula, either (10.6.1) or (10.6.2), to predict the pattern of cases. To this end, we track an HIV infected *cohort*, that is, a group of people infected about the same time, through the calculation.

Consider those infected over a two-year period, which we take to be $t = 0$ to $t = 2$. We will assume that there are 1000 cases in each of the two years; thus the HIV density we are interested in is

$$h(t) = \begin{cases} 1000 & \text{if } 0 \leq t \leq 2, \\ 0 & \text{otherwise.} \end{cases} \tag{10.6.3}$$

The total number of cases is $\int_0^2 h(s)ds = 2000$. With this choice for $h$, we can simplify the factor $h(t-s)$ in (10.6.1). By subtracting $t$ from the inequalities $0 \leq t - s \leq 2$ and multiplying by $-1$, we get the equivalent form $t - 2 \leq s \leq t$. In other words,

$$\text{if } t-2 \leq s \leq t, \quad \text{then } h(t-s) = 1000; \quad \text{otherwise, } h \text{ is } 0. \tag{10.6.4}$$

Therefore, the only contribution to the integral in (10.6.1) comes from the part of the $s$-axis between $t - 2$ and $t$.

There are three cases depending on the position of the interval $[t-2, t]$ relative to 0; see Figure 10.6.3. In (a), $t < 0$ and the interval is to the left of 0; in (b), $t - 2 < 0 < t$, the interval contains 0; and in (c), $0 < t - 2$, the interval is to the right of 0.

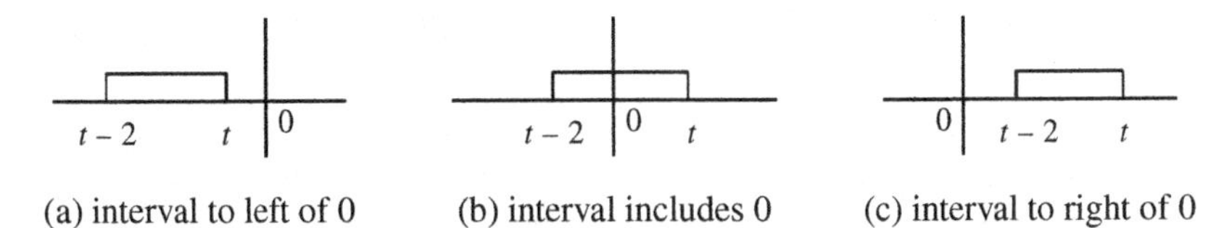


(a) interval to left of 0 (b) interval includes 0 (c) interval to right of 0

**Fig. 10.6.3.** Contributory subinterval of the $s$-axis.

Consider each case. If (a), $t \leq 0$, then $a(t) = 0$ from (10.6.2) and (10.6.3). If (c), $t \geq 2$, then $t - 2 \geq 0$ and (10.6.1) becomes, taking into account (10.6.4),

$$a(t) = 1000 \int_{t-2}^{t} p(s)ds, \quad t \geq 2.$$

Finally, for (b), $0 < t < 2$, the part of the interval to the left of $s = 0$ makes no contribution, and in this case (10.6.1) becomes

$$a(t) = 1000 \int_0^t p(s)ds, \quad 0 < t < 2.$$

Putting these three together, we have

$$a(t) = \begin{cases} 0, & t \le 0, \\ 1000 \int_0^t p(s)ds, & 0 < t < 2, \\ 1000 \int_{t-2}^t p(s)ds, & t \ge 2. \end{cases} \tag{10.6.5}$$

Because it is inconvenient to deal with a function defined by cases, such as $a(t)$ is defined by (10.6.5), a standard set of "cases"-type functions have been devised. One of these is the *Heaviside* function $H(t)$, and another is the *signum* function $S(t)$. The first is defined as

$$H(t) = \begin{cases} 0, & t < 0, \\ 1, & t \ge 0. \end{cases} \tag{10.6.6}$$

The signum function is just the *sign* of its argument, that is,

$$S(t) = \begin{cases} -1, & t < 0, \\ 0, & t = 0, \\ 1, & t > 0. \end{cases} \tag{10.6.7}$$

Actually, there is a relationship between the two, except for $t = 0$:

$$H(t) = \frac{1}{2}(S(t) + 1), \quad S(t) = 2H(t) - 1, \quad t \ne 0. \tag{10.6.8}$$

The Heaviside function $H(2 - t)$ cuts out at $t = 2$, while $H(t - 2)$ cuts in at $t = 2$, so in terms of Heaviside functions, (10.6.5) can be written as

$$a(t) = 1000H(2 - t) \int_0^t p(s)ds + 1000H(t - 2) \int_{t-2}^t p(s)ds. \tag{10.6.9}$$

For the simplest example, assume that the incubation density is the uniform distribution, $P_1$ above (Figure 10.6.1(a)):

$$P_1(t) = \begin{cases} \frac{1}{16} & \text{if } 2 \le t \le 18, \\ 0 & \text{otherwise.} \end{cases}$$

Substituting into (10.6.9) and integrating gives the onset distribution $a(t)$.

MAPLE
```
> restart:
> h:=t->1000*(Heaviside(2-t)-Heaviside(-t));
> plot(h(t),t=-3..3);
> int(h(t-s),s=2..18)/16;
```

```
> a:=unapply(int(h(t-s),s=2..18)/16,t);
> plot(a(t),t=0..20);

  MATLAB
  % make an m-file, casesOnset.m:
  % function a=casesOnset(t);
  %  if t<0  %    a=0;
  %  elseif t<2
  %    r=linspace(0,t,20);
  %    for i=1:20
  %      y(i)=incubationProfile(r(i));
  %    end
  %    a=1000*trapz(r,y);
  %  else
  %    r=linspace(t-2,t,20);
  %  for i=1:20
  %      y(i)=incubationProfile(r(i));
  %  end
  %    a=1000*trapz(r,y);
  %  end % end of casesOnset.m
  % recall t=linspace(0,20,100)
> for k=1:100
    a(k)=casesOnset(t(k)) % vector same size as t
> end
> plot(t,a)
> trapz(t,a) % integral 0 to 20
```

The output of this calculation is (in MAPLE syntax)

$$a(t) = \left(\frac{125}{4}t - \frac{1125}{2}\right) \text{signum}(18-t) + \left(\frac{125}{4}t - 125\right) \text{signum}(t-4) + \left(-\frac{125}{4}t + \frac{125}{2}\right) \text{signum}(2-t) + \left(-\frac{125}{4}t + 625\right) \text{signum}(20-t). \tag{10.6.10}$$

This provides an alternative realization for a formula for $a(t)$. Its form is different from that of (10.6.5). We can recover the previous one, however, by evaluating the signum function with various choices of $t$. To do this, suppose that $2 < t < 4$ or $4 < t < 18$ or $18 < t < 20$, respectively, and evaluate (10.6.10).

Eventually, all those infected will contract AIDS; therefore,

$$\int_0^\infty a(s)ds = \int_0^2 h(s)ds.$$

But the first integral reduces to the interval [0, 20]. That is, the total number of people who develop AIDS during the 20-year period is the same as the total number of people in the initial two-year cohort. This computation is done as

```
  MAPLE
> int(a(s), s=0..20);
```

This gives 2000.

Several observations should be made with the graph for each of the other distributions. There should be a gradual increase of the number of cases as the cohorts begin to develop symptoms of AIDS. Also, there should be a gradual decrease that may last past 20 years: those in the cohorts who were infected near the end of the second

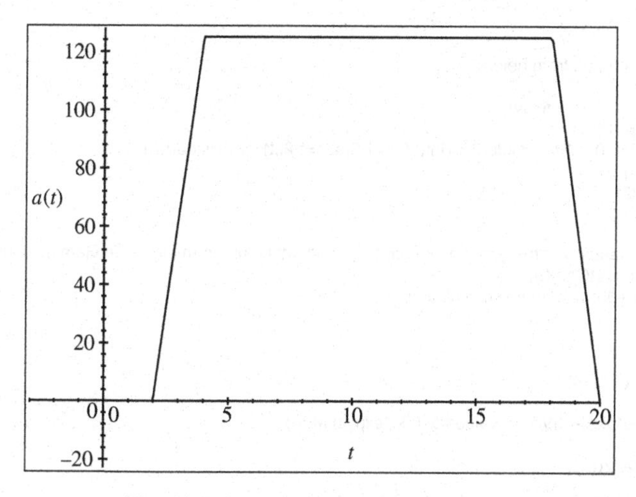


**Fig. 10.6.4.** Graph of $a(t)$ from (10.6.10).

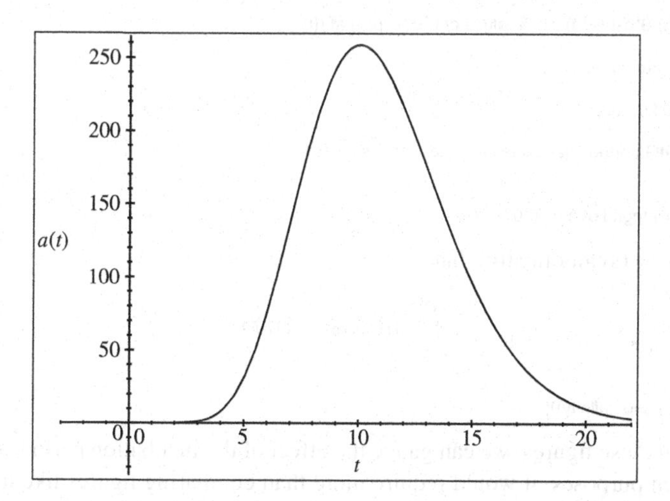


**Fig. 10.6.5.** Onset of AIDS cases for a two-year HIV cohort assuming gamma incubation.

year may not begin to show symptoms until the twenty-second year, depending on whether $P_2$, $P_3$, or $P_4$ is used.

We leave the computations for the other distributions to the exercises. However, Figure 10.6.5 shows the graph for the onset of AIDS cases for a two-year cohort assuming incubation as with the gamma distribution $P_3(t)$. The function $a(t)$ defined, evaluated, and plotted by

$$a(t) = \int_0^\infty h(t-s)P_3(s)ds$$

is evaluated with the following code:

```
MAPLE
#repeating P3 and c1 from before
> t:='t';
> c1:=int(t^9*exp(-t),t=0..infinity);
> P3:=t->t^9*exp(-t)/c1;
> int(1000*P3(s),s=0..t)*Heaviside(2-t)+int(1000*P3(s),s=t-2..t)*Heaviside(t-2):
> a:=unapply(%,t);
> plot(a(t),t=0..22);
```

```
MATLAB
% for onset of cases with the gamma incubation period, make up an m-file, onsetGam.m, containing
% function a=onsetGam(t);
% c1=3.6107e+05; % from text calculation
% if t<0
%  a=0;
% elseif t<2
%  r=linspace(0,t,20);
%  for i=1:20
%    y(i)=r(i)^9*exp(-r(i))/c1; %values of t.^9.*exp(-t) at r(i)
%  end
%  a=1000*trapz(r,y);
% else
%  r=linspace(t-2,t,20); % integral of y from 0 to t
%  for i=1:20
%    y(i)=r(i)^9*exp(-r(i))/c1; %values of t.^9.*exp(-t) at r(i)
%  end
%  a=1000*trapz(r,y);
% end
> t=linspace(0,20,100);
> for k=1:100
   a(k)=onsetGam(t(k)); % creates a vector same size as t
> end
> plot(t,a)
> trapz(t,a) % integral of a over 0 to 20
```

We verify (symbolically) that

$$\int_0^\infty a(s)ds = 2000.$$

```
MAPLE
> evalf(Int(a(s),s=0..infinity));
```

Comparing these figures, we can gauge the effect of the incubation period. Note that for research purposes, it would require more than comparing figures like these with AIDS epidemiologic data to determine the incubation distribution because one could not separate the AIDS cases into those stemming from a particular cohort; all cohort onsets are mixed together.

## Exercises/Experiments

1. Choose each of the two remaining hypothetical incubation densities of Figure 10.6.1 in turn. (The uniform and gamma have been done in the text.) Draw the graph of the number of AIDS cases expected to develop, $a(t)$, for the cohort of (10.6.2) with the assumption that the one you have chosen is correct.
2. We pose a *what if* experiment. Suppose that around 2010, a vaccine for HIV is developed, and while current cases cannot be cured, HIV is no longer transmitted. The number of reported new cases of HIV, $h(t)$, drops dramatically to zero by 2020. Model the reported cases of HIV with a hypothetical scenario such as

$$h(t) = \frac{t - 1980}{40}\left[1 - \frac{(t - 1980)^6}{40^6}\right].$$

(a) Draw a graph of $h(t)$. Observe that $h(1980) = 0$ and $h(2020) = 0$.

MAPLE
```
> restart;
> h:=t->((t-1980)/40)*(1-(t-1980)^6/40^6);
> plot(h(t),t=1980..2020,xtickmarks=0,ytickmarks=2);
> h(1980); h(2020);
```

MATLAB
```
% contents of m-file infectDensity.m:
% function h=infectDensity(t) % won't vectorize due to the conditionals
% if t<1980
%   h=0;
% elseif t<2020
%   h=((t-1980)/40)*(1-((t-1980)/40)^6);
% else
%   h=0;
% end
> t=linspace(1980,2020); % 100 values
> for k=1:100
    h(k)=infectDensity(t(k));
  end
> plot(t,h)
```

(b) Determine where the maximum value of $h$ occurs. This represents the time when the reported new cases of HIV-infected individuals peaks if this "optimistic scenario" were to happen.

MAPLE
```
> sol:=solve(diff(h(s),s)=0,s);
> evalf(sol[1]);
```

MATLAB
```
> hmax=max(h)
> m=0;
> for i=1:100
    if h(i)==hmax
      m=i;
    end
  end
> maxYear=t(m)
```

(c) Define a "later rather than sooner" hypothetical incubation density and draw its graph:

MAPLE
```
> c5:=int(1/16*(t-2)*(1-(1/16*(t-2))^2),t=2..18);
> P5:=t->1/16*(t-2)*(1-(1/16*(t-2))^2)/c5;
> plot([t,P5(t),t=2..18],t=0..20);
```

MATLAB
```
> s=linspace(2,18);
> incDen=((s-2)/16).*(1-((s-2)/16).^2);
> c5=trapz(s,incDen)
> incDen=incDen/c5;
> plot(s,incDen)
```

(d) Find $a(t)$ as in (10.6.1) associated with this distribution.

MAPLE
```
> a15:=t->int(h(t-r)*P5(r),r=2..t-1980);
> a25:=t->int(h(t-r)*P5(r),r=2..18);
> a:=t->a15(t)*Heaviside(2000-t)+a25(t)*Heaviside(t-2000);
> plot(a(t),t=1982..2020);
```

MATLAB
```
% the integral a(t)=int(h(t-s)*p(s)*ds)
% from 0 to infinity has 5 cases:
%     t<1982, a(t)=0,
% 1982<t<1998, a(t)= int(h(t-s)*p(s)ds from 2 to t-1998
% 1998<t<2022, a(t)= int(h(t-s)*p(s)ds from 2 to 18
% 2022<t, a(t)= int(h(t-s)*p(s)ds from t-2020 to 18
% Here's why. First case: suppose t=1981; since p(s)=0 unless s>2,
% t-s<1979 so h(t-s)=0.
% Second case: suppose t=1994;
% again p(s)=0 unless s>2, so the lower limit must be at least 2.
% For t=1994, t-1980=14, so s runs from 2 to 14 and t-s runs from 1992 down to 1980;
% after that h(t-s)=0.
% We leave the remaining cases for you.
%
% But all cases will be done automatically
% since we have defined infectDensity(t)=0 for t<1980 or t>2020
> t=linspace(1980,2050); % get the graph for 1980 to 2050
> for k=1:100 % k= time index
    T=t(k);
    for j=1:100 % j= s index
      S=s(j);
      y(j)=infectDensity(T-S)*incDen(j);
    end
    a(k)=trapz(s,y);
  end
> plot(t,a)
```

(e) Sketch the graphs for the hypothetical $h(t)$ and associated $a(t)$.

MAPLE
```
> plot({[t,h(t),t=1980..2020],[t,a(t),t=1982..2029.432]});
> int(h(t),t=1980..2020);
> int(a(t),t=1982..2029.432);
```

## Questions for Thought and Discussion

1. What are four suspected ways that HIV kills cells?
2. Why do viral mutations lead to the development of new antibodies by the immune system?
3. Describe the life cycle of HIV.
4. Why do we continue to get colds, year after year, but seldom get mumps more than once?
5. Describe clonal selection and clonal deletion.
6. How does the clonal deletion model explain the fact that a mouse injected prenatally with a virus never will raise antibodies against the virus after the mouse is born?
7. Describe three general immunologic mechanisms.
8. How does HIV infection result in the inactivation of both the humoral and cell-mediated immune responses?
9. Most DNA replication includes a proofreading function that corrects mismatched DNA nucleotides during DNA replication. The reverse transcriptase of HIV seems to lack this ability, which results in high mutation rates (as much as one or more per generation). Discuss this problem in terms of antibody production by a host's immune system.

## References and Suggested Further Reading

[1] BLOOD CELLS, IMMUNITY:
W. T. Keeton and J. L. Gould, *Biological Science*, 5th ed., Norton, New York, 1993.

[2] Immunity: Special issue on the immune system, *Sci. Amer.*, **269**-3 (1993).

[3] HIV AND AIDS:
What science knows about AIDS [full issue], *Sci. Amer.*, **259**-4 (1988).

[4] HIV AND AIDS:
M. A. Nowak and A. J. McMichael, How HIV defeats the immune system, *Sci. Amer.*, **273**-2 (1995), 58.

[5] HIV AND T CELLS:
A. S. Perelson, Modeling the interaction of the immune system with HIV, in C. Castillo-Chavez, ed., *Mathematical and Statistical Approaches to AIDS Epidemiology*, Lecture Notes in Biomathematics, Vol. 83, Springer-Verlag, New York, 1989, 350–370.

[6] HIV AND T CELLS:
A. S. Perelson, D. E. Kirschner, and R. J. De Boer, The dynamics of HIV infection of $CD4^+$ T cells, *Math. Biosci.*, **114** (1993), 81–125.

[7] HIV AND T CELLS:
K. E. Kirschner and A. S. Perelson, A model for the immune system response to HIV:AZT treatment studies, in O. Arino, D. E. Axelrod, M. Kimmel, and M. Langlais, eds., *Mathematical Population Dynamics: Analysis of Heterogeneity and the Theory of Epidemics*, Wuerz Publishing, Winnipeg, ON, Canada, 1995, 295–310.

[8] HIV AND T CELLS:
A. S. Perelson, Two theoretical problems in immunology: AIDS and epitopes, in G. Cowan, D. Pines, and D. Meltzer, eds., *Complexity: Metaphors, Models and Reality*, Addison–Wesley, Reading, MA, 185–197.

[9] THE IMMUNE RESPONSE:
W. C. Greene, AIDS and the immune system, *Sci. Amer.*, **269**-3 (special issue) (1993).

[10] MUTATIONS OF HIV:
M. A. Nowak, R. M. May, and R. M. Anderson, The evolutionary dynamics of HIV-1 quasi species and the development of immunodeficiency disease, *AIDS*, **4** (1990), 1095–1103.

[11] MUTATIONS OF HIV:
M. A. Nowak and R. M. May, Mathematical biology of HIV infections: Antigenic variation and diversity threshold, *Math. Biosci.*, **106** (1991), 1–21.

[12] CALCULATIONS OF THE TIME FROM HIV INFECTION TO AIDS SYMPTOMS:
P. Bacchetti, M. R. Segal, and N. P. Jewell, Backcalculation of HIV infection rates, *Statist. Sci.*, **8**-2 (1993), 82–119.

# 11

# Parasites and Their Diseases

## Introduction

In the first section of this chapter, we survey and briefly describe the parasites important to humans and the diseases they engender. In Section 11.2, we detail the life cycle of the parasites responsible for malaria. While there are four species of mosquitoes involved, the biggest threat is from *P. falciparum*. Next, we have a look at the complex interactions between parasites and their human hosts with an eye on potential lines of control of parasitic diseases. And in the last section, we introduce a mathematical model for malaria. The exercises invite the reader to use the model to explore some epidemiological scenarios for malaria.

## 11.1 Protozoan Parasites Cause Important Human Diseases

*Parasitic protozoa are a major cause of infectious disease worldwide. Parasite infections account for a higher incidence of morbidity and mortality than diseases produced by any other group of organisms. According to the World Health Organization, over* 300 *million people worldwide are affected by malaria alone and between* 1 *and* 1.5 *million people die from it every year.*

*Protozoans are classified according to their mode of locomotion.*

As noted in Chapter 4, *parasitism* describes a symbiotic relationship between two organisms in which one, the *parasite*, is benefited and the other, the *host*, is usually harmed by the interaction. The parasite is *obligate* if it can live only in association with a host. In contrast, *faculative* parasites are free-living organisms that are capable of becoming parasitic under favorable circumstances.

Parasites are typically small organisms, most of these members of several protozoan families and helminth (worm) families. Recall that protozoans are single-celled "animals" whose cell possesses a nucleus. Protozoan parasites, to which we will restrict ourselves here, are therefore quite small, being on the order of 10 to 40 micrometers ($\mu$m) in size; bacteria are typically 1 to 10 $\mu$m. Like viruses and bacteria,

R.W. Shonkwiler and J. Herod, *Mathematical Biology: An Introduction with Maple and Matlab*, Undergraduate Texts in Mathematics, DOI: 10.1007/978-0-387-70984-0_11,

protozoan parasites are called *microparasites*, while those of the helminthic phyla are *macroparasites*. What microparasites have in common is the ability to increase their numbers by a prodigious amount within the host. And if the parasite does live within the body of the host, it is termed an *endoparasite*; an *ectoparasite* lives on the body surface of its host, the tick being an example of the latter.

A primary way of recognizing and differentiating protozoan species is whether they are motile, and if so, by what means. All methods of single-cell locomotion are represented: via pseudopodia, cilia, and flagella. One phylum, Sporozoa, are not motile.

The *amoeboids* move by streaming cytoplasm in the direction of motion, forming a *pseudopod* extension of the cell. The remainder of the cell then pulls itself along in that direction. In addition to movement, this technique is used by the microorganism to feed. In what is called *phagocytosis*, the pseudopodia surround and engulf particles of food. The amoeboids largely constitute the subphylum *Sarcodina*.

The most widespread pathogenic disease caused by this group of organisms is *amebiasis* or *amebic dysentery*, which results from an infection of the protozoan *Entamoeba histolytica*. Infection occurs when cysts on fecally contaminated food or hands are ingested. The cyst is resistant to the gastric environment of the host and passes into the small intestine, where it decysts and multiplies. *E. histolytica* thrives best in oxygen free environments and does not possess mitochondria. The cycle of infection is completed when the organisms encyst for mitosis and are passed out of the gut with feces.

The *ciliates* move by means of *cilia*, which are short, hairlike projections emanating over most of the surface of the organism. As mentioned in Chapter 7, the beating of the cilia is synchronized in order that directed motion result.

The ciliates largely make up the subphylum *Ciliophora* and are the most complex of the protozoans. Ciliates have special organelles for processing food; food vacuoles move along a gullet from which digestible nutrients are absorbed by the cytoplasm. At the end of the gullet, the indigestible residue is eliminated at an anal pore. Ciliates also have two types of nuclei. The *macronucleus* controls metabolism, while the *micronucleus* controls reproduction. Two ciliates may even exchange genetic material in a process called *conjugation*.

Only one ciliophoran, *Balantidium coli*, infects humans. The *trophozoite*, or motile stage, of the organism inhabits the *caecum* and nearby regions of the intestinal tract of its host. This is at the upper end of the colon, where the small intestine empties. *B. coli* is the largest known protozoan parasite of humans, measuring between 50 and 130 $\mu$m. The organism is also unique among protozoan parasites in that it contains two prominent *contractile vacuoles* used for the control of osmosis. Depending on their environment, protozoans can gain water via osmosis. The excess water is forced into the vacuole, which consequently expands in size. At some point, the water is expelled outside the cell through a temporary opening in the plasma membrane, and at this time the vacuole rapidly contracts.

*B. coli* typically reproduce asexually by fission. However, conjugation also occurs in this species.

Transmission of the disease *balantidiosis* from one host to another is accomplished by the cyst stage of the organism. Encystation usually occurs in the large intestine of the host and is expelled with the feces. Decystation occurs in the small intestine of the newly infected host.

The *flagellates* possess one or more long, slender, whiplike protrusions from one end of their bodies used for locomotion. (The word *flagellum* is Latin for whip.) Flagellates belong to the phylum *Zoomastigina* (in some classification systems).

The flagella of these organisms enable them to swim and thus thrive in liquid media, which can include blood, lymph, and cerebrospinal fluid. With an elongate, torpedo-like shape, they are adapted to swim with reduced resistance. A flagellum achieves locomotion by beating in a regular rhythm. A series of bends propagates in wavelike fashion along the length of the flagellum, starting at its attachment point and proceeding to the free end. This movement is fueled by ATP.

The fine structure seen in cross-sections of flagella and cilia are identical; thus flagella may be considered as elongated cilia.

Several clinically important diseases are attributed to the flagellates. Those infecting the intestine or other spaces within the body (lumen) are *Giardia lamblia* and *Trichomonas vaginalis*. The former causes *giardiasis*, which is acquired by the ingestion of cysts, usually from contaminated water. Decystation occurs in the duodenum, and trophozoites colonize the upper small intestine, where they may swim freely or attach to the submucosal epithelium. The free trophozoites encyst as they move downstream, and mitosis takes place during the encystment. The cysts are passed in the stool. Man is the primary host, although beavers, pigs, and monkeys are also infected and serve as reservoirs (see below).

*Trichomoniasis* is the disease caused by *T. vaginalis*. The organism colonizes the vagina of women and the urethra and sometimes prostate of men. Infection occurs primarily via sexual contact, although nonvenereal infections are possible. The organism divides by binary fission, which is favored under low acidity (pH $> 5.9$; normal pH is 3.5–4.5). *T. vaginalis* does not encyst.

Those diseases in which flagellates inhabits the blood are *African trypanosomiasis* (sleeping sickness) and *leishmaniasis*. Sleeping sickness is caused by *Trypanosoma brucei gambiense*, mainly affecting Western and Central Africa, and *Trypanosoma brucei rhodesiense*, which is restricted to the Eastern third of the continent. *Tryponosoma* are the first protozoans we have discussed whose life cycle takes place in more than one host, with the consequential problem of transmission between hosts. Actually, this sort of life cycle is the rule rather than the exception, and we digress for a moment to discuss some of its general characteristics.

The host harboring the adult or reproductive stage of the parasite is called the *definitive host*. An *intermediate host* serves as a temporary but essential environment for the development or metamorphosis of the parasite short of sexual maturity. A *vector* is an organism that transmits the parasite to the definitive host. This is usually an intermediate host. Infected animals that serve as sources of the parasite for later transmission to humans are *reservoir hosts*. The reservoir host shares the same stage of the parasite as humans but is often better able to tolerate the infection.

Returning to trypanosomiasis, mammals, including cattle and humans, are the definitive hosts for the organism. Thus from a human being's point of view, cattle are reservoir hosts. The intermediate host and vector is the *tsetse fly*. Humans acquire the parasite when an infected fly takes a blood meal. The infective stage of the parasite makes its way to the salivary gland of the fly. As the fly bites, saliva serves to dilate the blood vessels and to prevent coagulation of the blood. At the same time, the parasite is transmitted with the saliva. Conversely, an infected human transmits a stage of the organism to any biting fly within the extracted blood.

*Leishmaniasis* is caused by any of the three subspecies *Leishmania donovani*, *L. tropica*, and *L. braziliensis*. These organisms are also multihost parasites, whose intermediate host and vector is the *sandfly*. Transmission of the parasite to the host is unique. The parasites, many hundreds of them, reside in the gut of the fly and are deposited in the skin of the victim when the sandfly feeds. Macrophages of the host quickly engulf the intruders. But remarkably, macrophages are the target cells of infection by the parasite! The invading parasite encases itself in a vacuole within the macrophage, where it lives and reproduces. Later, a biting sandfly will be infected via the blood of its meal.

The last group of protozoan parasites, the *sporozoans*, are essentially immotile, and every species of the group is parasitic. In humans, they give rise to such diseases as *malaria*, *babesiosis*, *toxoplasmosis*, and *Pneumocystis carinii pneumonia* (PCP), among others. While most members of the group are multihost, the agent causing PCP, *Pneumocystis carinii*, is an exception. In place of locomotor devices, members of this group possess structures known as the *apical complex*. This is a collection of readily identifiable organelles (microscopically) located beneath the plasma membrane at the anterior end of the organism. It is thought that the function of the complex is to secrete proteins facilitating the parasite's incorporation into the host cell.

Malaria is caused by four members of the genus *Plasmodium*: *P. vivax*, *P. falciparum*, *P. malariae*, and *P. ovale*. Malaria is one of the most prevalent and debilitating diseases afflicting humans. The World Health Organization estimates that each year 300–500 million cases of malaria occur worldwide and more than two million people die of malaria. *P. falciparum* (malignant tertian malaria) and *P. malariae* (quartan malaria) are the most common species of malarial parasite and are found in Asia and Africa. *P. vivax* (benign tertian malaria) predominates in Latin America, India, and Pakistan, whereas *P. ovale* (ovale tertian malaria) is almost exclusively found in Africa. The vector for malaria is the female mosquito of the genus *Anopheles*. The life cycle of the parasite proceeds through several stages, each of which is attended to by a different form of the organism. In the next section we will examine the life cycle in detail.

*Babesiosis* is a disease of recent appearance. It is caused by the protozoan *Babesia microti*, which is a natural parasite of the meadow vole and other rodents. The intermediate host and vector of the disease is the deer tick, *Ixodes dammini*. As is familiar by now, the parasite is injected into the definitive host through the saliva of an infected tick when it takes a blood meal. In the host, the organism lodges in erythrocytes (red blood cells), where it multiplies. In turn, the tick acquires the parasite by ingesting the blood of an infected host. Since the mortality rate for ticks is high, it is likely that an infected tick will not live to pass on the parasite. Therefore, the parasite has adapted to colonize the offspring of an infected female tick, thereby greatly increasing its chance of transmission to a definitive host. This is achieved when a form of the parasite invades the ovarian tissues of the tick. As a result, the newly hatched eggs are already infected.

*Toxoplasmosis* has worldwide distribution and a surprisingly high infection rate. In the United States, 50% of the population is seropositive to the causative organism, *Toxoplasma gondii*, meaning that they have been immunologically exposed to it. Normally infection is asymptomatic, but it does pose a serious threat in immunosuppressed individuals and pregnant females. As we will shortly see, cats are the definitive host. For this reason, women should avoid contact with litterbox filler during a pregnancy. With the spread of AIDS, toxoplasmosis has become much more serious.

The principal means for acquiring toxoplasmosis is by eating inadequately cooked meat or by contact with feral or domestic cats. A form of the organism, called an *oocyst*, is passed in large numbers with the feces of an infected cat to the soil. There they may be ingested by cattle, sheep, pigs, rodents, or even humans directly. The cycle of infection is completed when a cat eats an infected animal such as a mouse. In the intermediate host, the organism develops into a form called *pseudocysts* that may persist for years in muscle and especially nerve tissue. Eating undercooked meat containing pseudocysts can also result in infection.

The disease PCP has vaulted into prominence coincident with the increasing incidence of AIDS. Of the opportunistic diseases associated with AIDS, PCP is the most common cause of death, affecting an estimated 60% of AIDS patients in the United States. The disease is caused by *Pneumocystis carinii*, which is an extracellular

**Table 11.1.1.** Parasite summary.

| Disease | Parasite | Host multiplicity | Infection path | Infection site |
|---|---|---|---|---|
| amebiasis | *E. histolytica* | unihost | contaminated food | intestine |
| balantidiosis | *B. coli* | unihost | contaminated food | intestine |
| giardiasis | *G. lamblia* | unihost | contaminated water | intestine |
| trichomoniasis | *T. vaginalis* | unihost | sexually transmitted | sex organs |
| trypanosomiasis | *T. brucei gambiense*, *T. brucei rhodesiense* | multihost | vector (tsetse fly) | blood |
| leishmaniasis | *L. donovani/tropica*, *L. braziliensis* | multihost | vector (sandfly) | blood |
| malaria | *P. vivax/falciparum*, *P. malariae/ovale* | multihost | vector (mosquito) | blood |
| babesiosis | *B. microti* | multihost | vector (tick) | blood |
| toxoplasmosis | *T. gondii* | multihost | contaminated soil | nerve tissue |
| PCP | *P. carinii* | unihost | aerosol | lungs |

parasite found in the interstitial tissues of the lungs and within the alveoli. The mode of transmission from one human to another is thought to be via inhalation of cysts from the air.

We summarize the above in Table 11.1.1.

## 11.2 The Life Cycle of the Malaria Parasite

*The life cycle of* P. falciparum (*see Figure* 11.2.1) *is typical of that of many protozoan parasites. As we will see, it is quite complicated, much too complicated to be supported by the relatively meager genome of a virus or bacterium.*

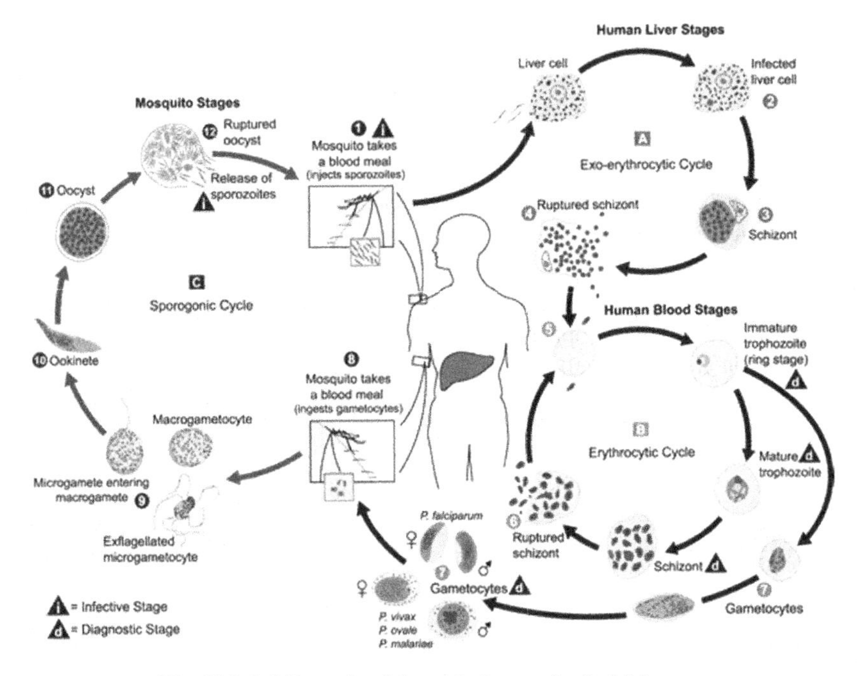


**Fig. 11.2.1.** Life cycle of the malaria parasite *P. falciparum*.

*Of course, the complication of their life cycle confers survival advantages. For example, if some natural disaster befell mosquitoes, the intermediate host, the pool of human infectees, would serve as a reservoir for the restoration of the parasite. A simple life cycle, such as that of the variola virus, which is responsible for smallpox, is vulnerable to eradication.*

*Although the need for living in an intermediate host does present the opportunity to attack the parasite at that source, eliminating the mosquito has proved impossible. In like fashion, with many stages to their life cycle, there are just as many opportunities for a drug against the organism, but in reality these stages have come to exist as a way for the parasite to thwart the host's immune system. The time the organism spends floating free in the bloodstream, where an opportunity for a vaccine exists, is brief.*

*Plasmodia have evolved a complicated life cycle with many specialized forms.*

Infection begins in humans with the bite of an infected female *Anopheles* mosquito. The male *Anopheles* lacks the mouth parts for penetrating human skin and instead feeds on plant juices. The parasite is in a slender, elongated form between 10 and 55 $\mu$m in length called a *sporozoite*. This form lasts only about one hour in the circulating blood. The parasite makes its way to the liver, where it enters a parenchymal (functional) cell of the liver. Here the sporozoite undergoes development into the *exoerythrocytic schizont* form of the organism and feeds on the cell's cytoplasm. The sporozoites of *P. vivax* and *P. ovale* are capable of becoming *hypnozoites*. This form remains dormant in the *hepatocyte (liver cell)* for anywhere between several months and up to four years. Upon emergence from their hibernation, the normal life cycle is resumed. This is the first of the parasite's two asexual reproduction cycles in man. The second is within a blood cell.

After one to two weeks and several cell divisions, each schizont ruptures, giving rise to thousands of *merozoites*, which enter the bloodstream. Merozoites are about 2.5 $\mu$m in length. In the bloodstream, they invade red blood cells, initiating the *erythrocytic schizogonic phase* of the infection. This is the stage responsible for the symptoms of fevers and chills experienced by victims of malaria. The attack upon red blood cells is mediated by surface antigens. For example, *P. vivax* requires the red blood cell surface to present the so-called *Duffy* blood group antigen for recognition. However, nearly all West Africans lack this antigen and are resistant to vivax malaria.

Inside the erythrocyte, the merozoite grows to the *early trophozoite*, or ring stage. This stage feeds on hemoglobin and develops into the mature, or *late trophozoite*, stage. These trophozoites reproduce by multiple fission events into schizonts. And once again, schizonts produce a new generation of merozoites, each of which is capable of infecting a new erythrocyte.

However, these second-generation merozoites may transform into the gametocyte stage of the parasite instead; *microgametocytes* are the male gametocytes, and *macrogametocytes* are female. Despite their names, the two are approximately the same size. Up to this point in the life cycle, reproduction has been asexual, but that will change in the next stage.

The sexual phase of development takes place in the mosquito. Gametocytes are ingested by the mosquito with a blood meal and conveyed to its gut. The gametocytes are unaffected by the insect's digestive juices and are, in fact, released from the erythrocyte as it is digested. Then the microgametocyte undergoes a maturation process, accompanied by cell division, called *exflagellation*, in which six to eight *microgametes* are formed. Each is equipped with flagella, with which microgametes seek out their counterpart, *macrogametes*. Of course, macrogametes have derived from macrogametocytes according to a separate development. The gametes fuse, giving rise to a diploid zygote.

Within 12 to 24 hours, the zygote elongates into a motile wormlike form known as an *ookinete*. Ookinetes penetrate the wall of the gut, where the next phase of development takes place. It changes to a round shape and further develops into a form called an *oocyst*. It is from the oocysts that the sporozoites will arise.

To begin this last phase of the life cycle, an oocyst grows four to five times its original size due to internal cell divisions. The cells division are by meiosis (sexual cell divisions), and the proliferating haploid cells are referred to as *sporoblasts*. Sporoblasts undergo numerous cell divisions themselves, producing thousands of sporozoites. Finally, within 10 to 24 days after the mosquito's blood meal, the sporozoite-filled oocysts rupture, releasing the sporozoites, which make their way to the ducts of the insects' salivary gland. The parasite is now ready to be injected into a new host.

A noteworthy postscript to the story of malaria infection is the resistance to the disease brought about by the *sickle-cell trait*. The gene responsible for it contains the substitution of a valine amino acid in place of glutamic acid in the $\beta$ chain of the hemoglobin molecule. Sickle-cell is invariably fatal to those homozygous for the trait, but in the heterozygous state it confers substantial protection against falciparum malaria. Red blood cells infected with the parasite are subject to low oxygen tensions and potassium leakage during the cell's passage through the capillaries. This kills the parasite.

## 11.3 Host–Parasite Interactions

*In order for an endoparasite to succeed, it must overcome two major obstacles: getting into the host's body and defeating the host's immune system. The host, of course, reacts to the invasion. There ensues a struggle between the host's immune system and the parasite's defenses. Over the ages, the parasite has won. It continues to exist and conduct its life cycle within an internal universe of one or more hosts. Further, it has evolved so as to keep its host alive, albeit debilitated, and thereby ensure itself a long reproductive life.*

*Gaining entrance to the host.*

Unihost parasites gain entrance to their host principally by utilizing an existing orifice. This admits the organism to the gut, the lungs, or the sex organs. This is also usually the site where they carry on their life cycle. From Table 11.1.1, we see that the normal mode of transmission of a parasite inhabiting the gut is by way of a cyst.

While multihost parasites can do the same, they can also exploit an alternative mechanism that admits them directly to the bloodstream. This is by hitching a ride on an organism, the vector, having the capability and practice of penetrating the host's body directly. The most convenient way to be at the right place at the right time is by living within or on such a vector, hence the evolution of multihost parasitism.

Of course, now the multihost parasite must also breech the vector's body. However, this is a solved problem: The simplest and most convenient mechanism is just to go back the way they came in, at the moment the vector is again feeding on the host. And so we have a tidy, closed cycle waiting to be exploited. As we have seen, multi-host parasites have many morphological and physiological adaptations to accomplish their life cycle in multiple hosts.

*Immunological response and counterresponse.*

From the moment a parasite breeches its host's body, it comes under immunological attack. The parasite must counter or be eliminated; as a result, a kind of arms race ensues in which each side attempts to overcome the defense of the other. Much is known about this struggle, especially with respect to malaria and several of the diseases we have examined above.

The body's defenses include antibodies, cytotoxic cells, lysosomal enzymes, toxic metabolites, and predatory phagocytes. Some of the principal players in mediating these defenses are immunoglobins (IgA, IgD, IgE, IgG, and IgM), lymphocytes, CD4+ helper T-cells, CD8+ cytotoxic T-cells, cytokines, macrophages, and granulocytes (see Section 10.2).

The role of antibodies is mainly confined to stages circulating in the bloodstream. Antibodies attack the merozoite stage of the malaria parasite. In addition, part of the resistance of adult humans against malaria in endemic regions is their antibody attack on the sporozoite stage. In another example, although *Giardia lamblia* is confined to portions of the small intestine, the antibodies IgA and IgM help control giardia infection in the submucosal epithelium.

For their part, plasma-based parasites respond by forging complicated life cycles alternating between the bloodstream and an intracellular subsistence. The time spent in the bloodstream is limited and they do so in various forms, thereby presenting differing surface antigens. Some species are even able to present a surface mimicking that of their host and thus be mistaken as "self" to the host's immunological system. This is called *molecular mimicry*. The parasite produces hostlike molecules on its body surface or its surface becomes covered with host molecules themselves. As a result, the host is duped into accepting the parasite as self.

However, a most remarkable feat is the ability of some species to present variable surface antigens. This has been observed in trypanosomes. If single parasite cells are cloned from different infected animals or patients, the surface coat is biochemically different—not just a bit different, but so different that the coat protein must come from the expression of different genes by trypanosomes in each animal. Moreover, if cells are taken from a defined wave of parasitemia in the same patient, it is found that all of the trypanosomes in that wave of organisms are expressing the same single-surface antigen, whereas in other waves, all of the parasites are expressing a single but completely different antigen. The organism is presenting variable surface antigens or variable surface glycoproteins (VSGs) in each wave.

In the laboratory, it has been observed that no antigen has been repeated even after hundreds of these waves. It follows that there must also be an equal number of VSG genes. In fact, there are probably 1000–2000 such genes. Thus 10% of the cell's genome is devoted to genes that express these surface molecules allowing the organism to be one step ahead of the host's immune response.

As the host mounts an attack to each new wave of trypanosome parasitemia, eventually the host's lymphoid organs become depleted of lymphocytes, and immunodepression sets in. As we have seen, this is a technique also exploited by the HIV virus.

Another tactic used by parasites to avoid destruction is the adaptation to leave the bloodstream and take up residence inside a cell of the host. As a bonus, the cell's cytoplasm can be used for nutrition as well! Although an intracellular habitat shields the invader from antibodies, it does invoke a new form of immunological attack.

An infected cell of the host will react to the presence of an intruder. First, an attempt is made to lyse, or break up, the intruder by the action of the cell's lysosomes. Lysosomes are the cells' garbage disposal system. A section of rough endoplasmic reticulum wraps itself around the intruder and forms a vacuole. Then vesicles containing lysosomal enzymes fuse with it. The pH becomes more acidic and this activates the enzymes that break up the contents. Lysosomes also degrade worn-out organelles such as mitochondria.

As expected, parasites have evolved to counter this threat. One means is by encapsulation within the cell. Failing to rid itself of the parasite in this way, the cell consigns itself to suicide. With the help of the endoplasmic reticulum, the cell presents antigens of the intruder on its own surface. T-cells respond to such antigens and mature into cytotoxic killers, combining with and lysing any cell expressing these antigens (cf. Section 10.2).

With respect to malaria, cytotoxic defense is mounted against parasites invading liver cells, the exoerythrocytic schizonts, but obviously not against those in red blood cells (erythrocytic schizonts), since the latter are not living cells.

Despite its attempts to do so, the body is not able to totally eliminate intracellular parasites; the most it can hope for is to keep them in check.

So why is the immune response not more effective at combating parasites? Some of the explanations that have been proposed are the following:

- Parasites show considerable antigenic diversity and variation—between species, between strains, between stages, and during the course of an infection.
- Parasites avoid immunity by hiding inside cells.
- Infection stimulates T-cell mediated immunity, but there is little T-cell memory.
- Parasites misdirect or suppress the immune response.

*Incidence and control of parasitic diseases.*

*Epidemiology* is the study of the factors responsible for the transmission and distribution of disease. Contagious diseases are those that are transmitted from person to person directly. The life cycle of the organisms responsible for such diseases can be quite simple. Indeed, contagious diseases are often caused by viruses or bacteria. These perpetrators are frequently highly virulent, resulting in acute diseases that are deadly. Of course, pathogens themselves die with the patient, but by then the disease has already been passed on, ensuring the survival of their genes.

*Common-source* diseases are those that are not transmitted directly but instead by some third party, or common source. The common source need not be another organism—for example, it can be the soil—but nevertheless it often is an organism. Utilizing a third party already requires a more complicated life cycle because the agent must survive in at least two different environments. As we have seen, parasitic diseases are usually common source, since they are acquired via a vector or some contaminated inanimate material. (PCP is an exception in that it can be caught from an infected person directly. Even in this case the organism must survive for a time outside the host.) Thus it is not surprising that there is a large overlap between higher organisms, protozoans and helminths, and parasitism. These organisms, with their more extensive genome, can accommodate complicated life cycles.

Parasitic diseases are usually chronic. If death comes, it is usually after a lengthy period of debilitation. Of course, this works to the favor of the parasite; it is not advantageous to kill its host. This suggests that parasitic disease have been around for some time and have evolved to prolong the life of their host as long as possible.

Especially as a result of the differing mechanisms for gaining entrance to the host, certain environments are conducive to the differing parasitic life cycles. The tropics and subtropics favor multihost parasitism among humans. Factors for this are the warm weather, allowing an insect vector to be active all year around; abundant bodies of stagnant water; more diverse ecosystems supporting greater numbers of species; and the fact that human hosts generally present more bare skin and sleep exposed.

The host's behavior can also be a factor in the incidence of parasitic diseases. Unsanitary conditions and inadequate cooking are major contributors to disease. The same can be said for certain social and ethnic customs such as communal bathing and the ritual consumption of undercooked meat.

It might seem that the multitude of forms and attendant multitude of proteins for bringing them about constitutes many opportunities for the control and even eradication of a given parasite species. And indeed, this is the hope and design of modern research efforts. For example, in the case of multihost parasites, their vector can be attacked and thereby the chain of infection broken. Many efforts have been directed at the *Anopheles* mosquito for the control of malaria. These efforts have been only partially successful. Between the 1940s and 1960s, malaria eradication was achieved in the USA, the USSR, Southern Europe, and most Caribbean Islands mainly by vector control. Much progress was also made in the Indian subcontinent and parts of South America.

It even appeared for a time that eradication would also be successful in major problem areas such as Nigeria. But ultimately vector eradication failed and malaria vengefully resurged; see Section 11.4. The reason is that in endemic regions, where transmission is high, people are continuously infected, so that they gradually develop a degree of immunity to the disease. Until they have acquired such immunity, children remain highly vulnerable. Initially, the eradication program greatly reduced the mosquito population and incidence of malaria was low for several years. People lost immunity, and a cohort of children grew up with no immunity. When the control measures failed and were discontinued, widespread incidence ensued.

The problem with mosquito control is that it is very hard to do. There are, in effect, an infinite number of breeding sites for them. In rural areas in the wet tropics, *Anopheles* may breed in every water-filled foot- and hoofprint, and larval control is an almost hopeless undertaking. Equally important, insect vectors can eventually gain resistance to chemical treatments.

Chemical eradication programs can have a very damaging effect on the environment, generally because of their broad-based activity. The insecticide most widely used for house spraying aimed at the adult *Anopheles* mosquito has been DDT. DDT has continued to be recommended for this purpose long after it was banned for agricultural use in the USA and many other countries. It is recommended because of its cheapness per unit weight and its durability, which allows programs to be based on spraying twice a year, or only once in areas with a short annual malaria mosquito season. However, unfortunately in low-income countries, it is almost impossible to prevent illicit diversion of insecticides intended for antimalaria use to farmers. The consequent insecticidal residues in crops at levels unacceptable for the export trade have been an important factor in recent bans of DDT for malaria control in several tropical countries.

One can also attempt to control the parasite directly either by a preinfection drug poised to kill the parasite immediately upon its entrance or by a postinfection one designed to work after infection. With respect to malaria, some preinfection drugs are Proguanil, chloroquine, mefloquine, doxycycline, and malarone. Chloroquine is also a postinfection remedy. Arteminisin is a postinfection drug effective against drug resistant *P. falciparum* infections.

The problem with attacking the parasite itself is that protozoa tend to develop resistance to drugs even faster than insects develop resistance to sprays. Chloroquine was hugely successful in combating malaria when launched in the 1950s, but the malaria parasite gradually became resistant. *Plasmodium falciparum* has proved extraordinarily adept at evolving to combat many of the drugs currently on the market.

Still, the multiplicity of forms of the organism does provide targets. Understanding the details of each form, including for example metabolic pathways, offers the possibility of drugs specific to the parasite. We will encounter a success of this very kind in the section on genomic medicine, Section 14.4.

## 11.4 Mathematical Models for Parasitic Diseases

*In Section* 11.2, *we studied the life cycle of the parasites causing malaria. Now we will take that information into account and formulate a mathematical model for the incidence of malaria. The model divides the population into three groups, or compartments: those susceptible to the disease, those infected with it, and those in some form of recovery.*

*We analyze the basic model and two refinements, time-dependent immunity and drug-resistant parasites. These models are due in part to Professor Sylvanus Aneke.*

*SIRS is a compartmental differential equations system.*

The most widely used model for studying malaria is a modified SIRS system of differential equations. This means that the population is assumed to consist of three groups: Susceptibles ($S$) are those at risk of contracting the disease; infectives ($I$) are those who have the disease and are spreading it; and the removed ($R$) are those recovered with immunity. The population moves from $S$ to $I$ to $R$ and back to $S$. The first main modification for malaria is that the susceptibles do not catch the disease from the infectives directly, and the removed are only partially removed, as we will explain. In the SIRS model for contagious diseases, there is a term of the form $-\alpha SI$ in the equation for $\frac{dS}{dt}$. This is a mass action term such as we have seen in our study of the Lotka–Volterra equation in Section 4.4, which provides for susceptibles getting sick from infectives. But there is no such term in a common source disease.

Instead, it is assumed that the susceptibles become infected at a rate in proportion to their numbers, $-hS$, for an infection rate parameter $h$. This choice is influenced by the fact that the feedback dynamics from mosquito to man and back to mosquito involve considerable delay, mostly due to the incubation periods of the several forms of the parasite. Also, in certain cases, it has been observed that the incidence of infected mosquitoes remains very close to 3 per cent under widely varying circumstances, thus constituting an approximately steady threat [2, 3].

Secondly, by the nature of malaria, a large fraction of those who recover from the disease are, in fact, only partially recovered. What we are calling infectives are those with severe symptoms. Most who overcome this state are still infected, having milder symptoms and being capable of passing gametocytes to mosquitoes. So the "removed" group here refers to this group, those partially recovered.

The movement between groups is shown in Figure 11.4.1 along with the rates, and the system model looks like this:

$$\begin{aligned}\frac{dx}{dt} &= -hx + \rho y + \beta(h)z,\\ \frac{dy}{dt} &= hx - \rho y - ry,\\ \frac{dz}{dt} &= ry - \beta(h)z.\end{aligned} \tag{11.4.1}$$

This is a compartmental model; those leaving one group do so by entering another. The behavior of such systems is well known; solutions tend to a unique globally asymptotically stable stationary point.

The model tracks the experience of a birth cohort, i.e., a group of people of about the same age, moving through time with $t$ representing their age. The variables $x$, $y$, and $z$, all functions of $t$, denote, respectively, the relative size of the susceptible, the infective, and the partially recovered groups. Since these are relative sizes, their sum is 1 or 100%:

$$x(t) + y(t) + z(t) = 1, \quad t \geq 0. \tag{11.4.2}$$

Implicit in this statement is that mortality acts approximately equally on all groups. Initially, $x = 1$ and $y = z = 0$.

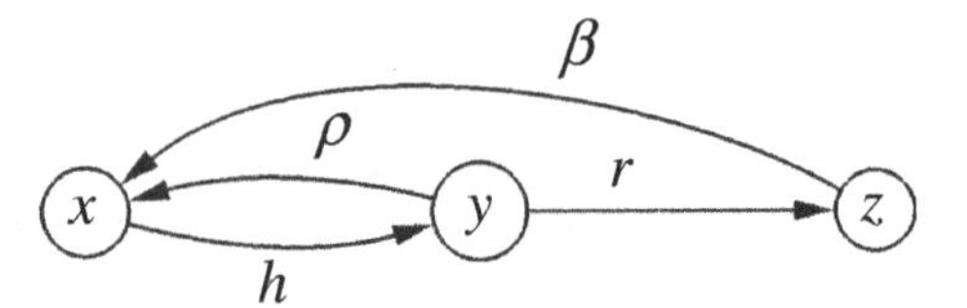


**Fig. 11.4.1.** Basic Aron–May malaria model.

As above, $h$ is the infection or "happening" rate. The parameter $r$ is the rate of (partial) recovery, that is, the transition rate from group $y$ to group $z$. Hence $\frac{1}{h}$ is the mean time until infection, and $\frac{1}{r}$ the mean time until recovery. The model postulates that the mildly symptomatic can arise only from the severely symptomatic. The terms $\rho y$ and $\beta(h)z$ are due to Aron and May [1, 2]. The *recovery rate* $\rho$ corresponds to how quickly parasites are cleared from the body. High recovery rates are associated with drug treatments.

*Note.* Greek symbols are not part of MAPLE or MATLAB, and so the parameters $\rho$ and $\beta$ will be coded as `rho` and `beta`, respectively, in the computer codes.

The term $\beta(h)y$ allows for a return path from the partially recovered back to the susceptibles. Furthermore, $\beta$ is taken as a function of $h$ in deference to the observation by many that the greater the endemicity of the disease, the greater the extent of immunity among the population. Thus $\beta$ should decrease with increasing $h$. The exact relationship is in terms of a parameter $\tau$ and takes the form

$$\beta(h) = \frac{he^{-h\tau}}{1 - e^{-h\tau}}, \tag{11.4.3}$$

where $\tau$ is mean duration of partial recovery; see Figure 11.4.2. If a person is reexposed before this time has elapsed, then another interval of duration $\tau$ without exposure is required before return to the susceptible group (probabilistically).

As a test of their models, and as a source of parameter estimation, mathematical epidemiologists often use real-life data. The Garki Project data and the Wilson data are well known and useful for these purposes.

In 1980, the World Health Organization (WHO) attempted to determine whether intensive spraying could eradicate malaria in and around Garki, Nigeria. As part of the project, WHO coordinated mass drug administration in 164 villages in addition to the spraying. The Garki Project had an enormous impact on the mosquito population in that area, reducing the biting rate of mosquitoes by 90%. But despite this dramatic decline, the prevalence of the malaria parasite among villagers did not significantly change. The vectorial capacity of the surviving mosquitoes was simply too high to be overcome by these extensive measures.

Figure 11.4.3 is a result of the Garki Project, showing that prevalence of the disease decreased at first but then rose to higher levels than controls upon cessation of treatment. That is, the prevalence curves crossed each other. This phenomenon is the motivation for the aforementioned infection-dependent recovery rate $\beta(h)$.

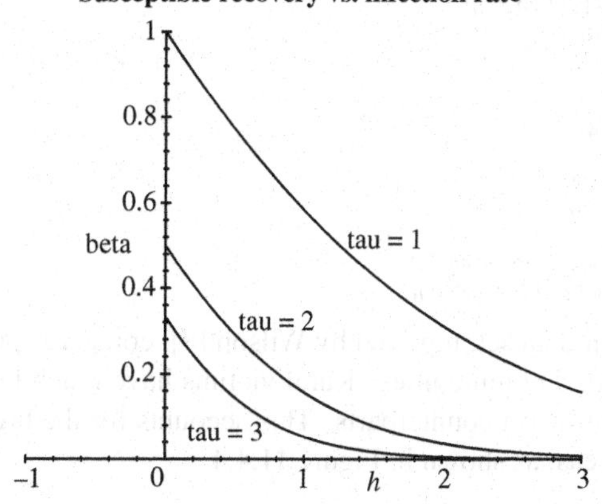


**Fig. 11.4.2.** Recovery rate $\beta$ vs. infection rate $h$ for various $\tau$.

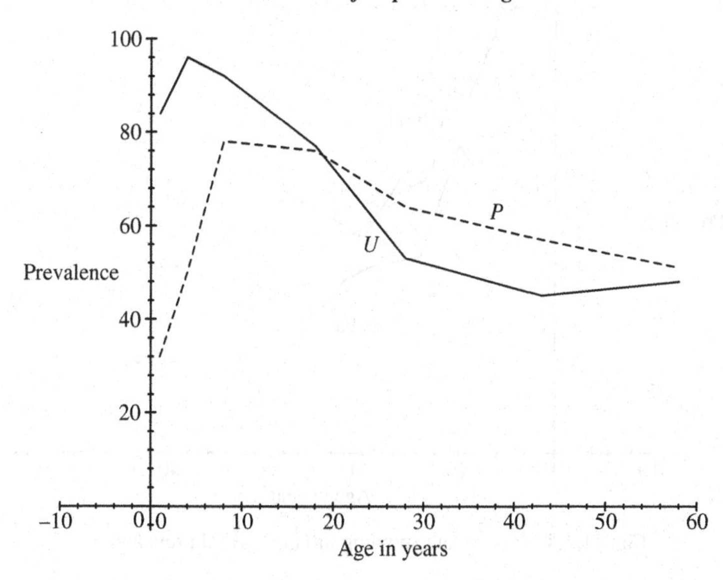


**Fig. 11.4.3.** Garki Project data. $U$ refers to the controls, $P$ to the treated.

The program for the Garki figure follows. This will be useful for comparing predicted incidence with actual prevalence:

```
MAPLE
> with(plots):
> controls:=[[1,84], [4,96], [8,92], [18,77], [28,53], [43,45], [58,48]];
> cPlot:=plot(controls,-10..60,-5..100,labels=['Age in years','Prevalence'],
        title='Prevalence of P. falciparum vs. age',titlefont=[COURIER,BOLD,16]):
> treated:=[[1,32], [4,50], [8,78], [18,76],[28,64], [43,57], [58,51]];
> tPlot:=plot(treated,linestyle=3):
```

```
> text:=textplot({[24,56,'U'],[38,63,'P']}):
> display({cPlot,tPlot,text});

MATLAB
> cx=[1,4,8,18,28,43,58];
> cy=[84,96,92,77,53,45,48];
> tx=[1,4,8,18,28,43,58];
> ty=[32,50,78,76,64,57,51];
> plot(cx,cy,tx,ty,[0,60],[0,0])
> axis([0,60,0,100])
> title('Prevalence of P. falciparum vs. age');
> xlabel('Age in years');ylabel('Prevalence');
```

Another relevant data set, reported by Wilson [8], compares prevalence in urban areas with that in rural communities. Rural victims have much less access to drugs as compared to their urban counterparts. This accounts for the higher prevalence of infection in rural areas, as shown in Figure 11.4.4.

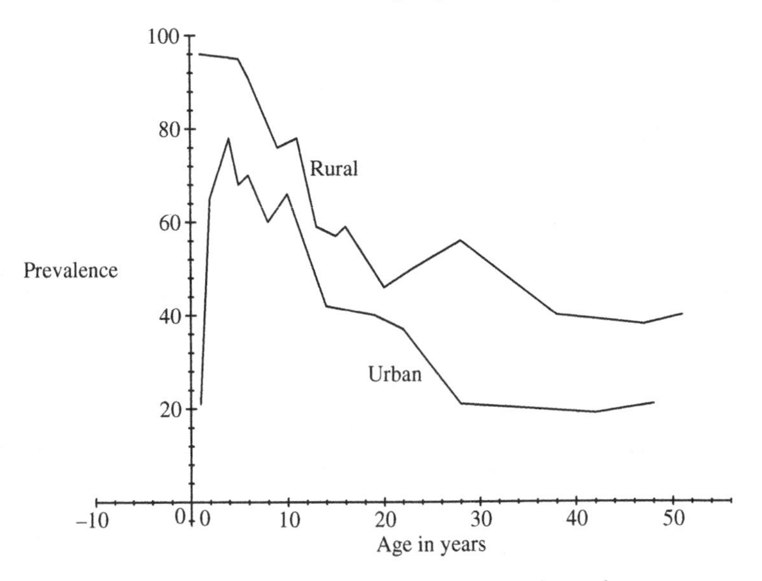


**Fig. 11.4.4.** Wilson data showing urban vs. rural prevalence.

Note the strong similarity between the untreated group in the Garki Project data and the rural group of the Wilson data.

We begin the mathematical analysis by finding the stationary point of the system. Setting the derivatives to zero gives

$$
\begin{aligned}
0 &= -hx + \rho y + \beta(h)z, \\
0 &= hx - \rho y - ry, \\
0 &= ry - \beta(h)z.
\end{aligned}
$$

Also take normalization, (11.4.2), into account. In MAPLE, we have the following:

```
MAPLE (symbolic, so no MATLAB equivalent)
> eSys:={-h*x+rho*y+beta*z=0, h*x-rho*y-r*y=0, r*y-beta*z=0, x+y+z=1};
> solve(eSys,{x,y,z});
```

So

$$y = \frac{1}{\left(\frac{\rho+r}{h} + 1 + \frac{r}{\beta}\right)}$$

and

$$x = \frac{\frac{\rho+r}{h}}{\left(\frac{\rho+r}{h} + 1 + \frac{r}{\beta}\right)}, \qquad z = \frac{\frac{r}{\beta}}{\left(\frac{\rho+r}{h} + 1 + \frac{r}{\beta}\right)}. \tag{11.4.4}$$

Next, we explore the predicted incidence itself by solving the Aron–May system (11.4.1). The following computer programs do this for three sets of parameter values.

**Code 11.4.1.**

```
MAPLE
> restart:
> sys:=diff(x(t),t)=-h*x(t)+rho*y(t)+b*z(t),
        diff(y(t),t)=h*x(t) - rho*y(t) - r*y(t),
        diff(z(t),t)=r*y(t) - b*z(t);
> sol:=dsolve({sys,x(0)=1,y(0)=0,z(0)=0},{x(t),y(t),z(t)}):
> ysol1:=unapply(subs(sol,y(t)),(h,rho,r,b,t)):
> tau:=2; rho:=1/6; r:=1/8;
> beta:=h->h*exp(-h*tau)/(1-exp(-h*tau));
> plot([subs(h=5,ysol1(h,rho,r,beta(h),t)),subs(h=5/10,ysol1(h,rho,r,beta(h),t)),
       subs(h=5/1000,ysol1(h,rho,r,beta(h),t))], t=0..50,0..1, color=[red,green,blue]);

MATLAB
% make up an m-file, beta.m, with
%   function b=beta(h,tau);
%   b=h*exp(-h*tau)/(1-exp(-h*tau));
% make up an m-file, parasConst.m, with
%   function parasParms = parasConst(t);
%   h=5; tau=2; b=beta(h,tau);
%   rho=.167; r=0.125; v=0; u=0; % v not used yet
%   parasParms=[h,rho,b,r,v,u];
% make up an m-file, aronmay.m, with
%   function SIRSprime=aronmay(t,X);
%   %X(1)=x, X(2)=y, X(3)=z
%   params=parasConst(t);
%   h=params(1); rho=params(2);b=params(3);
%   r=params(4);v=params(5);u=params(6);
%   SIRSprime=[-h*X(1)+rho*X(2)+b*X(3); h*X(1) - rho*X(2) - r*X(2); r*X(2) - b*X(3)];
> [t,X]=ode23('aronmay',[0 50],[1;0;0]);
> plot(t,X(:,2))
> hold on;
% change h=5 to h=0.5 in parasConst.m
> [t,X]=ode23('aronmay',[0 50],[1;0;0]);
> plot(t,X(:,2))
% change h=0.5 to h=0.005 in parasConst.m
> [t,X]=ode23('aronmay',[0 50],[1;0;0]);
> plot(t,X(:,2))
```

The results are plotted in Figure 11.4.5. Note that this model shows the crossover effect. In the exercises, we explore the possibilities further.

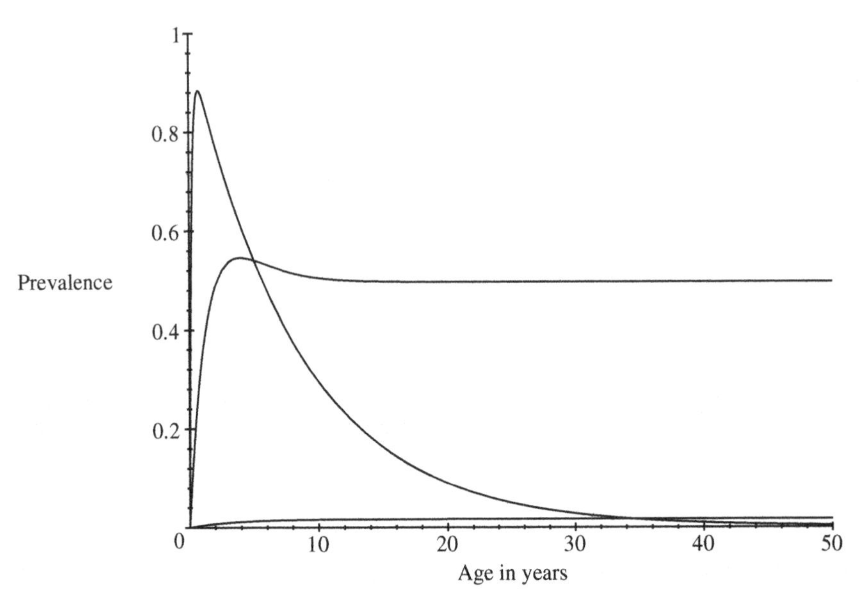


**Fig. 11.4.5.** Predicted prevalence of a cohort as a function of age for three sets of parameter values.

*Time-dependent immunity (TDI) model.*

Understanding that immunity is acquired and develops over time with exposure can be taken into account using a time-dependent immunity acquisition rate, $r = r(t)$. The assumptions are that immunity is initially nil, $r(0) = 0$; that upon exposure, there is a startup delay in acquiring immunity, $\dot{r}(0) = 0$; and that immunity tends asymptotically to a limiting value, say, $R$. These principles are captured in the simple differential equation

$$\frac{dr}{dt} = (\text{rate}) \cdot t \cdot (R - r), \quad r(0) = 0,$$

for some rate parameter. Let $v$, called *exposure*, denote twice the rate parameter. Then solving the differential equation gives

$$r = R(1 - e^{-vt^2}). \tag{11.4.5}$$

A plot of $r$ vs. $t$ for various $v$ is shown in Figure 11.4.6 using the following code:

```
MAPLE
> restart: with(plots):
> rsol:=dsolve({diff(r(t),t)=2*v*t*(R-r(t)),r(0)=0},r(t));
> r:=unapply(subs(rsol,r(t)),(v,t));
> R:=1;
> plot([r(1/1000,t),r(2/1000,t),r(3/1000,t)],t=0..50,view=[-10..50,-0.2..1],color=[blue,green,red]);
```

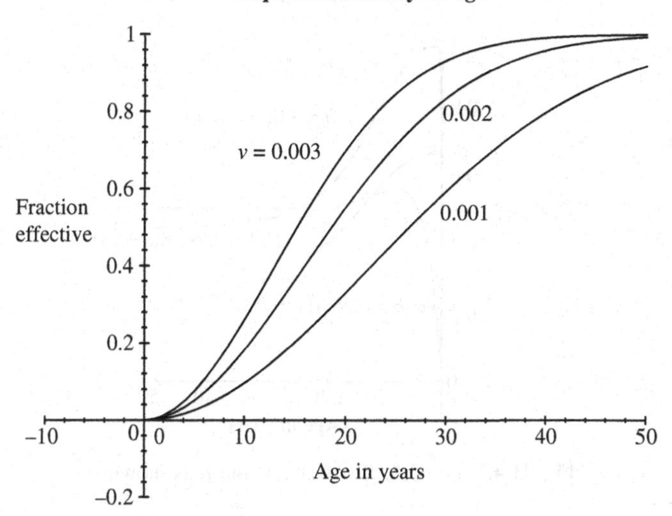


**Fig. 11.4.6.** Acquired immunity profile.

```
MATLAB
% make up an m-file, tdifcn.m, as
%  function r=tdifcn(R,v,t);
%  r=R*(1-exp(-v*t.^2));
> R=1;t=0:.1:50;
> v=.001; y1=tdifcn(R,v,t);
> v=.002; y2=tdifcn(R,v,t);
> v=.003; y3=tdifcn(R,v,t);
> plot(t,y1,t,y2,t,y3)
```

With this modification to system (11.4.1), we have the time-dependent immunity (TDI) model:

$$\begin{aligned}\frac{dx}{dt} &= -hx + \rho y + \beta(h)z,\\ \frac{dy}{dt} &= hx - \rho y - R(1-e^{-vt^2})y,\\ \frac{dz}{dt} &= R(1-e^{-vt^2})y - \beta(h)z.\end{aligned} \tag{11.4.6}$$

The modification has no effect on stationary values, since $r \to R$ as $t \to \infty$. So the stationary point is the same, only with $R$ replacing $r$ in (11.4.4).

Some runs with various parameter values are shown in Figure 11.4.7. The only change to the program listing code, Code 11.4.1, is to replace $r$ by the right-hand side of (11.4.5):

```
MAPLE
> R:=0.08; v:=0.01;
> sys:=diff(x(t),t)=-h*x(t)+rho*y(t)+b*z(t),
>       diff(y(t),t)=h*x(t)-rho*y(t)-R*(1-exp(-v*t^2))*y(t),
>       diff(z(t),t)=R*(1-exp(-v*t^2))*y(t)-b*z(t);
```

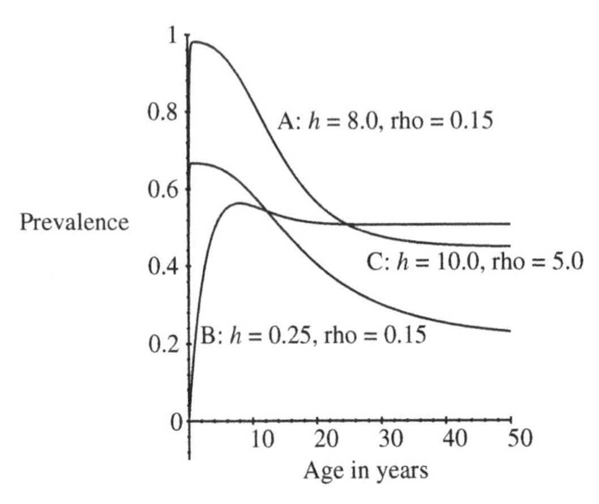


**Fig. 11.4.7.** $\tau = 0.6$, $R = 0.08$, $h$, and $\rho$ as shown.

```
> sol1:=dsolve({sys,x(0)=1,y(0)=0,z(0)=0},{x(t),y(t),z(t)},type=numeric,output=listprocedure);
> ysol1:=subs(sol1,y(t));
```

```
MATLAB
% make up an m-file, tdi.m, with
% function SIRSprime=tdi(t,X);
% %X(1)=x, X(2)=y, X(3)=z
% params=parasConst(t);
% h=params(1); rho=params(2);b=params(3);
% R=params(4);v=params(5);u=params(6);
% SIRSprime=[-h*X(1)+rho*X(2)+b*X(3); h*X(1)-rho*X(2)-R*(1-exp(-v*t^2))*X(2);
%              R*(1-exp(-v*t^2))*X(2)-b*X(3)];
% don't forget to set correct param values in parasConst.m.
% Also r in that file now plays the role of R.
```

Note that this time-dependent immunity model preserves the crossover phenomenon and the urban vs. rural phenomenon. Thus for $\rho = 0.15$, the curve for high $h$ crosses that for low $h$, curves A and B of the figure. In addition, a high value of $\rho$, $\rho = 5.0$, gives a profile matching the urban group of the Wilson data, curve C.

A weighted nonlinear least squares fit to the Garki Project data produces the parameters given in Table 11.4.1. The corresponding predicted prevalence curve is shown in Figure 11.4.8 superimposed on the $U$ Garki data.

**Table 11.4.1.** Time-dependent immunity.

| Parameter specification | Error | Stationary values |
|---|---|---|
| $h = 1.99$, $\rho = 0.074$,<br>$R = 0.113$, $\tau = 1.5$, $\upsilon = 0.0024$ | 0.003 | $\bar{x} = 0.04$, $\bar{y} = 0.46$, $\bar{z} = 0.50$ |

One would like to know how sensitive the predictions are to the various parameters. This is done by differentiating the variables with respect to the parameters

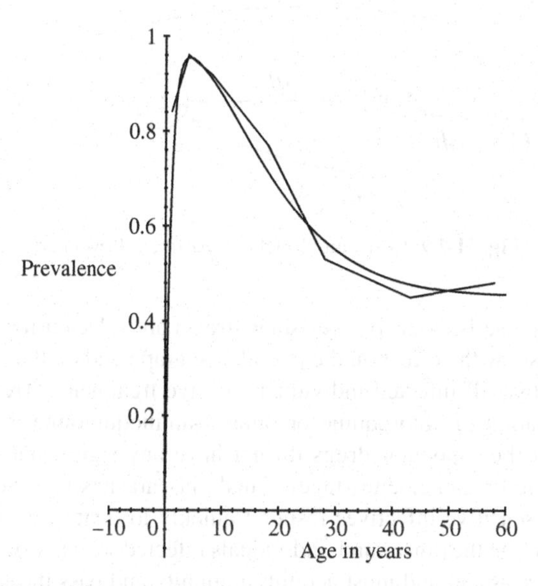


**Fig. 11.4.8.** Predicting the Garki urban data, parameters from Table 11.4.1.

themselves. Thus we compute the partial derivative of $x$ with respect to $h$, then with respect to $\rho$, then to $R$, and finally to $\tau$. Do the same for $y$ and $z$. The results can be presented in matrix form, called the *Jacobian*. For the parameter set in Table 11.4.1, the sensitivity at stationarity to parameter fluctuation is this:

$$J = \begin{bmatrix} -0.22 & 0.26 & 0.18 & -0.22 \\ 0.13 & -0.19 & -0.28 & -0.27 \\ 0.09 & -0.08 & 0.10 & 0.48 \end{bmatrix}.$$

The rows are ordered $x$, $y$, and $z$ and the columns $h$, $\rho$, $R$, and $\tau$. Thus $x$ is most sensitive to change in $\rho$, $y$ is most sensitive to change in $R$, and $z$ is most sensitive to change in $\tau$.

*Drug-resistant model.*

While the models above deal largely with the natural course of the disease, in this section we assume that the entire cohort at risk is treated with drugs to clear internal parasites. Treatment and control have become more difficult in recent years with the spread of drug-resistant strains of *P. falciparum* [3, 4, 9]. Drugs such as chloroquine, nivaquine, quinine, and fansidar are used for treatment. More recent and more powerful drugs include mefloquine and halofantrine.

To proceed, we distinguish resistant infectives from sensitive infectives; the former are those infected with parasites that are resistant to drugs. While this model cannot track the dynamics between sensitive and resistant strains of parasites—in

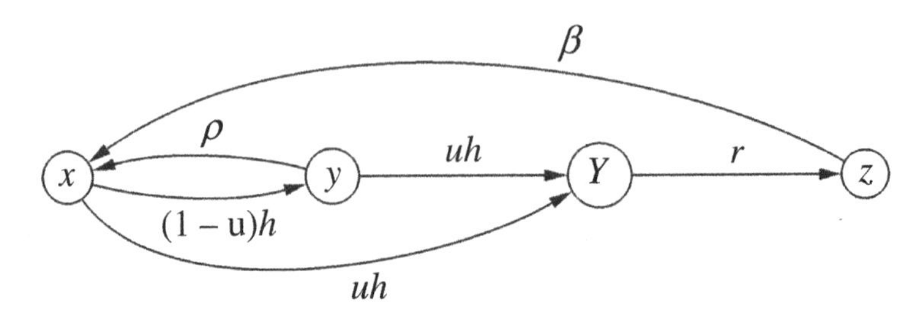


**Fig. 11.4.9.** Resistant model compartment flow chart.

areas where drug use is extensive, resistant strains are selected for and increase in number—it can show the effect on the prevalence graph and on the stationary point.

We assume that all infected individuals receive treatment. Treatment consists in the administration of chloroquine or other 4-aminoquinoline derivatives. It is well known that these popular drugs do not have any significant activity against the exoerythrocite or gametocite stage of malaria parasites (see Section 11.2). It is assumed that sensitive infectives respond quickly to treatment and return to the susceptible class. On the other hand, individuals infected with resistant strains do not respond to this treatment and must acquire immunity and pass through the partially recovered group before returning to the susceptibles.

Let $y$ represent the infectives stricken with sensitive parasites only and $Y$ those stricken with resistant parasites and, possibly, sensitive strains as well. Let $u$ be the probability that when an individual is infected, it is with a resistant strain of the parasite, with or without sensitive strains, and let $1 - u$ be the complementary probability, that is, the probability that an individual is infected with sensitive strains only. With the rest of the notation as before, we have the following sensitive–resistant strain model; Figure 11.4.9 illustrates the possibilities:

$$\begin{aligned}\frac{dx}{dt} &= -hx + \rho y + \beta z,\\ \frac{dy}{dt} &= (1-u)hx - \rho y - uhy,\\ \frac{dY}{dt} &= uhx + uhy - R(1-e^{-vt^2})Y,\\ \frac{dz}{dt} &= R(1-e^{-vt^2})Y - \beta z.\end{aligned} \tag{11.4.7}$$

As before, $\beta$ is a fuction of $h$ given by (11.4.3). By normalization,

$$x(t) + y(t) + Y(t) + z(t) = 1 \tag{11.4.8}$$

with initial condition $x(0) = 1$.

The modification to the computer code is again straightforward—the following, for example:

```
MAPLE
> beta:=h->h*exp(-h*tau)/(1-exp(-h*tau));
```

```
> tau:=.6; u:=0.8; R:=0.08; v:=0.0024; h:=.25; rho:=.15; b:=beta(h);
> sys:=diff(x(t),t)=-h*x(t)+rho*y(t)+b*z(t),
>       diff(y(t),t)=(1-u)*h*x(t)-rho*y(t)-u*h*y(t),
>       diff(Y(t),t)=u*h*(x(t)+y(t))-R*(1-exp(-v*t^2))*Y(t),
>       diff(z(t),t)=R*(1-exp(-v*t^2))*Y(t)-b*z(t);
> sol1:=dsolve({sys,x(0)=1,y(0)=0,Y(0)=0,z(0)=0},{x(t),y(t),Y(t),z(t)},type=numeric,output=listprocedure);
> ysol1:=subs(sol1,y(t));
> Ysol1:=subs(sol1,Y(t));
> ysum:=ysol1+Ysol1:
> plot([ysum,t,0..50]);
```

MATLAB

```
% make up an m-file, resis.m, with
% function SIRSprime=resis(t,X);
% %X(1)=x, X(2)=y, X(3)=Y, X(4)=z
% params=parasConst(t);
% h=params(1); rho=params(2); b=params(3); R=params(4);v=params(5); u=params(6);
% SIRSprime=[-h*X(1)+rho*X(2)+b*X(4); (1-u)*h*X(1)-rho*X(2)-u*h*X(2);
%            u*h*(X(1)+X(2))-R*(1-exp(-v*t^2))*X(3); R*(1-exp(-v*t^2))*X(3)-b*X(4)];
% don't forget to set correct param values in parasConst.m
% Also r in that file now plays the role of R.
> [t,X]=ode23('resis',[0 50],[1;0;0;0]);
> plot(t,X(:,2))
```

For stationarity, use the fourth equation of (11.4.7) to eliminate $Y$ and the second to eliminate $x$. The resulting equation (two times over) is

$$\frac{u(\rho+h)}{1-u}y-\beta z=0.$$

Solve this for $y$; the stationary point is given in terms of $z$ by

$$x=\frac{\beta(\rho+uh)}{uh(\rho+h)}z,\qquad y=\frac{\beta(1-u)}{u(\rho+h)}z,\qquad Y=\frac{\beta}{R}z.$$

As before, since solutions must sum to one, (11.4.8), we have a unique asymptotic stationary point.

MAPLE(symbolic, no MATLAB equivalent)

```
> restart;
> equi:=solve([-h*x+rho*y+beta*z=0, (1-u)*h*x-rho*y-u*h*y=0, u*h*x+u*h*y-R*Y=0, x+y+Y+z=1],{x,y,Y,z});
> ze:=simplify(subs(equi,z));
> xe:=simplify(subs(equi,x)/ze)*z;
> ye:=simplify(subs(equi,y)/ze)*z;
> Ye:=simplify(subs(equi,Y)/ze)*z;
> RHSmatrix:=matrix([[-h,rho,0,beta], [(1-u)*h,-rho-u*h,0,0], [u*h,u*h,-R,0], [0,0,R,-beta]]);
```

The effect of $u$ is illustrated in Figure 11.4.10 by plotting some prevalence profiles for the resistant infectives (the number of sensitive infectives is very small) for various values of $u$. In this figure, it is assumed that all parameters are as in the baseline TDI model, Table 11.4.1, except the recovery rate $\rho$, which is taken to be much higher. We see from the figure that the profiles are relatively insensitive to $u$ (provided it is not zero).

Next, we examine graphically some predictions of this model in relation to the previous ones. In order for our model to be of practical significance and describe certain hyperendemic situations, we use parameter values that are reported to be the situation in the Nsukka region of Nigeria: $\tau=0.6$, $u=0.8$, $h=0.5$, $\rho=0.8$, and $R=0.2$ (by personal communication from Professor Sylvanus Aneke).

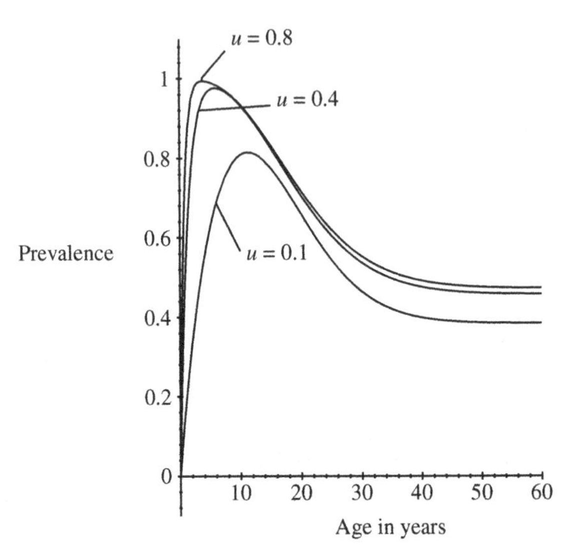


**Fig. 11.4.10.** Parameter values as in Table 11.4.1 except $\rho = 5$.

With these values the stationary point becomes

$$x = 0.250, \qquad y = 0.021, \qquad Y = 0.541, \qquad z = 0.188.$$

The population of the region is about 2 million.

Solution curves for both $y$ and $Y$ are shown in Figure 11.4.11 for three sets of parameter values as given in the figure. In the figure, $y_1$ goes with $Y_1$, $y_2$ with $Y_2$, and so on. The corresponding stationary points are given in Table 11.4.2.

It is seen from Figure 11.4.11 that in all cases, the principal infection is from resistant parasites. For low $\rho$, the resistant curves $Y_1$ and $Y_2$ resemble the corresponding curves in Figure 11.4.7. Thus in this setting, the first two sets of parameters could be seen as describing a situation in which infectives are treated initially with a drug that has very little impact on infectives. The third curve, $Y_3$, still shows that if $u$ is high, sensitive infectives diminish.

An observation that follows from the equilibrium equations is that for large $\rho$, meaning that treatment is widely administered and effective, most of the population will be in the partially recovered class. This is seen in Table 11.4.2. This occurs even if $h$ is high.

### Exercises/Experiments

1. Take as the baseline system the TDI model with these parameter values: $h = 2$, $\rho = 0.07$, $R = 0.1$, $\tau = 1.5$, and $v = 0.002$. Perform five experiments to test the effect of each parameter. Hold all parameters fixed except the one being tested and vary that parameter for a few values above and below the baseline value, say,

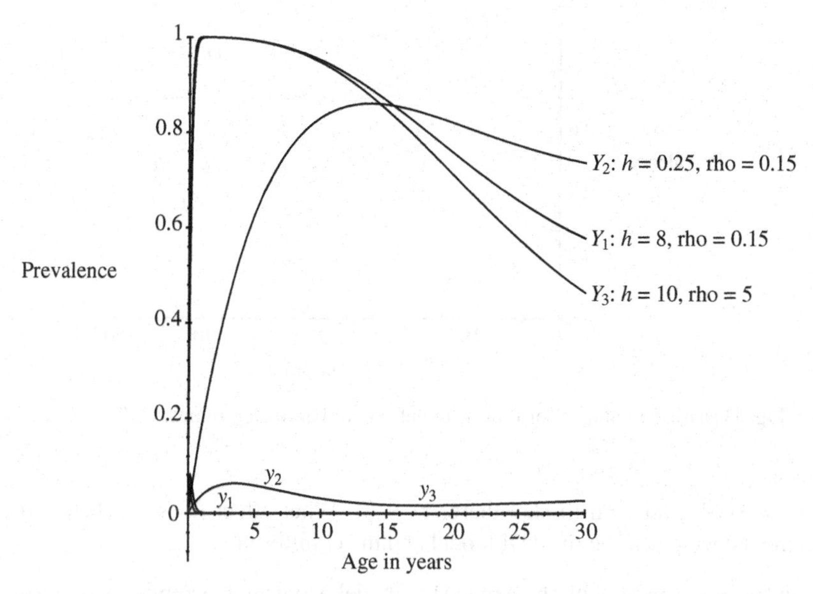


**Fig. 11.4.11.** $R = 0.08$, $\tau = 0.6$, $v = 0.0024$, $u = 0.8$, $h$ and $\rho$ as shown.

**Table 11.4.2.** Sensitive–resistant model.

| **Parameter specification** | **Asymptotic stationary values** |
|---|---|
| $h = 8$, $\rho = 0.15$ | $x = 0.01$, $y = 0$, $Y = 0.45$, $z = 0.54$ |
| $h = 0.25$, $\rho = 0.15$ | $x = 0.24$, $y = 0.03$, $Y = 0.69$, $z = 0.04$ |
| $h = 10$, $\rho = 5$ | $x = 0$, $y = 0$, $Y = 0.024$, $z = 0.76$ |

up to 50%. Plot the result of each test on a single plot along with the baseline. Altogether you will have five plots. Which parameter has the greatest effect on the disease according to your experiments?

**2.** According to the last two digits of your college registration number, pick graph A in Figure 11.4.12 if these digits are from 00 to 24, pick graph B if from 25 to 49, and so on. Experiment with the parameter values of the TDI model to match the selected graph as closely as you can. What are those values?

**3.** From the baseline case (see Exercise 1 above), suppose a cohort of individuals are born having very little propensity for acquiring immunity, $R = 0.01$, but for whom the tendency to surmount the disease is high, $\rho = 0.3$. What effect does this have on the prevalence profile and the equilibriums?

**4.** From the baseline case (see Exercise 1 above), suppose a cohort is born having the tendency to quickly pass into the acquired immunity stage, $R = 0.2$ and

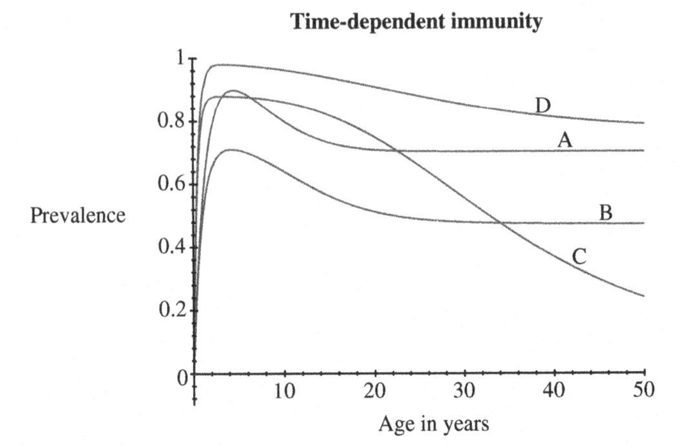


**Fig. 11.4.12.** Prevalence for four different sets of parameter values, $\tau$, $R$, $\rho$, $v$, $h$.

$v = 0.01$, and then to stay there, $\tau = 3$. What effect does this have on the prevalence profile in the TDI model and the equilibriums?

**5.** In this experiment with the Aron–May model, we want to suppose that mosquito control measures are quite effective at first, so $h = 0.2$ while a cohort is young, up to age 20, say, but then reverts to its baseline value $h = 2$. At the same time, put $v = 0.001$ to reflect that infecteds will be delaying acquired immunity. The modifications to the basic program are detailed below. What effect does this have on the prevalence?

```
MAPLE
> restart:
> h:=t->.2+1.8*Heaviside(t-20);
> beta:=h->h*exp(-h*tau)/(1-exp(-h*tau));
> tau:=1.5; R:=.1; rho:=.07; v:=.001;
> sys:=diff(x(t),t)=-h(t)*x(t)+rho*y(t)+beta*h(t)*z(t),
>      diff(y(t),t)=h(t)*x(t)-rho*y(t)-R*(1-exp(-v*t^2))*y(t),
>      diff(z(t),t)=R*(1-exp(-v*t^2))*y(t)-beta*h(t)*z(t);
> sol:=dsolve({sys,x(0)=1,y(0)=0,z(0)=0},{x(t),y(t),z(t)},type=numeric,output=listprocedure);
> ysol:=subs(sol,y(t));
> plot([ysol],0..50,-0.1..1.0);
```

```
MATLAB
% modify the m-file, parasConst.m, according to
% function parasParms = parasConst(t);
% tau= as desired;
% if t<20
%    h=.2;
% else
%    h=2.0;
% end
% b=beta(h,tau);
%    rho, r, v, u as desired
% parasParms=[h,rho,b,r,v,u];
> [t,X]=ode23('tdi',[0 50],[1;0;0]);
> plot(t,X(:,2))
```

6. In contrast to the previous problem, assume here that the baseline conditions prevail for a cohort until age 20, at which point the infection rate $h$ falls to 0.2. Leave the acquired immunity parameter $\upsilon$ at its baseline value.

**Questions for Thought and Discussion**

1. What would the world be like if malaria were conquered?
2. Discuss a possible evolutionary pathway leading to internal parasitism.
3. How can knowing the genomic sequence of *P. falciparum* help in controlling malaria?

## References and Suggested Further Reading

[1] J. L. Aron, Dynamics of acquired immunity boosted by exposure to infection, *Math. Biosci.*, **64** (1983), 249–259.
[2] J. L. Aron and R. M. May, The population dynamics of malaria, in R. M. Anderson, ed., *Population Dynamics of Infectious Disease*, Chapman and Hall, London, 1982, 139–179.
[3] N. T. J. Bailey, *The Biomathematics of Malaria*, Charles Griffin and Company, London, 1982.
[4] D. D. Despommier, R. W. Gwadz, and P. J. Hotez, *Parasitic Diseases*, 3rd ed., Springer-Verlag, Berlin, New York, Heidelberg, 1994.
[5] J. A. Jacquez and C. P. Simon, Qualitative theory of compartmental systems, *SIAM Rev.*, **35**-1 (1993), 43–79.
[6] L. Molineaux and G. Gramiccia, *The Garki Project*, World Health Organization, Geneva, 1980.
[7] G. Taubes, Malarial dreams, *Discover*, **109** (1998), 108–116.
[8] D. B. Wilson, Rural hyperendemic malaria in Tanganyika territory, *Trans. Roy. Soc. Tropical Med. Hygiene*, **29** (1936), 583–618.
[9] Division of Control of Tropical Diseases, World Health Organization, Geneva, April, 1997; available online from www.who.int/ctd.
[10] Disease Control Unit, Enugu State, Nigeria Health Management Board, Enugu, Nigeria, 1995 (unpublished).

# 12

# Cancer: A Disease of the DNA

## Introduction

Cancer is a group of diseases in which cells grow and spread unrestrained throughout the body. Cancers can arise in nearly any type of cell that retains the ability to divide. Although there are more than 100 forms of cancer, the basic processes underlying all of them are very similar. The process by which normal cells become cancerous is called *carcinogenesis*.

Cancers stem from mistakes and misapplication of cellular mechanisms, the cell's inability to heed normal growth and division controls or to undergo self-destruction, called apoptosis, when it detects that it is damaged. Normal cells are part of a cellular community and coordinate their activities with those of their neighbors especially regarding growth and division. Cancerous cells ignore cellular controls and even produce false signals for coercing their neighbors to help them. This errant behavior comes about due to the accumulation of small mutations, changes to the cellular genome that are perpetuated in cell reproduction. Two gene classes play major roles in choreographing the cellular life cycle: proto-oncogenes initiate cell growth and division, and tumor suppressor genes inhibit cell growth and division. When proto-oncogenes go awry and become oncogenes, they maintain continuous growth signals, like a car with the accelerator pedal stuck on. By contrast, dysfunctional tumor suppressor genes are like a car with no brakes. In order for a tumor to develop, mutations usually must occur in several genes.

## 12.1 Cell Growth and Division Is an Involved and Complicated Process

In Chapter 13, we consider the cell's growth and division cycle from a functional point of view. Here we must take a close look at the biochemistry of the sequence of steps leading to the creation of a new cell in some detail. We need to know the proteins involved and to understand their roles in the cycle.

R.W. Shonkwiler and J. Herod, *Mathematical Biology: An Introduction with Maple and Matlab*, Undergraduate Texts in Mathematics, DOI: 10.1007/978-0-387-70984-0_12,

*A cell cycle involves the orchestrated activity of intracellular enzymes.*

Each cell of the body lives as a member of a community of cells. The cell's community are the nearby cells of its particular tissue type. Together the community jointly controls the growth and division of its members. Cell division begins within a cell when it receives growth stimulatory signals transmitted by other cells of the community. These signals are in the form of protein *growth factors*, which move in the interstitial space between cells and bind to specific receptors on the surface of the cell. The receptors in turn signal proteins within the cell. The process is complicated in that several proteins of the cytoplasm are involved in a chain, or *stimulatory pathway*, of signaling, which ultimately ends in the nucleus of the cell.

In counterpoint to the stimulatory pathway, there is also an *inhibitory pathway*. Its signal transmitted to the cell nucleus is to hold off cell division.

The target of this activity in the nucleus is the *cell cycle clock*, also discussed in Section 13.1. The clock integrates all the stimulatory and inhibitory signals and decides whether the cell should undergo a division cycle, or *cell cycle*.

The cell cycle consists of four stages. To divide, the cell must double its genome; this stage is called the S, or *synthesis, phase*. Afterward, it must halve that genome in the *mitosis*, or M, phase. Between M and S is the *first gap phase*, or $G_1$, and between S and M is the *second gap phase*, or $G_2$. In $G_1$, the cell increases in size and at some point decides whether to divide. In $G_2$, the cell undergoes the necessary preparations for cell division, and in M the cell divides in half, sending a full complement of its chromosomes to each daughter cell. Each daughter cell immediately enters stage $G_1$ but normally does not proceed to S right away. Instead, it then enters a resting, or $G_0$, stage. Later it will reenter $G_1$ and decide to proceed on to S or return to $G_0$. The term *interphase* refers to any stage of the cell cycle other than M.

The cell cycle is quite complicated and is mediated at each step by a variety of molecules; principal among them are proteins called *cyclins* and *cyclin-dependent kinases* (CDKs). See Figure 12.1.1.

Cyclin D and CDK4 orchestrate $G_1$. Starting with $G_1$, a rising level of cyclin D binds to CDK4 and signals the cell to prepare the chromosomes for replication. It achieves this effect by nullifying the effect of the protein pRB, which exerts powerful growth-inhibitory control.

Cyclins A and E and CDK2 mediate phase S. A rising level of *S-phase promoting factor* (SPF), cyclin A bound to CDK2, enters the nucleus and prepares the cell to duplicate its DNA and its centrosomes. As DNA replication proceeds, cyclin E is destroyed, and the level of mitotic cyclins A and B begins to rise (in $G_2$).

In stage M, a complex of cyclins A and B and CDK1, called *M-phase promoting factor*, initiates several things: the assembly of the mitotic spindle, the breakdown of the nuclear envelope, and the condensation of the chromosomes.

These events take the cell to the metaphase of mitosis. At this point, the M-phase promoting factor activates the *anaphase-promoting complex* (APC), which

- allows the sister chromatids at the metaphase plate to separate and move to the poles (this is *anaphase*), completing mitosis;

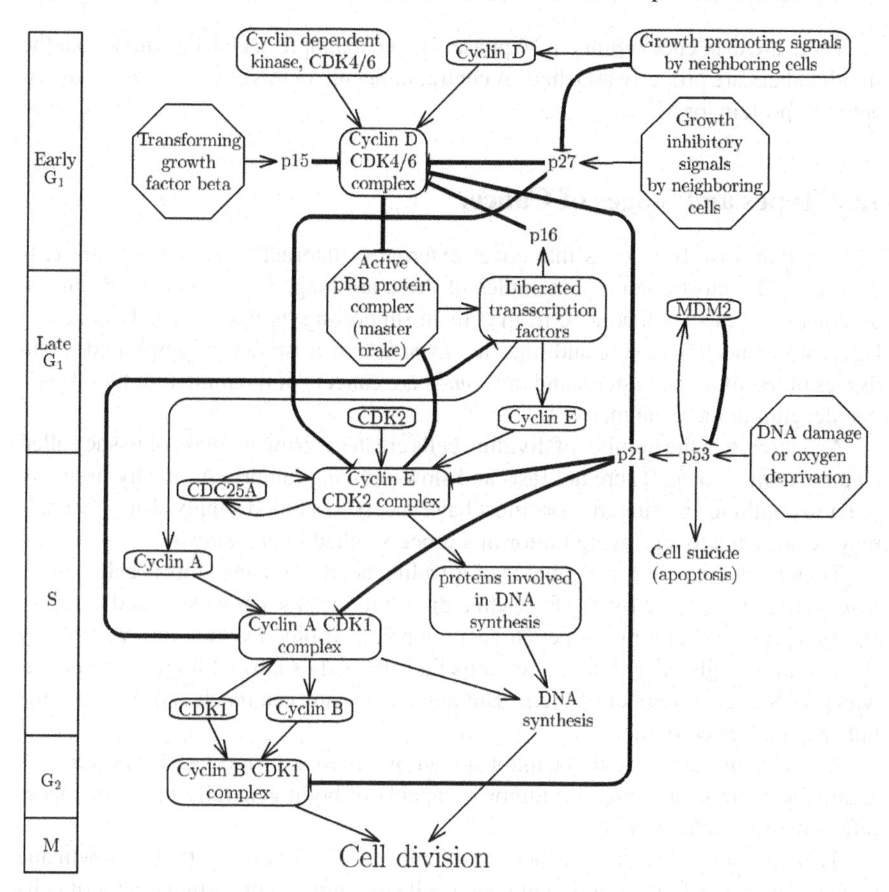


**Fig. 12.1.1.** Biochemistry of the cell cycle. An arrowhead path indicates a promoting signal; a blunthead path indicates an inhibiting signal.

- destroys cyclin B—it does this by attaching it to the protein *ubiquitin*, which targets it for destruction by proteasomes;
- turns on the synthesis of cyclin D for the next turn of the cycle;
- degrades *geminin*, a protein that has kept the freshly synthesized DNA in S phase from being rereplicated before mitosis.

In addition to these mechanisms for conducting the cell cycle, there are also checking mechanisms. There is a check of the DNA for errors in $G_1$ before the cell enters S phase. There is also a check that all the DNA has been copied before the cell can leave S phase. The principal protein involved in this is p53. The gene for p53 is recessive, since both copies must be defective for p53 to fail. The protein p53 is also an important mediator of *apoptosis*, leading defective cells to undergo self-destruction.

After S phase, in $G_2$, there is again a DNA check for damage. A molecule involved in this is *ataxia telangiectasia mutated*, or ATM.

There are also checkpoints in M phase. In one example, a check is made that the spindle fibers are properly attached. A central molecule in this check is *mitotic arrest deficient* protein, or MAD.

## 12.2 Types and Stages of Cancer

Cancers that arise from cells that cover external or internal body surfaces are *carcinomas*. The most common examples of these are lung, breast, and colon cancer. *Sarcomas* are cancers that arise from cells in supporting tissues such as bone, cartilage, fat, connective tissue, and muscle. *Lymphomas* arise in the lymph nodes and tissues of the immune system, and *leukemias* are cancers of the immature blood cells that develop in the bone marrow.

An increase in the number of dividing cells creates a growing mass of tissue called a *tumor* or *neoplasm*. There are fast- and slow-growing tumors. As in any tissue, in order to continue growing, tumors must have access to a blood supply. Blood vessels may be attracted to a growing tumor in a process called *angiogenesis*.

Tumor cells may become capable of invading nearby tissues—this is called *invasion*—or, in its extreme form, of invading distant tissues by a process called *metastasis*. In metastasis, tumor cells penetrate a lymphatic or blood vessel, into which they shed dividing cells. There the tumor cells face the body's normal immune response, which kills a large number of them. But any that escape can reach and invade some hitherto unaffected tissue.

A *benign* tumor is one that cannot spread by invasion or mestastsis and remains local. By contrast, a *malignant* tumor is capable of both; generally the term cancer refers to a malignant tumor.

There are several stages to cancer development. *Hyperplasia* is tissue growth due to an excessive cell division rate in which cell structure and the arrangement of cells are normal. Hyperplasia can be due to an irritating stimulus. *Dysplasia* is excessive cell proliferation along with the loss of normal tissue arrangement and normal cell structure to some degree. Sever dysplasia is referred to as a *carcinoma in situ* and is uncontrolled cell growth. Nevertheless, at this stage, the tumor is not invasive or metastatic.

In addition, tumors are graded for degree of progression. If the cells are highly abnormal in appearance and there is a large number of dividing cells, the grade is III or IV. Cancers less progressed have a grade of I or II.

Cancer, being a disease of the DNA, is amenable to study from the standpoint of genetics. In Section 14.4, we delve into this topic briefly and note some of the methods being tried to bring the disease under control.

## 12.3 The Role of Proto-Oncogenes and Tumor Suppressor Genes

As mentioned above, proto-oncogenes are responsible for the stimulatory pathways leading to cell division. When a proto-oncogene mutates, leading to an overexcitation of its function in the pathway, the mutated gene is termed an *oncogene*. All stages of the growth and division pathway can be subverted starting with the receptor proteins embedded in the cell surface. For example, the overactive receptor Erb-B2 is associated with breast cancer.

The next step in the pathway is the signal cascade operating in the cytoplasm. Here, too, oncogenes are responsible for overactive proteins. One example is the *ras* family. Proteins coded for by mutant *ras* genes are continuously stimulatory even in the absence of active growth factor receptors. Hyperactive *ras* proteins are found in one-fourth of all human tumors, including carcinomas of the colon, pancreas, and lung.

Oncogenes are implicated in growth and division pathologies in the nucleus as well. An example of this relates to the *myc* family of genes, coding for altered activity of various transcription factors. Such *myc* proteins are constantly at high levels despite the absence of their usual antecedents in the stimulatory pathway.

*Tumor suppressor genes are generally recessive.*

Equally troublesome, aberrant *tumor suppressor genes* (TSG) can be at fault. Normal TSGs produce proteins that restrain cell growth and division, but when a TSG is defective, this control is absent. Defective tumor suppressor genes differ from oncogenes in the important respect that they are recessive, and hence both genes of the chromosomal pair must be defective for the encoded protein to be dysfunctional.

In some cases, tumor suppressor proteins can hold the stimulatory effect of an oncogene in check, and cancer will develop only when some component of the suppressor system fails. In one example, the substance *transforming growth factor beta* (TGF-$\beta$) is a regulatory component in the growth of normal cells. However, some colon cancers overcome TGF-$\beta$ by inactivating the gene that codes for its surface receptor. In another example, this time in the cytoplasm, some cancers evade the effect of the protein produced by the DPC4 gene by inactivating that gene. The same goes for the p15 gene, which codes for a protein that works in concert with TGF-$\beta$. Normally, p15 works by blocking the binding of cyclin D with CDK4, thus preventing the cell going from $G_1$ to S. Should its gene become inactive, the cell loses this control.

The suppressor gene NF-1 is an example of a control that acts directly on a specific stimulatory protein. The suppressor produced by the NF-1 gene operates by blocking the growth signals sent by the *ras* protein. When NF-1's protein is dysfunctional, the stimulatory signal goes unchecked.

*p53 plays a central role in the health of the cell.*

One of the most prominent suppressor proteins is p53, which is multiroled. It checks the health of the entire cell generally: It monitors the integrity of the chromosomal DNA, it monitors the successful completion of the cell cycle, and it is the key protein involved in apoptosis. Half of all types of human cancers lack a functional p53 protein.

Regarding the effect of TSGs, it has been shown that the external introduction of a missing tumor suppressor protein can restore the control that that protein exhibits and hence arrest an incipient cancer. Thus it may be possible to treat cancers that arise from dysfunctional TSGs pharmacologically.

*Cell death and cell division limits are also subverted.*

If a cell's essential components become damaged, including the chromosomal DNA, then the cell will initiate events that lead to its own destruction; this is called *apoptosis*. As previously mentioned, the p53 protein generally monitors the health of the cell and triggers apoptosis if warranted. Among the DNA damage that can be detected is the conversion of a proto-oncogene to an oncogene or damage to a TSG.

Of course, if the gene coding for p53 itself becomes damaged, then apoptosis may not be possible for the cell. In addition, cancer cells devise ways of evading apoptosis. One of these is through mechanisms that inactivate the p53 encoding gene. Another is to make excessive amounts of the protein Bcl-2, which counters p53.

It was once thought that radiation therapy disrupted a cell so badly that it died. But now it is known that, in fact, only minor damage is inflicted on the cell, and it is apoptosis that kills the cell. The implication is that if p53 is absent or dysfunctional, then radiation therapy is far less effective.

A second mechanism that normally operates to kill cancer cells relates to the fact that cells may divide only a limited number of times, up to 50 to 60 depending on the type of cell. At that point the cell is said to be *senescent*; see also Section 5.1.

The mechanism for this has to do with segments on the ends of the chromosomal DNA strands known as *telomeres*. Each time a cell's chromosomes are replicated, the telomeres shorten a bit. Once the telomere length falls below a threshold, a signal goes out instructing the cell to enter senescence. If chromosomal duplication should proceed despite this signal, chromosomes fuse with one another, causing the cell to die.

Replication of a chromosome is accomplished by a protein that moves over the length of its surface and reads the base pairs. By way of analogy, think of a railroad car that is pulled over the rails to clean the track. When the cleaning train reaches the end of the line, the track is not cleaned to the very end because the locomotive is in the way. In the same way, the tip of the chromosomes cannot be read and duplicated, hence the function of the telomeres. The unread tip is the length by which the telomere shortens on each replication.

The proteins pRB and p53 are involved in detecting telomere length and signaling senescence. But cells that have mutations to the genes for either of these proteins will continue to divide anyway. Thus as explained above, eventually the telomeres disappear altogether and the cell dies. However, there is a way that a cell can circumvent this fate and become immortal.

The enzyme telomerase systematically replaces lost telomere segments, thereby enabling the chromosomes to be replicated endlessly. This enzyme is absent in most normal cells but present in almost all tumor cells. Thus not only does the tumor cell not die, but there is a second pathological effect. Immortality gives the cell time to accumulate other mutations.

Yet another way the necessary mutations can come about is through nonfunctional DNA repair proteins. The cell possesses proteins that move over the chromosomal

DNA and detect and repair copying mistakes and other abnormalities. As mentioned above, one of these is ATM. But these proteins may become dysfunctional, for example, by means of a mutation to their own genes. In this case, unrepaired mutations within the cell can accumulate very quickly. ATM is implicated in 10% of inherited breast cancers.

*It usually takes several mutations or "hits" for malignancy.*

Most malignancies are the product of tumor cells that have acquired several mutations. These may include several or all of the following:

- an overstimulated cell cycle pathway,
- a defective tumor suppressor pathway,
- the ability to circumvent apoptosis,
- the ability to produce telomerase,
- a mechanism for attracting blood vessels (angiogenesis),
- the ability to invade nearby tissues (invasion),
- the ability to metastasize.

Normally, it takes decades for an incipient tumor to collect enough mutations for malignant growth, but there is a mechanism that greatly accelerates the rate—an inherited mutation. If a germ cell harbors some mutation, then that mutation is part of every cell of the body. For example, suppose a germ cell has a defective copy of the TSG gene. Since these genes are recessive, the defect will not manifest itself so long as its twin on the homologous chromosome remains intact. But now every cell of the body is a target. Should a mutation occur to the remaining working gene in any one of them, the tumor suppression mechanism will now fail for that cell. Such an event is vastly more probable than accumulating two defects by chance to the same cell. We investigate this reality mathematically through a retinoblastoma study in Section 12.4.

Although inherited genetic alterations are a factor in some cases, nevertheless 80–90% of cancers occur in people with no family history of the disease.

The progression of cells to cancer is reminiscent of Darwinain evolution. Cells that acquire any of these mutations have the capacity for reproducing themselves to a greater extent than normal cells. This has the consequence that their kind outgrow neighboring cells with several detrimental consequences. The mildest of these is that these aberrant cells no longer participate fully in the intended function of the tissue while at the same time garnering more and more of the tissue's resources such as space and nutrients. Thus the normal cells of the tissue are at a competitive disadvantage. More seriously, they beget a colony of cells capable of further mutation.

## 12.4 A Model for Retinoblastoma

Retinoblastoma is a cancer of the retina and is one of the simplest cancers in terms of mechanism. It results when a dividing retinal cell lacks a tumor suppressor protein as a result of defects in the corresponding gene to both homologous chromosomes. Retinal cells grow and divide from early in fetal development and continue growing after birth until about age 2 or 3 years. Thus retinoblastoma is a childhood disease.

One way a retinal cell can suffer the two defects, or *hits*, to its tumor suppressor gene is by chance alone. Such a case is called *somatic* and is thought to account for 60% of all cases of the disease. Actually, this is a very rare occurrence with an incidence rate in the general population of about 30 per million.

The other possibility is for an individual to inherit a defective gene for the disease; such a person is called a *carrier*. In this event, every cell of the body, including all retinal cells, has one hit already. Consequently, if a mutation now occurs to any retinal cell at all during the growth phase, that cell will have two hits, and tumor growth ensues. These are called *germinal* cases and account for the balance of the cases, about 40%.

*Patient data.*

In his study of the disease, Knudson surveys cases of retinoblastoma and presents the data shown in Table 12.4.1. In fact, this purely mathematical study suggested the existence of tumor suppressor proteins, which led to their discovery. As reported in the table, retinoblastoma cases are observed as to sex, onset age, laterality (that is, which eye holds tumors), number of tumors, and family history of the disease. In addition, Knudson further presents the summary data of Table 12.4.2. Among this data is an estimate of the fraction of the cases that are germinal and, by complement, somatic. This estimate is made using the assumption that somatic cases have only one tumor, due to the unlikelihood of two hits; cf. statistic 13. A key item researched by Knudson is "carriers never affected," statistic 1, which can be used to determine the gene mutation rate $p$, as we will see below. Note that Knudson did not assess statistic 6, "carriers among the population," but we will be able to do so from our analysis.

We formulate a simple model to account for the data, namely that a retinal cell grows and matures for a period of time, after which it divides. In the process of division, a mutation randomly occurs to the tumor suppressor gene with probability $p$. We assume that there is no cell death. The two daughter cells are genetically the same as their shared parent, except that with probability $p$, one of them has mutated. (Probability $p$ is so small that we ignore the possibility that both daughter cells mutate.) A mutant cell then begets a clone of like cells.

To further simplify the model, we assume that all cells wait the same period of time, one unit, and hence we have a synchronous population. A more advanced model allows for random growing periods and therefore overlapping generations, but the results are essentially the same.

Our first task is to determine how many such cell divisions, $T$, retinal cells undergo. It is known that there are about 2 million retinal cells per eye, or $4 \times 10^6$ counting both eyes, all starting from a single cell. Therefore, there must be $T = 22$ division cycles, since $2^{22} = 4{,}194{,}304$. As mentioned above, this takes about two to three years on average. This result can also be derived by solving a simple recurrence

**Table 12.4.1.** Cases of retinoblastoma ($*$ indicates no information).

| (a) Bilateral cases | | | | | | (b) Unilateral cases | | | | |
|---|---|---|---|---|---|---|---|---|---|---|
| | | **Age at diagnosis** | **Number of tumors** | | | | | **Age at diagnosis** | | |
| **Case** | **Sex** | **(months)** | **left** | **right** | **Family history** | **Case** | **Sex** | **(months)** | **Number of tumors** | **Family history** |
| 1 | F | 8 | * | * | no | 24 | F | 48 | * | no |
| 2 | M | 3 | * | * | no | 25 | M | 22 | * | no |
| 3 | F | 11 | * | * | no | 26 | M | 33 | * | no |
| 4 | M | 2 | * | 1 | no | 27 | M | 38 | * | no |
| 5 | M | 60 | 1 | * | affected sib. | 28 | F | 47 | * | no |
| 6 | M | 22 | * | * | no | 29 | M | 50 | * | no |
| 7 | F | 4 | 3 | * | no | 30 | M | 32 | * | no |
| 8 | F | 18 | 2 | * | no | 31 | M | 28 | * | no |
| 9 | F | 30 | * | * | no | 32 | F | 31 | * | no |
| 10 | M | 3 | 2 | * | no | 33 | F | 29 | * | no |
| 11 | F | 6 | * | 1 | no | 34 | F | 21 | * | no |
| 12 | M | 7 | * | 2 | affected sib. | 35 | M | 46 | * | no |
| 13 | M | 9 | 3 | * | no | 36 | F | 36 | * | no |
| 14 | F | 4 | 5 | * | no | 37 | F | 73 | * | no |
| 15 | F | 13 | * | * | no | 38 | M | 29 | * | no |
| 16 | M | 18 | * | 4 | no | 39 | F | 15 | * | no |
| 17 | F | 24 | * | * | no | 40 | M | 52 | * | no |
| 18 | F | 44 | 1 | * | no | 41 | M | 24 | * | no |
| 19 | F | 5 | * | * | no | 42 | M | 8 | * | no |
| 20 | M | 12 | * | 1 | no | 43 | F | 19 | * | no |
| 21 | M | 3 | * | * | no | 44 | M | 36 | * | no |
| 22 | M | 12 | * | 1 | no | 45 | F | 34 | * | no |
| 23 | M | 15 | 1 | * | father | 46 | F | 27 | * | no |
| | | | | | | 47 | M | 10 | * | no |
| | | | | | | 48 | F | 8 | * | no |

relation (cf. Section 2.5). Let $N(t)$ be the number of retinal cells after $t$ cell divisions. Then $N(0) = 1$, and for any $t = 1, 2, \ldots,$

$$N(t) = 2N(t-1). \tag{12.4.1}$$

The solution is

$$N(t) = 2^t, \quad t = 0, 1, 2, \ldots;$$

try it!

*Germinal cases determine the mutation probability p.*

Consider a germinal case. Should a mutation occur to some retinal cell of a carrier, the result is a doubly mutated cell and hence a tumor. We are not interested in the number of such doubly mutated cells per se but rather in the number of tumors, which here is the same as the number of mutation events occurring to the growing tissue.

**Table 12.4.2.** Summary data.

| Statistic | Empirical value | Calculated using $p = 7.14 \times 10^{-7}$ |
|---|---|---|
| 1. carriers never affected, used for $p$ | 1–10% | 0.05 |
| 2. average number of tumors among germinal cases | 3 | 2.99 |
| 3. unilateral cases among germinal cases, $g_1$ | 25–40% | 0.337 |
| 4. bilateral cases among germinal cases | 60–75% | 0.663 |
| 5. bilateral cases among all cases, $B$ | 25–30% | 0.27 |
| 6. carriers among general population, $f$ | — | 27 per million |
| 7. incident rate of somatic cases, $u$ | 30 per million | 40 per million |
| 8. germinal cases among all cases | 35–45% | 0.387 |
| 9. somatic cases among all cases | 55–65% | 0.613 |
| 10. unilateral cases among all cases | 70–75% | 0.750 |
| 11. unilateral and hereditary cases among all cases | 10–15% | 0.137 |
| 12. germinal cases among unilateral cases | 15–20% | 0.183 |
| 13. unilateral cases among nonhereditary cases | 100% | assumption |

Let $X(t)$ be the random variable denoting the number of tumors at time $t$, and let $x_k(t)$ be the probability that there are $k$ tumors at time $t$, that is, $x_k(t) = \Pr(X(t) = k)$, $k = 0, 1, 2, \ldots$.

A powerful tool in the derivation of equations for the $x_k$ is the *probability-generating function* $G(s,t)$, a version of which will also be prominent in Section 13.7. In this case, $G(s,t)$ is defined as

$$G(s,t) = \sum_{0}^{\infty} x_k(t)s^k. \tag{12.4.2}$$

Evidently, $G$ is a function of two variables, $s$ and $t$, a polynomial in $s$ whose coefficients are the $x_k(t)$. The variable $s$ merely acts as a bookkeeping device here in that it manages the $x_k$, but it is a most useful bookkeeping device indeed. Even so, treating $s$ as a variable and setting $s = 1$, we get

$$G(1,t) = x_0(t) + x_1(t) + x_2(t) + \cdots = 1, \tag{12.4.3}$$

which sums to 1 because the $x$s account for all possibilities at time $t$: Either there is none, or 1, or 2, and so on. Also, we can recover all the $x_k$ from $G$ by differentiating with respect to $s$ and substituting $s = 0$. Thus

$$G(0,t) = x_0(t)$$

gives $x_0(t)$, and the partial derivative

$$\frac{\partial G}{\partial s} = x_1(t) + 2x_2(t)s + 3x_3(t)s^2 + \cdots,$$

along with the substitution $s = 0$, gives $x_1(t)$,

$$\left.\frac{\partial G}{\partial s}\right|_{s=0} = x_1(t).$$

Continuing to differentiate with respect to $s$ and substituting $s = 0$, one sees that

$$x_k(t) = \left.\frac{1}{k!}\frac{\partial^k G}{\partial s^k}\right|_{s=0}. \tag{12.4.4}$$

In addition to the $x_k$, we can also calculate expectations from $G(s,t)$. The expectation $E(X(t))$, or expected value, of $X(t)$ is defined as the sum over all its possible values weighted by the probability of each, so

$$E(X(t)) = 1 \cdot x_1(t) + 2 \cdot x_2(t) + 3 \cdot x_3(t) + \cdots.$$

But this is exactly the partial derivative of $G$ with respect to $s$ evaluated at $s = 1$,

$$\left.\frac{\partial G}{\partial s}\right|_{s=1} = x_1(t) + 2sx_2(t) + 3s^2x_3(t) + \cdots|_{s=1} = \sum_{k=1}^{\infty} kx_k(t). \tag{12.4.5}$$

One more observation on $G$. Its square is also an infinite series in $s$, whose coefficients take a special form,

$$\begin{aligned} G^2 &= (x_0 + x_1 s + x_2 s^2 + \cdots)(x_0 + x_1 s + x_2 s^2 + \cdots) \\ &= x_0^2 + (x_0 x_1 + x_0 x_1)s + (x_0 x_2 + x_1^2 + x_2 x_0)s^2 + \cdots \\ &= \sum_{k=0}^{\infty}\left(\sum_{k_1+k_2=k} x_{k_1} x_{k_2}\right) s^k. \end{aligned} \tag{12.4.6}$$

With these preliminaries behind us, we set our sights on finding $x_0(t)$. Remarkably, it is easier to calculate all the $x_k$ than just $x_0$! Start by decomposing on the possibilities at $t = 1$; either there is a mutation on this first cell division or not. "Given"—symbolized by |—there is a mutation, the probability that there will be $k$ mutations by time $t$ is exactly the probability that there are $k - 1$ in the remaining time, $t - 1$, so the contribution here is

$$\Pr(k \text{ tumors at } t \mid 1 \text{ at } t = 1) \Pr(\text{a mutation at } t = 1) = x_{k-1}(t-1)p.$$

On the other hand, if there is no mutation at time $t = 1$, then there must be $k$ mutations in the remaining time $t - 1$ stemming from one or both of the two daughter cells. If one of the daughter cells leads to $k_1$ tumors in the remaining time, where $k_1$ could be any of $0, 1, 2, \ldots$, the other must lead to $k_2 = k - k_1$. Thus the contribution here is

$$\begin{aligned} &\Pr(k \text{ tumors at } t \mid \text{none at } 1) \Pr(\text{no mutation at } t = 1) \\ &= \left(\sum_{k_1+k_2=k} x_{k_1}(t-1)x_{k_2}(t-1)\right)(1-p). \end{aligned}$$

Combining these two contributions, we have the equation

$$\begin{aligned}x_k(t) &= \Pr(k \text{ tumors at } t \mid \text{none at } 1)\Pr(\text{no mutation at } t=1)\\&\quad+\Pr(k \text{ tumors at } t \mid 1 \text{ at } t=1)\Pr(\text{mutation at } t=1)\\&=(1-p)\sum_{k_1+k_2=k} x_{k_1}(t-1)x_{k_2}(t-1)+px_{k-1}(t-1).\end{aligned} \tag{12.4.7}$$

Multiply both sides by $s^k$ and sum over $k$ to get

$$\begin{aligned}\sum_{k=0}^{\infty} x_k(t)s^k &= (1-p)\sum_{k=0}^{\infty}\left(\sum_{k_1+k_2=k} x_{k_1}(t-1)x_{k_2}(t-1)\right)s^k\\&\quad+ps\sum_{k=1}^{\infty} x_{k-1}(t-1)s^{k-1}.\end{aligned}$$

It remains only to recall that the double sum on the right is exactly $G^2(s,t-1)$. Thus we arrive at the fundamental equation

$$G(s,t) = (1-p)G^2(s,t-1) + spG(s,t-1). \tag{12.4.8}$$

As noted above, from the generating function we can feasibly calculate many of the properties of interest. One of these is incidence expectation. The expected number of tumors, $E(X(t))$, is given by

$$\begin{aligned}E(X(t)) &= \left.\frac{\partial G}{\partial s}\right|_{s=1}\\&=2(1-p)G(1,t-1)\frac{\partial G(1,t-1)}{\partial s}+pG(1,t-1)+ps\frac{\partial G(1,t-1)}{\partial s}.\end{aligned}$$

Remembering that $G(1,t)=1$ for all $t$ gives the following recurrence relation and initial value (the first and last terms combine):

$$E(X(t)) = (2-p)E(X(t-1))+p, \quad E(X(1)) = p.$$

This is easily solved to give

$$E(X(t)) = \frac{p}{1-p}\left((2-p)^t-1\right). \tag{12.4.9}$$

We may also calculate some individual probabilities.

Since $x_0(t) = G(0,t)$, substituting $s=0$ in (12.4.8) gives the following recurrence relation and initial value:

$$x_0(t) = (1-p)x_0^2(t-1), \quad x_0(1) = 1-p.$$

This, too, is easily solved, and yields the probability that no tumor will occur by time $t$,

$$x_0(t) = (1-p)^{2^t-1}. \tag{12.4.10}$$

In a similar fashion, one obtains the probability that there will be one tumor, $x_1(t)$. From the definition of the generating function, $x_1(t) = \frac{\partial G}{\partial s}|_{s=0}$. Differentiating (12.4.8), setting $s = 0$, and using (12.4.10) gives

$$\begin{aligned} x_1(t) &= 2(1-p)x_0(t-1)x_1(t-1) + px_0(t-1) \\ &= (1-p)^{2^{t-1}-1}[2(1-p)x_1(t-1)+p], \quad x_1(1) = p. \end{aligned} \tag{12.4.11}$$

While this equation and, to a greater extent, those for the larger values of $k$ are hard to solve in closed form (and we won't do that), they present no difficulty numerically. The reader can check that the next two equations for the $x_k$ are given by

$$\begin{aligned} 2x_2(t) &= 2(1-p)[2x_0(t-1)x_2(t-1) + x_1^2(t-1)] \\ &\quad + 2px_1(t-1), \qquad x_2(1) = 0, \end{aligned} \tag{12.4.12}$$

and

$$\begin{aligned} 3!x_3(t) &= 2(1-p)[3!x_1(t-1)x_2(t-1) + 3!x_0(t-1)x_3(t-1)] \\ &\quad + 3!px_2(t-1), \qquad x_3(1) = 0. \end{aligned} \tag{12.4.13}$$

These may be calculated in MAPLE and MATLAB with the following programs:

```
MAPLE
> Digits:=60:
> x0:=array(1..22): x1:=array(1..22):
> x2:=array(1..22): x3:=array(1..22):
> expTumors:=(p,t)->(p/(1-p))*((2-p)^t-1):
> x0calc:=proc(p)
  global x0; local i;
    for i from 1 to 22 do
      x0[i]:= (1-p)^(2^i-1);
    od;
  end:
> x1calc:=proc(p)
  global x0, x1; local i;
    x1[1]:=p;
    for i from 2 to 22 do
      x1[i]:=x0[i-1]*(2*(1-p)*x1[i-1]+p);
    od;
  end:
> x2calc:=proc(p)
  global x0, x1, x2; local i;
    x2[1]:=0;
    for i from 2 to 22 do
      x2[i]:=(1-p)*(2*x0[i-1]*x2[i-1]+x1[i-1]^2)+p*x1[i-1];
    od;
  end:
> x3calc:=proc(p)
  global x0, x1, x2, x3; local i;
    x3[1]:=0;
    for i from 2 to 22 do
      x3[i]:=2*(1-p)*(x1[i-1]*x2[i-1]+x0[i-1]*x3[i-1])+p*x2[i-1];
    od;
  end:
> p:=0.0000007;
> x0calc(p); x1calc(p); x2calc(p); x3calc(p);
```

```
MATLAB
% make an m-file, retinox0.m, with
% function x0=retinox0(p)
% for i=1:22
% x0(i)=(1-p)^(2^i-1);
% end
% make an m-file, retinox1.m, with
% function x1=retinox1(p,x0)
% x1(1)=p;
% for i=2:22
%    x1(i)=x0(i-1)*(2*(1-p)*x1(i-1)+p);
% end
% make an m-file, retinox2.m, with
% function x2=retinox2(p,x0,x1)
% x2(1)=0;
% for i=2:22
%    x2(i)=(1-p)*(2*x0(i-1)*x2(i-1)+x1(i-1)^2)+p*x1(i-1);
% end
% make an m-file, retinox3.m, with
% function x3=retinox3(p,x0,x1,x2)
% x3(1)=0;
% for i=2:22
%    x3(i)=2*(1-p)*(x1(i-1)*x2(i-1)+x0(i-1)*x3(i-1))+p*x2(i-1);
% end
> p=.0000007;
> x0=retinox0(p)
> x1=retinox1(p,x0)
> x2=retinox2(p,x0,x1)
> x3=retinox3(p,x0,x1,x2)
```

## 12.5 Application to the Retinoblastoma Data

In this section, we apply the model to the summary statistics. The predicted results are reported on the right side of Table 12.4.2. It can be seen that the model is consistent with the data. First, we use statistic 1 to calculate $p$.

From (12.4.10) in conjunction with statistic 1, carriers never affected, which should be about 5%, we get a value for $p$,

$$0.05 = x_0(22) = (1-p)^{2^{22}-1}, \qquad \log(1-p) = \frac{\log(0.05)}{2^{22}-1} = -0.0000007142.$$

So $(1-p) = 0.9999992858$ and $p = 7.14 \times 10^{-7}$. Thus the probability of a mutation occurring upon cell division that renders a tumor suppressor gene dysfunctional is about $7 \times 10^{-7}$. A very small value, but in the production of a full complement of retinal cells, there are about $4 \times 10^6$ cell divisions. (The last of the synchronous cell divisions involves $2 \times 10^6$ cells by itself.)

With this value of $p$, (12.4.9) gives the expected number of tumors in the germinal case, statistic 2, to be 2.99; compare with 3.

Denote by $g_1$ the unilateral cases among germinal cases, statistic 3. It is calculated by the infinite series

$$g_1 = x_1(22) + \frac{1}{2}x_2(22) + \frac{1}{2^2}x_3(22) + \cdots \tag{12.5.1}$$

for the following reason: There could be one tumor after all cell divisions; this is $x_1(22)$. If there are two tumors, with probability $\frac{1}{2}$ both are in the same eye. Similarly, if there are three, with probability $\frac{1}{4}$ all three are in the same eye because with probability $\frac{1}{2}$ the second is in the same eye as the first and with another probability $\frac{1}{2}$ the third is in that eye, too. And so on, but the remaining terms are small, so we stop with these three terms. Now step through the recurrence relations, (12.4.11), (12.4.12), (12.4.13), numerically with the value of $p$ found above to get $g_1 = 0.337$,

$$x_1(22) + \frac{1}{2}x_2(22) + \frac{1}{2^2}x_3(22) + \cdots \approx 0.337.$$

Of course, statistic 4, bilateral cases among germinal cases, is the complement of this at 0.663.

*The model also predicts carrier frequency.*

Let $f$ denote the fraction of carriers in the general population. Knowing $f$ would allow us to estimate the other statistics of Table 12.4.2. Or we can use one of them to find $f$ and then estimate the rest. The most reliable statistic is 5, the fraction of cases that are bilateral, about 27%; denote it by $B$:

$$B = \frac{\text{bilateral cases}}{\text{all cases}}.$$

Now by the assumption that all bilateral cases are germinal, statistic 13, the numerator will be

$$f(1 - g_1).$$

For the denominator, the overall incidence rate of retinoblastoma, counting both germinal and somatic cases, is given by $f(1 - x_0(22)) + (1 - f)u$, where, as above, $u$ is the probability of a somatic case. Hence we have for $B$,

$$B = \frac{f(1 - g_1)}{f(1 - x_0(22) + (1 - f)u}. \tag{12.5.2}$$

Solve this for $f$ to get

$$f = \frac{Bu}{1 - g_1 + Bu - B(1 - x_0(22))}. \tag{12.5.3}$$

Substituting the numerical values, including $u = 30 \times 10^{-6}$ from statistic 7, we get

$$f = 0.0000274,$$

or about 27 carriers per million in the general population.

Now the remaining statistics of Table 12.4.2 can be calculated. The fraction of germinal cases among all cases is given by

$$\frac{f(1 - x_0(22))}{f(1 - x_0(22)) + (1 - f)u} = 0.387. \tag{12.5.4}$$

Of course, the somatic cases are the complementary fraction.

The fraction of all cases that are unilateral is given by (explain)

$$\frac{f(x_1(22)+\frac{1}{2}x_2(22)+\frac{1}{2^2}x_3(22)+\cdots)+(1-f)u}{f(1-x_0(22))+(1-f)u}=0.750. \tag{12.5.5}$$

The fraction of unilateral and heredity cases among all cases is given by

$$f(x_1(22)+\frac{1}{2}x_2(22)+\frac{1}{2^2}x_3(22)+\cdots)f(1-x_0(22))+(1-f)u=0.137. \tag{12.5.6}$$

And finally, the fraction of unilateral cases that are germinal is

$$\frac{f(x_1(22)+\frac{1}{2}x_2(22)+\frac{1}{2^2}x_3(22)+\cdots)}{f(x_1(22)+\frac{1}{2}x_2(22)+\frac{1}{2^2}x_3(22)+\cdots)+(1-f)u}=0.183. \tag{12.5.7}$$

## 12.6 Persistence of Germinal Cases

It is natural to wonder how germinal cases arise: familial or new. In this section, we show that carriers are not persistent in society in that the genetic defect lasts at most two generations normally. Therefore, the condition is also the result of a chance mutation.

*Carriers can persist in the population.*

Let $q_k$ be the probability that a carrier zygote will survive and beget $k$ offspring who are also carriers, and let $F(s)=q_0+q_1s+q_2s^2+\cdots$ be the probability-generating function for the $q_k$. To compute the $q_k$, we also need the probabilities $c_i$ that a surviving carrier will beget $i$ offspring. Finally, let $p_0$ be the probability that a carrier will survive to adulthood, taken as 0.05 from Table 12.4.2. Assuming that a carrier mates with a noncarrier, we get, for $k>0$,

$$q_k=p_0\left(c_k\frac{1}{2^k}+c_{k+1}\binom{k+1}{k}\frac{1}{2^{k+1}}+c_{k+2}\binom{k+2}{k}\frac{1}{2^{k+2}}+\cdots\right),$$

since with probability $\frac{1}{2}$ an offspring is a carrier or a noncarrier. And

$$q_0=1-p_0+p_0\left(c_0+c_1\frac{1}{2}+c_2\frac{1}{2^2}+\cdots\right). \tag{12.6.1}$$

Standard theory provides that a trait will persist with probability $1-V$ and die out with probability $V$, where $V$ is the smallest fixed point of $F$, i.e., the smallest solution of $F(L)=L$; see also Section 13.7. Since already $F(0)=q_0>0.95$, $L$ must be very close to 1.

Actually, $L<1$ if and only if $F'(1)\geq 1$. But

$$F'(1) = q_1 + 2q_2 + 3q_3 + \cdots, \tag{12.6.2}$$

and since each term is multiplied by $p_0 = 0.05$, even with extraordinary high values of the $c_i$ for large $i$, $F'(1)$ will be less than 1. For example, suppose $c_5 = 1$, i.e., an average carrier has five children. Then $q_k = 0.05\frac{\binom{6}{k}}{2^6}$, $k = 1, \ldots, 6$ and $F'(1) = \frac{192}{1280}$.

## Exercises/Experiments

1. The first three statistics in Table 12.4.2 depend only on $p$. Plot these statistics as a function of $p$ varying, say, from $5 \times 10^{-7}$ to $8 \times 10^{-7}$. Recall that the code is given on p. 411 and $g_1$ is given by (12.5.1). Explain your findings.
2. It can be shown that $u$, the probability of a somatic case, can also be calculated from $p$; in fact, $u = 1 - h(1, 22)$, where $h(1,t)$ is given by the recurrence equation[1]
$$h(1,t) = 2px_0(t-1)h(1,t-1) + (1-2p)h^2(1,t-1).$$
As in Exercise 1 above, experiment with the behavior of $u$ as a function of $p$. What value of $p$ makes $u = 30 \times 10^{-6}$, 30 per million? What are the values of the first three statistics of Table 12.4.2 for this value of $p$?
3. As in Exercise 2 above, $u$ can be calculated knowing $p$. If also $B$ is given a value, statistic 5 of Table 12.4.2, then all the other statistics can be calculated from these two; see (12.5.3), (12.5.4), (12.5.5), (12.5.6), and (12.5.7). For each of the four extreme values of statistics 1 and 5 of Table 12.4.2, namely,
(a) carriers never affected = 1%, $B = 25\%$,
(b) carriers never affected = 1%, $B = 30\%$,
(c) carriers never affected = 10%, $B = 25\%$,
(d) carriers never affected = 10%, $B = 30\%$,
tabulate the other statistics of the table.
4. It is not necessarily the case that the retinal cell divisions proceed uniformly in time. Suppose the time $t$ in months from conception for the $i$th cell division, $0 \le i \le 22$, is given by $i = 22(1 - e^{-t/6})$.
(a) How many cell divisions have occurred by birth, $t = 9$?

---

[1] Let $r_{n,0}(t) = \Pr(n$ cells with a single mutation but no tumors at time $t)$. Decompose on the possibilities at $t = 1$,

$$\begin{aligned} r_{n,0}(t) &= \Pr(n \text{ with single mutation, no tumor at } t \mid \text{a mutation at } 1)2p \\ &\quad + \Pr(n \text{ with single mutation, no tumor at } t \mid \text{no mutation at } 1)(1-2p) \\ &= 2px_0(t-1)r_{n-2^{t-1},0}(t-1) + (1-2p)\sum_{n_1+n_2=n} r_{n_1,0}(t-1)r_{n_2,0}(t-1). \end{aligned}$$

Define the generating function $h(s,t) = \sum_n r_{n,0}(t)s^n$ and proceed as in the other derivations with generating functions.

(b) What is the chance the infant is born with a tumor if $p = 7.14 \times 10^{-7}$ and what is the expected number of tumors at that time? (*Hint*: In the computer codes, stop the `for` loops at the value of $i$ of part (a).)

(c) When does the 22nd cell division occur?

**5.** Referring to Section 12.6, assume that each carrier mates with a noncarrier and has exactly five children. What is the minimum value of $p_0$ that a carrier will survive to adulthood, in order that $F'(1) \geq 1$ (see (12.6.2)), i.e., that the single tumor suppressor gene mutation persists?

### Questions for Thought and Discussion

**1.** In what ways will knowing the human genome be a help in fighting cancer?

**2.** The retinoblastoma model measures time in terms of cell division generation. What observations are necessary to map this sense of time to real time?

**3.** The mechanism in the text offers an explanation as to why the telomeres are shorter on the duplicate chromosomes. What about the original chromosomes used as the templates?

## References and Suggested Further Reading

[1] B. Vogelstein and K. W. Kinzler, The multistep nature of cancer, *Trends Genet.*, **9**-4 (1993), 138–141.

[2] A. G. Knudson, Jr., Mutation and cancer: Statistical study of retinoblastoma, *Proc. Nat. Acad. Sci. USA*, **68**-4 (1971), 820–823.

[3] A. G. Knudson, Jr., H. W. Hethcote, and B. W. Brown, Mutation and childhood cancer: A probabilistic model for the incidence of retinoblastoma, *Proc. Nat. Acad. Sci. USA*, **72**-12 (1975), 5116–5120.

[4] H. W. Hethcote and A. G. Knudson, Jr., Model for the incidence of embryonal cancers: Application to retinoblastoma, *Proc. Nat. Acad. Sci. USA*, **75**-5 (1978), 2453–2457.

[5] G. M. Cooper, *Oncogenes*, 2nd ed., Jones and Bartlett, Boston, 1995.

# Part III

# Genomics

# 13

# Genetics

## Introduction

In this chapter, we will study the ways that genetic information is passed between generations and how it is expressed. Cells can make exact copies of themselves through asexual reproduction. The genes such cells carry can be turned off and on to vary the cells' behaviors, but the basic information they contain can be changed only by mutation, a process that is somewhat rare to begin with and usually kills the cell anyway.

Genetic material is mixed in sexual reproduction, but the result of such mixing is seldom expressed as a "blend" of the properties' expressions. Rather, the rules for the combination of genetic information are somewhat complex. Sexual reproduction thus results in offspring that are different from the parents. Much research shows that the ultimate genetic source of this variation is mutation, but the most immediate source is the scrambling of preexisting mutations.

The variations produced by sexual reproduction serve as a basis for evolutionary selection, preserving the most desirable properties in a particular environmental context.

## 13.1 Asexual Cell Reproduction: Mitosis

*Asexual reproduction of a cell results from the copying and equal distribution of the genetic material of a single cell. Each resultant daughter cell then possesses the same genes as the parent cell. If we are considering a single-celled organism, an environment for which the parent cell is suited should therefore also be suitable for the daughter cells. If we are considering a multicellular organism, the daughter cell may take on functions different from those of the parent cell by selectively turning genes off. This process creates the various tissues of a typical multicellular organism.*

*In this section, we will take a brief look at the mitosis cell division cycle, the process by which a cell reproduces an exact copy of itself. This is complementary to*

R.W. Shonkwiler and J. Herod, *Mathematical Biology: An Introduction with Maple and Matlab*, Undergraduate Texts in Mathematics, DOI: 10.1007/978-0-387-70984-0_13,

*the more detailed scrutiny of the cell cycle undertaken in Chapter 12 needed for the discussion of cancer.*

*Eukaryotic mitosis gives each of two daughter cells the same genes that the parent cell had.*

The actual process of eukaryotic mitosis is comparable to a movie, with sometimes complex actions flowing smoothly into one another, without breaks. For reference, however, mitosis is usually described in terms of five specific stages, named interphase, prophase, metaphase, anaphase, and telophase. It is important to remember, however, that a cell does not jump from one stage to the next. Rather, these stages are like "freeze-frames," or preserved instants; they are guideposts taken from the continuous action of mitosis (see [1] for further discussion).

Most of the time a cell's nucleus appears not to be active; this period is called *interphase*. If one adds to an interphase cell a stain that is preferentially taken up by nuclei and then examines the cell through a microscope, the nucleus appears to have no internal structure over long periods of time. This appearance is actually quite misleading, because, in fact, the nucleus is very active at this time. Its activity, however, is not reflected in changes in its outward appearance. For example, the addition of radioactive thymine to an interphase cell often leads to the formation of radioactive DNA. Clearly DNA synthesis takes place in interphase, but it does not change the appearance of the nucleus.

Biologists further subdivide interphase into three periods: $G_1$, during which preparations for DNA synthesis are made; S, during which DNA is synthesized; and $G_2$, during which preparations are made for actual cell division. (The "G" stands for "gap.") If we could see the DNA of a human skin cell during $G_1$ we would find 46 molecules. Each molecule, as usual, consists of two covalent polynucleotides, the two polymers being hydrogen-bonded to one another in a double helix. Genetic information is linearly encoded into the base sequence of these polynucleotides.

When we discussed DNA structure in Chapter 8, we associated a gene with the nucleic acid information necessary to code for one polypeptide. Thus a gene would be a string, not necessarily contiguous, of perhaps a few hundred to a few thousand bases within a DNA molecule. It is convenient to define a gene in another way, as a *functional unit of heredity*, a definition that has the virtue of generality. It can therefore include the DNA that codes for transfer RNA or ribosomal RNA, or it can just be a section of DNA that determines a particular observable property, such as wing shape or flower color. In this general definition, each DNA molecule is called a *chromosome*, where each genetic region, or *gene locus*, on the chromosome, determines a particular observable property.

In Figure 13.1.1, one chromosome is illustrated for a cell progressing through mitosis. (A human skin cell, for example, has 46 chromosomes, and each one behaves like the one in the figure.) The structure of the chromosome at $G_2$ cannot be seen in a microscope, so we must surmise its structure by its appearance in the next stage (prophase).

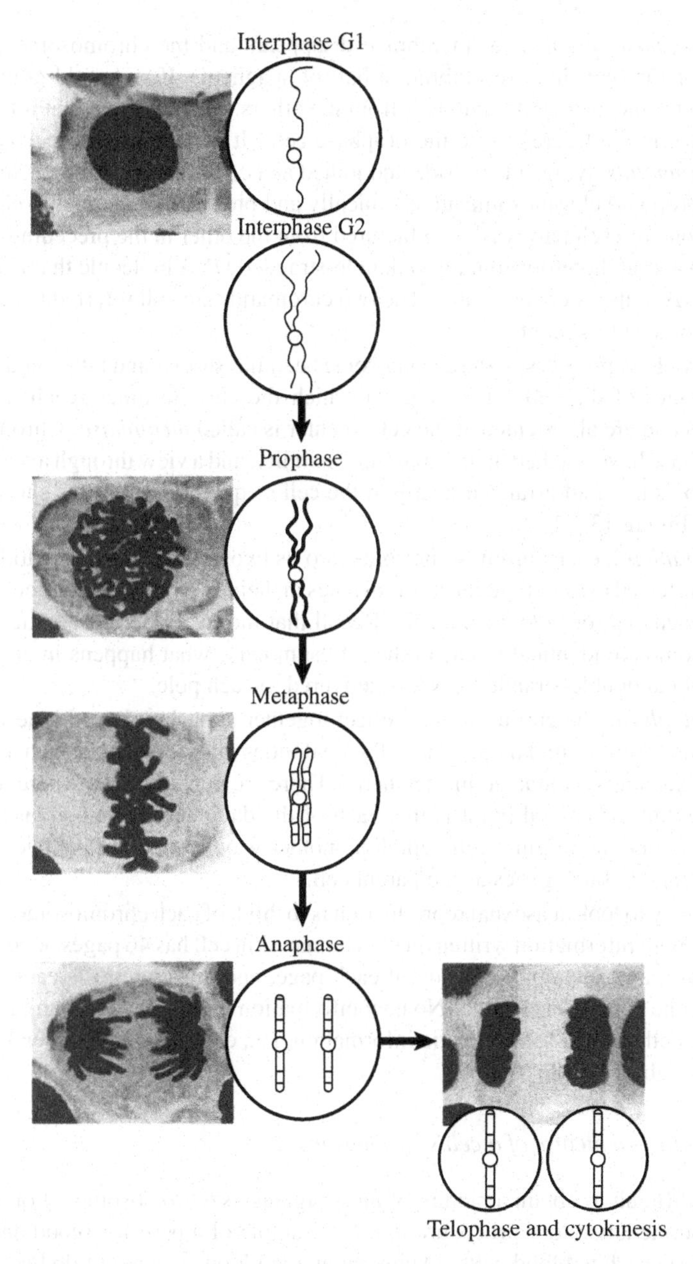


**Fig. 13.1.1.** The stages of mitosis. The figure shows actual photographs of a dividing cell's chromosomes. The line drawings show how the individual chromosomes are behaving during that stage of division. During mitosis, each chromosome replicates lengthwise and the two copies go to different daughter cells. Thus each daughter cell ends up with exactly the same genetic complement as the parent cell. (Photos of mitosis taken from *Radiation and Chromosomes Biokit*, item F6-17-1148, Carolina Biological Supply Company, Burlington, NC. Used with permission.)

At *prophase*, the nuclear membrane disappears and the chromosomes become visible for the first time, resembling a ball of spaghetti. If we could grab a loose end of a chromosome and separate it from the others, we would see that it looks like that shown in the figure beside the prophase cell. It consists of two halves, called sister *chromatids*, lying side by side and joined at a *centromere*. The two chromatids of each prophase chromosome are chemically and physically identical to each other because one of each pair was manufactured from the other in the preceding S phase. Each chromatid therefore contains a double-stranded DNA molecule that is identical to the DNA of its sister chromatid. The two chromatids are still referred to as a single chromosome at this point.

As prophase progresses, the chromosome becomes shorter and fatter, and it moves to the center of the cell. The stage at which the chromosomes reach maximum thickness and are all oriented at the cell's center is called *metaphase*. Chromosomes at metaphase have reached their maximum visibility, and a view through a microscope often shows them all arranged neatly in the cell's equatorial plane, as shown in the photo in Figure 13.1.1.

At *anaphase*, each chromosome splits into its two component chromatids, which are now referred to as individual chromosomes in their own right, and one copy moves toward each end, or *pole*, of the cell. Recall that the two sister chromatids of each chromosome are identical to each other. In summary, what happens in anaphase is that identical double-stranded DNA is delivered to each pole.

At *telophase*, the chromosomes collect together at each pole and a new nuclear membrane forms around them. The cell then divides its cytoplasm in such a way that one new nucleus is contained in each half.[1] There are now two cells where there was only one, but the crucial point is that each of the daughter cells now has the same DNA code that the original cell had. Put another way, two cells have been formed, each having the same genes as the parent cell.

One way to look at asexual reproduction is to think of each chromosome as a piece of paper, with information written on it. A human skin cell has 46 pages, labeled 1–46. At S phase, an exact copy is made of each page, and during mitosis each daughter cell gets one copy of each page. No new information is created, nor is any lost. Each daughter cell gets the same genetic information, i.e., each daughter cell ends up with 46 pages, labeled 1–46.

*A karyotype is a picture of a cell's chromosomes.*

It is not difficult to obtain a picture of most organisms' chromosomes. For example, it is a routine laboratory procedure to take a sample of a person's blood and isolate some of their white blood cells. (Mammalian red blood cells won't do because they lose their nuclei as they mature.) These white cells are then cultured in a test tube and their nuclear material is stained as they enter metaphase, which is when chromosomes are most easily visualized. The cell, with its chromosomes, is photographed through a microscope. The chromosomes are then cut out of the photograph and arranged

[1] The actual splitting of the cell is called *cytokinesis*.

in a row, according to size. This picture is a *karyotype*. An example is shown in Figure 13.1.2.

There are several interesting features of the illustrated karyotype:

1. These are metaphase chromosomes and therefore are lengthwise doubled, joined at a centromere. Each chromosome consists of two chromatids, a feature that sometimes confuses students. The problem is that the chromosomes must be photographed at metaphase because that is when they are most easily visible and distinguishable from one another. This is also the point at which they are in a duplex form. You may want to refer back to the discussion of Figure 13.1.1 to clarify the distinction between chromosome and chromatid.
2. There are 46 chromosomes in this cell. This is the number found in most of the cells of the human body, the exceptions being mature red blood cells, which lack nuclei, and certain cells of the reproductive system, called *germinal cells*, to be discussed later in this chapter. Any cells of our body that are not germinal are said to be *somatic* cells, a category that therefore includes virtually the entire bulk of our body: skin, blood, nervous system, muscles, the structural part of the reproductive system, etc. Our somatic cell chromosome number is thus 46.
3. The chromosomes in the karyotype seem to occur in identical-appearing pairs, called *homologous pairs*. Evidently, our human chromosomal complement is actually two sets of 23 chromosomes. It is very important to understand the difference between a homologous pair of chromosomes and the two chromatids of a single metaphase chromosome. The karyotype shows 23 homologous pairs; each member of each pair consists of two chromatids. Each chromatid contains a double-helical DNA molecule that is identical to the DNA of its sister chromatid, but which is different from the DNA of any other chromatid.

*Asexual reproduction can generate daughter cells that differ from each other.*

We could imagine an amoeba, a common single-celled eukaryote, dividing by mitosis to yield two identical amoebas. We could just as easily imagine a skin cell of a human, a multicellular eukaryote, dividing by mitosis to give two identical human skin cells. Indeed, this is the way that our skin normally replaces those cells that die or are rubbed off. In both cases, the daughter cells have the same DNA base sequence that the parent cell had, and that is reflected in the identical physiology and appearance of the daughter cells.

There is another possibility: consider a single fertilized human egg. It divides by mitosis repeatedly to form a multicellular human, but the cells of a developed human are of many sizes, shapes, and physiological behaviors. Liver cells look and behave one way, nerve cells another, and muscle cells still another. Mitosis seems not to have been conserved. How could cells that have exactly replicated their DNA in mitosis and then partitioned it out equally have yielded different progeny cells?

One possibility is that cells in each unique kind of tissue of a multicellular organism have lost all their genes except those essential to the proper functioning of that particular tissue. Thus liver cells would have retained only those genes needed for

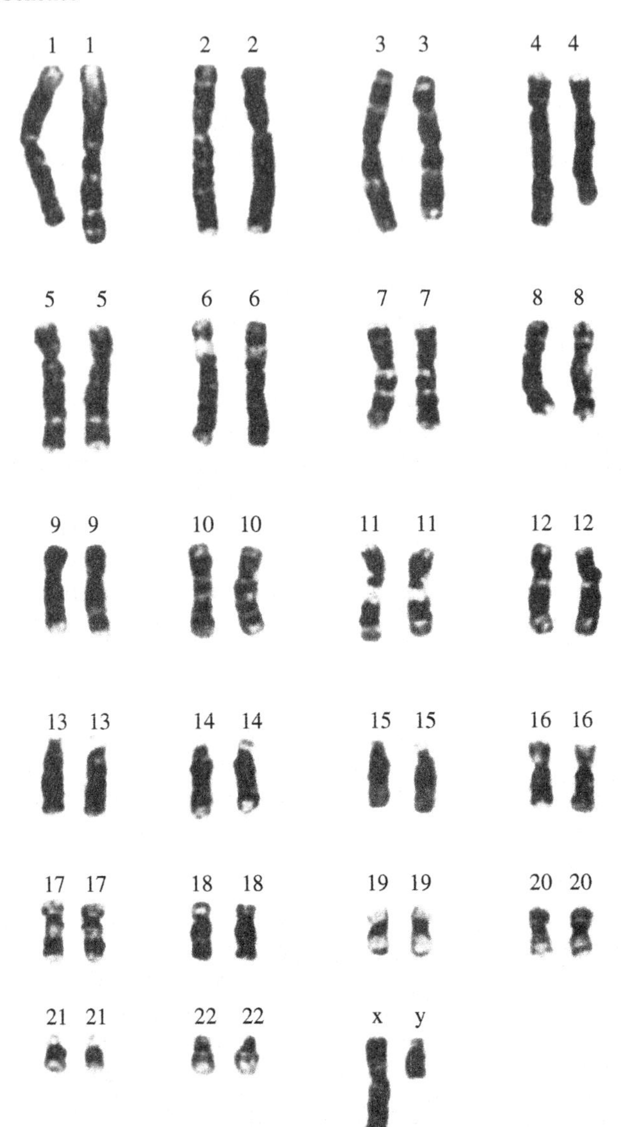


**Fig. 13.1.2.** A karyotype of a normal human male. The chromosomes were photographed at metaphase and images of the individual chromosomes were then cut out and arranged by size. The result is a group of 22 chromosome pairs, called homologs, each pair of which is matched by length, centromere location, and staining pattern. Because this is a male's karyotype, the 23rd pair of chromosomes (sex chromosomes; X and Y) do not match each other. Each of the chromosomes shown in the figure consists of two identical daughter chromatids, but they are so closely associated that they are often indistinguishable at metaphase. However, note the right-hand homolog of number 18; the two chromatids can be distinguished. (Photo of karyotype arranged from *Human Karyotypes, Normal Male*, item F6-17-3832, Carolina Biological Supply Company; Burlington, NC. Used with permission.)

liver functioning and muscle cells would have retained only those genes needed for muscle functioning. This possibility is easy to reject by a simple experiment: In the cells of a plant stem the genes necessary for stem growth and function are obviously active, and there is no evidence of genes involving root formation. If the stem is broken off and the broken end inserted into soil, within a few weeks the plant will often start to grow roots at the broken stem end. Clearly the genes for root growth and function were in the cells of the stem all along, but were reversibly turned off. A similar experiment has been done on a vertebrate, in which a nucleus from a specialized somatic tissue, the intestinal lining of a tadpole, has been used to grow a whole tadpole and the subsequent toad. We can conclude that mitosis generates different tissues of multicellular organisms when selected genes are turned off or on in the course of, or in spite of, asexual cell division.

The process by which unspecialized cells of a multicellular organism take up specialized roles—liver, nerve, skin, etc.—is called *differentiation*. Differentiation is not restricted to embryos, but can occur all our lives, e.g., in bone marrow, where unspecialized stem cells can become specialized blood cells. Differentiation is only one part of *development*, which includes all the changes in an organism in its life, from conception to death. Other aspects of development include tissue growth and deterioration, as described in Section 9.2.

*Some cell types rarely divide.*

Certain cells of multicellular organisms seem to have a very limited, even nonexistent, capacity for division. For example, muscle cells don't divide; the muscle enlargement associated with exercise comes from cellular enlargement. Fat cells get larger or smaller, but their numbers stay the same (which is why cosmetic liposuction works—the lost fat cells can't be replaced). Cells of the central nervous system don't divide, which explains the seriousness of spinal injuries. Liver cells rarely divide unless part of the liver is cut away—in which case the liver cells undergo division to replace those removed. Note the implication here: Genes controlling liver cell division haven't been lost. They were shut off, and can be reactivated.

## 13.2 Sexual Reproduction: Meiosis and Fertilization

*Sexual reproduction involves the creation of an offspring that contains genetic contributions from two parents. A type of cell division called meiosis halves the chromosome number of germinal cells to produce sperms or eggs. A sperm and an egg then combine in fertilization to restore the double chromosome number. The new offspring now has genetic information from two sources for every characteristic. The ways that these two sets of information combine to produce a single property are complex, and this is the subject of the study of classical genetics.*

*Sexual reproduction provides variation upon which evolutionary selection can act.*

Recall the Darwinian model: More organisms are born than can survive, and they exhibit variability. Those with favored characteristics survive and may pass the favored properties to their offspring. It is tempting to credit genetic mutation with this variability and let it go at that. The fact is that all of the ten (nontwin) children in a hypothetical large family look different and virtually *none* of the variations among them are the result of mutations in their, or their parents', generation. This fact, surprising at first, seems more reasonable when we consider the accuracy of DNA base pairing, the "proofreading" capability of some kinds of DNA polymerase and the existence of repair mechanisms to correct DNA damaged by such mutagens as radiation. Thus DNA sequences tend to be conserved over many generations. We can therefore conclude that most of the variations among the ten children of the same family are the result of scrambling of existing genes, not the result of recent mutation. The cause of this shuffling of the genetic cards is sexual reproduction. Of course, the variant genes *originated* through mutation, but virtually all of them originated many generations earlier (see [2] for further discussion).

*Sexual reproduction involves the combination of genetic material from two parents into one offspring.*

Refer back to the karyotype in Figure 13.1.2. The human chromosome complement consists of 23 homologous pairs or, put another way, of two sets of 23 each. The sources of the two sets of 23 can be stated simply: we get one set from each of our parents when a sperm fertilizes an egg. What is not so simple is how the genetic material in those 46 chromosomes combines to make each of us what we are. The rules for combination will be the subject of Section 13.3. Our more immediate concern, however, is the means by which we generate cells with 23 chromosomes from cells having 46.

*Meiosis halves the chromosome number of cells.*

A special kind of reductional cell division, called *meiosis*, creates *gametes* having half the number of chromosomes found in somatic cells.[2]

The chromosomes are not partitioned at random however; rather, every gamete winds up *with exactly one random representative of each homologous pair*, giving it one basic set of 23 chromosomes. Such a cell is said to be *haploid*. A cell that has two basic sets of chromosomes is said to be *diploid*. We see that somatic cells are diploid and germinal cells are haploid. Thus meiosis in humans converts diploid cells, with a chromosome number of 46, to haploid cells with a chromosome number of 23.

Meiosis is diagramed in Figure 13.2.1 for a hypothetical organism having two homologous pairs; its diploid number is 4. Each chromosome is replicated in interphase and thus contains two identical chromatids joined at a centromere. In a departure from meiosis, homologs bind together, side by side, in a process called *synapsis*, to form *tetrads* consisting of two chromosomes (four chromatids). The homologs then separate to end the first meiotic division. Next, the chromatids separate to complete

[2] Gametes are often called *germinal cells* to distinguish them from somatic cells.

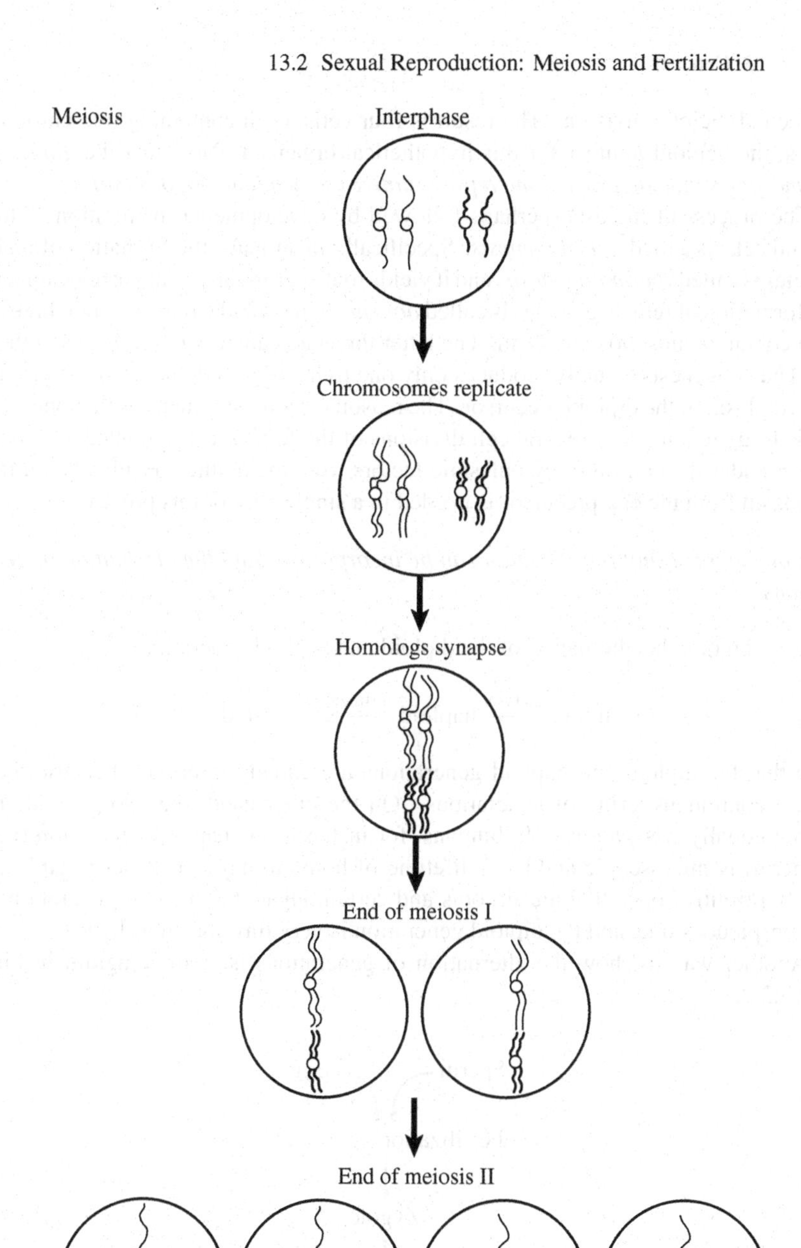


**Fig. 13.2.1.** The stages of meiosis. The cell shown has two homologous pairs. Each chromosome replicates lengthwise to form two chromatids, synapses to its homolog, and then two cell divisions ensue. The daughter cells each end up with exactly one representative of each homologous pair. Thus a diploid cell at the start of meiosis results in four haploid cells at the end of meiosis.

the second meiotic division. The result is four cells, each containing two chromosomes, the haploid number for this hypothetical organism. Note that the *gametes' chromosomes include exactly one representative of each homologous pair*.

The process of meiosis (perhaps followed by developmental maturation of the haploid cell) is called *gametogenesis*. Specifically in animals, the formation of male gametes is called *spermatogenesis*, and it yields four sperms, all similar in appearance. The formation of female gametes is called *oogenesis* and yields four cells, but three of them contain almost no cytoplasm. The latter three are called *polar bodies*, and they die. Thus oogenesis actually produces only one living egg, and that one contains all the cytoplasm of the diploid precursor. The reason for this asymmetry is that once the egg is fertilized, the first several cell divisions of the fertilized egg (called a *zygote*) remain under the control of cytoplasmic factors from the mother. Evidently, all the cytoplasm from the egg precursor is needed in a single egg for this process.

*The concept of sexual reproduction can be incorporated into the alternation of generations.*

We can diagram the alternation of the diploid and haploid generations:

$$\cdots \longrightarrow \text{diploid} \stackrel{\text{meiosis}}{\longrightarrow} \text{haploid} \stackrel{\text{fertilization}}{\longrightarrow} \text{diploid} \longrightarrow \cdots .$$

Note that the diploid and haploid generations are equally important because they form a continuous string of generations. On the other hand, the two generations are not equally *conspicuous*. In humans, for instance, the haploid generation (egg or sperm) is microscopic and has a lifetime of hours to days. In other organisms, mainly primitive ones like mushrooms and certain algae, the haploid generation is the conspicuous one, and the diploid generation is very tiny and short lived.

Another way to show the alternation of generations is in the diagram in Figure 13.2.2.

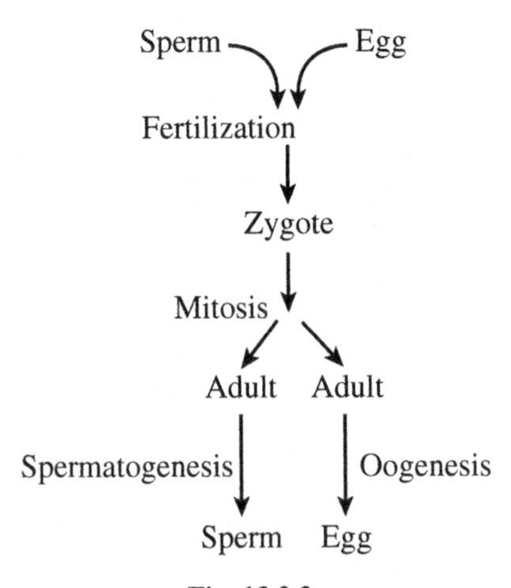


**Fig. 13.2.2.**

## 13.3 Classical Genetics

*Classical genetics describes the many ways that the genetic material of two parents combines to produce a single observable property. For instance, a red-flowered plant and a white-flowered plant usually produce an offspring with a single color in its flower. What that color will be is not predictable unless a geneticist has already studied flower colors in that plant—because there are about a dozen ways that parental genes can combine. We describe many of those ways in this section.*

*Classical genetics describes the result of interactions in genetic information.*

A diploid human cell carries 23 homologous pairs of chromosomes: One member of each pair comes from a sperm cell of the male parent, and the other member comes from an egg cell of the female parent. Other diploid organisms may have chromosome numbers ranging from a few up to hundreds, but the same principle about the origin of homologous pairs holds. What we will consider now is how the genetic information from the two parents combines to produce the characteristics that appear in the offspring and why the latter are so variable. Let us first examine a chromosome at $G_1$ phase, because that is the usual condition in a cell.

Genes, defined generally as functional units of heredity, are arranged linearly along the chromosome (Figure 13.3.1). Each gene locus affects some property, say, flower color or leaf shape in a plant. The order in which these loci appear is the

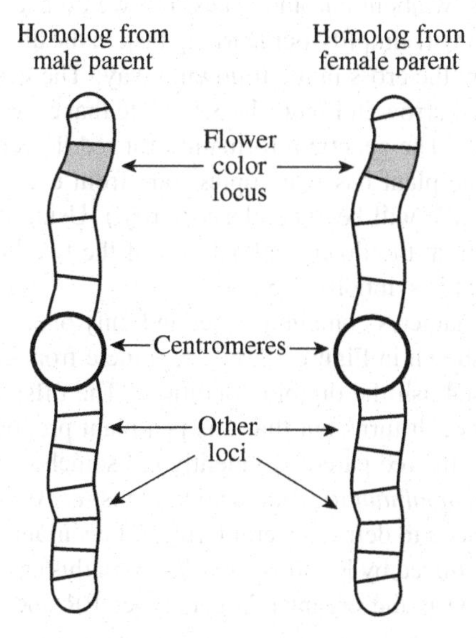


**Fig. 13.3.1.** This shows a simple model of the chromosome. The genes are lined up along the length of the chromosome, like beads on a string. A hypothetical flower color locus is labeled.

same on each member of the homologous pair. Thus it is common to refer to the "flower color" locus, meaning the section of either member of a homologous pair that is the gene that determines flower color. Clearly, each property is determined by two such sections, one on each homolog. *Each parent, then, contributes to each genetic property in the offspring.*

*The behavior of chromosomes provides a basis for the study of genetics.*

The pioneering geneticist was Gregor Mendel, who studied the genetics of peas, a common flowering plant. Peas, like many flowering plants, have male and female reproductive structures in the same flower. The male part makes pollen that is carried to the female part of that or another plant; the pollen then produces a sperm cell and fertilizes an egg. It a straightforward matter to dissect out the male part of a flower to prevent the plant from self-pollinating. Further, it is simple to use pollen from the male part of one plant to fertilize an egg of another plant and thus to make controlled matings. The seed that results from fertilizing an egg can be planted and the appearance of the offspring studied. The principles of chromosomal behavior and gene interaction in peas are the same as for humans.

Mendel had two groups, or populations, of plants that were *true breeding*. A population is true breeding if its freely interbreeding members always give rise to progeny that are identical to the parents, generation after generation. Members of a population of true-breeding red-flowered peas fertilize themselves or other members of the population for many generations, but only red-flowered plants ever appear. Mendel made a cross between a plant from a true-breeding red-flowered population and one from a true-breeding white-flowered population.

Mendel did not know about chromosomes, but we do and we will make use of that knowledge, which will simplify our learning task in the discussion to follow. We will therefore represent the cross in the following way: The gene for flower color is indicated by the labeled arrow in Figure 13.3.1. Note that each of the two homologs has such a gene locus.[3] The genetic information for red flower color is symbolized by the letter R, and the plant has two copies, one from each parent. (The reason for the copies being alike will become clear shortly.) Using the same convention, the genetic information at the flower color locus of the two homologs in the other (white-flowered) parent is symbolized by w.

Meiosis produces gametes containing one, and only one, representative of each homologous pair, as shown in Figure 13.3.2. A gamete from each parent combines at fertilization to reestablish the diploid condition. The offspring has flower color genetic information Rw. It turns out that this pea plant produces only red flowers, indistinguishable from the red parent. Evidently red somehow masks white; we say that red information is *dominant* to white, and white is *recessive* to red.

At this point, we need to define several terms. The variant forms of information for one property, symbolized by R and w, are *alleles*, in this case flower color alleles. The allelic composition is the organism's genotype; RR and ww are *homozygous*

[3] For learning purposes, we will ignore all other chromosomes, as if they do not have loci that affect flower color. In actual fact, this may not be true.

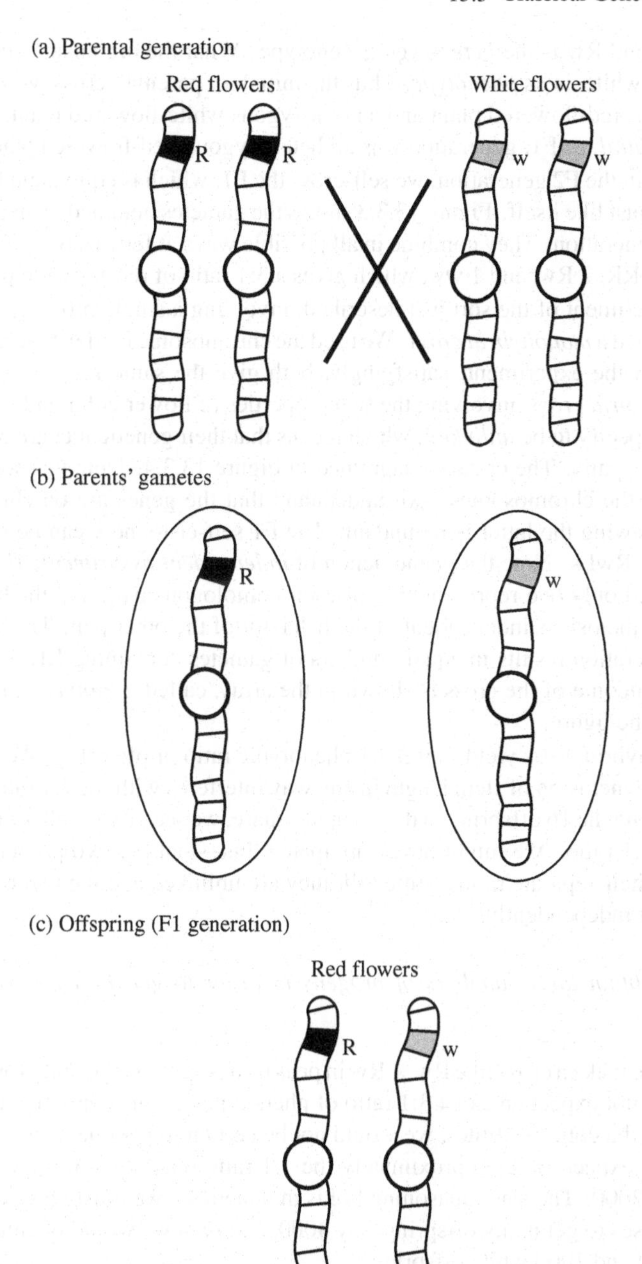


**Fig. 13.3.2.** The behavior of chromosomes and their individual loci during a cross between two homozygous parents. The parents (RR and ww) each contribute one chromosome from the homologous pair to form gametes. The gametes combine in fertilization to restore the diploid number of two. The offspring's flowers will be red.

genotypes and Rw is the *heterozygous* genotype. What the organism actually looks like, red or white, is its *phenotype*. Thus the initial, or parental, cross, was between a homozygous red-flowered plant and a homozygous white-flowered plant. The result in the first *filial*, or F1, generation was all heterozygous, red-flowered plants.

To obtain the F2 generation, we self-cross the F1, which is equivalent to crossing it with one just like itself. Figure 13.3.3 shows the gametes obtained from each parent in the F1 generation. They combine in all possible ways at fertilization. The result is a ratio of 1 RR, 2 Rw, and 1 ww, which gives a 3:1 ratio of red-to-white phenotypes.

An experiment of the sort just described, involving a single property like flower color, is called a *monohybrid cross*. We used the chromosome model, whereas Mendel actually ran the experiment; satisfyingly, both give the same results. Let us now make a *dihybrid cross*, involving the two properties of flower color and stem length, which we specify to be *unlinked*, which means that their genetic loci are on different homologous pairs. The cross is diagramed in Figure 13.3.4. Note that we have quit drawing in the chromosomes—we understand that the genes are on chromosomes and that drawing the latter is redundant. The F1 self-cross now can be represented as RwLs × RwLs. Note the phenomenon of *independent assortment*: Each gamete gets one and only one representative of each homologous pair, and the behavior of one pair in meiosis is independent of the behavior of the other pair. Thus meiosis in the F1 generation results in equal numbers of gametes containing RL, Rs, wL, and ws. The outcome of the cross is shown in the array, called a *Punnett square*, at the bottom of the figure.

The dihybrid cross yields a 9:3:3:1 phenotypic ratio of offspring. We should ask whether the inclusion of stem length in any way interferes with the 3:1 ratio of flower color. Among the 16 offspring in the Punnett square, we see 12 red and 3 white, which gives the 3:1 ratio. We might have anticipated this—that the two properties would not affect their separate ratios—after all, they are unlinked and the two homologous pairs assort independently.

*We must obtain large numbers of progeny in order to get the expected ratios of offspring.*

Suppose we make a cross like Rw × Rw in peas (red × red) and get only four progeny. We should not expect an exact 3:1 ratio of phenotypes in this experiment. After all, if we flipped a coin two times, we would not be certain to get one head and one tail. Rather, we expect to get approximately the 1:1 ratio only if we flip the coin many times, say 2000. The same reasoning holds in genetics—we must make enough Rw × Rw crosses to get many offspring, say 4000, and *then* we would obtain very close to 3000 red and 1000 white offspring.

The ratios 3:1 and 9:3:3:1 are often called *Mendelian ratios*, because they are what Mendel reported. There is a bit of a problem here: Statisticians have examined Mendel's data and some have concluded that the experimental data are too good, i.e., consistently too close to the 3:1 and 9:3:3:1 expected ratios. For the sample sizes Mendel reported, it would be expected that he would have gotten somewhat larger deviations from "Mendelian" ratios.

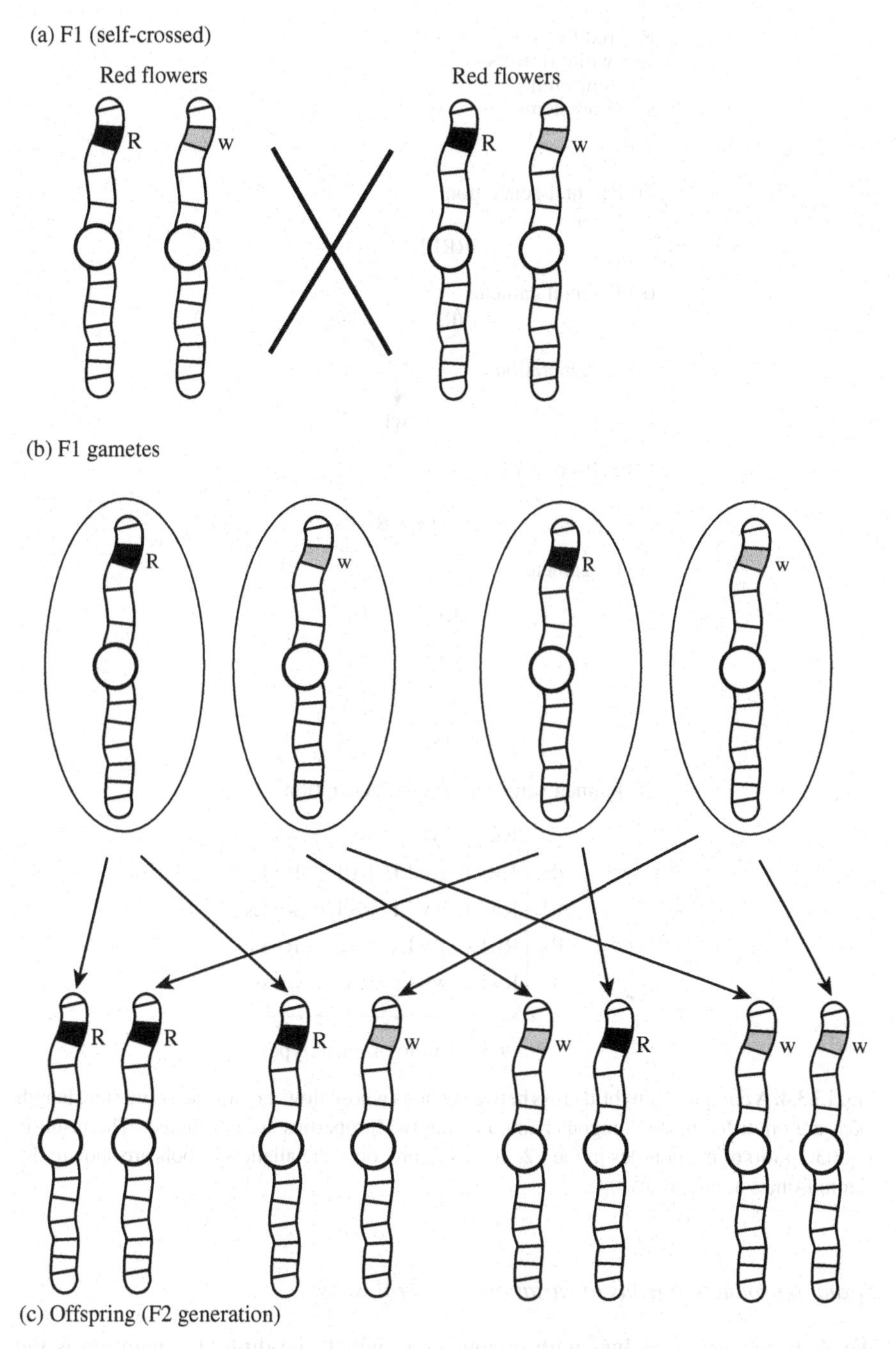


**Fig. 13.3.3.** A cross between two heterozygotes. Each F1 from Figure 13.3.2 makes gametes having the genes $R$ and $w$ with equal probability. When the gametes combine to make the F2 generation, the result is offspring of genotypes RR, Rw, and ww in the ratio 1:2:1.

R = red flowers
w = white flowers
L = long stems
s = short stems

(a) Parental generation

RRLL × wwss

(b) Parental gametes

(c) F1 generation

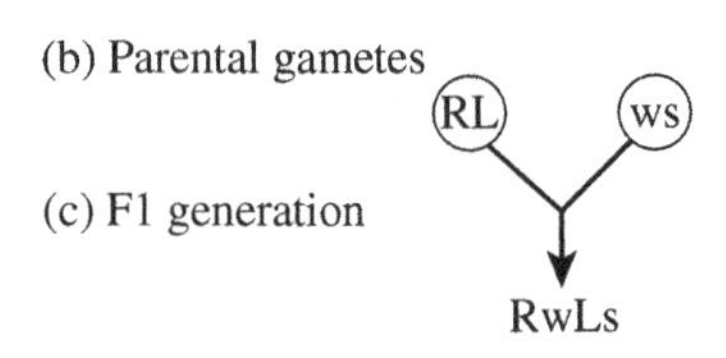


(d) Self-cross F1

RwLs × RwLs

(e) F1 gametes

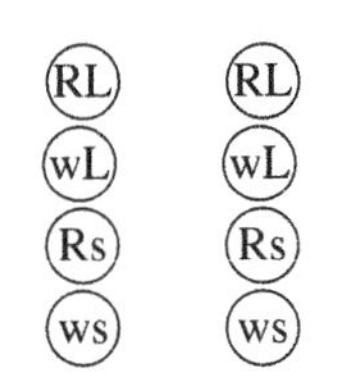


(f) Punnett square to give F2 generation

| | RL | wL | Rs | ws |
|---|---|---|---|---|
| RL | RRLL | RwLL | RRLs | RwLs |
| wL | RwLL | wwLL | RwLs | wwLs |
| Rs | RRLs | RwLs | RRss | Rwss |
| ws | RwLs | wwLs | Rwss | wwss |

9:3:3:1 ratio of phenotypes

**Fig. 13.3.4.** A complete dihybrid cross between plants whose flower color locus and stem length locus are on different homologous pairs, i.e., the two properties are not linked. The result is a 9:3:3:1 ratio of phenotypes in the F2. In this figure, only the allelic symbols are shown; the chromosomes are not drawn.

*Sexual reproduction leads to variation in several ways.*

We shall concern ourselves with organisms in which the diploid generation is the most conspicuous, e.g., humans, and we will examine the variations introduced into the diploid organism by sexual reproduction. It should always be borne in mind, however, that haploid organisms are under genetic control also.

Earlier it was pointed out that while mutation is the ultimate cause of genetic variation, there is only a very small chance that a given locus will mutate between two generations, will be unrepaired, and will not kill the cell. In spite of this, there are great variations among even the offspring of a single mating pair. We are now in a position to understand the sources of this immediate variation. First, look at the Punnett square of the dihybrid cross in Figure 13.3.4. Note that the F1 (RwLs) yields the gametes RL, Rs, wL, and ws, and yet the gametes of the parental generation were RL and ws. Thus two new combinations have turned up in the gametes of the F1. The reason is that the flower color locus and the stem length locus are unlinked—they are on different homologous pairs—and every homologous pair assorts independently of every other pair. Thus in the gametes of the F1, R paired up with L as often as R paired up with s. There were therefore $2^2 = 4$ combinations of chromosomes in the gamete. A human has 23 homologous pairs, all of which assort independently; thus a person can produce $2^{23}$ different combinations of chromosomes in their gametes, using independent assortment alone!

Second, when homologous chromosomes synapse they can exchange pieces in a process called *crossing over*. Let us cross two true-breeding parents, AABB $\times$ aabb, as shown in Figure 13.3.5. Notice that the two gene loci are *linked*, i.e., on the same chromosome. The F1 genotype is AaBb, and we *test-cross* it.[4] Some of the gametes of the F1 are the expected ones, AB and ab, but as the figure shows, crossing over, in which the homologs break and rejoin in a new way, produces gametes with two new allelic combinations, Ab and aB. These two new kinds of gametes, called *recombinant* gametes, are different from the gametes of either members of the parental generation. When the various gametes are paired up with the ab gametes in the test-cross, the following *phenotypes* appear in the F2 generation: Ab, aB, AB, and ab. The last two of these are the same phenotypes as the parental generation, and the first two are recombinant offspring, having phenotypes not seen in the previous crosses. We see that crossing over rearranges genetic material and presents novel phenotypes upon which selection can act.

How often does such crossing over occur? Actually, it is not unusual to find at least one example in every tetrad. Furthermore, crossing over is predictable: The farther apart two loci are, the more likely crossing over is to occur between them. The frequency of crossing over, measured by the frequency of recombinant offspring, is used by geneticists as a measure of the distance between two loci.

Note that we can account for an immense number of allelic combinations just using independent assortment and crossing over, without a mention of mutation. Independent assortment and crossing over account for virtually all the phenotypic variation seen in members of a single family generation. This variation, in the main, is what Darwinian selection works on.

A final point is worth mentioning here: Self-fertilization might be considered to be a limiting form of sexual reproduction.[5] Suppose that allele A is completely

---

[4] A test-cross is a cross with a homozygous recessive individual.

[5] Think of it this way: A self-cross is just like a cross between two separate, but genetically identical, parents.

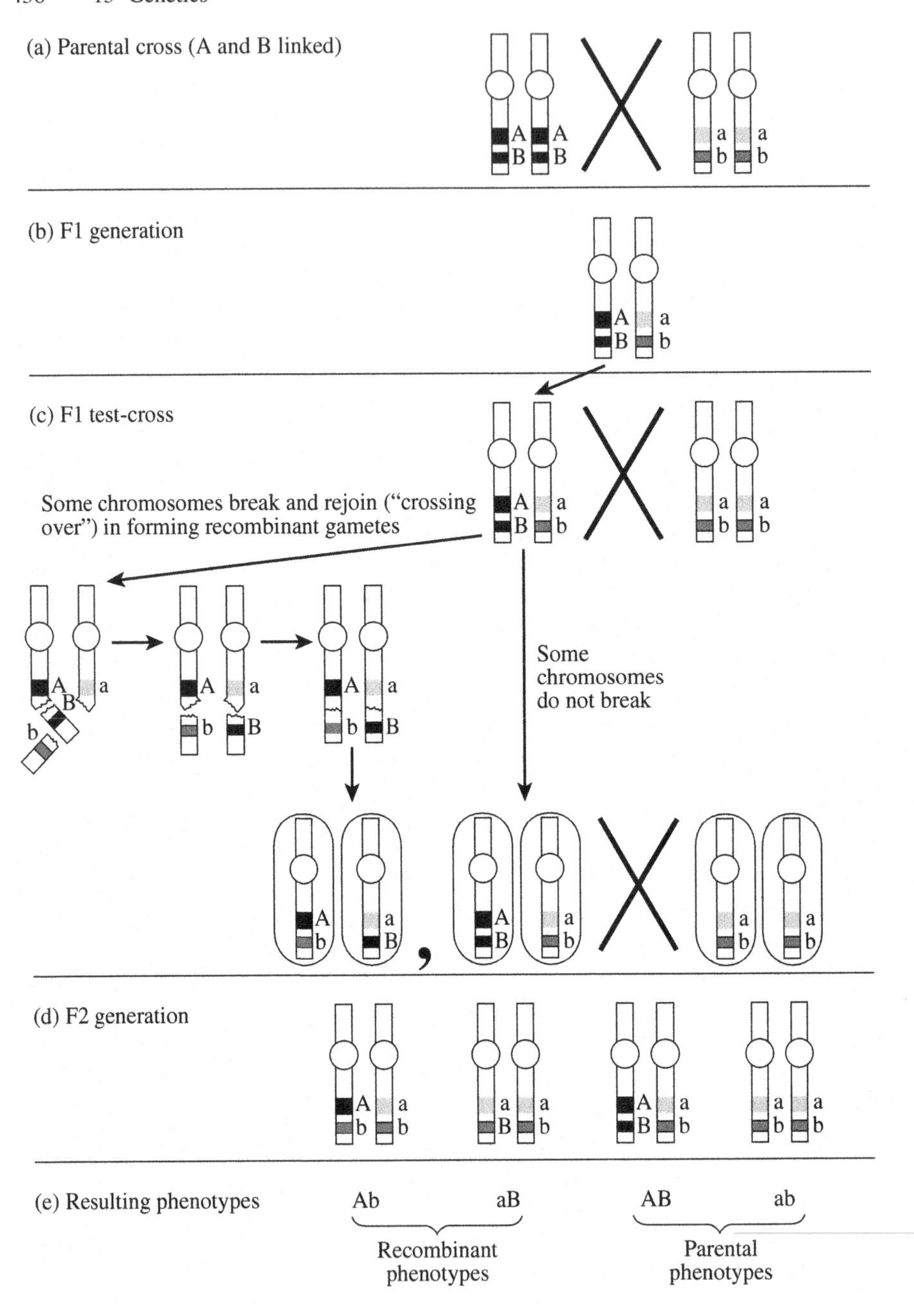


**Fig. 13.3.5.** A complete dihybrid cross, in which loci A and B are on the same chromosome, i.e., the two properties are linked. The results are predictable until the F1 test-cross at (c), when the chromosomes may break, yielding new combinations of the two loci. Notice that the resulting phenotypes at (e) include two (Ab and aB) that are unlike either of the two original parents.

dominant to allele a: If we self-cross an individual of genotype Aa, variant offspring appear in the ratio of 3:1, a mark of sexual reproduction. Asexual reproduction in the same organism yields only one kind of offspring—Aa. Where self-fertilizing organisms might run into evolutionary problems is in *continued* self-fertilization, which minimizes variation. This is shown by the following example: Take a population that is 100% heterozygotes (Aa) and self-cross all individuals. Note that the result is 50% heterozygotes and 50% homozygotes. Now self-cross all of that generation and note that 75% of the next generation will be homozygotes. After a few more generations of self-fertilization, virtually the entire population will be homozygous, either AA or aa. This can create problems for the population in two ways: First, suppose that the recessive allele is an unfavorable one that is usually masked by the dominant allele. As shown above, self-fertilization increases homozygosity, and homozygous recessive individuals would be selected out. Second, when homozygotes fertilize themselves, independent assortment and crossing over can occur, but they cannot generate variation. (You should verify this statement by schematically working out the cross.)

Here is an idea to think about: We sometimes hear about the "rescue" of a species that is near extinction. The last few members of the species are brought together to be bred in a controlled environment, free from whatever forces were causing the extinction in the first place. Suppose now that a particular species has been depleted until only one male and one female are left. This mating pair must serve to reestablish the species. It is to be expected that each member of this pair would be heterozygous for at least a few unpleasant recessive genes. In light of the information in the preceding paragraph, what unique problems will the reconstituted species face?

*A group of questions for practice and for extending Mendelian genetics.*

1. Refer to the definition of "true breeding" two sections back. In the discussion of the monohybrid cross and Figures 13.3.2 and 13.3.3, "true breeding" was asserted to mean "homozygous." Suppose for a moment that a member of a supposedly true-breeding population were a heterozygote. Show that being heterozygous is inconsistent with the definition of true breeding.

2. Suppose you are given a red-flowered pea. A *test-cross* will enable you to determine whether this dominant phenotype is a heterozygote (Rw) or a homozygote (RR). Cross it with a homozygous recessive individual (ww); the cross is therefore either RR × ww or Rw × ww. Note the different results obtained, depending on the genotype of the dominant phenotype. How do we know that a white-flowered plant is homozygous?

3. The red-flower allele in peas completely masks the white-flower allele, i.e., red is *completely dominant* to red. If we cross a true-breeding red-flowered snapdragon with a true-breeding white-flowered one, the F1 offspring are all pink. We say that dominance is *incomplete*, or *partial*, for snapdragon flower color; partial dominance is a very common phenomenon. Cross two pink snapdragons to get offspring with a phenotypic ratio of 1 red:2 pink:1 white.

4. Foxes with platinum fur have the genotype Pp and silver foxes are pp. The genotype PP kills the fetus right after conception, i.e., it is *lethal*. Evidently, the gene locus for fur color controls other properties as well, among them at least one very basic metabolic process. Show that a cross of two platinum foxes gives a 2:1 phenotypic ratio of offspring.

5. There is a notable exception to the statement that every chromosome in a mammalian diploid cell has an exact homolog. Mammalian males have one chromosome called an X chromosome and one called a Y chromosome. Females have two Xs and no Ys. These *sex chromosomes* carry a number of genes having to do with gender and many others that do not. Despite the fact that they are not homologous, the X and Y chromosomes in a male can synapse over a portion of their length to facilitate meiosis. A well-known recessive gene on the X chromosome is for hemophilia, a blood-clotting disorder. Let us represent a heterozygous ("carrier") female as $X^h X^+$, where "X" indicates X-linkage, "h" indicates the hemophilia allele, and "+" represents the normal allele. Note that a male of genotype $X^hY$ will show the disorder because there is no possibility of a dominant allele on his Y chromosome to mask the hemophilia allele on his X chromosome. Cross a carrier female with a hemophilic male to show that a female can get hemophilia. Cross a carrier female with a normal male to show that no daughters and half the sons would be affected.

6. Often there are more than two choices for the alleles for a property, a phenomenon called *multiple alleles*. The presence of certain molecules on red blood cells is determined by the alleles A, B, and O. For example, the genotypes AA and AO yield the A molecule, the genotypes BB and BO yield the B molecule, the genotype OO yields neither molecule, and the genotype AB yields both the A and B molecules. The latter case, expression of both alleles, is called *codominance*. Cross an AB parent with an O parent; what ratio of offspring is obtained? Could an O-type man be the parent of an AB child? Is it possible that a particular A-type man is the father of an A-type child by an A-type mother?

7. The expression of some genes is determined by the environment. The gene for dark pigmentation in Siamese cats is expressed only in cool parts of the cat's body—nose, ears, and tail tip. The expression of the gene for diabetes mellitus, a deficiency in sugar metabolism, is affected by diet and the person's age. As an example, environmental effects might cause a dominant allele not to be expressed under certain conditions, and an individual with genotype AA or Aa might show the recessive phenotype. How might you determine that such an individual is actually of the dominant genotype?

## 13.4 Genetic Drift

Natural selection is not the only mechanism at work in evolution. The other important mechanism is *genetic drift*. Genetic drift is a random or stochastic process in which the allelic fractions of a population change from generation to generation due to chance alone.

At one level chance works on mating itself. In a pair of diploid sexually reproducing parents (such as humans), not all of the parent's alleles will be passed on to their progeny due to chance assortment of chromosomes at meiosis. This is called *sampling error*. Extended over an entire population, sampling error will be mitigated by the law of large numbers but not completely eliminated. Thus population-wide, frequencies of alleles change from generation to generation.

In addition to that, not all offspring of a new generation survive and reproduce. Of course, natural selection works on this principle when the underlying cause is differential fitness. But fitness is not always the issue; random events can intervene with the result that even the most fit individuals fail to reproduce.

Genetic drift is not self-correcting. The population does not have a genetic memory and is not urged back to some previous genetic state. The changes that arise in the allelic frequencies of the previous generation become the basis of the new gene pool. As more and more generations pass, allelic frequencies can range far from where they started. Surprising as it may seem, over time genetic drift moves the fraction of every allele (not subject to natural selection) to either 1 or 0. This will even happen due to sampling error alone.

Genetic drift is an example of the mathematical process known as a *random walk*. In a simplified version, suppose that the frequency of some allele, say $A$, can change by the amount $s$, the step size, in each generation; assume $s = \frac{1}{10}$. Right now $A$ comprises the fraction $f$ of the gene pool for its trait; suppose $f = 60\%$. From generation to generation, $f$ will move up by 10% or down by 10% with some probability $p > 0$. The walk will go back and forth along the points $0\%, 10\%, \ldots, 90\%, 100\%$, but eventually $f$ will become 0% or 100% and be trapped there. In random walk terminology, the walk has been *absorbed*. The larger the value of $s$ or $p$, the fewer expected number of generations until absorption.

The effects of genetic drift are accentuated on small populations. Loss of alleles due to limited mating outcomes and random environmental occurrences represent a larger fractional allelic change to the population when it is small. So the step size $s$ of the previous paragraph is larger. There are two extreme situations in which genetic drift is of primary importance due to small population size. The first is when some catastrophic event occurs to a population and its numbers fall to a very low level; this is called the *bottleneck effect*. The second is when a subpopulation of the whole becomes reproductively isolated; this is the *founders effect*.

An example of the bottleneck effect occurred to the northern elephant seal. This animal was hunted almost to extinction. By 1890, there were fewer than 20 animals remaining. Although it now numbers around 30,000, there is very little genetic variation in this population. As a result the population is highly vulnerable to extinction, for example from disease. In any case, the present elephant seal population is sure to have large differences in allelic frequencies from its pre-1890 counterpart.

Founder effect occurs when a small subpopulation of a species becomes reproductively isolated. These are the founders of the isolated population and their allelic frequencies will be its norm. But these frequencies will most certainly be much different from those of the parent population for many traits. Native American Indians constitute an example in that their ancestors crossed the Bering Strait in small numbers to found societies in the Western Hemisphere. Unlike other races, American Indians lack the blood group B, in all probability due to the absence of its allele among the founders.

The founders effect often results in the high prevalence of normally rare diseases. The Amish people of Pennsylvania constitute a closed population stemming from a small number of original German immigrants, about 200. But the Amish carry unusual concentrations of gene mutations that cause a number of otherwise rare inherited disorders. One of these, *Ellis–van Creveld syndrome*, involves dwarfism, polydactyl abnormalities (extra fingers or toes), and, in about half of the afflicted, a hole between the two upper chambers of the heart.

Since founders and their progeny are genetically isolated and must interbreed, their recessive genes will pair up much more frequently than occurs in the parent population. But recessive genes are often defective, hence the increased incidence of these kinds of genetic disorders. In the Amish, Ellis–van Creveld syndrome has been traced back to a single couple who came to the area in 1744.

Obviously, genetic drift is highly important as a mechanism of evolution. Arguments over how important as compared to natural selection is an unsettled issue among geneticists. It is known that genetic drift can overcome natural selection if the selection pressure is weak. The fixation of less fit alleles is an integral feature of the evolutionary process.

## 13.5 A Final Look at Darwinian Evolution

*We close out our discussion of biology with a last look at the Darwinian model of evolution, which we introduced in Section* 3.1. *Fitness is measured by the persistence of a property in subsequent generations. If a property cannot be inherited, it cannot be selected. Thus acquired properties like facelifts cannot be selected, nor can genetic properties of sterile individuals, like a mule's hardiness.*

*Populations evolve; individuals do not. An individual is born with a fixed set of genes; mutations in somatic cells are not transmitted to offspring, and mutations in germinal cells can be seen only in the offspring.*

*Some organisms do not exhibit sexual reproduction but rather reproduce only asexually. Their only source of variation is therefore mutation. Nevertheless, such organisms have long evolutionary histories.*

*Fitness is measured by the ability to project genes into subsequent generations.*

---

Common phrases like "struggle for survival" and "survival of the fittest" can be very misleading because they bring to mind vicious battles to the death between two contestants. The fact is that, except arguably among humans, violence is rarely the route by which Darwinian fitness is achieved in the biological world. Even the noisy, aggressive encounters between male animals seen on television nature programs seldom result in serious injury to participants. We must look to much more subtle interactions as a source of fitness.

One group of organisms may be slightly more able than another to tolerate heat, to thrive on available food, or to elude predators. Subtle pressure is the norm in evolution; it works slowly, but there is no hurry. *Drosophila*, a common fruit fly, is used in many genetic experiments because it is easy to raise, has a short life span, and has many simple physical properties, such as eye color, whose modes of genetic transmission are easy to follow. If a large number of red-eyed and white-eyed *Drosophila* are put together in an enclosure and left to their own devices, the fraction of flies with white eyes will decrease steadily for several tens of generations and finally reach zero. Close observation reveals the reason: A female *Drosophila*, either red-eyed or white-eyed, will generally choose not to mate with a white-eyed male if a red-eyed male is available. Thus there is a definite selection for the red-eye genetic trait.

Humans are not excluded from such subtle pressures: "Personals" ads in newspapers contain wish lists of traits people prefer in a mate. Height and affluence (control of territory?) are prized male traits, and hourglass figures and youth (ability to bear children?) are valued female traits.

Regardless of the strength of the selective pressure or the nature of the properties being selected, there is really only one way to measure the evolutionary value of a trait, and that is the degree to which it is propagated into future generations. A shy, ugly person who has lots of fertile children has a high degree of fitness. We see that one generation of propagation is not enough; the trait must be persistent. For example, mules are known for their hardiness, but they are sterile offspring of horses and donkeys. As a result, the hardiness of a mule cannot confer any evolutionary advantage.[6]

*Populations evolve; individuals do not.*

A *population* is a group of organisms of the same species, living in the same area. As before, we will restrict our discussion here to populations of organisms for which the diploid generation is most conspicuous, e.g., humans.

If we observe a population over many generations, the "average" phenotypic property will change, in keeping with our earlier discussion of species formation and genetic drift. Thus the average height may increase, or the typical eye color may darken. We now ask and answer two questions: At what points in the alternation of generations do the changes occur, and what kinds of changes are relevant to evolution?

The Darwinian model stipulates that favored properties may be transmitted to offspring; in any case, they certainly must be *capable* of transmission for the model to apply. A diploid individual is conceived with a set of genes that are relatively fixed for that individual's lifetime. Exceptions to this statement might involve mutations in somatic cells and infection by lysogenic viruses (see Chapter 10). As long as these changes do not occur in germinal cells or germinal cell precursors, they cannot be transmitted to the next generation and thus have no evolutionary effect. In addition,

[6] There is a peculiar example of a noninheritable trait—a desire for a large family—that might be passed from one generation to another by teaching and which could have a strong positive selective value. This was discussed in Section 4.1.

there are many phenotypic properties that favor reproduction but that, because they are not of a genetic nature, cannot be transmitted to offspring. Examples are suntans, exercise-strengthened bodies, and straightened teeth.

Genetically transmissible variations must originate via one of at least three routes, all of which require sexual reproduction (in other words, an *intervening haploid generation*) for their expression:

1. independent assortment;
2. crossing over;
3. mutation in a sperm, or an egg, or in their precursors in a parent prior to conception, or in a zygote at a very early stage of development—the altered genetic material in any one of these cases should turn up in those cells of the reproductive system that undergo meiosis to form the next generation of gametes.

We can conclude that because the Darwinian model requires changes that are inheritable, and because the observation of inheritable changes requires the observation of more than one generation, *it is the population that evolves*. Changes restricted to the somatic cells of individuals are not genetically transmitted to offspring; thus in terms of evolution, an individual is fixed. Over a period of time, however, the average, or typical, characteristics of the population evolve.

*Some organisms do not exhibit sexual reproduction.*

Sexual reproduction is unknown (and probably nonexistent) in several kinds of organisms, for example, most bacteria, blue-green algae, and some fungi. In those cases, all reproduction is asexual, which would seem to limit severely the possibilities of variation. Nonetheless, these organisms seem to have gotten along fine over long periods of history. We must conclude that some combination of three things applies: Either these organisms have not been exposed to large fluctuations in their environments, or they possess an innate physiological flexibility that permits them to get along in different environments, or their spontaneous mutation rates are sufficiently high to generate the variation necessary for adapting to new environmental situations.[7]

## 13.6 The Hardy–Weinberg Principle

*Diploidism and sexual reproduction complicate the calculation of inheritance probabilities. But remarkably, the results are the same as if alleles were balls selected for combination from an urn. This is the Hardy–Weinberg principle. Although its veracity depends on random mating, among other properties, it continues to provide good approximations in many other situations as well.*

*Mendelian inheritance follows the laws of probability.*

We will be concerned with probabilities associated with Mendelian inheritance for a

[7] There is now good evidence that bacteria, including asexual ones, can pass small pieces of DNA, called plasmids, to other bacteria.

diploid organism. As explained in Section 13.2, meiosis produces four haploid cells of two different kinds, each equally likely to participate in fertilization. Then the probability is $\frac{1}{2}$ that a given kind of gamete will do so.

Consider first a single locus for which there are only two alleles, say A and a. Hence there are three distinct genotypes, the homozygotes AA and aa, and the heterozygote Aa (or aA). If one parent is AA and the other Aa, then the possible zygote genotypes resulting from a mating may be represented by an *event tree* as follows.

Let the first branch point in Figure 13.6.1 correspond to the allele donated by the first parent, AA. There are two possible alleles, and so here the diagram will have two branches. But for the parent AA, both branches lead to the same result, namely, the contribution of allele *A* to the offspring. Let the second branch point correspond to the allele donated by the second parent. Again there are two possibilities, but this time the outcomes are different as indicated.

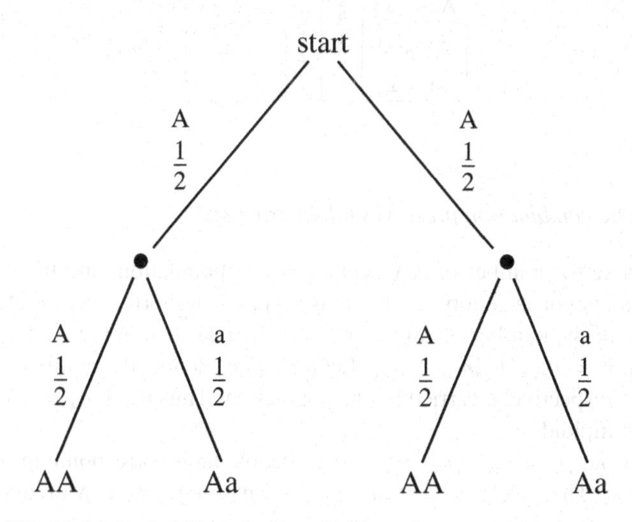


**Fig. 13.6.1.** Probabilities for the offspring of an AA with Aa mating.

Nhe resulting probabilities may be calculated in several ways. Since all the legs, or *edges*, of the diagram are equally likely, so are the resulting outcomes, each having probability $\frac{1}{4}$. Hence

$$\Pr(\mathrm{AA}) = \frac{1}{2} \quad \text{and} \quad \Pr(\mathrm{Aa}) = \frac{1}{2}.$$

Alternatively, starting at the top node, the *root node*, and traversing the two edges to the left leading to AA gives a probability of $\frac{1}{4}$ for this outcome by multiplying the probabilities along each edge of the path ($\frac{1}{2} \cdot \frac{1}{2}$). This way of calculating the probabilities is the method of *conditional probabilities*, since the probabilities along the branches leading away from any node are conditioned on the sequence of events leading to the node. Altogether, the probability of an AA zygote by this method is $\frac{1}{2}\frac{1}{2} + \frac{1}{2}\frac{1}{2} = \frac{1}{2}$, since AA can occur in two different ways according to the tree.

Finally, the probabilities can be calculated by the principle of independence (see Section 2.8). The selection of a gamete from the AA parent will result in an A with probability 1. The selection of a gamete from the Aa parent is independent and will result in an A with probability $\frac{1}{2}$. Therefore, the probability of an AA zygote is $1 \cdot \frac{1}{2}$.

The complete list of Mendelian inheritance probabilities is given in Table 13.6.1.

**Table 13.6.1.** Mendelian inheritance probabilities.

| Parent | Zygote genotypes | | |
|---|---|---|---|
| genotypes | AA | Aa | aa |
| AA × AA | 1 | | |
| AA × Aa | $\frac{1}{2}$ | $\frac{1}{2}$ | |
| Aa × Aa | $\frac{1}{4}$ | $\frac{1}{2}$ | $\frac{1}{4}$ |
| Aa × aa | $\frac{1}{2}$ | $\frac{1}{2}$ | |
| aa × aa | 1 | | |

*Random allelic combination preserves allelic fractions.*

Let $n_{\text{AA}}$ denote the number of AA genotypes in a population, and likewise let $n_{\text{aa}}$ denote the number of aa genotypes. For reasons that will shortly become clear, let $n_{\text{Aa}}$ denote one-half the number of Aa genotypes. Then the size of the entire population $N$ is the sum $N = n_{\text{AA}} + 2n_{\text{Aa}} + n_{\text{aa}}$. Let $n_{\text{A}}$ and $n_{\text{a}}$ denote the number of A alleles and a alleles, respectively, carried by the population. Thus $n_{\text{A}} + n_{\text{a}} = 2N$, since the population is diploid.

Similarly, let $p_{\text{AA}}$, $p_{\text{Aa}}$, $p_{\text{aa}}$, $p_{\text{A}}$, and $p_{\text{a}}$ denote their corresponding fractions of the population. Then $p_{\text{A}} + p_{\text{a}} = 1$ and $p_{\text{AA}} + 2p_{\text{Aa}} + p_{\text{aa}} = 1$. Moreover,

$$p_{\text{A}} = \frac{n_{\text{A}}}{2N} = \frac{2n_{\text{AA}} + 2n_{\text{Aa}}}{2N} = p_{\text{AA}} + p_{\text{Aa}},$$

and similarly

$$p_{\text{a}} = p_{\text{Aa}} + p_{\text{aa}}.$$

Now imagine that all the *alleles* of the population are pooled and two are selected at random from the pool to form a pair. The selection of an A happens with probability $p_{\text{A}}$, while the selection of an a happens with probability $p_{\text{a}}$. (We assume that the pool is so large that the removal of any one allele does not appreciably change the subsequent selection probability.) Then, for example, the probability of forming an AA pair is $p_{\text{A}}^2$, since we assume that the selections are made independently. In the same way, the other possible pair selections are calculated, with the results shown in Table 13.6.2. As always, these probabilities are also the (approximate) fractions of the various outcomes in a large number of such pairings.

**Table 13.6.2.** Mendelian inheritance probabilities.

| **Female gametes (frequencies)** | **Male gametes (frequencies)** | |
|---|---|---|
| | **A ($p_A$)** | **a ($p_a$)** |
| A ($p_A$) | AA ($p_A^2$) | Aa ($p_A p_a$) |
| a ($p_a$) | aA ($p_A p_a$) | aa ($p_a^2$) |

From the table we can calculate the fraction, $p'_A$, of A alleles among the resultant pairs. Each pair of type AA contributes two A alleles, and while each Aa pair contributes only one, there are twice as many such pairs. Hence

$$p'_A = \frac{2p_A^2 + 2p_A p_a}{2} = p_A(p_A + p_a) = p_A.$$

In this it is necessary to divide by 2 because each pair has two alleles. Thus the fraction of A alleles among a large number of pairings is the same as their fraction in the original gene pool, $p_A$. The same is (consequently) true for the a allele, $p'_a = p_a$.

Of course, the process of gene maintenance for bisexual diploid organisms is much more complicated than the simple random pairing of alleles selected from a common pool that we have explored here (see Section 13.3). Nevertheless, we will see in the next subsection that the results are the same if mating is random.

*Random mating preserves allelic fractions.*

Again consider a one-locus, two-allele system and suppose mating is completely random. Then the probability of an AA × Aa mating, for example, is $2p_{AA}(2p_{Aa})$, since the first parent could be AA and the second Aa or the other way around. Altogether, there are six different kinds of matings; their probabilities are listed in Table 13.6.3.

**Table 13.6.3.** Mendelian inheritance probabilities.

| **Genotype mating** | **Probability** |
|---|---|
| AA × AA | $(p_{AA})^2$ |
| AA × Aa | $2p_{AA}(2p_{Aa})$ |
| AA × aa | $2p_{AA}p_{aa}$ |
| Aa × Aa | $(2p_{Aa})^2$ |
| Aa × aa | $2(2p_{Aa})p_{aa}$ |
| aa × aa | $(p_{aa})^2$ |

Now apply the Mendelian inheritance laws to calculate the probability of the various possible zygotes, for example, an AA zygote. First, an AA results from an AA × AA parentage with probability 1. Next, an AA results from an AA × Aa parentage with probability $\frac{1}{2}$ (see Figure 13.6.1), and finally an AA results from an Aa × Aa cross with probability $\frac{1}{4}$. Now, by the method of conditional probabilities as discussed at the beginning of this section, we have

$$\begin{aligned}\Pr(\mathrm{AA}) &= p_{\mathrm{AA}}^2 \cdot 1 + 2p_{\mathrm{AA}}(2p_{\mathrm{Aa}}) \cdot \frac{1}{2} + (2p_{\mathrm{Aa}})^2 \cdot \frac{1}{4} \\ &= p_{\mathrm{AA}}^2 + 2p_{\mathrm{AA}}p_{\mathrm{Aa}} + p_{\mathrm{Aa}}^2 \\ &= (p_{\mathrm{AA}} + p_{\mathrm{Aa}})^2 = p_{\mathrm{A}}^2.\end{aligned}$$

Similarly, we leave it to the reader to show that

$$\Pr(\mathrm{aa}) = (p_{\mathrm{aa}} + p_{\mathrm{Aa}})^2 = p_{\mathrm{a}}^2$$

and

$$\Pr(\mathrm{Aa}) = 2(p_{\mathrm{AA}} + p_{\mathrm{Aa}})(p_{\mathrm{Aa}} + p_{\mathrm{aa}}) = 2p_{\mathrm{A}}p_{\mathrm{a}}.$$

But this shows that the fractions of alleles A and a are again $p_{\mathrm{A}}$ and $p_{\mathrm{a}}$, respectively, among the offspring just as among their parents, assuming that the various genotypes are equally likely to survive. This is the same result we calculated in the last section. In other words, the effect of random genotype mating is indistinguishable from that of random gamete recombination. This is the *Hardy–Weinberg principle*.

**Hardy–Weinberg principle.** *Under the condition that mating is random and all genotypes are equally fit, the fractions of alleles will stay the same from generation to generation.*

A consequence of the Hardy–Weinberg principle is that after at most one generation, the fractions of genotypes also stabilize and at the values

$$\begin{aligned}p'_{\mathrm{AA}} &= p_{\mathrm{A}}^2, \\ 2p'_{\mathrm{Aa}} &= 2p_{\mathrm{A}}p_{\mathrm{a}} \\ p'_{\mathrm{aa}} &= p_{\mathrm{a}}^2.\end{aligned}$$

For example, suppose that initially 70% of a population is AA and the remaining 30% is aa. Then the fractions of alleles in subsequent generations are also 70% and 30% for A and a, respectively. Therefore, after one generation, the fractions of genotypes will be

$$\begin{aligned}\mathrm{AA}: &\quad (0.7)^2 = 0.49, \\ \mathrm{Aa}: &\quad 2(0.7)(0.3) = 0.42, \\ \mathrm{aa}: &\quad (0.3)^2 = 0.09.\end{aligned}$$

In some cases the Hardy–Weinberg principle is applicable even when mating is not random. Mating would fail to be random for example if the homozygote for a recessive gene is impaired or unviable. But in fact, the homozygotes in these cases are so rare that the induced error is very small. Keep in mind that for a recessive gene a, the homozygote AA and heterozygote Aa are indistinguishable, so that random mating among them is a reasonable assumption.

The Hardy–Weinberg principle breaks down when there is migration, inbreeding, or nonrandom mating, that is, phenotypes are selected for some attribute.

*Sex-linked loci give rise to different rates of expression between males and females.*

In the event that males (or females) have one or more nonhomologous chromosomes, the foregoing derivations must be modified. One consequence of nonhomologous chromosomes is that there can be a large difference in expression of a sex-linked character between males and females. For definiteness, suppose the male has the nonhomologous pair XY, while the female has the homologous pair XX.[8] For this case, fractions of alleles for genes on either the X or the Y chromosome are identical to genotype fractions for the male. For example, suppose a recessive sex-linked allele occurs with frequency $p$ among a population. Then $p$ is also the rate at which the allele will occur in males. However, the rate at which the homozygous condition will occur in females is $p^2$.

An example of such an allele is color blindness in humans. Through various studies, it is believed that the frequency of the recessive allele is 8% as derived from the incidence rate in males. Therefore, the incidence rate in females ought to be $(0.08)^2 = 0.0064$ or 0.6%. Actually, the female incidence of the disease is about 0.4%. The discrepancy is an interesting story in its own right and stems from the fact that there are four different kinds of color blindness, two of which are red blindness and the other two green blindness. The bearer of defective genes for different types, such as these two, can still see normally.

Another possibility that can arise relative to sex-linked genes is that the allelic fractions are different between males and females. This can happen, for instance, when males and females of different geographical backgrounds are brought together. Let $F$ be the fraction of allele A in the females and let $M$ be its fraction in males. Then $f = 1 - F$ is the fraction of a in females and $m = 1 - M$ is its fraction in males. Assuming an equal number of males and females, the population frequencies work out to be

$$p_{\mathrm{A}} = \frac{M+F}{2} \quad \text{and} \quad p_{\mathrm{a}} = \frac{m+f}{2} = 1 - p_{\mathrm{A}},$$

and these will remain constant by the Hardy–Weinberg principle. However, the values of $M$ and $F$ will change from generation to generation.

To follow these fractions through several generations, we need only keep track of $F$ and $M$, since $f$ and $m$ can always be found from them. Let $F_n$ and $M_n$ refer to generation $n$ with $n = 0$ corresponding to the initial fractions.

Since a male gets his X chromosome from his mother, the allelic frequencies in males will always be what it was in females a generation earlier; thus

$$M_{n+1} = F_n.$$

On the other hand, the frequency in females will be the average of the two sexes in the preceding generation, since each sex contributes one X chromosome; hence

$$F_{n+1} = \frac{1}{2}M_n + \frac{1}{2}F_n.$$

[8] This is a mammalian property. In birds, the situation is reversed.

In matrix form, this can be written

$$\begin{bmatrix} M_{n+1} \\ F_{n+1} \end{bmatrix} = \begin{bmatrix} 0 & 1 \\ \frac{1}{2} & \frac{1}{2} \end{bmatrix} \begin{bmatrix} M_n \\ F_n \end{bmatrix}.$$

In the exercises, we will investigate where this leads.

Before we leave this example, there is another observation to be made. We used matrix $T$ above,

$$T = \begin{bmatrix} 0 & 1 \\ \frac{1}{2} & \frac{1}{2} \end{bmatrix},$$

in conjunction with multiplication on its right to update the column of male/female fractions $M_n$ and $F_n$. But in this example there is a biological meaning to left multiplication on the matrix $T$. In each generation, there will be a certain fraction of the alleles on the X chromosome in males that originally came from the females. It is possible to track that distribution.

To fix ideas, suppose that a ship of males of European origin runs aground on a South Sea island of Polynesian females. Further, suppose (hypothetically) that the alleles for a gene on the X chromosomes of the Europeans, the $E$-variant, are slightly different from those of the Polynesians, the $P$-variant, in, say, two base pairs. So the distribution of $E$-variant and $P$-variant chromosomal alleles of the emigrating males can be described by the (row) pair

$$(1\ 0),$$

where the first element is the fraction originating with the males and the second is the fraction originating with the females. The distribution of these fractions can be traced through the generations by a matrix calculation similar to that above, only this time using matrix multiplication on the left. In the first generation, we have

$$(1\ 0) \begin{bmatrix} 0 & 1 \\ \frac{1}{2} & \frac{1}{2} \end{bmatrix} = (0\ 1),$$

showing that all the alleles in the males in this generation come from the females. In the second generation, the fraction works out to

$$(0\ 1) \begin{bmatrix} 0 & 1 \\ \frac{1}{2} & \frac{1}{2} \end{bmatrix} = (\tfrac{1}{2}\ \tfrac{1}{2}),$$

or 50–50. Of course, the calculation can be continued to obtain the fractions for any generation.

The same calculation can give the female ratios, by starting with the initial female ratio of $(0\ \ 1)$.

## 13.7 The Fixation of a Beneficial Mutation

*A beneficial mutation does not necessarily become a permanent change in the gene pool of its host species. Its original host individual may die before leaving progeny, for example. Under the assumption that such a mutation is dominant (rather than recessive) and that individuals with the mutation behave independently, it is possible to derive the governing equations for calculating the fixation probability. One way of measuring the value of a vital factor is the expected number of surviving offspring, beyond self replacement, an adult will leave. For an r-strategist, the chance that a beneficial mutation will become permanent is about twice the overreplacement value of the mutation to its holder.*

*Probability of fixation of a beneficial mutation is the complement of the fixed point of its probability-generating function.*

Let $p_k$ be the probability that a chance mutation appearing in a zygote will subsequently be passed on to $k$ of its offspring. A convenient method of organizing a sequence of probabilities, such as $p_k$, $k = 0, 1, \ldots$, is by means of the polynomial

$$f(x) = p_0 + p_1 x + p_2 x^2 + \cdots,$$

in which the coefficient of $x^k$ is the $k$th probability. This polynomial is called the *probability-generating function* for the sequence $p_k$. The probability-generating function is purely formal, that is, it implies nothing more than a bookkeeping device for keeping track of its coefficients. Note that $f(1) = 1$. And $f(0) = p_0$ is the probability that the mutation disappears in one generation. Also note that the expected number of offspring to have the mutation is given (formally) by

$$\sum_{k=1}^{\infty} k p_k = f'(1)$$

(see Section 2.8). To say that the mutation is beneficial is to say that this expectation is greater than 1, that is,

$$\sum_{k=1}^{\infty} k p_k = 1 + a > 1$$

for some value a, which is a measure of the benefit in terms of overreplacement in fecundity.

Now if two such individuals with this mutation live and reproduce independently of each other (as in a large population), then the probability-generating function for their combined offspring having the mutation is

$$p_0^2 + 2p_0 p_1 x + (2p_0 p_2 + p_1^2)x^2 + (2p_0 p_3 + 2p_1 p_2)x^3 + \cdots, \tag{13.7.1}$$

which is proved by considering each possibility in turn. There will be no mutant offspring only if both parents leave none; this happens with probability $p_0^2$ by independence. There will be one mutant offspring between the two parents if one leaves none and the other leaves exactly one; this can happen in two ways. There will be two mutant offspring if the first leaves none while the second leaves two, or they both leave one, or the first leaves two while the second leaves none; this is $(2p_0 p_2 + p_1^2)$. The other terms of (13.7.1) may be checked similarly.

But note that (13.7.1) is exactly the polynomial product $f^2(x)$,

$$(p_0 + p_1x + p_2x^2 + \cdots)(p_0 + p_1x + p_2x^2 + \cdots)$$
$$= p_0^2 + 2p_0p_1x + (2p_0p_2 + p_1^2)x^2 + \cdots .$$

More generally, $m$ independent individuals with the mutation as zygotes will pass on the mutation to their combined offspring with probability-generating function given by the $m$th power $f^m(x)$.

Now start again with one mutant zygote and consider the probability-generating function $f_2$ for generation 2. Of course, the outcome of generation 2 depends on the outcome of generation 1. If there are no mutants in generation 1, and this occurs with probability $p_0$, then there are none for certain in generation 2. Hence this possibility contributes

$$p_0 \cdot 1$$

to $f_2$. On the other hand, if the outcome of generation 1 is one, then the probability-generating function for generation 2 is $f(x)$; so this possibility contributes

$$p_1 f(x).$$

If the outcome of generation 1 is two mutant individuals (and they behave independently), then the probability-generating function for generation 2 is, from above, $f^2(x)$; so this possibility contributes

$$p_2 f^2(x).$$

Continuing this line of reasoning yields the result that the probability-generating function for generation 2 is the composition of the function $f$ with itself, $f_2(x) = f(f(x))$,

$$f_2(x) = p_0 + p_1 f(x) + p_2 f^2(x) + p_3 f^3(x) + \cdots = f(f(x)).$$

More generally, the probability-generating function for generation $n$, $f_n(x)$, is given as the composition $f \circ f \circ \cdots \circ f$ of $f$ with itself $n$ times, or

$$f_n(x) = \underbrace{f(f(\cdots f(x)\cdots))}_{n \text{ times}}.$$

Now the probability that the mutation dies out by the $n$th generation is the constant term of $f_n(x)$ or $f_n(0)$. Hence the probability that the mutation dies out or vanishes some time is the limit

$$V = \lim_{n\to\infty} \underbrace{f(f(\cdots f(0)\cdots))}_{n \text{ times}}.$$

Applying $f$ to both sides of this equality shows that $V$ is a fixed point of $f$,

$$f(V) = V.$$

The fixed point of $f(x)$ is where the graphs $y = f(x)$ and $y = x$ intersect (see Figure 13.7.1). Since $f'(x)$ and $f''(x)$ are nonnegative for $x > 0$ (having all positive or zero coefficients), and since $f(1) = 1$, we see that there can be either zero or one fixed point less than $x = 1$. If there is a fixed point less than one, then $V$ is that value; otherwise, $V = 1$.

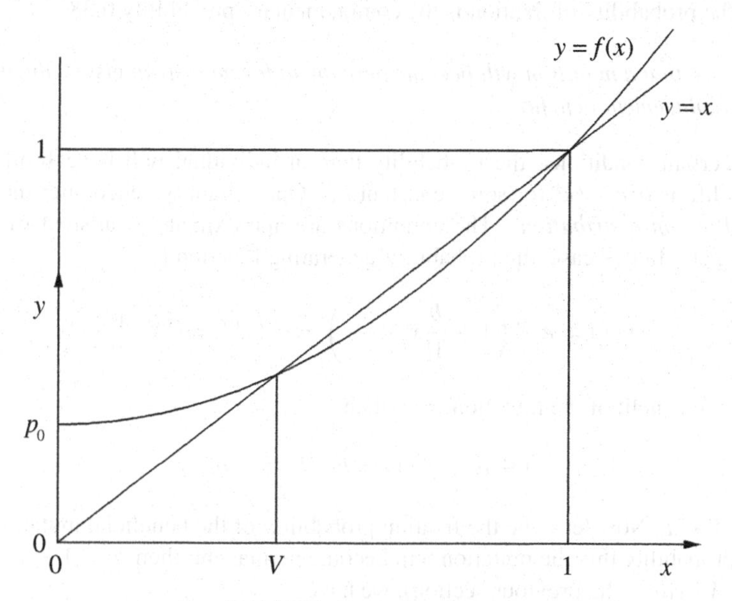


**Fig. 13.7.1.** $V$ is the fixed point of $f(x)$.

For example, suppose that a mutation arose on the X-chromosome of a human female about the time that "Lucy" walked the earth (2 million years ago). Further suppose that the following probabilities of producing surviving (female) offspring pertained to the holder of such a mutation:

- probability of leaving no female offspring, $p_0 = 0.35$;
- probability of leaving one female offspring, $p_1 = 0.25$;
- probability of leaving two female offspring, $p_2 = 0.20$;
- probability of leaving three female offspring, $p_3 = 0.1$;
- probability of leaving four female offspring, $p_4 = 0.1$,
- and zero probability of leaving more than four female offspring.

Then the probability-generating function is

$$f(x) = 0.35 + 0.25x + 0.2x^2 + 0.1x^3 + 0.1x^4.$$

Its fixed points can be found by solving the roots of the fourth-degree polynomial

$$0.1x^4 + 0.1x^3 + 0.2x^2 + (0.25 - 1)x + 0.35 = 0.$$

With the following code, the appropriate root is found to be 0.62:

MAPLE
```
> f:=.1*x^4+.1*x^3+.2*x^2+(.25-1)*x+.35;
> fsolve(f,x,0..1);
```

MATLAB
```
> p=[.1 .1 .2 (.25-1) .35];
> min(roots(p))
```

Hence the probability of fixation is the complementary probability 0.38.

*The chance that a mutation will become permanent for an $r$-strategist is about twice its overreplacement benefit.*

Under certain conditions, the probability that an individual will have $k$ offspring over its life is $b^k e^{-b}/k!$ for some constant $b$.[9] Once again we encounter the ubiquitous *Poisson distribution*. The conditions are approximately satisfied by many $r$-strategists. In this case, the probability-generating function is

$$f(x) = e^{-b}\left(1 + \frac{b}{1!}x + \cdots\right) = e^{-b}e^{bx} = e^{b(x-1)}.$$

Let the benefit of the mutation be $a$; then

$$1 + a = f'(1) = be^{b(1-1)} = b,$$

so $b = 1 + a$. Now let $F$ be the fixation probability of the beneficial mutation, that is, the probability that the mutation will become permanent; then $F = 1 - V$. Since $V = f(V)$ (from the previous section), we have

$$1 - F = e^{-(1+a)F}.$$

Taking logarithms,

$$(1 + a)F = -\ln(1 - F) = F + \frac{F^2}{2} + \frac{F^3}{3} + \cdots.$$

The infinite series is the Taylor series for the middle term. Divide by $F$ and subtract 1 to get

$$a = \frac{F}{2} + \frac{F^2}{3} + \cdots.$$

If $a$ is small, then approximately

$$a \approx \frac{F}{2},$$

so the fixation probability is about $2a$.

[9] The conditions are (a) the probability of an offspring over a short period of time $\Delta t$ is proportional to $\Delta t$; (b) the probability of two or more offspring over a short period of time is essentially zero; and (c) offspring occur independently of one another. The distribution would also apply if offspring occurred in batches; the $k$ counts batches.

## Exercises/Experiments

**1.** In this problem, assume a diploid organism having three loci per homologous chromosomal pair and two alleles per locus.

(a) If the organism has only one such chromosomal pair, how many different genotypes are possible?

(b) Same question if there are two chromosomal pairs.

(c) Suppose there are two chromosomal pairs with genes $\alpha$, $\beta$, and $\gamma$ on one of them, while genes $\delta$, $\epsilon$, and $\phi$ lie on the other. How many different haploid forms are there?

(d) For a given genotype as in (c), how many different gametes are possible? That is, suppose that a particular individual has the homologous chromosomes (1) $(A, b, C)$ and $(A, B, C)$ and (2) $(d, e, F)$ and $(D, e, F)$. How many haploid forms are there?

(e) What is the maximum number of different offspring possible from a mating pair of organisms as in (d)? What is the minimum number? How could the minimum number be achieved?

(f) Work out a graph showing how the number of haploid forms varies with (i) number of chromosomal pairs or (ii) number of genes per chromosomal pairs. Which effect leads to more possibilities?

**2.** For a given diploid two-allele locus, the initial fractions of genotypes are AA : $p$, Aa : $q$, and aa : $r$ (hence $p+q+r=1$). Recall that the frequencies in the next generation will be $p_{\mathrm{AA}} = x$, $p_{\mathrm{Aa}} = y$, and $p_{\mathrm{aa}} = z$, where

$$x = \left(p + \frac{1}{2}q\right)^2, \qquad y = 2\left(p + \frac{1}{2}q\right)\left(r + \frac{1}{2}q\right), \qquad z = \left(r + \frac{1}{2}q\right)^2.$$

Under the assumption that the various genotypes are selected neither for nor against, show that these ratios will be maintained in all future generations; i.e., show that

$$x = \left(x + \frac{1}{2}y\right)^2,$$
$$y = 2\left(x + \frac{1}{2}y\right)\left(z + \frac{1}{2}y\right),$$
$$z = \left(z + \frac{1}{2}y\right)^2.$$

Hence when the Hardy–Weinberg principle holds, genotype frequencies stabilize in one generation,

```
MAPLE
> x:=(p+q/2)^2;
> y:=2*(p+q/2)*((1-p-q)+q/2);
> z:=1-x-y;
> X:=(x+y/2)^2;
```

```
> simplify(X-x);
```

MATLAB

```
> p=.7; q=0; r=.3;
> x=(p+q/2)^2
> y=2*(p+q/2)*(r+q/2)
> z=(r+q/2)^2
> X=(x+y/2)^2
> Y=2*(x+y/2)*(z+y/2)
> Z=(z+y/2)^2
```

etc.

**3.** In this problem, we want to see how many homozygous recessives for a trait result from homozygous parents and how many result from heterozygous parents (see Figure 13.7.2). The question is, given an aa progeny, what is the probability the parents were aa × aa?

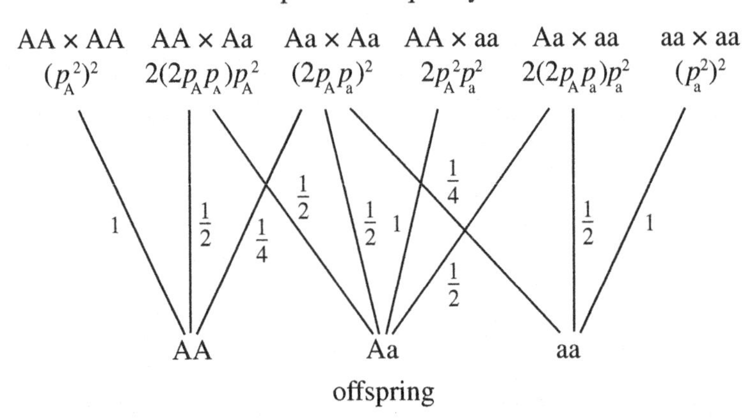


**Fig. 13.7.2.**

Since the progeny is known to be aa, the universe for this problem is the paths of the tree leading to aa; its frequency is given by

$$u = (2p_A p_a)^2 \cdot \frac{1}{4} + 2(2p_A p_a)p_a^2 \cdot \frac{1}{2} + (p_a^2)^2 \cdot 1.$$

So the relative frequency in which this occurs via aa × aa parents is

$$\frac{(p_a^2)^2 \cdot 1}{u}.$$

(a) Calculate the probable parentage of an aa progeny via Aa × Aa genotypes and Aa × aa genotypes.

(b) Make three graphs of these probable parentages over the range of frequencies of allele a from 0.25 to 0.001, say.

(c) If $p_a = 0.01$, then what is the chance that an aa individual had heterozygous parents? Same question for at least one heterozygous parent.

For part (a):

```
MAPLE
> u:=(2*pA*pa)^2*(1/4)+2*(2*pA*pa)*pa^2*(1/2)+(pa^2)^2;
> pA:=1-pa;
> aaxaa:= pa->(pa^2)^2/u;
> #similarly for AaxAa and Aaxaa
```

For part (b):

```
MAPLE
> plot(aaxaa(pa),pa=0.001..0.25);

MATLAB
> pa=(.001:.001:.25);
> pA=1-pa;
> u=(2*pA.*pa).^2*(1/4)+2*(2*pA.*pa).*pa.^2*(1/2)+(pa.^2).^2;
> aaxaa=pa.^4./u;
> plot(pa,aaxaa)
```

**4.** This problem refers to the sex-linked loci subsection of Section 13.6.

(a) Let the starting fraction of allele a in males be $M_0 = 0.1$ and in females be $F_0 = 0.3$. By performing the matrix calculation

$$\begin{bmatrix} M_{t+1} \\ F_{t+1} \end{bmatrix} = \begin{bmatrix} 0 & 1 \\ \frac{1}{2} & \frac{1}{2} \end{bmatrix} \begin{bmatrix} M_t \\ F_t \end{bmatrix}$$

repeatedly, find the limiting fractions $M_\infty$ and $F_\infty$. What is the ratio $\frac{M_\infty}{F_\infty}$?

(b) Do the same for the starting ratios $M_0 = 1$ and $F_0 = 0$. What is the limiting ratio $\frac{M_\infty}{F_\infty}$?

(c) Let $T$ be the matrix in part (a):

$$T = \begin{bmatrix} 0 & 1 \\ \frac{1}{2} & \frac{1}{2} \end{bmatrix}.$$

Show that $T$ satisfies

$$T \begin{pmatrix} 1 \\ 1 \end{pmatrix} = \begin{pmatrix} 1 \\ 1 \end{pmatrix}.$$

We say that this column vector, with both components 1, is a right eigenvector for $T$ with eigenvalue 1.

(d) As in part (a), iterate the calculation

$$(0\ 1) = (1\ 0) \begin{bmatrix} 0 & 1 \\ \frac{1}{2} & \frac{1}{2} \end{bmatrix},$$

this time multiplying the matrix on the left by the vector, to obtain the limit. This will represent the ultimate distribution of the original male vs. female alleles. Show that

$$\begin{pmatrix} \frac{1}{3} & \frac{2}{3} \end{pmatrix}$$

is a left eigenvector for $T$. What is the eigenvalue?

MAPLE
```
> with(LinearAlgebra):
> T:=Matrix([[0,1],[1/2,1/2]]);
> v:=Vector([1/10,3/10]);
> #the next to see the trend
> for n from 0 to 10 do
    evalf((T^n).v);
  od;
> #Now get the eigenvalue and eigenvector
> Eigenvectors(T);
```

MATLAB
```
> T=[0 1; .5 .5]
> x=[1 0]
> for k=1:20
> x=x*T
> end
> y=x % print eigenvector
> lambda = y(1)/x(1) % and evalue
```

In the output, the first item is the eigenvalue (as above), the second is its multiplicity (how many times repeated, should be 1 here), and the third is the eigenvector. Eigenvectors may be multiplied by any constant, so if $\binom{1}{1}$ is an eigenvector, so is $\binom{3}{3}$.

**5.** Two hypotheses that explain the greater incidence of early baldness in males than in females are (1) an autosomal dominance that is normally expressed only in males and (2) an X-linked recessive. If the first is correct and $Q$ is the frequency of the gene for baldness, what proportion of the sons of bald fathers are expected to be bald? What proportion are from nonbald fathers? What are the corresponding expectations for the X-linked recessive hypothesis.

Data gathered by Harris[10] found that 13.3% of males in the sample were prematurely bald. Of 100 bald men, 56 had bald fathers. Show that this is consistent with the sex-limited dominance hypothesis but not the sex-linked recessive. (Note that it is easier to get data about the fathers of bald sons than it is to wait for the sons of bald fathers to grow up to get data about bald sons.)

**6.** Suppose an organism that is capable of both sexual and asexual reproduction reproduces $c$ fraction of the time asexually (by cloning) and $1 - c$ fraction of the time sexually with random mating. Let $P_t$ be the fraction of the genotype AA in generation $t$ and let $p$ be the frequency of allele A. Assumethat $c$ is independent of genotype, and consequently $p$ will remain constant from generation to generation. However, the frequency of genotype AA can change. Using the Hardy–Weinberg principle, show that the change in this fraction is given by

$$P_{t+1} = cP_t + (1 - c)p^2.$$

Find the limiting fraction $P_\infty$.

---

[10] H. Harris, The inheritance of premature baldness in men, *Ann. Eugenics*, **13** (1946), 172–181.

```
MAPLE
> restart;
> # first try
> F:=x->c*x+(1-c)*p^2;
> x:=0; y:=F(x); w:=F(y);
> simplify(%);
> restart;
> F:=x->c*x+(1-c)*p^2;
> y[0]:=0;
> for n from 1 to 10 do
    y[n]:=F(y[n-1]):
  od:
> simplify(y[10]);

MATLAB
> c=1/3; P(1)=.7;
> p=.8
> for k=1:30
>   P(k+1)=c*P(k)+(1-c)*p^2;
> end
> P
> c=.9;
> for k=1:30
>   P(k+1)=c*P(k)+(1-c)*p^2;
> end
> P % how does P(infinity) depend on c?
> p=.3
> for k=1:50
>   P(k+1)=c*P(k)+(1-c)*p^2;
> end
> P % compare P(infinity) with p^2
```

**7.** Suppose the frequency of a recessive allele is $p$ (equal to $\frac{1}{1000}$, say); therefore, the frequency of homozygotes under the hypothesis of random mating will be $p^2$. But what if mating is not random? In this problem we want to investigate this somewhat.

First, suppose the species is capable of self-fertilization. Then clearly the offspring of a homozygous adult will again be homozygous. On the other hand, the heterozygous Aa will produce A and a haploid cells in 50–50 mix as before. Hence as before, an offspring will be AA with $\frac{1}{4}$ chance, aa with $\frac{1}{4}$ chance, and Aa with $\frac{1}{2}$ chance. We record these observations in the following $3 \times 3$ matrix:

$$T = \begin{bmatrix} 1 & 0 & 0 \\ \frac{1}{4} & \frac{1}{2} & \frac{1}{4} \\ 0 & 0 & 1 \end{bmatrix}.$$

In this, the rows correspond to the genotypes AA, Aa, and aa in that order and so do the columns.

Next, suppose we start out with a mix of genotypes, say, their fractions are $p$, $q$, and $r$, respectively, $p + q + r = 1$. Then after one generation, the new fractions $p'$, $q'$, and $r'$ will be given by the matrix product

$$(p' \; q' \; r') = (p \; q \; r)T.$$

(a) Using specific values for the starting fractions, find the limiting fractions after many generations.

Next, consider parent/child matings and calculate the probability that a homozygous recessive aa will be the result. First, condition on the parent (sketch a tree diagram) that from the root node, there will be three edges corresponding to the possibilities that the parent is AA, Aa, or aa. The AA branch cannot lead to an aa grandoffspring, so there is no need to follow that edge further. The Aa parent occurs with frequency $2p(1-p)$, as we have seen, and the aa parent with frequency $p^2$.

Next, condition on the genotype of the child. Use the Hardy–Weinberg principle for probabilities of alleles A and a. Starting from the Aa node, the possibilities are AA with probability $\frac{1}{2}(1-p)$, Aa with probability $\frac{1}{2}p + \frac{1}{2}(1-p) = \frac{1}{2}$, and finally aa with probability $\frac{1}{2}p$. You do the possibilities from the aa node.

Now assign the offspring probabilities using Mendelian genetics. From the Aa node along the path from root through the Aa parent, the probability of an aa offspring is $\frac{1}{4}$. From the aa node through the Aa parent, the probability is $\frac{1}{2}$, and so on.

(b) Altogether, the result should be

$$P(\text{aa offspring}) = \frac{1}{2}p\left(\frac{3}{4} + \frac{3}{2}p - p^2\right).$$

Finally, consider sibling matings. As in part (a) above, we want to investigate the trend of the population toward homozygosity. Starting with the parents, there are six possible matings by genotype, AA × AA, AA × Aa, and so on through aa × aa. Consider the AA × Aa parents. Their offspring are AA and Aa both with frequency $\frac{1}{2}$. Therefore, the sibling mating possibilities are AA × AA with frequency $\frac{1}{4}$, AA × Aa with frequency $\frac{1}{2}$, and Aa × Aa with frequency $\frac{1}{4}$.

Justify the rest of Table 13.7.1. The corresponding transition matrix $T$ is

**Table 13.7.1.**

| **Parent** | **Sibling mating frequencies** | | | | | |
|---|---|---|---|---|---|---|
| **genotypes** | **AA x AA** | **AA x Aa** | **Aa x Aa** | **AA x aa** | **Aa x aa** | **aa x aa** |
| **AA x AA** | 1 | 0 | 0 | 0 | 0 | 0 |
| **AA x Aa** | $\frac{1}{4}$ | $\frac{1}{2}$ | 0 | $\frac{1}{4}$ | 0 | 0 |
| **Aa x Aa** | $\frac{1}{16}$ | $\frac{1}{4}$ | $\frac{1}{8}$ | $\frac{1}{4}$ | $\frac{1}{4}$ | $\frac{1}{16}$ |
| **AA x aa** | 0 | 0 | 1 | 0 | 0 | 0 |
| **Aa x aa** | 0 | 0 | 0 | $\frac{1}{4}$ | $\frac{1}{2}$ | $\frac{1}{4}$ |
| **aa x aa** | 0 | 0 | 0 | 0 | 0 | 1 |

$$T = \begin{bmatrix} 1 & 0 & 0 & 0 & 0 & 0 \\ \frac{1}{4} & \frac{1}{2} & 0 & \frac{1}{4} & 0 & 0 \\ \frac{1}{16} & \frac{1}{4} & \frac{1}{8} & \frac{1}{4} & \frac{1}{4} & \frac{1}{16} \\ 0 & 0 & 1 & 0 & 0 & 0 \\ 0 & 0 & 0 & \frac{1}{4} & \frac{1}{2} & \frac{1}{4} \\ 0 & 0 & 0 & 0 & 0 & 1 \end{bmatrix}.$$

(c) Make up an initial distribution of genotypes $(p\ q\ r\ s\ t\ u)$, track the change in distribution over a few generations, and find the limiting distribution.

**Questions for Thought and Discussion**

1. Discuss the concept of fitness as it is used in the Darwinian model. What kinds of selection factors might be involved in the case of humans?
2. A woman with type A blood has a child with type O blood. The woman alleges that a certain man with type B blood is the father. Discuss her allegation and reach a conclusion, if possible.
3. In *Drosophila*, females are XX and males are XY. On the X chromosome, there is an eye-color gene such that red is dominant to eosin and to white, and eosin is dominant to white. What is the result of crossing an eosin-eyed male with a red-eyed female whose mother had white eyes?
4. Mitosis is a conservative form of cell replication, because each daughter cell gets an exact copy of the genetic material that the parent cell had. How can we explain the fact that most of our tissues were formed by mitosis and yet are different?
5. Suppose there is an organism that reproduces only by self-fertilization, which is the highest degree of inbreeding. Start with a heterozygote for a single property and let it and its descendants reproduce by self-fertilization for three generations. Note how the fraction of homozygotes increases with each generation. What implication does this have if the recessive allele is harmful? Or suppose it is not harmful?
6. Combining the concepts of the central dogma of genetics with that of meiosis, trace the path of hereditary control of cellular chemistry from one generation to another.
7. In a hypothetical laboratory animal, a solid-color allele is dominant to striped, and long hair is dominant to short hair. What is the maximum number of phenotypes that could result from the mating of a long, solid animal with a short, striped animal?

## References and Suggested Further Reading

[1] CELL DIVISION AND REPRODUCTION:
W. S. Beck, K. F. Liem, and G. G. Simpson, *Life: An Introduction to Biology*, 3rd ed., Harper–Collins, New York, 1991.

[2] GENETICS:
D. T. Suzuki, A. J. F. Griffiths, J. H. Miller, and R. C. Lewontin, *An Introduction to Genetic Analysis*, W. H. Freeman, 3rd ed., New York, 1986.

[3] SEXUAL SELECTION IN HUMANS:
D. M. Buss, The strategies of human mating, *Amer. Sci.*, **82** (1994), 238–249.

[4] MATHEMATICAL GENETICS:
J. F. Crow and M. Kimura, *An Introduction to Population Genetics Theory*, Harper and Row, New York, 1970.

[5] MATHEMATICAL GENETICS:
J. B. S. Haldane, A mathematical theory of natural and artificial selection V: Selection and mutation, *Cambridge Philos. Soc. Proc.*, **23**-VII (1927), 838–844.

# 14

# Genomics

## 14.1 The Promise of Genomics

In Chapter 8, we learned that DNA holds the information on how to construct a living organism and make it work. This is achieved by instructing the cell how to make proteins characteristic for that organism, one cell at a time. So biologists became interested in determining the precise sequence of the nucleotides in the DNA.

At first, this was hard to do and took a great deal of time even for small segments. Simple strands of DNA were the first to be attempted, those of viruses and the plasmids of bacteria. In time, new techniques were invented and sequencing became faster. A big step was made with the invention of PCR, *polymerase chain reaction*, for making millions of copies of a strand of DNA.

Still, sequencing the *genome* of an organism, its entire complement of DNA, was beyond reach, except possibly that of a virus. And so it was a wildly ambitious plan when, in 1990, the U.S. government launched the *Human Genome Project* (HGP). The project was to sequence the entire human genome of 3 billion base pairs to an accuracy of one error in 10,000 bases.[1] The science of genomics was born, and with it, biology is forever changed.

Genomics is the science of understanding the information contained in the genomes of the world's organisms. Genomics has engendered whole new branches of biology: proteomics, comparative genomics, genomic medicine, pharmacogenomics, and structural genomics among them. Besides the biology, an essential ingredient in genomics is computation. Genomics has been called "computational biology."

*How is DNA sequenced?*

At the time it was announced, the Human Genome Project seemed an insurmountable task that would require at least 20 years to complete. But in fact, the job was done in less than 10. This was possible due to enormous advances in the technology of

[1] The original plan of one error in 100,000 bases, a somewhat arbitrary figure, was eased to reduce costs.

R.W. Shonkwiler and J. Herod, *Mathematical Biology: An Introduction with Maple and Matlab*, Undergraduate Texts in Mathematics, DOI: 10.1007/978-0-387-70984-0_14,

nucleotide sequencing. Today automated machines can sequence 600 to 800 bp (base pairs) in an hour and issue the results.

The basis for DNA sequencing today is the approach discovered by Frederick Sanger. The key element in this approach is the use of DNA polymerase, which adds complementary bases to a single-strand DNA template, and dideoxynucleotide, ddNTP, which stops the addition of bases. There is a different dideoxynucleotide for each base ddATP, ddTTP, ddCTP, and ddGTP.

Neglecting technical issues, such as the use of primers to get the reaction started, the essentials of the method are these. Four reaction vessels are used, one for each of the four bases. Consider the "A" reaction vessel. It includes complementary single strands of the DNA to be sequenced, a supply of the four bases, and a small amount of ddATP. By the action of DNA polymerase, bases are added one by one to these strands, forming partially complete double-stranded DNA of growing length. If by chance a ddATP is added where an A should have gone, then the chain is "frozen": No more bases will be added, and the length will not increase.

At the end of the reaction, the entire content of the vessel is now separated by electrophoresis in capillary gel tubes. Shorter segments will move farther than longer ones, and they all will end with a ddATP. For example, if the first complementary base in the sequence is A (so the original base is T), then the double strand of primer followed by the T–ddATP pair will move to the end of the capillary tube.

By lining up the four capillary tubes from each reaction and noting the tube of the segment that migrated farthest, say the "A" tube, and the tube of the segment that migrated second farthest, say the "G" tube, and the third farthest, maybe "G" again, and so on, the complimentary sequence can be read, in this case

AGG....

Originally, the positions of the migrating fragments were determined by radioactive labeling, but today fluorescent dyes are used instead, a different color for each base. The DNA sequencing machine can determine the colors as the four capillary gels pass by the reading window.

Of course, only fragments of a genome can be sequenced this way in any one run. Ultimately the fragments must be stitched together. There are two main approaches to this step, *map-based sequencing* and the *whole genome shotgun* (WGS) technique.

In the first, a chromosome to be sequenced is divided into large segments of about 150 kilobases. These segments have generous overlap. The whole collection is called a library. The segments are then cloned using *bacterial artificial chromosomes* (BACs), a man-made piece of DNA that can be replicated in bacteria. Before sequencing the segments, their order must first be worked out. As an aid in this step, each segment is "fingerprinted," or given a unique tag. With their order known, these segments are then sequenced. For this they are broken up again into subsegments of about 1,500 bases long. These too have overlap. The subsegments are sequenced to a length of about 500 bp from each end. This generates thousands of A, T, C, G sequences, which are pieced together by computer algorithms that make essential use of the overlaps.

In the whole genome shotgun approach, the entire chromosome is randomly shredded into segments about 2,000 bp long. Meanwhile, separately, a copy is broken into segments 10,000 bp long. All segments are sequenced to a length of 500 bp from each end; this generates millions of sequences. Then, as above, very powerful parallel computers fit all the pieces together.

In both approaches, sequencing to 500 base pairs would seem to leave gaps. But due to the overlap, most of these are filled in. Any remaining gaps are flagged in the finishing process and further sequencing is expended to fill them.

## 14.2 Structural Genomics

The aim of structural genomics is to determine the three-dimensional structure of a protein from its amino acid sequence. This effort extends to RNA as well, since some of the genome codes for RNA as the end result. Structural genomics has some overlap with *proteomics*, which studies the function of proteins in addition to elucidating protein structure. Of course, the function of a protein is closely related to and determined by its structure. Proteins that serve as enzymes depend on having a carefully crafted topography and electronic presentation along their active surfaces. Those functioning as cell membrane pores rely on their array of hydrophobic and hydrophilic parts, and those intended as structural materials typically have alpha-helices or beta-pleated sheets.

A protein starts out as a linear sequence of amino acids. This is called its *primary structure*. Even before the protein is completely synthesized, the partially assembled sequence begins changing shape to bring about its ultimate, biologically active conformation. This process is called *protein folding*. The final arrangement in space of all its atoms is the *tertiary* structure of a protein. Protein *secondary structure* refers to the three-dimensional form of substructures within the protein, such as alpha-helices and beta-pleated sheets.

With every possible configuration of a molecule, proteins included, there is an associated energy. The graph of this energy vs. configuration is its *energy landscape*; it is a multidimensional graph. Important factors determining the energy landscape are hydrogen-bonding, dipolar interactions, Van der Waals attraction, and disulfide bonds between the amino acid side chains. The energy landscape also depends on the molecule's environment. For a protein, this usually means the temperature, pH, and solvent in which it is located.

For simple molecules the energy landscape is a simple bowl-shaped, or monomodal, surface. Immediately upon formation, such a molecule assumes a configuration at or near the bottom of the bowl, its energy minimum. For complicated molecules the energy landscape is multimodal, meaning it has multiple bowl-shaped regions. The energy of one basin will generally be different from that of another. The correct tertiary structure of a protein is that of its lowest energy basin.

The most successful method of determining protein structure has been *X-ray crystallography*. This method requires high-quality crystals of the protein subject. These are then placed in the path of an X-ray source, and the resulting diffraction pattern is recorded. Using a great deal of skill and insight, the shape of the protein can be worked out. The structure of myoglobin, published in 1959, took 20 years for the job. With modern methods, heavily computational, the time is much shorter.

X-ray crystallography is limited by its need for high quality crystals and for an X-ray source. With regard to the second of these, today that source is usually a synchrotron, and they are in limited supply. The first requirement can be a bigger hurdle, since it is very hard to crystallize proteins that function in a hydrophobic environment.

Another method makes use of NMR technology. This relies on the spin of the nuclei of hydrogen, carbon, and nitrogen atoms. A protein being tested is in aqueous solution and is therefore closer to its *in vivo* condition. The drawbacks of this method are its poor resolution; the data obtained are very difficult to analyze, and this, in turn, imposes a severe limit on protein size; and finally, the sample must be at high concentration.

The ideal method would allow the structure to be worked out from its amino acid sequence and environment alone, computationally. This is the *ab initio* method. One line of attack is to simulate the events of protein folding. The physics of the simulation are well understood; the problem is that the computational requirement for anything but the very smallest protein is beyond present-day capability. Another possibility is to combine computational techniques with experimentally derived substructures. So far, the computational method has had only limited success.

The Protein Data Bank (PDB) is a repository for three-dimensional protein structures. This is a consortium of the San Diego Supercomputing Center, Rutgers University, and the National Institute of Standards and Technology. Every protein in the repository has been validated by the staff of PDB for physical plausibility. These data are available online at http://www.rcsb.org/pub/.

*Structure-based drug design.*

Understanding the binding properties of proteins has become an integral part of modern drug discovery. The goal is to produce a molecule that will bind, or *dock*, with an active site on the target protein. To succeed, a thorough understanding of the molecular recognition between the active site and an interacting molecule is required. The active site of the protein is a space to be filled with a molecule that complements it in terms of shape, charge, and other binding components.

An important tool in the design process is software that visualizes how the drug interacts with these components three-dimensionally. This knowledge can then guide the chemistry, called *computational chemistry*. Varieties of promising compounds are made in parallel and checked by means of some sort of activity or binding assay. This high-throughput method rapidly sorts through all these possibilities to find out which ones really work.

Prior to the use of structural genomics, combinatorial chemistry by itself was found to be an ineffectual strategy. This is because the number of molecules that can be made by the technique is still infinitesimally small compared to all the possibilities. Of the huge numbers of compounds made, nearly all were irrelevant.

Likewise, designing drugs *de novo*, purely computationally, turned out to be equally ineffective. Docking software suffered from an incomplete understanding of the energy, electron densities, and thermodynamics of the interacting molecules. The molecules derived from such studies did not conform to the expectations in potency when made.

Structure-based drug design is at its most powerful when coupled with combinatorial techniques. One of the most effective uses of structure is in optimization. An understanding of structure is key to devising methods for modifying the drug molecule to get the desired properties. Potential compounds can be screened on the computer in a very short period of time.

The binding mode of synthesized compounds to the target is often verified using X-ray crystallography and NMR. Three-dimensional structures produced by X-ray crystallography have become an integral part of the drug discovery process because of changes that have increased throughput. Synchrotron beam lines have made it possible to run a large number of crystals all at once because crystallographic information can be obtained from smaller crystals than was previously possible. The verification results are feedback into the prediction simulations to improve the process.

Speeding up drug discovery is only one of the goals of structure-based drug design. Another is to improve the quality of the drug, to determine molecules that have better pharmacological properties.

## 14.3 Comparative Genomics

Comparative genomics is the science of understanding and making use of similar genomic segments between two, several, or a group of organisms. Such segments can help locate gene coding and regulatory regions, clarify the importance of gene expression, elucidate protein–protein interactions, and help unravel gene evolution. As a warmup to decoding the human genome, the Human Genome Project tested and refined their techniques by decoding the genomes of brewer's yeast, *Saccharomyces cerevisiae*, the roundworm *Caenorhabditis elegans*, the plant *Arabidopsis thaliana* of the mustard family, and the fruit fly *Drosophila melanogaster*. Thus began the science of comparative genomics.

*Synteny* refers to the preserved order of genes, on a chromosome, between related organisms. Synteny can be derived from genetic or physical maps of the genome. A genetic map provides a distance between two genes on the same chromosome by noting the frequency of their recombination. During meiosis, chromosomes become paired, and in the process of duplication, they may break. It can then happen that when the chromosome is put back together, a broken segment is swapped with the matching segment of the homologous chromosome. This form of recombination is called *crossing over*. The greater the distance between two genes on the same chromosome, the more likely crossing over will occur between them. In turn, these recombination rates are used to construct a genetic map along the chromosome.

A physical map gives the location of identifiable landmarks along a chromosome measured in base pairs. Examples of landmarks are restriction enzyme cutting sites and the location of genes. Physical maps are constructed by cutting the DNA segment between landmarks into overlapping fragments called *contigs*. These are then inserted into a *vector*, an agent for cloning the DNA, such as a phage virus that can accommodate up to 45 kilobases, or as a *bacterial artificial chromosome* (BAC) spliced into *Escherichia coli* cells. These can accommodate up to 300 kb. The contigs are now sequenced, as for example by the Sanger–Hood method.

Synteny appears to be common. Long stretches of the human and mouse genomes show a remarkable degree of synteny. As a result, a human gene having a function similar to a gene of the mouse can be approximately located within the human genome if the analogue's location in the mouse genome is known.

But comparison between genomes is much more useful than that. DNA is subject to background mutation or random nucleotide drift. Several lines of research point to a gene mutation rate of about $10^{-7}$ per replication (as we obtained in the retinoblastoma study, Section 12.4). The rate is approximately the same even across biological divisions: Archaea, Bacteria, and Eukaryota. However, most gene mutations are detrimental to the organism. As a result, two gene sequences that started out the same tend to be conserved even over millions of years of independent evolution. Their coding sequences as well as their regulatory regions, everything functionally important, resist random drift.

Not so for the other segments of the genome. Noncoding DNA that does not have a structural or regulatory function tends to diverge much more rapidly than coding DNA. By comparing genomes, especially across related species, much can be learned about the location of exons, introns, and the regulatory segments of genes (recall Section 8.3 for the definitions of these terms). When the genome of *Fugu* (a type of blowfish) was compared with the human genome, over 1,000 new human gene candidates were discovered.

In genomics, *homology* refers to the similarity in DNA or protein sequences between individuals of the same species or among different species. In Section 14.6, we will learn how to search genomes for matching segments. This is a powerful tool for finding homologies. A surprising discovery resulting from such a homology search is that the sea squirt *Ciona intestinalis* has a homolog with the gene endoglucanase of the plant *Arabidopsis*. This gene is believed to have an important role in cellulose synthesis. The search was undertaken because the coat of *Ciona* is made primarily of cellulose.

In the science to which comparative genomics has given rise, the term *ortholog* refers to genes found in two species having a common ancestor. On the other hand, *paralogs* are similar genes found within a single species, copies having been created by duplication events. A challenge of comparative genomics is to discern true orthologs and not just similar DNA segments.

The utility of comparative genomics extends to gene expression as well. Transcription, the synthesis of mRNA, involves three main events: initiation, elongation, and termination. The main work is done by RNA polymerase. In the elongation step, RNA polymerase adds complementary nucleotides to the $3'$ end of the growing RNA chain. Termination occurs when RNA polymerase encounters a termination signal within the gene. The initial mRNA transcript is processed by splicing out introns and ligating exons into the mature mRNA. Gene expression therefore is primarily controlled by initiation, which is a fairly involved process.

It begins with the binding of RNA polymerase to the double-stranded DNA at a site called the *promoter*. This is usually a short distance upstream of the start of transcription and contains the nucleotide sequence TATA, called the *TATA-box*. The DNA becomes single-stranded at this site. Note that a TA-rich area is more easily unwound than a CG area, since T–A is doubly hydrogen-bonded, while C–G is triply bonded.

Besides RNA polymerase, the help of other proteins, called *transcription factors*, is also required for initiation. Gene expression may also be regulated by repressors that must vacate their DNA binding sites before transcription can begin. Identifying regulatory sequences in a sequenced genome is a challenging problem.

But as we saw above, because the regulatory sites are part of the functional apparatus of the gene, they are likely to be conserved across species. By aligning orthologs from two or more species and noting matching intergenic sequences, candidate regulatory segments can be uncovered. These conserved segments are called *phylogenic footprints*.

Comparative genomics can be used to to identify proteins that interact with each other. The method exploits an insight that if multiple protein-coding segments in one species are coded as a single protein in another, then it is likely those proteins interact. The technique has been effective in finding interacting proteins in the three bacteria species *Escherichia coli*, *Haemophilus influenzae*, and *Methanococcus jannaschii*.

Probably the biggest use made of comparative genomics is in elucidating both biological and gene evolution. Before the human genome was available, it was thought that the number of human genes would be around 100,000. It did not turn out that way; it is now believed there are only 30,000 human genes. This number is not very different from those of other organisms such as chimpanzees and even the plant *Arabidopsis*. Furthermore, humans' and chimps' genomes are about 98% similar. But humans and chimps have significant differences. So how can these facts be reconciled?

One way this is possible is by differences in gene expression. In fact, it has been shown that there is over a fivefold difference in gene expression related to brain tissue between chimps and humans. This shows that gene expression is an important factor in biological evolution and that coopting existing genes is as important as creating new ones.

## 14.4 Genomics in Medicine

The human genome codes for about 30,000 genes. Of course, a base pair mutation in any one of them has the potential for causing disease. Hence we see that there can be a very large number of genetic diseases. Table 14.4.1 lists some of these diseases along with the gene whose mutation causes it. As of the year 2000, there were 1,000 known heredity disorders.

**Table 14.4.1.** Genes and their diseases (when mutated). (Source: www.ornl.gov/sci/techresources/Human_Genome/medicine/genetest.shtml.)

| APKD | Adult polycystic kidney disease |
|---|---|
| AAT | Alpha-1-antitrypsin deficiency |
| ALS | Amyotrophic lateral sclerosis (Lou Gehrig's disease) |
| APOE | Alzheimer's disease |
| AT | Ataxia telangiectasia |
| CMT | Charcot–Marie–Tooth |
| CAH | Congenital adrenal hyperplasia |
| CF | Cystic fibrosis |
| DMD | Duchenne–Becker muscular dystrophy |
| DYT | Dystonia |
| FA | Fanconi anemia group C |
| FVL | Factor V-Leiden |
| FRAX | Fragile X syndrome |
| GD | Gaucher disease |
| HEMA, HEMB | Hemophilia A and B |
| HFE | Hereditary hemochromatosis |
| CA | Hereditary nonpolyposis colon cancer |
| HD | Huntington's disease |
| BRCA 1, BRAC 2 | inherited breast and ovarian cancer |
| MD | Myotonic dystrophy |
| NF1 | Neurofibromatosis type 1 |
| PKU | Phenylketonuria |
| PW/A | Prader–Willi–Angelman syndromes |
| SS | Sickle-cell disease |
| SMA | Spinal muscular atrophy |
| SCA1 | Spinocerebellar ataxia |
| TS | Tay–Sachs disease |
| THAL | Thalassemias |

Genomics offers the possibility of curing heredity diseases. First, the cause of the disease must be worked out. This is easiest for *monogenic diseases*, or those attributable to a single faulty gene. The victim either lacks or has too little of some crucial protein, or has a flawed version, or some protein is made to excess. The initial task is discovering the offending protein. Sometimes this can be done via the genome itself. Another possibility is to identify the responsible protein directly and reverse the transcription, translation process. This approach works through the intermediary mRNA, which is another clue to the gene. This is the method of *expressed sequence tags* (ESTs). Complicating the EST approach is that genes are seldom contiguous, and in any case it will not find initiators or promoters.

Along with the base pair sequence, the location of the gene within the genome must also be found—on which chromosome and where along the chromosome. This step is greatly aided by having the complete human genome sequence and is mostly a software task. Knowing the location includes knowing the exons, introns, promoters, and initiators. Then the task of deducing the problem with the faulty gene can begin.

This process has been worked through for Huntington's disease, a disease of progressive degeneration of brain cells in certain areas of the brain. The gene, HD, is on the fourth chromosome and is dominant. In the normal form, the trinucleotide CAG is repeated within the sequence up to 26 times. In Huntington's patients, the trinucleotide is repeated 36 to 125 times. With repetition between 36 to 39 times, the individual may or may not contract the disease, but with repeats over 39, disease is certain and symptoms occur sooner with increasing number of repeats.

Another genetic disease for which the responsible protein and corresponding gene has been worked out is *cystic fibrosis* (CF). It is caused by a defect in a gene called the cystic fibrosis transmembrane conductance regulator (CFTR) gene. In 1989, scientists identified the CF gene on chromosome 7. About 70% of the mutations observed in CF patients result from deletion of the three base pairs CTT (and GAA on the complementary strand) in CFTR's nucleotide sequence. This deletion causes loss of the amino acid phenylalanine (even though CTT is not a codon for phenylalanine, the error is between frames) located at position 508 in the CFTR protein; as a result, this mutation is referred to as *delta F*508.

The normal CFTR protein product is a chloride channel protein found in membranes of cells that line passageways of the lungs, liver, pancreas, intestines, reproductive tract, and skin. It controls the movement of salt and water in and out of these cells. In people with CF, the gene does not work effectively, resulting in a sodium and chloride ion imbalance. This in turn creates a thick, sticky mucus layer that cannot be removed by cilia and traps bacteria, resulting in chronic infections.

The faulty gene is autosomal (not sex-linked) recessive, so one must have two defective genes to have the disease. It also means that individuals with one defective gene are carriers.

How can these and other genetic diseases be cured? At the present time, genetic tests are available for the diseases listed in Table 14.4.1. Consequently, prospective parents can ensure that their child will be free of these diseases. If they do not have an aberrant gene themselves, then neither will their child.

On the other hand, through genetic counseling, couples can explore options. One of these is to forgo having children. Another is adoption. But there is also another possibility, *in vitro* fertilization and genotyping of embryos. In the case of an autosomal dominant disease, such as Huntington's, each embryo has a 50% chance of being free of the disease if one parent is afflicted and 25% chance even if both are. Similarly for an autosomal recessive disease, such as cystic fibrosis, even if both parents are carriers. This procedure was first performed in 1992 for cystic fibrosis carriers.

It should be noted that some people have ethical issues with *in vitro* fertilization, since several human embryos are created and genetically tested. Unused embryos are discarded.

Remedies are also sought for those with existing disease. Most of these are in the research stage of development. It may be possible to create a drug for a missing or ineffective protein. A more daring possibility is *gene therapy*. The goal here is to deliver a functional gene to the cells of the body that need it. At this time, the most effective means for delivering therapeutic genes have been viruses. Adenoviruses are frequently used because they are associated with relatively mild illnesses or with no illness at all. Retroviruses have also been used.

Although there have been some successes with gene therapy, there have also been some tragic failures. In one case, a healthy volunteer died as a result of an exaggerated immune response to an adenovirus. In another, a child being treated for SCID developed leukemia following gene therapy with a retrovirus. It is believed the virus activated a cancer-causing gene.

Genomics is a major tool in cancer research. In one application, EST sequencing is undertaken for both normal and tumor cells. The EST method of sequencing, via mRNA, targets gene expression. Once in the database, the sequences can be analyzed using nucleotide matching and other bioinformatics software, a topic we will take up later in this chapter.

Another approach to studying cancer is by means of *microarrays*. A microarray is used to determine, in a single experiment, the expression levels of hundreds or thousands of genes. The array is made up using robotic equipment to spot cDNA samples on a microscope slide or other substrate. *Complementary DNA*, or *cDNA*, is synthesized from an mRNA template or, turning this around, a segment of DNA used to bind to mRNA as here. Fluorescent labels are attached to DNA or mRNA taken from a cell under study and allowed to bind to the cDNA. The expression levels of these probes can be determined by measuring the amount bound to each site on the array. With the aid of a computer, that amount is precisely measured, generating a profile of gene expression in the cell.

The gene expression profile of metastasizing medulloblastoma (a largely childhood cancer of the brain) was determined by microarray experiments. One of the genes enhanced under metastasis was one coding for *platelet derived growth factor receptor alpha* (PDGFR$\alpha$). In this way, the experiment pointed to candidate drug targets for combating metastasis.

Genomics is helping to fight microbial diseases too. In 2003, sequencing of the genome of *P. falciparum* was completed. It took an unusually long time, six years, since it contains a high A+T content, which caused problems for the software that joins overlapping segments. The organism has 5,300 protein-coding segments, which can now be studied in detail genetically. It is hoped that this will result in opportunities for new drug targets and vaccine candidates.

In fact, this has already been done and has resulted in the discovery that an existing drug, *DOXP reductoisomerase*, is effective in curing malaria in rats. The discovery began by searching for homologies between *P. falciparum*'s genes and those in other organisms. It was learned that the parasite has a homolog to the enzyme DOXP reductoisomerase found in bacteria that is part of the biosynthetic pathway synthesizing isopentenyl diphosphate, a vital precursor chemical. Animals use a different pathway from DOXP for this synthesis, providing an opportunity for a drug with few side effects. By chance, the DOXP reductoisomerase drug was already on the market as an antibacterial drug.

Alternatively, the disease might be overcome through manipulation of its mosquito vector. The genome of *Anopheles gambiae* was sequenced in 2002. A genetically modified mosquito was subsequently created with the ability to block the transmission of the parasite to humans. This was done by inserting a gene into the mosquito to produce a 12-amino-acid peptide that blocks the oocysts' entry into the mosquito's salivary glands; see Section 11.2. This is truly a novel and unforeseen application of genomics and a hallmark of good science.

## 14.5 Protein Substitution Matrices

Eventually, mutations occur in the genome leading to mistakes in the encoded amino acid sequence. But these mistakes have varying effects. Most will render the protein dysfunctional and compromise its host organism. However, some amino acid substitutions are effectively harmless. And there is a spectrum of possibilities between these two extremes.

This is possible because, with respect to specific attributes, subgroups of the 20 amino acids are similar to each other. Thus with respect to size, glycine, alanine, serine, and cysteine are alike in that they are small. The peptides serine, threonine, and tyrosine are hydroxyl. And so on for several other attributes. The main antithetic pairs are large/small, hydrophobic/hydrophilic, aliphatic/aromatic, acidic/basic, and sulphur-containing/non-sulphur-containing. An amino acid substitute, sufficiently similar to the original, might not be disruptive and could even be beneficial.

To quantify this phenomenon, Margaret Dayhoff analyzed 1572 substitutions between closely related proteins. Frequencies of substitutions were noted and tabulated as shown in Table 14.5.1. Along the left side of the table are the amino acids denoted by their standard three-letter designation. Along the top are those same amino acids denoted by their single-letter designations. (These were also introduced by Dayhoff. To find the familiar three-letter protein for "D," say, move down the D column to the dash and then across to "asp." Note that Table 14.6.1, given later on in this chapter, gives the correspondence directly and includes "wildcards.") In the body of the table are the frequencies of substitution. The table is triangular, since it is assumed that the substitutions can go both ways just as well, e.g., methionine for leucine just as well as leucine for methionine.

This exchange-count table has a profound significance. With some mathematical processing, it can become the basis of a model for protein evolution. This is because the table's frequency data expresses the extent to which residues change as a result of accepted mutations as compared with purely random exchanges.

The model is *Markovian*, meaning it is based on assumptions that certain events, as given next, are independent and not affected by the other events of the model. The first, neighbor independence, is that each residue mutates independently; its mutation rate is uncoupled from that of the other residues. The second is positional independence, the probability that residue $i$ mutates to residue $j$ depends only on $i$ and $j$ and not on where $i$ appears in the sequence. And the third, historical independence, means that mutation is memoryless, the probability of mutation at each site does not depend on how the present peptide sequence came about.

**Table 14.5.1.** Dayhoff exchange counts.

| | G | A | V | L | I | M | C | S | T | N | Q | D | E | K | R | H | F | Y | W | P |
|---|---|---|---|---|---|---|---|---|---|---|---|---|---|---|---|---|---|---|---|---|
| gly | — | | | | | | | | | | | | | | | | | | | |
| ala | 58 | — | | | | | | | | | | | | | | | | | | |
| val | 10 | 37 | — | | | | | | | | | | | | | | | | | |
| leu | 2 | 10 | 30 | — | | | | | | | | | | | | | | | | |
| ile | | 7 | 66 | 25 | — | | | | | | | | | | | | | | | |
| met | 1 | 3 | 8 | 21 | 6 | — | | | | | | | | | | | | | | |
| cys | 1 | 3 | 3 | | 2 | | — | | | | | | | | | | | | | |
| ser | 45 | 77 | 4 | 3 | 2 | 2 | 12 | — | | | | | | | | | | | | |
| thr | 5 | 59 | 19 | 5 | 13 | 3 | 1 | 70 | — | | | | | | | | | | | |
| asn | 16 | 11 | 1 | 4 | 4 | | | 43 | 17 | — | | | | | | | | | | |
| gln | 3 | 9 | 3 | 8 | 1 | 2 | | 5 | 4 | 5 | — | | | | | | | | | |
| asp | 16 | 15 | 2 | | 1 | | | 10 | 6 | 53 | 8 | — | | | | | | | | |
| glu | 11 | 27 | 4 | 2 | 4 | 1 | | 9 | 3 | 9 | 42 | 83 | — | | | | | | | |
| lys | 6 | 6 | 2 | 4 | 4 | 9 | | 17 | 20 | 32 | 15 | | 10 | — | | | | | | |
| arg | 1 | 3 | 2 | 2 | 3 | 2 | 1 | 14 | 2 | 2 | 12 | 9 | | 48 | — | | | | | |
| his | 1 | 2 | 3 | 4 | | | 1 | 3 | 1 | 23 | 24 | 4 | 2 | 2 | 10 | — | | | | |
| phe | 2 | 2 | 1 | 17 | 9 | 2 | | 4 | 1 | 1 | | | | | 1 | 2 | — | | | |
| tyr | | 2 | 2 | 2 | 1 | | 3 | 2 | 2 | 4 | | | 1 | 1 | | 4 | 26 | — | | |
| trp | | | | 1 | | | | 2 | | | | | | | 3 | | 1 | 1 | — | |
| pro | 5 | 35 | 5 | 4 | 1 | | 1 | 27 | 7 | 3 | 9 | 1 | 4 | 4 | 7 | 5 | 1 | | | — |

The importance of the Markovian assumption is this: Given a mutation matrix $M$ of probabilities $m_{ij}$ that amino acid $i$ will mutate to amino acid $j$ over some evolutionary time period, then in two such time periods the mutation probabilities are given by $M^2$, and in three time periods by $M^3$, and so on.

The time period chosen by Dayhoff is that required to give an average amino acid mutation rate of 1%. The corresponding matrix is called the *PAM* 1 *matrix* (point accepted mutation). It is possible to compute the PAM1 matrix starting from the exchange matrix of Table 14.5.1, but one must first construct the missing diagonal elements; these are the frequencies of an amino acid not changing. Then, too, the entries above the diagonal have to be added, for example, by assuming that amino acid $j$ changes to $i$ with the same rate that $i$ changes to $j$, which is the assumption that is, in fact, made. After that, to be a probability matrix, the exchange matrix has to be normalized so its rows sum to 1. For brevity, we will instead start with the matrix in Table 14.5.2, for which these steps have already been done.

*Constructing PAM* 1.

We start by figuring the average mutation rate. Each diagonal entry is the probability that its residue does not mutate, so the difference from 1 is the probability that it does. These have to be weighted according to the frequency of the amino acid. These are given in Table 14.5.3 and compiled by counting the number of occurrences of the various amino acids in biological proteins (from Jones, Taylor, and Thornton [2]).

**Table 14.5.2.** Assumed mutation matrix $M$.

| | A | C | D | E | F | G | H | I | K | L | M | N | P | Q | R | S | T | V | W | Y |
|---|---|---|---|---|---|---|---|---|---|---|---|---|---|---|---|---|---|---|---|---|
| A | .44 | .00 | .02 | .04 | .00 | .09 | .00 | .01 | .01 | .02 | .00 | .02 | .05 | .01 | .00 | .12 | .09 | .06 | .00 | .00 |
| C | .00 | .96 | .00 | .00 | .00 | .00 | .00 | .00 | .00 | .00 | .00 | .00 | .00 | .00 | .00 | .02 | .00 | .00 | .00 | .00 |
| D | .02 | .00 | .68 | .13 | .00 | .02 | .01 | .00 | .00 | .00 | .00 | .08 | .00 | .01 | .01 | .02 | .01 | .00 | .00 | .00 |
| E | .04 | .00 | .13 | .68 | .00 | .02 | .00 | .01 | .02 | .00 | .00 | .01 | .01 | .06 | .00 | .01 | .00 | .01 | .00 | .00 |
| F | .00 | .00 | .00 | .00 | .89 | .00 | .00 | .01 | .00 | .03 | .00 | .00 | .00 | .00 | .00 | .01 | .00 | .00 | .00 | .04 |
| G | .09 | .00 | .02 | .02 | .00 | .72 | .00 | .00 | .01 | .00 | .00 | .02 | .01 | .00 | .00 | .07 | .01 | .02 | .00 | .00 |
| H | .00 | .00 | .01 | .00 | .00 | .00 | .86 | .00 | .00 | .01 | .00 | .03 | .01 | .04 | .02 | .00 | .00 | .00 | .00 | .01 |
| I | .01 | .00 | .00 | .01 | .01 | .00 | .00 | .77 | .01 | .04 | .01 | .01 | .00 | .00 | .00 | .00 | .02 | .10 | .00 | .00 |
| K | .01 | .00 | .00 | .02 | .00 | .01 | .00 | .01 | .73 | .01 | .01 | .05 | .01 | .02 | .07 | .03 | .03 | .00 | .00 | .00 |
| L | .02 | .00 | .00 | .00 | .03 | .00 | .01 | .04 | .01 | .78 | .03 | .01 | .01 | .01 | .00 | .00 | .01 | .05 | .00 | .00 |
| M | .00 | .00 | .00 | .00 | .00 | .00 | .00 | .01 | .01 | .03 | .91 | .00 | .00 | .00 | .00 | .00 | .00 | .01 | .00 | .00 |
| N | .02 | .00 | .08 | .01 | .00 | .02 | .03 | .01 | .05 | .01 | .00 | .65 | .00 | .01 | .00 | .07 | .03 | .00 | .00 | .01 |
| P | .05 | .00 | .00 | .01 | .00 | .01 | .01 | .00 | .01 | .01 | .00 | .00 | .82 | .01 | .01 | .04 | .01 | .01 | .00 | .00 |
| Q | .01 | .00 | .01 | .06 | .00 | .00 | .04 | .00 | .02 | .01 | .00 | .01 | .01 | .77 | .02 | .01 | .01 | .00 | .00 | .00 |
| R | .00 | .00 | .01 | .00 | .00 | .00 | .02 | .00 | .07 | .00 | .00 | .00 | .01 | .02 | .81 | .02 | .00 | .00 | .00 | .00 |
| S | .12 | .02 | .02 | .01 | .01 | .07 | .00 | .00 | .03 | .00 | .00 | .07 | .04 | .01 | .02 | .47 | .11 | .01 | .00 | .00 |
| T | .09 | .00 | .01 | .00 | .00 | .01 | .00 | .02 | .03 | .01 | .00 | .03 | .01 | .01 | .00 | .11 | .64 | .03 | .00 | .00 |
| V | .06 | .00 | .00 | .01 | .00 | .02 | .00 | .10 | .00 | .05 | .01 | .00 | .01 | .00 | .00 | .01 | .03 | .69 | .00 | .00 |
| W | .00 | .00 | .00 | .00 | .00 | .00 | .00 | .00 | .00 | .00 | .00 | .00 | .00 | .00 | .00 | .00 | .00 | .00 | .99 | .00 |
| Y | .00 | .00 | .00 | .00 | .04 | .00 | .01 | .00 | .00 | .00 | .00 | .01 | .00 | .00 | .00 | .00 | .00 | .00 | .00 | .92 |

**Table 14.5.3.** Amino acid frequencies.

| | | | |
|---|---|---|---|
| L 0.091 | E 0.062 | R 0.051 | Y 0.032 |
| A 0.077 | T 0.059 | P 0.051 | M 0.024 |
| G 0.074 | K 0.059 | N 0.043 | H 0.023 |
| S 0.069 | I 0.053 | Q 0.041 | C 0.020 |
| V 0.066 | D 0.052 | F 0.040 | W 0.014 |

It is listed in order of most to least frequent; thus leucine makes up over 9% of all proteins.

So the average mutation rate for matrix $M$ is

$$\begin{aligned}\text{average mutation rate} &= \sum_i f_i(1 - m_{ii}) \\ &= 0.077 * (1 - 0.44) + \cdots + 0.032 * (1 - 0.92) = 0.247,\end{aligned}$$

or 24.7%.

But for PAM1, the average mutation rate must be 1%. Since powers of PAM1 give mutations over multiples of the basic time period, we must find $k$ such that

$$(\text{PAM1})^k = M, \quad \text{or} \quad \text{PAM1} = M^{1/k}.$$

This shows that PAM1 is some appropriate root of $M$ for which the average mutation rate is 1%, a computationally intensive calculation but otherwise not a problem. The result is shown in Table 14.5.4.

**Table 14.5.4.** PAM1 matrix.

| | A | C | D | E | F | G | H | I | K | L | M | N | P | Q | R | S | T | V | W | Y |
|---|---|---|---|---|---|---|---|---|---|---|---|---|---|---|---|---|---|---|---|---|
| A | 9867 | 3 | 10 | 17 | 2 | 21 | 2 | 6 | 2 | 4 | 6 | 9 | 22 | 8 | 2 | 35 | 32 | 18 | 0 | 2 |
| C | 1 | 9973 | 0 | 0 | 0 | 0 | 1 | 1 | 0 | 0 | 0 | 0 | 1 | 0 | 1 | 5 | 1 | 2 | 0 | 3 |
| D | 6 | 0 | 9859 | 53 | 0 | 6 | 4 | 1 | 3 | 0 | 0 | 42 | 1 | 6 | 0 | 5 | 3 | 1 | 0 | 0 |
| E | 10 | 0 | 56 | 9865 | 0 | 4 | 2 | 3 | 4 | 1 | 1 | 7 | 3 | 35 | 0 | 4 | 2 | 2 | 0 | 1 |
| F | 1 | 0 | 0 | 0 | 9946 | 1 | 2 | 8 | 0 | 6 | 4 | 1 | 0 | 0 | 1 | 2 | 1 | 0 | 3 | 28 |
| G | 21 | 1 | 11 | 7 | 1 | 9935 | 1 | 0 | 2 | 1 | 1 | 12 | 3 | 3 | 1 | 21 | 3 | 5 | 0 | 0 |
| H | 1 | 1 | 3 | 1 | 2 | 0 | 9912 | 0 | 1 | 1 | 0 | 18 | 3 | 20 | 8 | 1 | 1 | 1 | 1 | 4 |
| I | 2 | 2 | 1 | 2 | 7 | 0 | 0 | 9872 | 2 | 9 | 12 | 3 | 0 | 1 | 2 | 1 | 7 | 33 | 0 | 1 |
| K | 2 | 0 | 6 | 7 | 0 | 2 | 2 | 4 | 9926 | 1 | 20 | 25 | 3 | 12 | 37 | 8 | 11 | 1 | 0 | 1 |
| L | 3 | 0 | 0 | 1 | 13 | 1 | 4 | 22 | 2 | 9947 | 45 | 3 | 3 | 6 | 1 | 1 | 3 | 15 | 4 | 2 |
| M | 1 | 0 | 0 | 0 | 1 | 0 | 0 | 5 | 4 | 8 | 9874 | 0 | 0 | 2 | 1 | 1 | 2 | 4 | 0 | 0 |
| N | 4 | 0 | 36 | 6 | 1 | 6 | 21 | 3 | 13 | 1 | 0 | 9822 | 2 | 4 | 1 | 20 | 9 | 1 | 1 | 4 |
| P | 13 | 1 | 1 | 3 | 1 | 2 | 5 | 1 | 2 | 2 | 1 | 2 | 9926 | 8 | 5 | 12 | 4 | 2 | 0 | 0 |
| Q | 3 | 0 | 5 | 27 | 0 | 1 | 23 | 1 | 6 | 3 | 4 | 4 | 6 | 9876 | 9 | 2 | 2 | 1 | 0 | 0 |
| R | 1 | 1 | 0 | 0 | 1 | 0 | 10 | 3 | 19 | 1 | 4 | 1 | 4 | 10 | 9913 | 6 | 1 | 1 | 8 | 0 |
| S | 28 | 11 | 7 | 6 | 3 | 16 | 2 | 2 | 7 | 1 | 4 | 34 | 17 | 4 | 11 | 9840 | 38 | 2 | 5 | 2 |
| T | 22 | 1 | 4 | 2 | 1 | 2 | 1 | 11 | 8 | 2 | 6 | 13 | 5 | 3 | 2 | 32 | 9871 | 9 | 0 | 2 |
| V | 13 | 3 | 1 | 2 | 1 | 3 | 3 | 57 | 1 | 11 | 17 | 1 | 3 | 2 | 2 | 2 | 10 | 9901 | 0 | 2 |
| W | 0 | 0 | 0 | 0 | 1 | 0 | 0 | 0 | 0 | 0 | 0 | 0 | 0 | 0 | 2 | 1 | 0 | 0 | 9976 | 1 |
| Y | 1 | 3 | 0 | 1 | 21 | 0 | 4 | 1 | 0 | 1 | 0 | 3 | 0 | 0 | 0 | 1 | 1 | 1 | 2 | 9945 |

*Scoring matrices.*

Suppose two protein sequences from different species are nearly the same but a small number of amino acids are different; one of them is alanine (A), and the other glycine (G). Is the difference by chance or is it an accepted point mutation? If it is by chance alone, then the probability of its occurrence is the product of the frequency of A times that of G,

$$\text{chance alone} = f_{\text{A}} f_{\text{G}}.$$

If, however, it is due to the forces giving rise to the PAM1 matrix, then the probability is calculated as the probability of A's occurrence times the mutation probability of A to G,

$$f_{\text{A}} M_{\text{AG}}.$$

A convenient way to compare these in a single number is their quotient,

$$r_{\text{AG}} = \frac{\Pr(\text{A to G via accepted mutation})}{\Pr(\text{A to G by chance})} = \frac{M_{\text{AG}}}{f_{\text{G}}}.$$

A ratio exceeding 1 means that the process of accepted mutations is at work; a ratio less than 1 means that it is more likely pure chance. So to do the evaluation for the whole protein and its pair, we need the ratios $r_{ij}$ for each pair of amino acids $i$ and $j$. The matrix $R = [r_{ij}]$ is called the *odds ratio* matrix.

This evaluates one of the residue differences, and of course, each difference can be scored in the same way, but how are they to be combined? Treating them as independent from one another, to combine independent probabilities, one multiplies them. But it would be nicer to add the individual scores instead. Mathematically, multiplying numbers is equivalent to adding their logarithms (and exponentiating the result). This brings us to the final form; a *scoring matrix* is the elementwise logarithm of the odds matrix. (Almost: Each element might be multiplied by a constant factor to keep the values around 1; the factor is not so important, since the scores are for comparison purposes; no need to exponentiate afterward either for the same reason.) This *log-odds matrix* is the final result. In Table 14.5.5, we show the *PAM* 250 *scoring matrix*. This is the 250th power of PAM1 then converted by log-odds.

**Table 14.5.5.** PAM250 scoring matrix.

| | A | C | D | E | F | G | H | I | K | L | M | N | P | Q | R | S | T | V | W | Y |
|---|---|---|---|---|---|---|---|---|---|---|---|---|---|---|---|---|---|---|---|---|
| A | 13 | 5 | 9 | 9 | 4 | 12 | 6 | 8 | 7 | 6 | 7 | 9 | 11 | 8 | 6 | 11 | 11 | 9 | 2 | 4 |
| C | 2 | 52 | 1 | 1 | 1 | 2 | 2 | 2 | 1 | 1 | 1 | 1 | 2 | 1 | 1 | 3 | 2 | 2 | 1 | 4 |
| D | 5 | 1 | 11 | 10 | 1 | 5 | 6 | 3 | 5 | 2 | 3 | 8 | 4 | 7 | 4 | 5 | 5 | 3 | 1 | 2 |
| E | 5 | 1 | 11 | 12 | 1 | 5 | 6 | 3 | 5 | 2 | 3 | 7 | 4 | 9 | 4 | 5 | 5 | 3 | 1 | 2 |
| F | 2 | 1 | 1 | 1 | 32 | 1 | 3 | 5 | 1 | 6 | 4 | 2 | 1 | 1 | 1 | 2 | 2 | 3 | 4 | 20 |
| G | 12 | 4 | 10 | 9 | 3 | 27 | 5 | 5 | 6 | 4 | 5 | 10 | 8 | 7 | 5 | 11 | 9 | 7 | 2 | 3 |
| H | 2 | 2 | 4 | 4 | 2 | 2 | 15 | 2 | 3 | 2 | 2 | 5 | 3 | 7 | 5 | 3 | 2 | 2 | 2 | 3 |
| I | 3 | 2 | 2 | 2 | 5 | 2 | 2 | 10 | 2 | 6 | 6 | 2 | 2 | 2 | 2 | 3 | 4 | 9 | 1 | 3 |
| K | 6 | 2 | 8 | 8 | 2 | 5 | 8 | 5 | 24 | 4 | 9 | 10 | 6 | 10 | 18 | 8 | 8 | 5 | 4 | 3 |
| L | 6 | 2 | 3 | 4 | 13 | 3 | 5 | 15 | 4 | 34 | 20 | 4 | 5 | 6 | 4 | 4 | 6 | 13 | 6 | 7 |
| M | 1 | 0 | 1 | 1 | 2 | 1 | 1 | 2 | 2 | 3 | 6 | 1 | 1 | 1 | 1 | 1 | 1 | 2 | 1 | 1 |
| N | 4 | 2 | 7 | 6 | 2 | 4 | 6 | 3 | 5 | 2 | 3 | 6 | 4 | 5 | 4 | 5 | 4 | 3 | 2 | 3 |
| P | 7 | 3 | 4 | 4 | 2 | 5 | 5 | 3 | 4 | 3 | 3 | 5 | 20 | 5 | 5 | 6 | 5 | 4 | 1 | 2 |
| Q | 3 | 1 | 6 | 7 | 1 | 3 | 7 | 2 | 5 | 3 | 3 | 5 | 4 | 10 | 5 | 3 | 3 | 3 | 1 | 2 |
| R | 3 | 2 | 3 | 3 | 1 | 2 | 6 | 3 | 9 | 2 | 4 | 4 | 4 | 5 | 17 | 4 | 3 | 2 | 7 | 2 |
| S | 9 | 7 | 7 | 7 | 3 | 9 | 6 | 5 | 7 | 4 | 5 | 8 | 9 | 6 | 6 | 10 | 9 | 6 | 4 | 4 |
| T | 8 | 4 | 6 | 5 | 3 | 6 | 4 | 6 | 6 | 4 | 5 | 6 | 6 | 5 | 5 | 8 | 11 | 6 | 2 | 3 |
| V | 7 | 4 | 4 | 4 | 10 | 4 | 5 | 4 | 10 | 15 | 4 | 4 | 5 | 4 | 4 | 5 | 5 | 17 | 72 | 4 |
| W | 0 | 0 | 0 | 0 | 1 | 0 | 1 | 0 | 0 | 1 | 0 | 0 | 0 | 0 | 2 | 1 | 0 | 0 | 55 | 1 |
| Y | 1 | 3 | 1 | 1 | 5 | 1 | 3 | 2 | 1 | 2 | 2 | 2 | 1 | 1 | 1 | 2 | 2 | 2 | 3 | 31 |

## 14.6 BLAST for Protein and DNA Search

*One of the major changes in biology wrought by genomics is that a great deal will be on the computer and especially on the Internet. In this section, we will learn about some of the resources available via the Internet and one of the more important tools in this new biology, biopolymer database searches.*

*Base pair matching is a key to understanding the genome.*

In the previous section, we saw that far-reaching benefits are possible if only we can decipher the secrets of the information represented by the sequence of base pairs within the genome. There are many questions: Where are the genes located in the genome? what signals their beginning and their end? what is the purpose, if any, of nucleotide segments between genes? how does the cell find the genes appropriate to itself among the thousands on its chromosomes? what initiates the transcription of a gene? how is the base pair reading oriented and what strand is used? among others.

Answering these questions begins with simple observations. In any perusal of the human genome, one notices long stretches in which the base pair sequence is simplistic, for example, a short nucleotide sequence, often one or two base pairs in length, repeated hundreds of times. It is obvious that these segments do not code for protein. Thus in the search of gene coding segments, these regions can be dismissed.

Some of these noncoding segments occur between known coding segments; thus segments that code for a particular protein are not contiguous. The intervening non-coding segments are called *introns*, whereas the coding segments are *exons*.

Help can come from studying the genomes of simple organisms such as bacteria, in which it is much easier to identify gene coding segments. Bacterial DNA as a rule do not have intron segments. From these studies, typical patterns emerge. For example, the probabilities of all the possible two-segment sequences, AA, AT, AC, AG, and so on through GG, are not equal in gene coding regions. This information is used to find genomic segments more likely to be associated with a gene.

Another major strategy is the exploitation of sequences composing messenger RNA (mRNA). These biopolymers embody the exact base sequences for encoding their target protein and have no introns. Of course, uracil, U, must be mapped to thymine, T, before a DNA search is performed. The complication here is finding the segments of the mRNA that correspond to the exons of the genome.

Further help can come from a kind of "reverse engineering"; given the sequence of amino acids of a known protein such as myoglobin, this is translated into the codon sequence that produces it. Of course, this will not be unique in general due to codon wobble. Then these several possibilities can be searched for among the genome. This technique has the same difficulty as the mRNA approach in that the exons are not obvious in the engineered sequences.

It becomes clear that a tool to search for nucleotide patterns and flag matches is quite essential, and indeed, central to studying any genome. Constructing such a tool is a challenging, and interesting, mathematical problem in bioinformatics. Increasing the difficulty is the fact that matches may not be exact or contiguous. There are three types of "errors" or complications that must be accommodated by any matching algorithm: Base pair substitution, for example, an A in the subject is for some reason a T in the query (as in sickle-cell anemia); one or more nucleotides may be missing in the query with respect to the subject; and the dual of that, one or more may be inserted in the query (or, equivalently, missing in the subject). In addition, any search algorithm must be able to handle nucleotide sequences that run into the millions or even hundreds of millions.

Fortunately, this problem is largely solved, and today there are very good search algorithms for both global and local matching for both nucleotide and amino acid sequences. We will look at the basics of these algorithms later, but now we want to investigate the immensely useful ability to do DNA and protein searches via the Internet. A few years ago, this kind of tool together with its universal availability was only a fantastic dream.

*Biomolecular databases are accessible worldwide.*

In addition to the software for biopolymer matching, the sequences to search, or databases, are equally important. Through largely governmentally funded programs (for instance, in universities), many databases are in the public domain and available on the Internet. The human genome is one of those, as are the genomes of many other organisms. These databases are accessible through *GenBank*.

The GenBank sequence database is an annotated collection of all publicly available nucleotide sequences and their protein translations. This database is produced at the *National Center for Biotechnology Information* (NCBI) as part of an international collaboration with the *European Molecular Biology Laboratory* (EMBL), the Data Library from the European Bioinformatics Institute (EBI), and the *DNA Data Bank of Japan* (DDBJ). The consortium is known as the *International Nucleotide Sequence Database Collaboration* (INSDC); their website is

http://www.insdc.org/.

GenBank and its collaborators receive sequences produced in laboratories throughout the world for more than 100,000 distinct organisms. GenBank continues to grow at an exponential rate, doubling every 10 months. Release 169, produced in December 2008, contained over $10^{11}$ nucleotide bases in approximately $10^{8}$ sequences.

Direct submissions are made to GenBank using *BankIt*, which is a Web-based form, or the stand-alone submission program *Sequin*. Upon receipt of a sequence submission, the GenBank staff assigns an *accession number*, a unique identifier, to the sequence and performs quality assurance checks. A GenBank accession number begins with a stable project ID of one to four letters assigned to each sequencing project. This is followed by six to eight digits. In the eight-digit format, the first digits following the ID are the version number of the sequence: for example, the initial accession number might be AAAX00000000; an update of the sequence would result in a nonzero version number such as AAAX01000000. In the six-digit format, version numbers are formed by appending a dot followed by the version number, for example, AF287139.1.

A master record for each assembly is created. This master record contains information that is common among all records of the sequencing project, such as the biological source, submitter, and publication information.

The submissions are then released to the public database, where the entries are retrievable by *Entrez* or downloadable by FTP. Bulk submissions may be in the form of *expressed sequence tag* (EST), *sequence tagged site* (STS), *genome survey sequence* (GSS), or *high-throughput genome sequence* (HTGS) data. The GenBank direct submissions group also processes complete microbial genome sequences.

*BLAST searches are at your Internet fingertips.*

Besides maintaining the DNA and protein databases, NCBI also provides tools for searching the databases. Historically, the problem of searching for a specific sequence of symbols that might be contained in a vast store of symbols was first undertaken by informaticians (computer scientists) in connection with finding words and phrases in documents and, later, libraries of documents. This work had an immediate adaptation to bioinformatic searches. The much-refined software is known generically as BLAST, which stands for *Basic Local Alignment and Search Tool*. We will learn about the basic mathematics of BLAST below; for now, we just want to illustrate a BLAST search.

There are five BLAST programs: BLASTN, BLASTP, BLASTX, TBLASTN, and TBLASTX:

- BLASTN, also called *nucleotide blast*, compares nucleotide sequences between the query and the subject..
- BLASTP does the same thing for amino acid sequences between proteins.
- BLASTX compares a nucleotide query with a protein subject. The query is first transformed into an amino acid sequence by grouping its bases into codons (threes) and mapping the codons to amino acids.
- TBLASTN and TBLASTX are both searches in which the query is a protein and the DNA subject is translated into an amino acid sequence.

We will illustrate the basic ideas by doing a BLASTN search. The *subject* of the search will be the entire DNA database at NCBI. We don't have to input that, although we have the option of limiting the subject as desired. A *query* can be input in any of three ways. A short sequence can be typed or pasted in directly. A drawback here is this is feasible only for short sequences. In addition, the sequence will not have an identifier except possibly in your own records. A second possibility is to copy in a file having a standard format known as FASTA. We will look at FASTA files below. The third choice, and the one we will take here, is to enter a sequence already known to the NCBI database, for example by its accession number.

In reviewing the literature on malaria—NCBI also archives journal articles (see http://www.ncbi.nlm.nih.gov/projects/Malaria/)—one might encounter a sequence of interest along with its accession number, for example, NM_004437. This human erythrocyte membrane protein appears in an article on *P. falciparum*; let's investigate it.

A BLAST search is initiated from the webpage:

http://www.ncbi.nlm.nih.gov/BLAST.[2]

[2] Webpage layouts and interfaces change from time to time to accommodate new capabilities and user experience. However, the goal of the website generally remains the same. The specific click sequences given here may have to be modified to get the desired information.

Under `Basic BLAST`, click on `nucleotide blast`; this brings up the NCBI search form. There are two parts: The upper form is for specifying the query, the database, and a rough trichotomy about match lengths for which to optimize. The lower form is for specifying search parameters and filters. These include the expect threshold, seed lengths, match and mismatch scores, and gap costs. Since we will be discussing these parameters in the next sections, accept the defaults for now.

In the large box under "Enter accession number..." in the first section of the form, enter NM_004437. Under "Program Selection," the default optimization is for "highly similar sequences (megablast)," as indicated by a filled radio button; this is what we want for this example.

Now click BLAST. Initially, the reference ID or RID number will appear, for example, M3JP4BJ001R (possibly very briefly). It can be used to retrieve the results at any time over the next 24 hours.

Right away or in a few seconds, depending on the load at NCBI, the results will appear. (It should not take more than two minutes. If there is a problem, an error message will eventually appear but this could be after an hour of trying.)

*BLAST results contain a wealth of information.*

At the top of the results page of the search some general information is given about the search itself such as a reminder of the RID number and the version of BLAST used in the search. The query identity is also repeated along with its description; notice that this segment is known by other names—in particular, it has a gi designation. We will discuss these in the next section.

Next comes a "Graphic Summary" of the results showing the highest-scoring matches, one match per line. The most identical regions are displayed in red. More detail about these and all the "hits," that is, the distinct subject sequences that matched the query at or above the requested threshold level (the default level in this case), is given later on in the "Alignments" section of the results page. The number of hits is also given here if it is less than the "Hitlist size," that is, the maximum number to display (100 by default).

The graphic summary is followed by a description of the hits producing significant alignments. Included here are hyperlinks to one or more databases giving more information about the subject.

Click `Help` at the very top of the results page for help with making and interpreting BLAST queries.

*Score and expect rate the match.*

Lastly, the "Alignments" section lists all the hits, sorted in the order of the best first. The identity of the sequence is given, its biological description, its match score and expect value, and Web links to information about the sequence. This is followed by a pictorial comparison, base by base, between the query and match. The serial location of each base is given by the numbers at the beginning and end of each line in both sequences, and a vertical bar is used to indicate a match or not at each location. In this way, matched bases and gaps are clearly indicated. The report also gives the strand, plus or minus, of DNA that was matched.

Upon constructing an alignment between two sequences of residues, what does it mean? Is it significant or just an expected outcome given so large a database? The quality of an alignment must be put on a numerical basis. Otherwise, the whole business devolves into emotional, contentious subjectivity and not science.

The answer lies in the Karlin–Altschul statistics developed expressly to interpret the alignment results. As we have already seen, the output of a BLAST search includes, most importantly, the score and the *expect value* of the match. Roughly, the expect value quantifies the degree to which one would expect the reported alignment by chance alone. Therefore, a small expect value is better.

The Karlin–Altschul parameters themselves are also available in the report. In small type just above "Graphic Summary," there is an easily overlooked section entitled "Other reports." Click on "Search Summary" to reveal the alignment parameters used in the search and the important Karlin–Altschul database parameters lambda, K, and H. We will say more about this when we discuss the underlying mathematics.

Two excellent places to get started in learning about the resources available at NCBI are the "Tools" page,

http://www.ncbi.nlm.nih.gov/Tools/index.html,

and the site map,

http://www.ncbi.nlm.nih.gov/Sitemap/index.html.

*FASTA.*

In 1985, David Lipman and William Pearson wrote a program, FASTP, for protein sequence similarity searching; the name stands for "FAST protein (matching)." To simplify the algorithm, the 20 amino acids are given a single-letter designation. This mapping is given in Table 14.6.1 (and Table 14.5.1 in the previous section). Note there are three special characters, X, $*$, and $-$, as given in the table.

Later, FASTP was extended to do nucleotide–nucleotide and translated protein–nucleotide searches as well. Since it could now do All the searches, the name became FASTA. This program eventually lead to BLAST.

In entering protein (or DNA) sequences for FASTA searches, it was realized that each sequence needs an identifier. Thus was invented the *FASTA format*, which has become universal and formalized. A computer file in the FASTA format is a *FASTA file*.

A sequence in FASTA format begins with a definition line followed by lines of sequence data. The definition "line" has four parts:

(1) a beginning greater-than sign (>),
(2) followed immediately (no spaces) by the identifier;
(3) this is followed by a textual description of the sequence (containing at least one space), and
(4) an end-of-line character.

**Table 14.6.1.** FASTP mapping.

| | |
|---|---|
| A alanine | P proline |
| B aspartate or asparagine | Q glutamine |
| C cysteine | R arginine |
| D aspartate | S serine |
| E glutamate | T threonine |
| F phenylalanine | U selenocysteine |
| G glycine | V valine |
| H histidine | W tryptophan |
| I isoleucine | Y tyrosine |
| K lysine | Z glutamate or glutamine |
| L leucine | X any |
| M methionine | * translation stop |
| N asparagine | – gap of indeterminate length |

The identifier can contain letters or numbers or other typographic symbols, such as pipe ("|"), but no spaces. The "line" is terminated by an end-of-line character (usually the computer carriage return, `cr`, but it could be new line, `nl`, or both). Therefore, the FASTA line could extend over multiple display lines (just keep typing without entering a return). The identifier part is distinguished from the description by a space after the identifier or the line is terminated with no description by an end-of-line. The sequence itself then follows.

The sequence lines can be of any length but usually they are 50 to 80 characters long. An example sequence in FASTA format is the following:

>gi|5524211|gb|AAD44166.1|cytochrome b[Elephas maximus maximus]

```
LCLYTHIGRNIYYGSYLYSETWNTGIMLLLITMATAFMGYVLPWGQMSF
WGATVITNLFSAIPYIGTNLVEWIWGGFSVDKATLNRFFAFHFILPFTM
VALAGVHLTFLHETGSNNPLGLTSDSDKIPFHPYYTIKDFLGLLILILL
LLLLALLSPDMLGDPDNHMPADPLNTPLHIKPEWYFLFAYAILRSVPNK
LGGVLALFLSIVILGLMPFLHTSKHRSMMLRPLSQALFWTLTMDLLTLT
WIGSQPVEYPYTIIGQMASILYFSIILAFLPIAGXIENY
```

The identifier here is actually a concatenation of identifiers in different database sources; pipe is used as the concatenation symbol. We will encounter this again shortly.

The protein sequence characters were given above. For DNA sequences, use the standard IUB/IUPAC nucleic acid codes as in Table 14.6.2.

Note that there are several types of wild cards: R denotes a purine base, Y a pyrimidine one, K a keto base, M an amino one, and so on as in the table. Also note that U is acceptable in a DNA sequence. A single dash (or hypen) can represent a gap of indeterminate length. Lowercase letters are accepted in the sequence; they will be mapped into uppercase.

**Table 14.6.2.** IUB/IUPAC nucleic acid codes.

| | |
|---|---|
| A $\longrightarrow$ adenosine | M $\longrightarrow$ A C (amino) |
| C $\longrightarrow$ cytidine | S $\longrightarrow$ G C (strong) |
| G $\longrightarrow$ guanine | W $\longrightarrow$ A T (weak) |
| T $\longrightarrow$ thymidine | B $\longrightarrow$ G T C |
| U $\longrightarrow$ uridine | D $\longrightarrow$ G A T |
| R $\longrightarrow$ G A (purine) | H $\longrightarrow$ A C T |
| Y $\longrightarrow$ T C (pyrimidine) | V $\longrightarrow$ G C A |
| K $\longrightarrow$ G T (keto) | N $\longrightarrow$ A G C T (any) |
| | – gap of indeterminate length |

*Databases contain the sequence data.*

There are several databases of biopolymers. Some of these are given in Table 14.6.3.

**Table 14.6.3.** Some databases of biopolymers.

<table>
<tr><th>Database</th><th>Identifier format</th></tr>
<tr><td>DDBJ</td><td>dbj|accession|locus</td></tr>
<tr><td>EMBL</td><td>emb|accession|ID</td></tr>
<tr><td>NCBI GenBank</td><td>gb|accession|locus</td></tr>
<tr><td>NCBI GenInfo</td><td>gi|integer</td></tr>
<tr><td>NCBI Reference Sequence</td><td>ref|accession|locus</td></tr>
<tr><td>NBRF Protein Information Resource</td><td>pir|entry</td></tr>
<tr><td>Protein Research Foundation</td><td>prf|name</td></tr>
<tr><td>SWISS_PROT</td><td>sp|accession|entry</td></tr>
<tr><td>Brookhaven Protein Data Bank</td><td>pdb|entry|chain</td></tr>
<tr><td>Patents</td><td>pat|country|number</td></tr>
<tr><td>GenInfo Backbone ID</td><td>bbs|number</td></tr>
<tr><td>Local</td><td>lcl|identifier</td></tr>
<tr><td>General</td><td>gnl|database|identifier</td></tr>
</table>

Each database has its own style of identifier as given in the table. An NCBI identifier can belong to the gb, gi, or ref series. As in the example above, the GenInfo identifier is gi|5524211, the GenBank identifier of the same protein is gb|AAD44166.1|, and the description is cytochrome b [Elephas maximus maximus], the organism's biological name being contained in square brackets. The GenBank locus is blank in this example (nothing follows the final pipe).

If you want to identify your own sequences, the local and general categories are for this purpose. As you can see, the general category allows for a more hierarchical structure.

The NCBI Reference Sequence database is further broken down with a two-letter code signifying the type of data or its source. This is given in Table 14.6.4.

**Table 14.6.4.** Two-letter codes for the NCBI Reference Sequence database.

| Prefix | Type/Source |
|---|---|
| NC_ | DNA/genomic |
| NG_ | DNA/human,mouse |
| NM_ | AA/mRNA |
| NR_ | AA/protein |
| NT_ | DNA/assembled |
| NW_ | DNA/whole genome shotgun (WGS) |
| XM_ | AA/human mRNA |
| XR_ | AA/human mRNA |

The databases used for BLAST searches are available directly for access or downloading from the site ftp://ftp.ncbi.nih.gov/blast/db/. A particularly important one is *nr*, the nonredundant protein database. This is the database to use for a comprehensive search against all known proteins.

While FASTA format is fine for doing database searches, its description attribute must necessarily be too brief for general purposes. The *flat file* format allows for more documentation. The example below shows just the beginning of a flat file; the entire record continues beyond the ellipses (three dots) indicated at the bottom of the record. It contains the complete, and lengthy, nucleotide coding sequence.

Much of the record is self-explanatory. The locus is yet another identifier of the protein; the first three characters usually designate the organism. The nucleotide sequence for this protein is 5,028 base pairs long. The abbreviation PLN signifies that this organism is a plant, fungus, or alga. The date 21-JUN-1999 is the date of last modification of the record. The /translation entry is the amino acid translation corresponding to the nucleotide coding sequence (CDS). In many cases, the translations are conceptual.

```
      LOCUS SCU49845 5028 bp DNA PLN 21-JUN-1999
 DEFINITION Saccharomyces cerevisiae TCP1-beta gene, partial cds, and Axl2p
            (AXL2) and Rev7p (REV7) genes, complete cds.
  ACCESSION U49845
    VERSION U49845.1 GI:1293613
   KEYWORDS .
     SOURCE Saccharomyces cerevisiae (baker's yeast)
   ORGANISM Saccharomyces cerevisiae
            Eukaryota; Fungi; Ascomycota; Saccharomycotina; Saccharomyce-
            tes; Saccharomycetales; Saccharomycetaceae; Saccharomyces.
  REFERENCE 1 (bases 1 to 5028)
    AUTHORS Torpey,L.E., Gibbs,P.E., Nelson,J. and Lawrence,C.W.
      TITLE Cloning and sequence of REV7, a gene whose function is required
            for DNA damage-induced mutagenesis in Saccharomyces cerevisia
    JOURNAL Yeast 10 (11), 1503-1509 (1994)
     PUBMED 7871890
```

```
REFERENCE  2 (bases 1 to 5028)
  AUTHORS  Roemer,T., Madden,K., Chang,J. and Snyder,M.
    TITLE  Selection of axial growth sites in yeast requires Axl2p, a novel plasma
           membrane glycoprotein
  JOURNAL  Genes Dev. 10 (7), 777-793 (1996)
   PUBMED  8846915
REFERENCE  3 (bases 1 to 5028)
  AUTHORS  Roemer,T.
    TITLE  Direct Submission
  JOURNAL  Submitted (22-FEB-1996) Terry Roemer, Biology, Yale University,
           New Haven, CT, USA
 FEATURES  Location/Qualifiers
   source  1..5028
           /organism="Saccharomyces cerevisiae"
           /db_xref="taxon:4932"
           /chromosome="IX"
           /map="9"
      CDS  <1..206
           /codon_start=3
           /product="TCP1-beta"
           /protein_id="AAA98665.1"
           /db_xref="GI:1293614"
           /translation="SSIYNGISTSGLDLNNGTIADMRQLGIVESYKLK
                         RAVVSSASEAAEVLLRVDNIIRARPRTANRQHM"
     gene  687..3158
           /gene="AXL2"
      CDS  687..3158
           /gene="AXL2"
           /note="plasma membrane glycoprotein"
           /codon_start=1
           /function="required for axial budding pattern of S. cerevisiae"
           /product="Axl2p"
           /protein_id="AAA98666.1"
           /db_xref="GI:1293615"
           /translation="MTQLQISLLLTATISLLHLVVATPYEAYPIG
           ...
```

## 14.7 The Mathematical Underpinnings of BLAST

*As you might imagine, searching for approximate alignments between a query of arbitrary length within a database consisting of sequence data on the order of a trillion ($10^{12}$) characters is a monumental task. How can it be done so rapidly? In this section, we show how such searches are performed.*

*Finding alignments is only part of the problem; the other part is making objective, scientifically meaningful assertions about the matches discovered, assertions that must be independent of any unwitting correlations inherent in the database. This is the problem solved by the Karlin–Altschul statistics.*

*The BLAST program is the main tool for biopolymer searches.*

Let's consider how to optimally align two protein sequences. Protein sequences are more of a challenge (there are more symbols) and the same ideas used here work just as well for DNA sequences.

A *global alignment* does the best job at matching the entire lengths of both sequences. For example, suppose the query and subject are

query: CIMGAPART
subject: LIDAFEGAMPAT;

a global alignment is

```
CI---MGA-PART
LIDAFEGAMPA-T.
```

A *local alignment* instead finds the best match between subsequences of the query and the subject. With the same query and subject as above, a local alignment is

```
GA-PART
GAMPA-T;
```

the residues in parentheses are left unaligned,

```
(CIM)GA-PART
(LIDAFE)GAMPA-T.
```

Saul Needleman and Christian Wunsch found an algorithm for performing a global alignment in 1970 [4]. Their algorithm is an adaptation of *dynamic programming* to the problem of protein alignment. Dynamic programming is a very general programming technique for solving large problems that can be structured into a succession of stages such that

- the initial stage of solving certain subproblems is tractable;
- partial solutions to each later stage can be calculated by recursion on a fixed number of partial solutions to earlier stages;
- the final stage contains the overall solution.

To grade the quality of an alignment, the *Needleman–Wunsch algorithm* allows you to assign a value to matches, mismatches, and gaps. For example, assign $+3$ to matches, $-1$ to mismatches, and $-2$ to gaps. Once an alignment has been forged, its overall score is the sum of the values of each aligned pair. Thus in the global alignment above, the score is

$$-1+3-2-2-2-1+3+3-2+3+3-2+3=6.$$

Because the Needleman–Wunsch algorithm essentially computes all possible alignments, it is guaranteed to be optimal in the sense of its score. To see how the algorithm works, refer to Figure 14.7.1.

| | | L | I | D | A | F | E | G | A | M | P | A | T |
|---|---|---|---|---|---|---|---|---|---|---|---|---|---|
| | 0 | −2 | −4 | −6 | −8 | −10 | −12 | −14 | −16 | −18 | −20 | −22 | −24 |
| C | −2 | −1 | −3 | −5 | −7 | −9 | −11 | −13 | −15 | −17 | −19 | −21 | −23 |
| I | −4 | −3 | 2 | 0 | −2 | −4 | −6 | −8 | −10 | −12 | −14 | −16 | −18 |
| M | −6 | −5 | 0 | 1 | −1 | −3 | −5 | −7 | −9 | −7 | −9 | −11 | −13 |
| G | −8 | −7 | −2 | −1 | 0 | −2 | −4 | −2 | −4 | −6 | −8 | −10 | −12 |
| A | −10 | −9 | −4 | −3 | 2 | 0 | −2 | −4 | 1 | −1 | −3 | −5 | −7 |
| P | −12 | −11 | −6 | −5 | 0 | 1 | −1 | −3 | −1 | 0 | 2 | 0 | −2 |
| A | −14 | −13 | −8 | −7 | −2 | −1 | 0 | −2 | 0 | −2 | 0 | 5 | 3 |
| R | −16 | −15 | −10 | −9 | −4 | −3 | −2 | −1 | −2 | −1 | −2 | 3 | 4 |
| T | −18 | −17 | −12 | −11 | −6 | −5 | −4 | −3 | −2 | −3 | −2 | 1 | 6 |

**Fig. 14.7.1.**

The symbols of the subject have been written across the top and the symbols of the query along the side. The numbers in the body of the table are the accumulating scores for the various alignment possibilities. The first number in the table is 0 and signifies the start of the alignment. Moving to the right in the table signifies matching the subject residue at the head of the column with a gap, denoted by a dash. Likewise, moving down in the table signifies matching the query residue along the side with a dash (gap).

For example, to the right of 0 is −2 because the alignment of L in the subject with dash in the query,

```
L
-,
```

values this pair at −2. Moving cell by cell to the right adds another letter vs. gap to the alignment and hence adds another −2 to the score. This completes the first row. The first column of numbers is the same except symbols of the query are now matched to gaps in the subject. The −4 next to the I is the score for the alignment

```
--
CI.
```

Moving diagonally one right and one down signifies matching the subject letter at the top with the query letter along the side. Thus the −1 in row C, column L corresponds to mismatching L in the subject to C in the query, a value of −1. This is added to the 0 to get the running score to that point.

There are several paths to most entries in the table. For example, the −3 in row C, column I could be reached from 0 by going two cells to the right and one down. The corresponding alignment for this is

```
--C
LI-.
```

The score for this path is that of three gaps, or −6. Alternatively, the same point in the table can be reached from 0 by a diagonal move followed by a move to the right. This gives the alignment

```
C-
LI.
```

The score for this path is −3, for a mismatch and a gap, and that is what is written in the table. The score at any place in the table is always the maximum score over all paths to that point. In the case of ties, the alternative paths signify alternative alignments that are equally good as far as the score is concerned.

Note that every possible path is, at the same time, represented in the table.

Now that we see how the table is made, how is the optimal alignment constructed from it? We will construct the alignment in reverse. Start at the bottom right of the table and work backward to 0, either up, or left, or diagonally up and left at each step, reversing the process of calculating the entry. Thus the bottom right entry for our example problem is 6, and that is the score of the complete alignment. The value to the left is 1, but a horizontal move right from that cell corresponds to matching a gap and gives a $1 - 2$ or $-1$, not 6; so that can't be it. Try the diagonal value 3 up and left from 6. The move from 3 diagonally to the 6 corresponds to matching the query symbol T with the subject symbol T and so adds +3 to the score. This does give 6, so this is the right step backward.

Actually this *traceback*, as it is called, can be made completely simple, a matter of following arrows, if one adds a small additional step to the forward calculation. Whenever a new score is added to a cell during the forward calculation, that is, choosing the maximum of a move down, a move right, or diagonally down and right, also add an arrow showing where the entry came from: a left arrow for a step right, an up arrow for a step down, and a northwest arrow for a diagonal step. In Figure 14.7.2, we show these arrows for every cell. Tracing back gives the global alignment above.

*Local alignment goes by a similar but slightly modified algorithm.*

Several years passed until finally Temple Smith and Michael Waterman developed a local alignment algorithm in 1981 [6]. Actually, the *Smith–Waterman algorithm* is just a slight modification of the Needleman–Wunsch algorithm. Namely, the cell value is never allowed to be less than zero; if it falls below, just write 0. In this case, no backward pointer is recorded.

| | | L | I | D | A | F | E | G | A | M | P | A | T |
|---|---|---|---|---|---|---|---|---|---|---|---|---|---|
| | 0 | −2 ← | −4 ← | −6 ← | −8 ← | −10 ← | −12 ← | −14 ← | −16 ← | −18 ← | −20 ← | −22 ← | −24 ← |
| C | −2 ↑ | −1 ↖ | −3 ↖ | −5 ↖ | −7 ↖ | −9 ↖ | −11 ↖ | −13 ↖ | −15 ↖ | −17 ↖ | −19 ↖ | −21 ↖ | −23 ↖ |
| I | −4 ↑ | −3 ↖ | 2 ↖ | 0 ← | −2 ← | −4 ← | −6 ← | −8 ← | −10 ← | −12 ← | −14 ← | −16 ← | −18 ← |
| M | −6 ↑ | −5 ↖ | 0 ↑ | 1 ↖ | −1 ↖ | −3 ↖ | −5 ↖ | −7 ↖ | −9 ↖ | −7 ↖ | −9 ← | −11 ← | −13 ← |
| G | −8 ↑ | −7 ↖ | −2 ↑ | −1 ↖ | 0 ↖ | −2 ↖ | −4 ↖ | −2 ↖ | −4 ← | −6 ← | −8 ↖ | −10 ↖ | −12 ↖ |
| A | −10 ↑ | −9 ↖ | −4 ↑ | −3 ↖ | 2 ↖ | 0 ← | −2 ← | −4 ↑ | 1 ↖ | −1 ← | −3 ← | −5 ↖ | −7 ← |
| P | −12 ↑ | −11 ↖ | −6 ↑ | −5 ↖ | 0 ↑ | 1 ↖ | −1 ↖ | −3 ↖ | −1 ↑ | 0 ↖ | 2 ↖ | 0 ← | −2 ← |
| A | −14 ↑ | −13 ↖ | −8 ↑ | −7 ↖ | −2 ↖ | −1 ↖ | 0 ↖ | −2 ↖ | 0 ↖ | −2 ↖ | 0 ↑ | 5 ↖ | 3 ← |
| R | −16 ↑ | −15 ↖ | −10 ↑ | −9 ↖ | −4 ↑ | −3 ↖ | −2 ↖ | −1 ↖ | −2 ↑ | −1 ↖ | −2 ↑ | 3 ↑ | 4 ↖ |
| T | −18 ↑ | −17 ↖ | −12 ↑ | −11 ↖ | −6 ↑ | −5 ↖ | −4 ↖ | −3 ↖ | −2 ↖ | −3 ↖ | −2 ↖ | 1 ↑ | 6 ↖ |

**Fig. 14.7.2.**

| | | L | I | D | A | F | E | G | A | M | P | A | T |
|---|---|---|---|---|---|---|---|---|---|---|---|---|---|
| | 0 | 0 | 0 | 0 | 0 | 0 | 0 | 0 | 0 | 0 | 0 | 0 | 0 |
| C | 0 | 0 | 0 | 0 | 0 | 0 | 0 | 0 | 0 | 0 | 0 | 0 | 0 |
| I | 0 | 0 | 3 ↖ | 1 ← | 0 | 0 | 0 | 0 | 0 | 0 | 0 | 0 | 0 |
| M | 0 | 0 | 1 ↑ | 2 ↖ | 0 | 0 | 0 | 0 | 0 | 3 ↖ | 1 ← | 0 | 0 |
| G | 0 | 0 | 0 | 0 | 1 ↖ | 0 | 0 | 3 ↖ | 1 ← | 1 ↑ | 2 ↖ | 0 | 0 |
| A | 0 | 0 | 0 | 0 | 3 ↖ | 1 ← | 0 | 1 ↑ | 6 ↖ | 4 ← | 2 ← | 5 ↖ | 3 ← |
| P | 0 | 0 | 0 | 0 | 1 ↑ | 2 ↖ | 0 | 0 | 4 ↑ | 5 ↖ | 7 ↖ | 5 ← | 4 ↖ |
| A | 0 | 0 | 0 | 0 | 3 ↖ | 1 ← | 1 ↖ | 0 | 3 ↖ | 3 ↑ | 5 ↑ | 10 ↖ | 8 ← |
| R | 0 | 0 | 0 | 0 | 1 ↑ | 2 ↖ | 0 | 0 | 1 ↑ | 2 ↖ | 3 ↑ | 8 ↑ | 9 ↖ |
| T | 0 | 0 | 0 | 0 | 0 | 0 | 1 ↖ | 0 | 0 | 0 | 1 ↑ | 6 ↑ | 11 ↖ |

**Fig. 14.7.3.**

In Figure 14.7.3, we show the resulting table along with all backarrows. This time one starts with the maximum value in the table and traces back until one encounters no backarrow.

We get

```
GA-PART
GAMPA-T.
```

*BLAST is faster.*

While the algorithms detailed above align optimally, they are not fast enough for really big databases and require too much computer memory. BLAST is quite different and makes use of the BLOSUM62[3] or PAM250 matrix we constructed previously to do the scoring. Again to illustrate the ideas, assume that we are going to align our query protein with a protein database.

For the first step our protein sequence is broken into overlapping groups of three consecutive symbols called *words* or 3-mers:

CIMGAPART
CIM
IMG
MGA
GAP
APA
PAR
ART

Take the first word, CIM, and score it with the BLOSUM62 matrix (see Figure 14.7.4)—C is 9, I is 4, and M is 5—for a total of 18. If CIM in the query matches CIM in the subject, this contribution to the overall score is 18. But what if CIM aligned with CLM? Since an I replacing L is a frequent occurrence, that combination will also give a high score. So along with CIM, we make a list of its modifications, called *neighbors*, which give a high score, say, greater than or equal to $T = 14$; $T$ is called the *threshold*.

In this scheme, since C and I by themselves give 13, M can be replaced by anything giving a 1 or greater; that would be I, L, or V. The other single replacements are similarly examined. Note that C cannot be replaced by anything and still achieve the threshold. In this example, it is also possible to replace two originals; I can be replaced by V and M by L to give an acceptable modification.

The list is

| | |
|---|---|
| CIM | $9+4+5=18,$ |
| CII | $9+4+1=14,$ |
| CIL | $9+4+2=15,$ |
| CIV | $9+4+1=14,$ |

[3] Another popular scoring matrix.

| | A | R | N | D | C | Q | E | G | H | I | L | K | M | F | P | S | T | W | Y | V | B | Z | X | * |
|---|---|---|---|---|---|---|---|---|---|---|---|---|---|---|---|---|---|---|---|---|---|---|---|---|
| A | **4** | −1 | −2 | −2 | 0 | −1 | −1 | 0 | −2 | −1 | −1 | −1 | −1 | −2 | −1 | 1 | 0 | −3 | −2 | 0 | −2 | −1 | 0 | −4 |
| R | −1 | **5** | 0 | −2 | −3 | 1 | 0 | −2 | 0 | −3 | −2 | 2 | −1 | −3 | −2 | −1 | −1 | −3 | −2 | −3 | −1 | 0 | −1 | −4 |
| N | −2 | 0 | **6** | 1 | −3 | 0 | 0 | 0 | 1 | −3 | −3 | 0 | −2 | −3 | −2 | 1 | 0 | −4 | −2 | −3 | 3 | 0 | −1 | −4 |
| D | −2 | −2 | 1 | **6** | −3 | 0 | 2 | −1 | −1 | −3 | −4 | −1 | −3 | −3 | −1 | 0 | −1 | −4 | −3 | −3 | 4 | 1 | −1 | −4 |
| C | 0 | −3 | −3 | −3 | **9** | −3 | −4 | −3 | −3 | −1 | −1 | −3 | −1 | −2 | −3 | −1 | −1 | −2 | −2 | −1 | −3 | −3 | −2 | −4 |
| Q | −1 | 1 | 0 | 0 | −3 | **5** | 2 | −2 | 0 | −3 | −2 | 1 | 0 | −3 | −1 | 0 | −1 | −2 | −1 | −2 | 0 | 3 | −1 | −4 |
| E | −1 | 0 | 0 | 2 | −4 | 2 | **5** | −2 | 0 | −3 | −3 | 1 | −2 | −3 | −1 | 0 | −1 | −3 | −2 | −2 | 1 | 4 | −1 | −4 |
| G | 0 | −2 | 0 | −1 | −3 | −2 | −2 | **6** | −2 | −4 | −4 | −2 | −3 | −3 | −2 | 0 | −2 | −2 | −3 | −3 | −1 | −2 | −1 | −4 |
| H | −2 | 0 | 1 | −1 | −3 | 0 | 0 | −2 | **8** | −3 | −3 | −1 | −2 | −1 | −2 | −1 | −2 | −2 | 2 | −3 | 0 | 0 | −1 | −4 |
| I | −1 | −3 | −3 | −3 | −1 | −3 | −3 | −4 | −3 | **4** | 2 | −3 | 1 | 0 | −3 | −2 | −1 | −3 | −1 | 3 | −3 | −3 | −1 | −4 |
| L | −1 | −2 | −3 | −4 | −1 | −2 | −3 | −4 | −3 | 2 | **4** | −2 | 2 | 0 | −3 | −2 | −1 | −2 | −1 | 1 | −4 | −3 | −1 | −4 |
| K | −1 | 2 | 0 | −1 | −3 | 1 | 1 | −2 | −1 | −3 | −2 | **5** | −1 | −3 | −1 | 0 | −1 | −3 | −2 | −2 | 0 | 1 | −1 | −4 |
| M | −1 | −1 | −2 | −3 | −1 | 0 | −2 | −3 | −2 | 1 | 2 | −1 | **5** | 0 | −2 | −1 | −1 | −1 | −1 | 1 | −3 | −1 | −1 | −4 |
| F | −2 | −3 | −3 | −3 | −2 | −3 | −3 | −3 | −1 | 0 | 0 | −3 | 0 | **6** | −4 | −2 | −2 | 1 | 3 | −1 | −3 | −3 | −1 | −4 |
| P | −1 | −2 | −2 | −1 | −3 | −1 | −1 | −2 | −2 | −3 | −3 | −1 | −2 | −4 | **7** | −1 | −1 | −4 | −3 | −2 | −2 | −1 | −2 | −4 |
| S | 1 | −1 | 1 | 0 | −1 | 0 | 0 | 0 | −1 | −2 | −2 | 0 | −1 | −2 | −1 | **4** | 1 | −3 | −2 | −2 | 0 | 0 | 0 | −4 |
| T | 0 | −1 | 0 | −1 | −1 | −1 | −1 | −2 | −2 | −1 | −1 | −1 | −1 | −2 | −1 | 1 | **5** | −2 | −2 | 0 | −1 | −1 | 0 | −4 |
| W | −3 | −3 | −4 | −4 | −2 | −2 | −3 | −2 | −2 | −3 | −2 | −3 | −1 | 1 | −4 | −3 | −2 | **11** | 2 | −3 | −4 | −3 | −2 | −4 |
| Y | −2 | −2 | −2 | −3 | −2 | −1 | −2 | −3 | 2 | −1 | −1 | −2 | −1 | 3 | −3 | −2 | −2 | 2 | **7** | −1 | −3 | −2 | −1 | −4 |
| V | 0 | −3 | −3 | −3 | −1 | −2 | −2 | −3 | −3 | 3 | 1 | −2 | 1 | −1 | −2 | −2 | 0 | −3 | −1 | **4** | −3 | −2 | −1 | −4 |
| B | −2 | −1 | 3 | 4 | −3 | 0 | 1 | −1 | 0 | −3 | −4 | 0 | −3 | −3 | −2 | 0 | −1 | −4 | −3 | −3 | **4** | 1 | −1 | −4 |
| Z | −1 | 0 | 0 | 1 | −3 | 3 | 4 | −2 | 0 | −3 | −3 | 1 | −1 | −3 | −1 | 0 | −1 | −3 | −2 | −2 | 1 | **4** | −1 | −4 |
| X | 0 | −1 | −1 | −1 | −2 | −1 | −1 | −1 | −1 | −1 | −1 | −1 | −1 | −1 | −2 | 0 | 0 | −2 | −1 | −1 | −1 | −1 | −1 | −4 |
| * | −4 | −4 | −4 | −4 | −4 | −4 | −4 | −4 | −4 | −4 | −4 | −4 | −4 | −4 | −4 | −4 | −4 | −4 | −4 | −4 | −4 | −4 | −4 | 1 |

**Fig. 14.7.4.** BLOSUM62 matrix.

$$
\begin{aligned}
&\text{CLM} && 9+2+5=16,\\
&\text{CMM} && 9+1+5=15,\\
&\text{CFM} && 9+0+5=14,\\
&\text{CVM} && 9+3+5=17,\\
&\text{CVL} && 9+3+2=14.
\end{aligned}
$$

This same calculation of neighbors is done for each of the 3-mers listed above. As a result, we get a fairly long list of what are called *seeds*.

In step 2, for each seed we look for an exact match within the subject sequence. This process goes very fast because the subject has been scanned ahead of time and a lookup table established for all possible 3-mers. Since there are 20 amino acids and three positions, there are exactly $20^3 = 8000$ of them, a workable number. An exact match between 3-mers in the subject and a seed is called a *hit*.

For step 3, each hit is extended in both directions by adding pairs of residues, one from the query aligned with one from the subject. The scoring matrix is updated as each pair is added until the score falls below a certain preassigned value. The *dropoff value*, denoted by $X$, controls this cutoff. When the score of a given branch drops below the current best score minus the $X$-dropoff, the exploration of this branch stops. The resulting aligned segment is called a *high-scoring segment pair*, or HSP.

We will use our continuing example to demonstrate step 3 of the method. (The other steps have already been illustrated.) Because the example is small, we will take word size to be 2. This, in fact, is the word size used by the FASTA algorithm, which preceded BLAST by two years. Also, we take the dropoff value to be 4. Among the seeds produced in step 1 is GA from residue 4 in the query; this will produce a hit in the subject at residue 7 with a result score of $6 + 4 = 10$. Working upstream one residue, we match M in the query with E in the subject, a mismatch and therefore from the BLOSUM62 matrix a penalty of 2. Since the dropoff limit is 4, we continue matching I in the query with F in the subject, again a mismatch with penalty of 0. Continuing upstream, the next lineup is C with A, again a mismatch penalty of 0. Since we are at the end of the query string, we stop matching and trim back to the maximum value encountered, which was 10 with the original GA.

Now work downstream, matching P in the query with M in the subject. This gives a penalty of 2. Next match A with P, penalty of 1; then R with T, penalty of 1, so stop. Trim back to the maximum value and arrive at the HSP of GA. The original BLAST did not allow for gaps.

*Improvements to BLAST allow for gaps.*

Improvements to BLAST occur frequently. It was improved by its creators themselves in 1997 to gapped BLAST. Unfortunately, we must ask the reader to research gapped BLAST due to the constraints of the scope of this text.

*Karlin–Altschul statistics are independent of scoring matrix.*

Attached to each alignment is its score in bits and its expect value. We mentioned earlier that expect refers to the likelihood that the score occurs by chance, so small expect values are better. Here we discuss the basis on which score and expect are calculated.

The mathematical theory underlying the statistics of an alignment was worked out by Samuel Karlin and Stephen Altschul in 1990. It draws on several branches of mathematics such as the theory of information, extreme value theory, and *Poisson probability distributions*. A technical discussion is beyond the scope of this text, but we hope to give an overview sufficient for understanding the issues.

As mentioned above, clicking on "Search Summary" in the NCBI report provides additional numerical information pertaining to the run. The information there includes the size of the database and query, the "effective" lengths of the subject and query (accounting for end effects), the number of database hits, and so on. Our focus here is on the parameters lambda, K, and H. Lambda, or $\lambda$, and $K$ are the *Karlin–Altschul parameters* and $H$ is the *relative entropy*. These parameters are concerned with the calculation of the bit scores and expect values and are central to the statistical underpinnings of the report.

Certainly, it is necessary that bit scores and expect values have universal meaning; in particular, the results cannot be dependent on the scoring matrix. BLOSUM62 is just one among many possible scoring matrices, and individual researchers are free to use their own. The statistical results must compensate for whatever matrix is used. This is the function of the Karlin–Altschul parameters. Lambda functions as a normalization factor and K compensates for interresidue dependencies that may be built in to the matrix, that is, the lack of probabilistic independence between the residues.

All scoring matrices, $S_{ij}$, have built within them an implicit probability distribution, $q_{ij}$, that residue $j$ can substitute for residue $i$ (and conversely). For example, we saw in the previous section how the PAM1 matrix was constructed from the Dayhoff data derived from empirical substitution probabilities. The implicit probabilities can be recovered from the scoring matrix using the defining equation for lambda,

$$\lambda S_{ij} = \log\left(\frac{q_{ij}}{p_i p_j}\right), \tag{14.7.1}$$

where the $p_i$ are the *background probabilities*. For the $i$th residue, $p_i$ is the frequency of its occurrence over a large number of protein sequences. Equivalently, it is the probability that it will appear at any particular position in a randomly constructed sequence. These probabilities are assumed to be independent of each other.

Solving for $q_{ij}$, we get

$$q_{ij} = p_i p_j e^{\lambda S_{ij}}, \tag{14.7.2}$$

provided $\lambda$ is known. But since the $q_{ij}$ are probabilities, their sum over the entire scoring matrix must be 1; hence we have

$$\sum_{i,j} p_i p_j e^{\lambda S_{ij}} = 1. \tag{14.7.3}$$

Since all the variables in this equation are known except $\lambda$, it can be used to find $\lambda$.

Finding $K$ is harder and is estimated by statistical inference methods. For most of the scoring matrices in common use, $K$ is on the order of 0.1.

With $\lambda$ and $K$ in hand, the *Karlin–Altschul equation* is used to find expect,

$$E = Kmne^{\lambda S}. \tag{14.7.4}$$

In this, $m$ is the length of the subject sequence and $n$ is the length of the query sequence. The equation gives the expected number of HSPs in an alignment that will have a score of at least $S$ just by chance. The equation makes intuitive sense because doubling the length of either sequence should double the *number* of HSPs attaining a given score. On the other hand, in order to double a given *score*, an HSP must attain that score twice in a row, so $E$ should decrease exponentially with score, and so it does. As we have seen, small values of $E$ are highly significant.

In reality, the Karlin–Altschul equation is an approximation. Attempts to improve its accuracy center on making adjustments for the values of $m$ and $n$ and for the presence of gaps. Using the actual lengths $m$ and $n$ of the sequences is not completely accurate because of end effects; for example, the middle of the query cannot be matched with the first few residues of the subject, since the initial part of the query would then align with nothing. So there are techniques for figuring *effective lengths*.

With respect to gaps, the derivations culminating in Karlin–Altschul statistics are theoretically exact for gapless alignments. There is no exact theory when gaps are included as a possibility. As a result, various empirical techniques have been derived to adjust for this.

Returning to the output parameters, we turn our attention to relative entropy $H$. It is defined by

$$H = -\sum_{i,j} q_{ij} \lambda S_{ij}.$$

Thus $H$ is the negative of the expected value of the normalized score. This will always be a positive value because the expected score for aligning a random pair of amino acids must necessarily be negative. Were this not the case, long alignments would tend to have high score independently of whether the aligned segments were related, and the statistical theory would break down.

As its name implies, $H$ is also an entropy in the information theory sense, and information theory is the basis for Karlin–Altschul statistics. Its interpretation with respect to a scoring matrix is that low-entropy matrices are more specialized (i.e., have more structure) and high-entropy matrices are more general.

*Bit scores compensate for the scoring system used.*

Once an alignment has been constructed, the raw scores of its HSPs are available. But these scores are not useful by themselves since they depend on the scoring matrix. However, once $\lambda$ and $K$ are known, the raw score can be adjusted to calculate a *bit score*, denoted by $S'$. The bit score is given by the equation

$$S' = \frac{\lambda S - \log K}{\log 2} \tag{14.7.5}$$

and is comparable to bit scores calculated by other scoring matrices.

By combining (14.7.5) with (14.7.4), expect can be calculated in terms of bit score as

$$E = mne^{-\lambda S + \log K} = mne^{-S' \log 2},$$
$$E = mn2^{-S'}.$$

Thus bit scores embody the statistical essence of the scoring system employed. From the bit score, one only needs to know the size of the search space in order to calculate the expect.

*P-values derive from the Poission distribution.*

The number of random HSPs with score $\geq S$ is described by a Poisson distribution. This means that the probability of finding no HSPs with score $\geq S$ is $e^{-E}$, so the probability of finding at least one such HSP is $P = 1 - e^{-E}$.

This is the $P$-value associated with the score $S$. The BLAST programs report $E$-values rather than $P$-values because it is easier to understand the difference between, for example, $E$-values of 5 and 10 than $P$-values of 0.993 and 0.99995. However, when $E < 0.01$, $P$-values and $E$-values are nearly identical.

## Exercises/Experiments

*General notes.* The layout of websites and the location of links change from time to time. You may find that the specific link following sequences below are not completely accurate. Please try to work your way to the destination link overcoming any such changes. These exercises are intended to get you started in the use of Internet resources in the new era of genomics biology. We will barely scratch the surface of what's available on the Web.

1. In this exercise, we want to study the map of a specific chromosome of some organism. Go to NCBI, click "Map Viewer" under "Hot Spots" (also accessible via Entrez and the nucleotide database). In the search box at the top, pick an organism (e.g., *Plasmodium falciparum*), click "Go!," and click on a chromosome (e.g., chromosome 3). Explain what information is presented on this page directly and through its links.

2. One may research proteins as well. From NCBI, click "Entrez Home"; click "Protein: sequence database"; under "Additional protein information," click "structure"; in the "Search Entrez Structure/MMDB" box, enter the protein to be studied, (e.g., tryrosine kinase); click "go"; click on the desired type of resource (e.g., 2COI for Src Family Kinase…). Explain what information is available on this page and through the links on the page. A downloadable helper application for browsers, Cn3D, is available for viewing the three-dimensional structure of the protein (click "Structure" in the toolbar across the top).

3. As noted in the chapter, *orthologous* genes between two or more species are those with a high degree of sequence similarity and usually code for proteins having similar function. They are assumed to have evolved from a common ancestral gene. The proteins coded for by such genes are also called orthologous. By comparing the genomes of the species sequenced up to the present time, several groups or clusters of orthologous proteins have been discovered. These clusters of orthologous groups are known as COGs for prokaryotic proteins and KOGs for eukaryotic ones. From the NCBI webpage, click "Clusters of orthologous groups" under "Hot Spots," click Eukaryotic Clusters, then click 16 in the first line under KOGs; finally, click KOG1019, Retinoblastoma pathway protein. Explain what information is presented on this page and its links. Note that we will discuss phylogeny diagrams in Chapter 15.

4. The objective this time is to use the Entrez reference resources to see what is available on a topic and to use BLAST to research the associated genomics. Feel free to substitute a subject of your own interest for toxoplasmosis in the example here. Use the Entrez database at NCBI (from the NCBI homepage, select "Entrez Home") to search for toxoplasmosis: enter "toxoplasmosis" in the search box and execute the search. There are many possibilities from this point; for example, click "Nucleotide," scroll down, and select "BD495032." After reading about this nucleotide sequence, do a blastn search on it as described in the chapter. Experiment with the parameters of the search, especially the expect, word size,

and database. Report on the information available and the differences you found by varying the parameters.

**5.** This exerise is similar to the one above, except this time do a tblastn search. Research your own example or take the following. In Exercise 3 above, we came across a reference to "At5g27610" (note it is a "gee" after 5, not a nine) in connection with KOGs. Enter this into an Entrez search and look under "proteins." One of these is NP_198113. Do a tblastn search on this protein and experiment with the search parameters. In the results page, just below the "Distribution of...BLAST hits," there is a list of "sequences producing significant alignments" along with their $E$-values. To the right of that are boxes labeled "U" or "E" or "G." What information is available on these links?

### Questions for Thought and Discussion

What changes in biology have been brought about by genomics (knowing the complete genome of organisms) in the fields of

**1.** genetics;
**2.** taxonomy (classifying organisms);
**3.** evolution;
**4.** enzyme activity?

## References and Suggested Further Reading

[1] P. Benfey and A. Protopapas, *Genomics*, Prentice–Hall, Englewood Cliffs, NJ, 2004.
[2] D. T. Jones, W. R. Taylor, and J. M. Thornton, The rapid generation of mutation data matrices from protein sequences, *Comput. Appl. Biosci.*, **8**-3 (1992), 275–282.
[3] I. Korf, M. Yandell, and J. Bedell, *Blast*, O'Reilly, Cambridge, UK, 2003.
[4] S. B. Needleman and C. D. Wunsch, A general method applicable to the search for similarities in the amino acid sequence of two proteins, *J. Molecular Biol.*, **48**-3 (1970), 443–453.
[5] G. Smith, *Genomics Age*, AMACOM, New York, 2005.
[6] T. F. Smith and M. S. Waterman, Identification of common molecular subsequences, *J. Molecular Biol.*, **147**-1 (1981), 195–197.
[7] E. Ukkonen, Algorithms for string matching, *Inform. Control*, **64** (1985), 100–118.

# 15

# Phylogenetics

## Introduction

One of the purposes of this chapter is to introduce the reader to the new mathematical field of algebraic statistics; cf. [5]. Among the many topics in biology in which algebraic statistics is making an impact, we have chosen phylogenetics as the vehicle for showcasing this new discipline. Our reasons are that

- phylogeny and cladistics are important semiclassical fields in biology (with beginnings in the mid-1950s) quite different from anything we have studied up to now;
- postgenomics phylogeny makes extensive use of algebraic statistics and demonstrates more of its techniques than other branches of biology;
- phylogeny draws heavily on genomic searches, which we studied in the last chapter, and hence reinforces what we investigated there; and
- phylogeny is related to several of the new fields of biology that have arisen with genomics that we outlined in the first section of the genomics chapter, Section 14.1.

Algebraic statistics, as mentioned above, is a new branch of mathematics arising out of the many needs and uses of mathematics in genomics. Not surprisingly, the basic mathematics of algebraic statistics originates in the fields of algebra and statistics, but already new mathematics, inspired by the biology, has been created in the discipline.

This chapter will take us to a higher level of mathematical abstraction, skill, and reasoning than in the other chapters of the book and is likewise more demanding. As in the earlier parts of the book, we make every effort to explain the mathematics we need from first principles, principles that one would encounter in two years of a college mathematics curriculum, one that includes linear algebra. Still, very little abstract algebra makes its way to this level, and so we pay extra attention to illustrate the ideas and terms with examples.

Phylogenetic trees contain a great deal of biological and evolutionary information. Taxa closer together on the tree signify a greater degree of shared evolutionary novelties. The tree shows ancestral relationships among taxa and indicates the geological time the process of evolution has taken step by step. We will see that trees

R.W. Shonkwiler and J. Herod, *Mathematical Biology: An Introduction with Maple and Matlab*, Undergraduate Texts in Mathematics, DOI: 10.1007/978-0-387-70984-0_15,

are constructed using several lines of observation including ontologic, morphologic, physiologic, the fossil record, and finally genomic.

## 15.1 Phylogeny

*Phylogenetics elucidates the history of evolution.*

*Phylogenetics* is the study of the evolutionary relatedness among various groups of organisms, for example among species. Derived from the Greek—*phylon* means tribe or race and *genetikos* means relative to birth—phylogenetics attempts to reconstruct and explain the pattern of events that have led to the distribution and diversity of life as it exists at the present time. The results of a phylogenetic study is a tree diagram graphically depicting ancestor–descendant relationships over evolutionary time. An example, the tree of life, showing all three domains of cellular life, is presented in Figure 15.1.1. This tree was pieced together by Carl Woese and colleagues by comparing base pair sequences of the 16S ribosomal RNA gene.

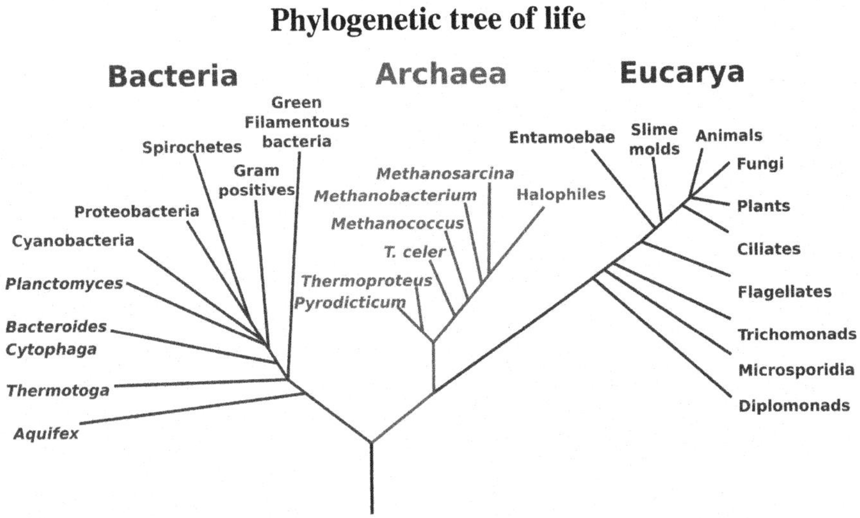

**Fig. 15.1.1.** The tree of life.

In addition to the ancestor–descendant relationships given in a phylogenetic tree, the tree also implies sets of shared, nested attributes, called *characters*, possessed by organisms farther along each branch of the tree. As a result, constructing a phylogenetic tree resolves two sets of problems. By far the more difficult of the two, without the use of genomics, is establishing the ancestor–descendancy relationship. The science of *cladistics* separates these two problems and concentrates solely on organizing groups of organisms according to nested sets of characters. Similar to phylogenetic trees, *cladists* portray their results in *cladograms*, which are also tree diagrams. The fundamental units of comparison in cladograms are *taxa*. These are collections of organisms sufficiently distinct from other sets to be given formal names and placed in a Linnaean hierarchy. Often taxa are taken to be species, but, as in the tree of life, they can be more inclusive sets as well.

In this section, we discuss the methodology for constructing and testing cladograms using classical, nongenomic, characters. However, since the end result of the analyses of phylogenetics and cladistics is presented in tree diagrams, it is important to understand how they are used.

*Tree diagrams contain a wealth of information.*

A *tree* (diagram) is a graph consisting of *nodes* or *vertices* and *edges* or *branches*. The nodes of a tree are labeled in some fashion, for example by the positive integers $1, 2, \ldots, N$. The *size* of a tree is the number of its nodes, $N$ in this case. Each edge connects two nodes, for example an edge might connect nodes 1 and 2; this is indicated by the notation $(1, 2)$. The edge $(1, 2)$ is *incident* on both node 1 and node 2. A *path* from node $a$ to node $b$ is a chain of edges, $(a, a_1), (a_1, a_2), \ldots, (a_n, b)$ connecting $a$ and $b$. A tree is *connected*; this means that every pair of distinct nodes is connected by a path. There are no circular paths in a tree, that is, paths beginning and ending on the same node. Some examples of trees are shown in Figure 15.1.2. Note that while trees (b) and (c) of the figure appear to be different, they are actually equivalent, since they have the same incidence structure.

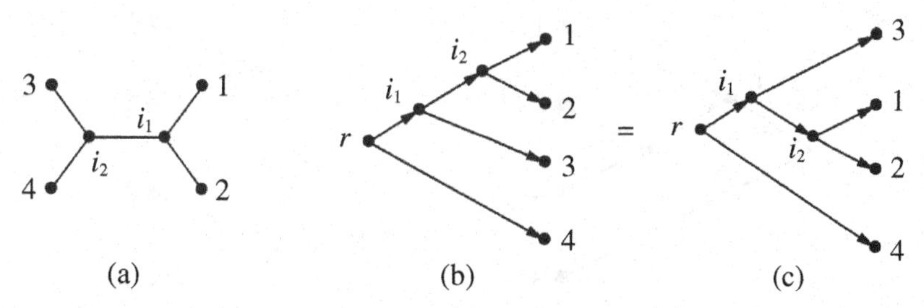


**Fig. 15.1.2.** (a) Undirected four-leaf tee. (b) Directed four-leaf tree. (c) Tree equivalent to (b).

Given a subset $S$ of nodes of a directed tree, the *subtree containing* $S$ is the part of the tree starting from the nearest common ancestor $r$ to $S$ and including all descendants of $r$. This concept is termed *monophyletic* in phylogenetics; a monophyletic group is a taxon and all of its descendants.

The number of edges incident at a node is its *degree*. Nodes having degree 1 are *leaf* nodes (except for the root node of a directed tree; see next). Nodes that are not leaf nodes are *interior nodes*. Nodes 1, 2, 3, and 4 are leaf nodes of all three trees in Figure 15.1.2; nodes $i_1$ and $i_2$ are interior nodes.

Frequently, a tree indicates the passage of time. In this case the edges are *directed* with the forward direction that of forward in time. The forward direction of a directed tree is indicated by means of arrows. However, when the direction is clear from the context, we will omit the arrows. Trees (b) and (c) of Figure 15.1.2 are directed.

By definition, in a directed tree, only one edge can lead into any given node of the tree. Thus there is a unique node with no edge leading into it; this is the *root node*. A (forward) path in a directed tree must follow the edge directions. A tree having exactly two edges branch out from each interior node is said to be *binary*. Both directed trees of Figure 15.1.2 are binary; the root node is indicated by $r$.

The leaf nodes of a tree represent taxa for which we have data. The interior nodes of a phylogenetic tree represent hypothetical ancestors. If Figure 15.1.2(b) represents a phylogenetic tree, then $i_2$ is the nearest common ancestor of taxa 1 and 2. Likewise, $i_1$ is the nearest common ancestor of 1, 2, and 3. The root node of a phylogenetic tree represents the ancestor of all taxa of the tree.

In contrast, the vertices of a cladogram only represent sister taxa with respect to some shared character. If Figure 15.1.2(b) represents a cladogram, its message is that taxa 1 and 2 share some evolutionarily novel attribute that taxa 3 and 4 do not possess. Likewise, taxa 1, 2, and 3 share some character that 4 does not possess. The term *synapomorphy* means sharing a derived character from an immediate common ancestor. Thus a cladogram expresses a series of synapomorphies.

A tree may be represented in text using nested parentheses to enclose all descendants of a node. The tree shown in Figure 15.1.3(a) has the representation $(((2,4),1),3)$, while that in (b) is written $((1,3),(2,4))$.

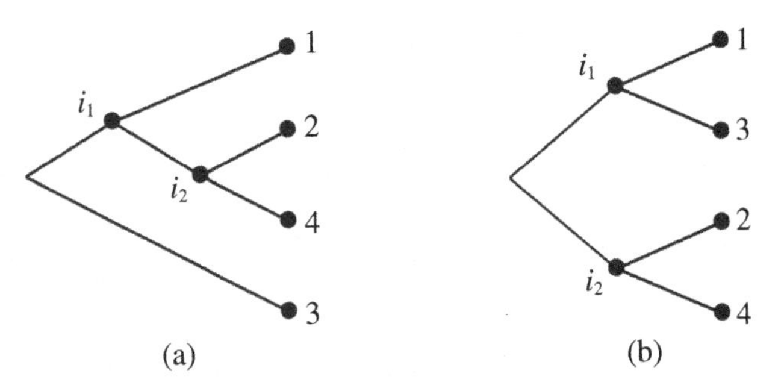


**Fig. 15.1.3.** (a) $(((2,4),1),3)$ tree. (b) $((1,3),(2,4))$ tree.

In addition to direction, the edges of a tree can depict other facts about taxa. The length of an edge can show the relative distance, in some sense, separating two taxa, for example, protein sequence distance. This is shown in Figure 15.1.4. Such a tree is called an *additive tree*, and it defines a *tree metric* in that the sum of the branch lengths along the unique path (not necessarily always forward) connecting two nodes gives the distance between them. In this way, we may construct the following distance matrix for the leaf nodes of Figure 15.1.4:

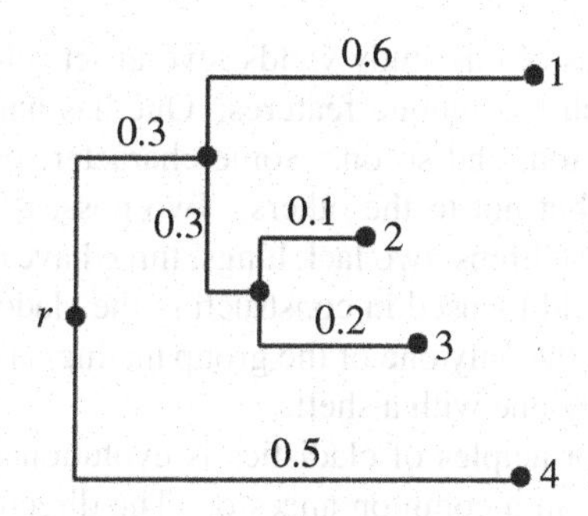


**Fig. 15.1.4.** Branch lengths give the relative distance between nodes.

$$\begin{array}{c} \\ 1 \\ 2 \\ 3 \\ 4 \end{array}\begin{array}{c} \begin{array}{cccc} 1 & 2 & 3 & 4 \end{array} \\ \begin{pmatrix} - & 1.0 & 1.1 & 1.4 \\ 1.0 & - & 0.3 & 1.2 \\ 1.1 & 0.3 & - & 1.3 \\ 1.4 & 1.2 & 1.3 & - \end{pmatrix} \end{array}.$$

Note that the root $r$ of Figure 15.1.4 is superfluous and may be omitted from such a graph.

*Cladograms express nested evolutionary novelties.*

William Hennig founded the field of cladistics in 1950 based on the principle that the evolutionary process produces, as an expectation, a nested set of evolutionary novelties. To illustrate the ideas and methods of cladistics, we begin with an example—construct a cladogram for the following organisms:

beaver, dolphin, salamander, shark, trout, and turtle.

Because these are extant organisms, many features and lines of evidence are available for distinguishing their similarities and differences. These include ontogenesis and anatomical, physiological, developmental, biochemical, and behavorial characteristics. These are called *intrinsic* characteristics of organisms. *Extrinsic* characteristics are their distribution in space and time. Of course, since the advent of genomics, DNA and protein sequence comparisons are also available, but our emphasis here is the use of nongenomic evidence. We postpone consideration of sequence matching data to the subsequent sections of this chapter.

The first task is to find general similarities shared by all the organisms in the study that could be used to define a group containing all of them. These characteristics are called the *universal set* of the comparison. In our example, these include physical symmetry, possession of an endoskeleton, appendages, a chambered heart, a dorsal nerve cord, a notochord, and visceral or gill pouches at some stage of the life cycle. Establishing a universal sets of characteristics helps in selecting outgroups used to fix evolutionary subgroups within the organisms of the study. We will see how this works below.

Examining the species of this study yields several sets of similarities and differences among them. Each has unique features: One has hair, one has a shell, one has a cartilaginous skeleton, and so on. Some characteristics are common to two or three of the subjects but not to the others: Two possess mammary glands, two have fleshy fins, three have limbs, two lack lungs, three have an amniotic egg, and so on. What principles should be used in constructing the cladogram? Maybe beavers should branch off first as the only one of the group having hair; maybe turtles should branch off first as the only one with a shell.

One of the guiding principles of cladistics is evolutionary descent: Organisms are related by descent from a common ancestor. The direction of descent is called *polarity*. Hence cladograms are constructed according to nested sets of evolutionary novelties, that is, synapomorphies, with more inclusive characters appearing nearer the root and more recent novelties shown nearer the leaves. Among the characteristics cited above, how does one decide on their evolutionary descent?

*Developmental processes help in deciding polarity.*

One line of evidence occurs during *ontogeny*, the development of an organism from a fertilized egg. During development, a trait or attribute similar to that of an ancestral species may be observed for a time, only to have it disappear at a later stage. This is known as *recapitulation.*

The first to state generalized rules of ontogeny, based on detailed studies in the 1820s, was Karl von Baer. They are the following:

1. In development from the egg, the general characteristics appear before the special characteristics.
2. From the more general characteristics, the less general and finally the special characteristics are developed.
3. During its development, an animal departs more and more from the form of other animals.
4. The young stages in the development of an animal are not like the adult stages of other animals lower down on the scale, but are like the young stages of those animals.

An example is the appearance of proto-pharyngeal gill pouches in almost all mammalian embryos at early stages of development. In particular, this applies to our example organisms. Accordingly, organisms in our group having gills, sharks and trout, should be placed toward the root of the cladogram, while lungs will be considered an evolutionary novelty.

The same rationale for ontogeny applies as well to developmental morphology. An evolutionary novelty is often the modification of some preexisting feature within the universal set. During preadult development, the modification might be repeated among derived organisms. In fact, this is the case for our example organisms with respect to skeletal composition. All six start out with cartilaginous skeletons, but except for the shark, the others replace much of this with bone prior to adulthood. Consequently, the bony members of our study are deemed a later subgroup within the group.

Our researches so far take the following form:

(universal set : shark + (bony skeleton : trout
+ (beaver, dolphin, salamander, turtle))).

This leaves four taxa to resolve.

Another striking developmental modification is seen in the salamander. As larvae, salamanders respire with gills; meanwhile, the lungs they will need as terrestrial adults are under development. Since the taxa we have already placed on the cladogram also have gills and the remaining taxa breathe with lungs, we are led to regard lungs as a derived evolutionary novelty and place salamanders next in sequence. The term used by cladists for a derived evolutionary characteristic is *apomorphy*, from the Greek *apo*, meaning away, and *morphy* for form. The term for the opposite is *plesiomorphy* (Greek: close form), meaning a primitive characteristic relative to the study group and therefore more widely shared.

Continuing in this way using ontogeny and developmental morphology, we formulate the cladogram of Figure 15.1.5.

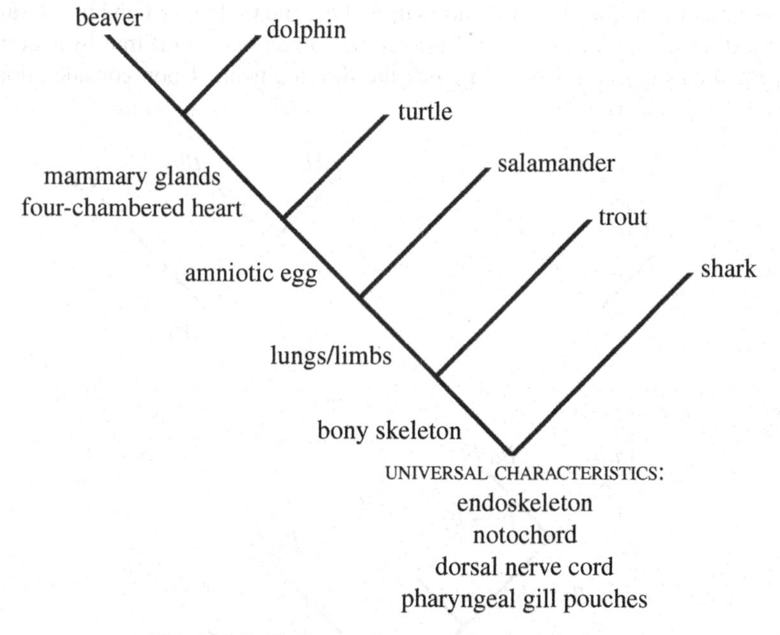


**Fig. 15.1.5.** Cladogram for example set of six taxa.

*Outgroup comparison also provides evidence.*

Ontogeny and developmental morphology are not the only criteria used in constructing and corroborating a cladogram. Outgroup comparison is another major source of clues for determining polarity. An *outgroup* is a taxon related to and believed to be more primitive than the organisms being classified. The perfect outgroup would be a close ancestor of these taxa and would have the primitive form of the characteristics being compared. In the real world, it is impossible to be certain that an outgroup chosen for cladistic analysis is actually the ancestor, or whether its traits are all truly primitive. The existence of *convergent evolution* is a primary complication in cladistics, since it confuses the identification of primitive characteristics. We briefly digress to explain.

In our example, dolphins have been included within the set of animals having evolved terrestrial limbs. But instead, they possess appendages similar to those of sharks. Their assignment therefore has an inconsistency. But the inconsistency is resolved if it were the case that the immediate line of ancestors of the dolphin, having taking up life in the sea, gradually modified their limbs into the finlike structures of today's animal. In fact, this is what cladists believe. *Convergence* is the independent evolution of similar structures to solve the same biological problem, in this case movement through water. Note that despite having returned to the sea, the ancestors of the dolphin did not revert to oxygen exchange via gills but retained their lungs.

To see how outgroups can help, consider the diagrams in Figure 15.1.6. Three taxa $A$, $B$, and $C$ present variation with respect to two different traits, $a$ and $b$. Taxon $A$ shows traits $a_2$ and $b_2$, taxon $B$ shows $a_2$ and $b_1$, and $C$ shows $a_1$ and $b_2$. If trait $a_2$ is derived from $a_1$, then Figure 15.1.6(a) is the correct one, but if trait $b_2$ is derived from $b_1$, then Figure 15.1.6(b) captures the development. Upon consideration of

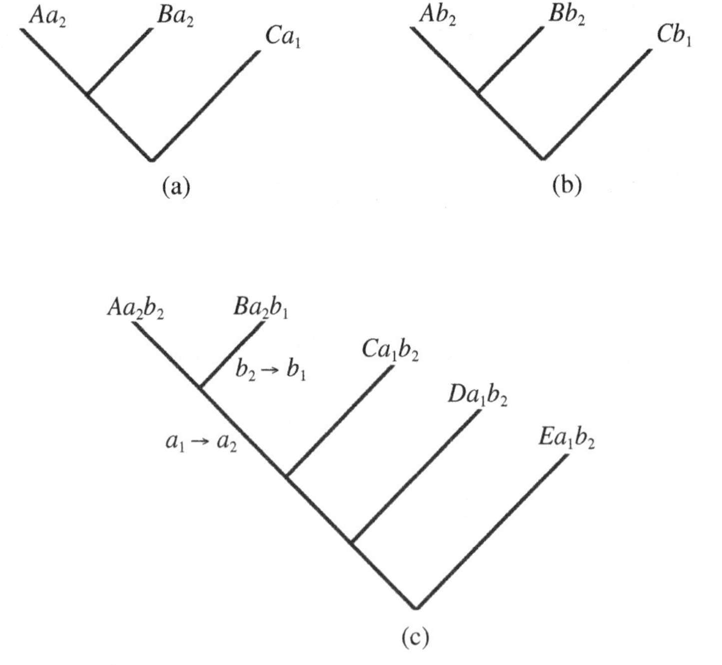


**Fig. 15.1.6.** Outgroups show that traits $a_1$ and $b_2$ are primitive.

outgroups $D$ and $E$, we find that they possess characteristics $a_1$ and $b_2$. Hence these are taken as the primitive forms and we get Figure 15.1.6(c).

Following the discussion above, as an outgroup we seek taxa that could be close ancestors of our study group. This is the purpose of establishing a universal set of characteristics. Evidently, organisms possessing these are the most likely to meet the ancestor criteria. In this example, the lamprey can serve as an outgroup for us, since they possess all the characteristics of the universal set. Since the lamprey does not have a bony skeleton at any stage of its life, we judge that this is a primitive trait. This corroborates our placement of sharks as the first branch of the cladogram.

Having made this determination, we may now include sharks as another outgroup in deciding derived traits for the remaining set: trout, salamander, turtle, dolphin, beaver. Among these, what evolutionary novelties are shared by all but one? We notice that trout lack lungs and fleshy limbs, and, moreover, these are not present in either outgroup, lamprey and sharks. We thus regard lungs as an apomorphy. Continuing to apply these principles, we again arrive at Figure 15.1.5.

*Cladograms help in constructing phylogenetic trees.*

As previously mentioned, phylogenetic trees go beyond cladograms in that they assert ancestor–descendant relationships. Since biological species are delineated according to the potential of its members to interbreed, species are necessarily the units of evolution.[1]

One can speak of one species as being the ancestor of another. Properly then, the taxa underlying a phylogenetic tree are species. Nevertheless, phylogenetic trees are constructed for higher biological units, for example reptiles as descending from amphibians. The inference is that some amphibian species—*Seymouria* has been cited—is the particular ancestor of the line leading to the reptiles.

Since a cladogram has less information than a phylogenetic tree, each subgraph of a cladogram can be explained by any one of several trees. For example, consider the cladogram shown in Figure 15.1.7. The information here is that taxa $A$ and $B$ share some derived characteristic, a synapomorphy, not possessed by $C$. Figure 15.1.8 shows six possible ways this could happen.

In the first of these, $A$ and $B$ derive from a common ancestor, and that ancestor and $C$ do likewise. This explains how $A$ and $B$ can have a shared characteristic while $C$ does not have it. In (b), $C$ is itself the ancestor of the common ancestor of $A$ and $B$. Again, the synapomorphy could derive from this nearest common ancestor and thus not from $C$. In the other diagrams, one of $A$ or $B$ is the ancestor of the other; for example, $A$ is the ancestor of $B$ in (f). The assumption is that the characteristic in question first appeared, with respect to the diagram, in $A$ and was passed on to $B$. And so once again, $A$ and $B$ share it, while C does not, as consistent with the cladogram.

[1] Formulating a testable definition of species is a challenge; how can one show that two organisms widely separated in space or time are or were capable of interbreeding? For an in-depth discussion, see [3].

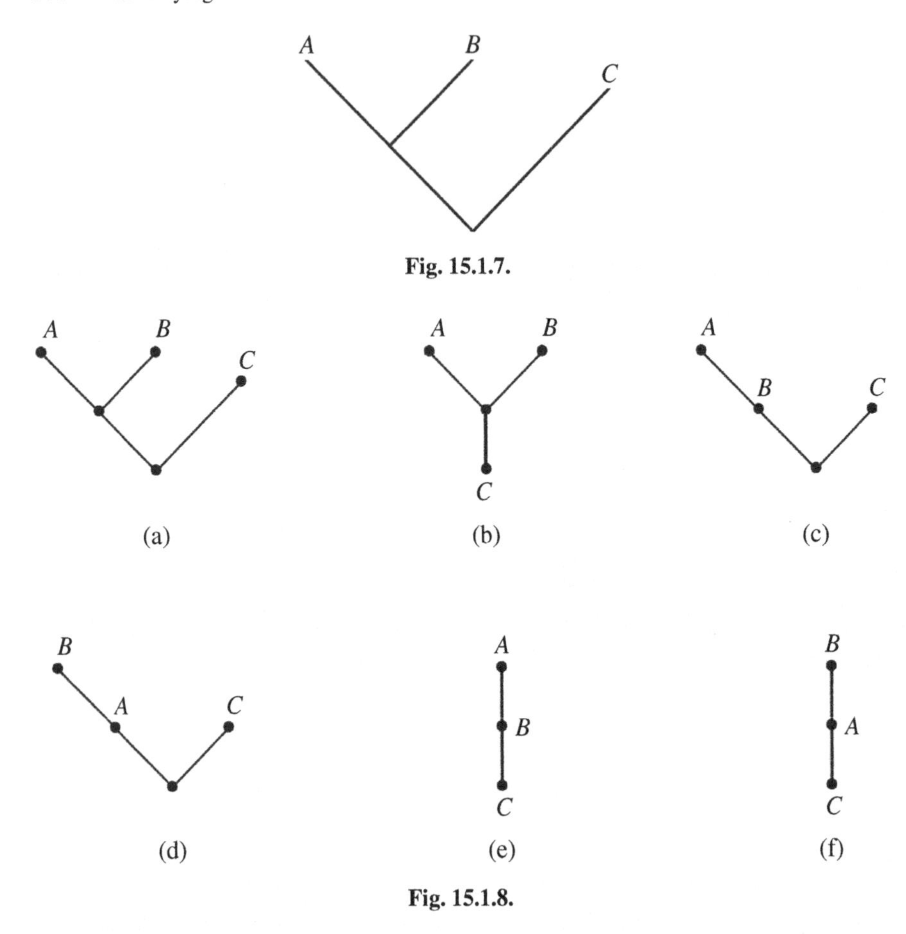


**Fig. 15.1.7.**

**Fig. 15.1.8.**

Therefore, additional information is needed to promote a cladogram to a phylogenetic tree. That information is the evolutionary history of speciation events occurring to the organisms being studied. But the sought-after ancestors are, in most cases, extinct organisms. As a result, until the advent of genomics, the information had to come from the fossil record. Continuing to focus on pregenomic methods here, we inquire into the nature of speciation.

In fact, the way in which new species arise is still not well understood; see [6]. We content ourselves with mentioning some of the possibilities. The *transformational* mechanism of speciation is that over time, a species transforms by accumulating modifications, either gradually or all at once (*saltationism*), emerging as another species. If the transformation is gradual and the species is a type that leaves fossils, then the fossil record should contain the multitude of intermediates through time. But the fossil record is generally incomplete, and what we see are merely "snapshots" separated by gaps in the record. Hence there is no need to speculate as to when the new species emerged during the gradual change; the incomplete fossil record has performed this task.

In the saltation theory, evolution proceeds by sudden jumps. Of course, paleontologically, a jump could mean a few hundreds of thousand years. Several mechanisms have been proposed for saltation. For example, a genetic mutation could produce a bifurcation in the apparatus producing hair and instead produce a proto-feather. The mutation does not interfere with interbreeding. In time, those with the mutation do better than those without and come to dominate the gene pool. Eventually, the original gene disappears altogether. There is fossil evidence for this mechanism; see [1].

Transitional speciation, whether gradually or as a jump, produces a phylogenetic tree as exhibited in Figure 15.1.9(a).

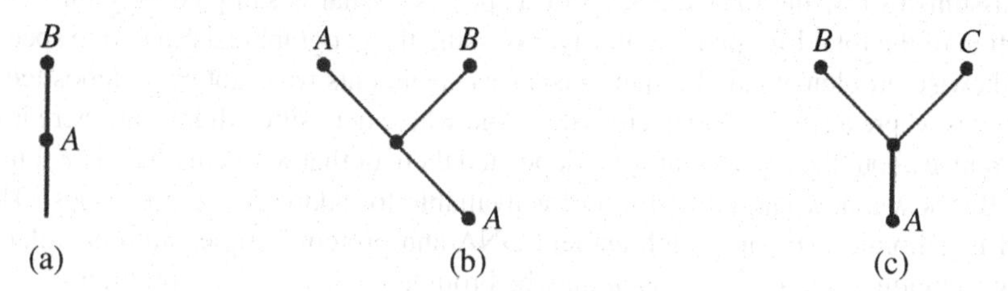


**Fig. 15.1.9.** Speciation possibilities.

The transitional theory does not produce diversification; the species count remains the same. But the other major theory of speciation, that of splitting, does increase the count. The idea here is that a portion of the population becomes reproductively isolated. This could be the result of geographical circumstances—a mountain range develops, a canyon widens, the subpopulation is transported to a new continent or an island, and so on. Over time, differences accumulate between the subpopulation and the main population. This could stem from natural selection or just by genetic drift. Such an isolated subpopulation already starts off genetically different from the main body. Every genetic trait has a specific distribution of its alleles among an interbreeding population. But any given subpopulation will have profound allelic differences for some subset of traits just by chance alone. Eventually, the subpopulation becomes reproductively incompatible with the original population, and a new species has thus been created.

Reproductive isolation strictly enforced by some geographical barrier is called *allopatric speciation* of type A. In allopatric speciation of type B, the isolation stems from the fact that the population is large and widely separated in distance (relative to the organism). As a result, individuals on the margins never have the opportunity to breed with other parts of the population. As above, these units proceed toward establishing a distinct species. Speciation by splitting results in a phylogenetic tree as exemplified in Figure 15.1.9(b) or (c). In (c), the ancestral species itself produces a new species by transition.

Once a tree has been constructed, its assertions and predictions must stand up to testing. Of course, a phylogenetic tree must be consistent with any cladogram depiction of its taxa. A cladogram that shows species *B* to have derived characteristics with respect to species $A$ cannot have $A$ as a descendant of $B$. Thus the evidence used in cladistics is applicable to phylogenetic trees as well.

Paleontology can be a source of falsification, or support, as well. The data here will be extrinsic, that is, information about the distribution of the organisms in space and time. The assertion that one species is the descendant of another would presuppose that fossils of the ancestor should predominately be found in older rock and those of the descendant in younger rock. Or in another case, a speciation event postulated to have taken place at a point in time would suppose the geographical distribution of the ancestor to be more widespread and their numbers to be larger. However, the nature of the fossil record is not so precise. Observed stratigraphic ranges cannot be assumed to be the total life span of a species. What is sampled may only be a portion of the total life span. Similarly, assessing the geographical range of a species is likewise problematical. In many instances, sediments were not even deposited in all areas where a species had been living. And where deposition did occur, there is no guarantee about the fossils that were deposited there or that will remain and be found.

But now a new and powerful tool is available for addressing these issues. That tool is genomics. Even though ancient DNA and protein samples are unavailable, new techniques and lines of attack may be brought to bear on these problems.

## 15.2 Branch Lengths Estimate the Separation of Species

In this section, we find that natural assumptions about the rate of molecular evolution leads to a quantitative description of the phenomenon via mathematical semigroups. Moreover, in order to estimate mutation rates, we employ the widely used technique of maximum likelihood estimation. We thus begin our first encounter with algebraic statistics. Throughout the remainder of this chapter, we closely follow some of the topics in the groundbreaking text by Pachter and Sturmfels [5]. In the next section, we take time out from our study of phylogenetics to introduce all the additional mathematics that will be needed. For more detail and to learn about the full scope of algebraic statistics applications in genomics, see Pachter and Sturmfels [5].

*The molecular clock assumption asserts that molecular evolution is constant over time.*

In 1965, Emile Zuckerkandl and Linus Pauling proposed the theory of a *molecular clock*, which states that the rate of molecular evolution is approximately constant over time for all the proteins in all lineages. According to this theory, any time of divergence between genes, proteins, or lineages can be dated simply by measuring the number of changes between sequences. Soon afterward, in 1969, Thomas Jukes and Charles Cantor (1969) proposed a stochastic model for DNA substitution in which all nucleotide substitutions occur at an equal rate, and when a nucleotide is substituted, any one of the other nucleotides is equally likely to be its replacement.

In this section, DNA alignment data is used to compute branch lengths under the assumption of a Markov model for point mutations. These assumptions are patterned after the molecular clock theory:

1. mutations occur at random, dependent only on a mutation rate;
2. mutations occur independently at different sites;
3. (continuous time assumption) at any instant in time, there is a nonzero probability that a mutation will occur.

As in all good science, the assumptions may be oversimplified, but they do capture the essence of the phenomenon and form the basis of a starting point for studying the subject.

While the Markov assumptions may apply to molecular evolution from ancestral species well enough, the requirement of DNA sequence data limits these methods to taxa for which there is such data. Hence at present these methods give rise to unrooted phylogenetic trees among existing species.

Although the model applies to point mutation phenomena in general, for example, protein evolution, we will specialize to DNA mutation. Thus our indices run over the set of bases $\Sigma = \{A, C, G, T\}$ taken in alphabetical order.

*The problem of calculating branch lengths.*

In consequence of the independence assumption, we can consider the mutation at each site one by one. At any single site, there is the probability $\theta_{ij}(t)$ that base $i$ will have changed to base $j$ after time $t$. The path of the change is not considered: $i$ may have changed directly to $j$, or may have changed to some intermediate $k$ that changed to $j$, or any other of many possibilities. Let $\theta(t)$ denote the $4 \times 4$ matrix of these probabilities. It represents the cumulative effect of changes over a time period $t$ and is called the *substitution matrix*. Thus $\theta(0) = I$, the identity matrix.

A consequence of the Markov assumptions is that the process is "memoryless": Over a period of time $s + t$, and decomposing on base $k$ the process visited at time $s$ (should a base be impossible at time $s$, then the probability of the transition from $i$ to $k$ will be 0), we have

$$\Pr(i \to j \text{ over time } s+t) = \sum_{k \in \Sigma} \Pr(i \to k \text{ over time } s) \cdot \Pr(k \to j \text{ over time } t).$$

In terms of matrix multiplication, this is exactly

$$\theta(s+t) = \theta(s)\theta(t), \quad s \geq 0, \quad t \geq 0. \tag{15.2.1}$$

A family of matrices, indexed by $t$, satisfying (15.2.1) is a mathematical *semigroup*. In turn, this implies the existence of an *infinitesimal generator* or *rate matrix* $Q$ having the properties

$$\theta(t) = e^{Qt} = \sum_{n=0}^{\infty} \frac{1}{n!} Q^n t^n,$$

$$\theta'(t) = \theta(t)Q = Q\theta(t), \quad t \geq 0, \quad ' \text{ signifying the derivative},$$

$$\theta^{(k)}(0) = Q^k, \quad k \geq 0, \quad (k) \text{ signifying the } k\text{th derivative}.$$

The off-diagonal elements of $Q$ are the transition rates between bases per unit time. Thus $q_{ij}$, with $i \neq j$, is the (average) instantaneous rate at which base $i$ mutates into base $j$. Mathematically, the rows of $Q$ must sum to 0,

$$q_{ij} \geq 0, \quad i \neq j, \qquad q_{ii} < 0,$$
$$\sum_{j \in \Sigma} q_{ij} = 0 \quad \text{for all } i \in \Sigma.$$

Two widely used rate matrices are the *Jukes–Cantor*,

$$Q_{\text{JC}} = \begin{bmatrix} -3\alpha & \alpha & \alpha & \alpha \\ \alpha & -3\alpha & \alpha & \alpha \\ \alpha & \alpha & -3\alpha & \alpha \\ \alpha & \alpha & \alpha & -3\alpha \end{bmatrix}, \tag{15.2.2}$$

and the *Kimora*-80,

$$Q_{\text{K80}} = \begin{bmatrix} -(\alpha+2\beta) & \beta & \alpha & \beta \\ \beta & -(\alpha+2\beta) & \beta & \alpha \\ \alpha & \beta & -(\alpha+2\beta) & \beta \\ \beta & \alpha & \beta & -(\alpha+2\beta) \end{bmatrix}. \tag{15.2.3}$$

In Jukes–Cantor, the first row says that the rates at which adenine (A) mutates into cytosine (C) or guanine (G) or thymine (T) are the same and equal $\alpha$ (a parameter of the model). Likewise, the other rows say the analogous thing for cytosine, guanine, and thymine.

In Kimora-80, the rate for a purine to a purine or a pyrimidine to a pyrimidine base is $\alpha$, while from a purine to a pyrimidine or conversely is $\beta$. This more accurately reflects actual mutation rates at the expense of introducing a second parameter, $\beta$.

Using the Jukes–Cantor rate matrix, the substitution matrix can be computed by MAPLE:

```
MAPLE (symbolic, no MATLAB equivalent)
> with(LinearAlgebra):
> assume(a>0):
> Q:=Matrix([[-3*a,a,a,a],[a,-3*a,a,a],[a,a,-3*a,a],[a,a,a,-3*a]]);
> MatrixExponential(Q)
```

Regarding $a = \alpha t$, this gives

$$\theta(t) = e^{Qt} = \frac{1}{4}\begin{bmatrix} 1+3e^{-4\alpha t} & 1-e^{-4\alpha t} & 1-e^{-4\alpha t} & 1-e^{-4\alpha t} \\ 1-e^{-4\alpha t} & 1+3e^{-4\alpha t} & 1-e^{-4\alpha t} & 1-e^{-4\alpha t} \\ 1-e^{-4\alpha t} & 1-e^{-4\alpha t} & 1+3e^{-4\alpha t} & 1-e^{-4\alpha t} \\ 1-e^{-4\alpha t} & 1-e^{-4\alpha t} & 1-e^{-4\alpha t} & 1+3e^{-4\alpha t} \end{bmatrix}. \tag{15.2.4}$$

Under the Markov assumptions enumerated on p. 508 above, the course of the ensuing mutation is an instance of a mathematical process known as a *Poisson process*. This means that the distribution of mutation events is given by

$$\Pr(k \text{ events in time } t) = \frac{(\lambda t)^k}{k!} e^{\lambda t},$$

where $\lambda$ is the Poisson event rate. In turn, this is the average mutation rate imposed by $Q$. The event rate for base $i$ is the sum of its mutation rates to the other bases; but this is exactly the negative of the diagonal element, $-q_{ii}$. Therefore, the average event rate is

$$\lambda = -\frac{1}{4}\,\text{trace}(Q),$$

where the *trace* of a matrix is the sum of its main diagonal elements. The expected number of events over time $t$ of a Poisson process is $\lambda t$, and so the expected number of mutations over time $t$ is

$$\text{branch length} = -\frac{1}{4}\,\text{trace}(Q)\cdot t. \tag{15.2.5a}$$

As indicated, the expected number of events is taken as the branch length of the phylogenetic tree. For the Jukes–Cantor model, we get

$$\text{branch length} = 3\alpha t. \tag{15.2.6}$$

By the diagonalization theorem of Section 2.6, (2.6.2), one can show that $\log\det(e^Q) = \text{trace}(Q)$; hence in terms of the substitution matrix directly,

$$\text{branch length} = -\frac{1}{4}\log\det(\theta(t)). \tag{15.2.5b}$$

*Estimating branch lengths leads to the method of maximum likelihood.*

Now suppose that we are given an alignment between two DNA sequences and we want to estimate the branch length between them with respect to a rooted tree; see Figure 15.2.1.

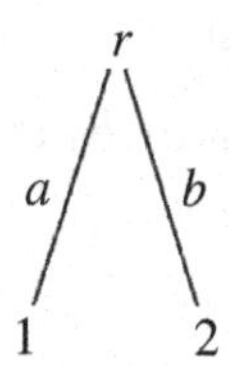


**Fig. 15.2.1.** Two-claw tree.

The problem is solved by the following theorem.

**Theorem 1.** *Given an alignment of two sequences of length n, with k differences between their bases, the maximum likelihood estimate of the branch length under the Jukes–Cantor rate model is*

$$\text{branch length} = -\frac{3}{4}\log\left(1-\frac{4k}{3n}\right). \tag{15.2.7}$$

*In terms of $c = n - k$, the number of identities between the two sequences, this is*

$$\text{branch length} = -\frac{3}{4}\log\left(\frac{4c}{3n}-\frac{1}{3}\right). \tag{15.2.8}$$

*Maximum likelihood estimation*, as mentioned in the theorem, is the most widely used method for parameter estimation in statistics. It means choosing the value of any unknown parameter in such a way as to make the outcome that was actually observed the most probable. Here is how that works out for branch length estimation.

Let $a$ be the substitution matrix along the left branch of the tree and $b$ that along the right branch. From (15.2.4), these matrices have only two distinct elements: the diagonal elements and the nondiagonal elements. Let

$$a_0 = \frac{1}{4}\left(1 + 3e^{-4\alpha t}\right) \quad \text{and} \quad a_1 = \frac{1}{4}\left(1 - e^{-4\alpha t}\right). \tag{15.2.9a}$$

Similarly, let

$$b_0 = \frac{1}{4}\left(1 + 3e^{-4\beta t}\right) \quad \text{and} \quad b_1 = \frac{1}{4}\left(1 - e^{-4\beta t}\right). \tag{15.2.9b}$$

Then by (15.2.6), the branch length we want to calculate is

$$\begin{aligned} \text{branch length} &= (\text{branch length 1 to } r) + (\text{branch length } r \text{ to } 2) \\ &= 3(\alpha + \beta)t. \end{aligned} \tag{15.2.10}$$

With the abbreviations defined above, we can write

$$a = \begin{bmatrix} a_0 & a_1 & a_1 & a_1 \\ a_1 & a_0 & a_1 & a_1 \\ a_1 & a_1 & a_0 & a_1 \\ a_1 & a_1 & a_1 & a_0 \end{bmatrix}, \qquad b = \begin{bmatrix} b_0 & b_1 & b_1 & b_1 \\ b_1 & b_0 & b_1 & b_1 \\ b_1 & b_1 & b_0 & b_1 \\ b_1 & b_1 & b_1 & b_0 \end{bmatrix}.$$

There are four unknown parameters, $a_0, a_1, b_0$, and $b_1$. But since the rows of stochastic matrices must sum to 1, we have

$$a_0 + 3a_1 = 1, \qquad b_0 + 3b_1 = 1. \tag{15.2.11}$$

This is automatically satisfied with the $a$s and $b$s taken according to (15.2.9).

Next, we calculate the probability that two bases at the leaves 1 and 2 of the tree will be the same. Let $p_{\text{AA}}$ be the probability that they are both A; then

$$p_{\text{AA}} = \frac{1}{4}a_0 b_0 + \frac{3}{4}a_1 b_1. \tag{15.2.12}$$

This is seen as follows: If the original base at the root is A, with probability $\frac{1}{4}$, then we will have A at leaf 1 if there is no change, and this happens with probability $a_0$ according to the edge matrix $a$. Similarly, the A at leaf 2 remains unchanged with probability $b_0$. This gives the first term in (15.2.12). But if the original base is not A, and this happens with $\frac{3}{4}$ probability—say it is C—then both leaves will be A if C mutates to A along both edges. The mutation from C to A along the left edge happens with probability $a_1$ according to the substitution matrix, and along the right edge it is $b_1$. So the probability that both leaves will be A when the root was not is $(\frac{3}{4})a_1b_1$. And this is the second term.

The probability that both leaves are C or G or T is the same as for A, and so the probability that both leaves are the same is given by

$$\theta = p_{\text{same}} = a_0b_0 + 3a_1b_1; \tag{15.2.13}$$

denote this by $\theta$. By a similar calculation, the probability that the bases at the two leaves are different works out to be (there are 12 ways they could be different)

$$p_{\text{dif}} = 12\left(\frac{1}{4}a_0b_1 + \frac{1}{4}a_1b_0 + \frac{1}{2}a_1b_1\right) = 3a_0b_1 + 3a_1b_0 + 6a_1b_1 = 1 - \theta.$$

As noted, this equals $1 - \theta$, since it is the complementary event to the leaves being the same.

Now return to our original problem; we have two DNA sequences of length $n$ differing in $k$ places. We apply the results derived above to each place. The probability that $k$ bases are different out of $n$ is like getting $k$ heads out of $n$ tosses of a weighted coin (from Sections 2.8), so we have

$$L(\theta) = \Pr(k \text{ differences out of } n) = \binom{n}{k}(1-\theta)^k\theta^{n-k}. \tag{15.2.14}$$

This is called the *likelihood function* for the model, and for emphasis we show it to be a function of $\theta$.

Now suppose that there were $n = 100$ bases and $\frac{3}{4}$ of them, or 75, remained the same, leaving $k = 25$ to mutate. What value of $\theta$ would make this outcome the most likely? It would be $\theta = \frac{3}{4}$. For example, if $\theta$, being the probability that a base remains unchanged, were $\frac{1}{2}$, it would be very unlikely to get $\frac{3}{4}$ of 100 bases unchanged by chance; see Figure 15.2.2.

This is how maximum likelihood works. To maximize (15.2.14), we set its derivative to zero and solve for $\theta$. Alternatively, we could take the logarithm of (15.2.14) and set its derivative to zero. Often likelihood functions are products of factors to various powers, and working with the log likelihood function is easier. The calculation is

$$\begin{aligned} 0 &= \frac{d\log L(\theta)}{d\theta} = \frac{d}{d\theta}\left(\log\binom{n}{k} + k\log(1-\theta) + (n-k)\log\theta\right) \\ &= \frac{-k}{1-\theta} + \frac{n-k}{\theta}. \end{aligned} \tag{15.2.15}$$

The solution is $\theta = \frac{n-k}{n}$. From (15.2.13), this gives

$$\frac{n-k}{n} = a_0b_0 + 3a_1b_1.$$

Now substitute (15.2.9) into this to get

$$\frac{n-k}{n} = \frac{1}{4} + \frac{3}{4}e^{-4(\alpha+\beta)t}.$$

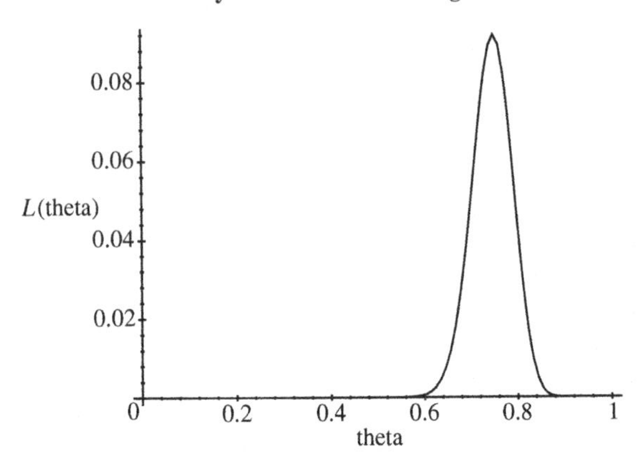


**Fig. 15.2.2.** Maximizing outcome probability for $n = 100$ and $k = 25$.

Solve this for $\alpha + \beta$ and, remembering (15.2.10), we get the conclusion of Theorem 1,

$$\text{branch length} = -\frac{3}{4}\log\left(1 - \frac{4k}{3n}\right).$$

Keep in mind that, branch length is taken to be the expected number of mutation events over the time between the two observations and is a pure number. If the time between the observations is known, then a mutation rate estimation can be worked out. Conversely, if the mutation rate is known, then the time can be estimated.

We make a last observation on this model. Note that we have not made a determination of $\alpha$ and $\beta$ individually, only their sum. Likewise, the $a$s and $b$s occur only combined, as in

$$a_0b_0 + 3a_1b_1 \quad \text{and} \quad a_0b_1 + a_1b_0 + 2a_1b_1. \tag{15.2.16}$$

This is because we took the states of the root to be equally likely; we have information only about differences along the combined link from node 1 through $a$ to $r$ then through $b$ to node 2. The combined edges are governed by the matrix product

$$ab = \begin{bmatrix} a_0b_0 + 3a_1b_1 & a_0b_1 + a_1b_0 + 2a_1b_1 & \dots \dots \\ a_0b_1 + a_1b_0 + 2a_1b_1 & a_0b_0 + 3a_1b_1 & \dots \dots \\ \vdots & \vdots & \ddots \end{bmatrix}.$$

## 15.3 Introduction to Algebraic Statistics

In order to avoid interrupting the flow of ideas in subsequent sections, we gather together here the concepts and tools of algebra that we will need in those sections.

*It all takes place in the ring of polynomials.*

A *mathematical ring* is a set $S$ together with two operations, denoted by $+$ and $*$, satisfying the following basic arithmetic laws:

- Associativity:

$$(a+b)+c = a+(b+c), \qquad (a*b)*c = a*(b*c), \quad a,b,c \in S.$$

- Commutativity for $+$ (not necessarily for $*$):

$$a+b = b+a, \quad a,b \in S.$$

- Distributivity:

$$a*(b+c) = a*b + a*c, \quad a,b,c \in S,$$

  and

$$(b+c)*a = b*a + c*a, \quad a,b,c \in S.$$

- Existence of an additive identity, 0:

$$0+a = a+0 = a, \quad a \in S.$$

- Existence of an additive inverse: For $a \in S$, there is an inverse, denoted by $-a$, such that

$$a+(-a) = (-a)+a = 0.$$

Of course, the familiar number systems—the integers $\mathbb{Z}$, the real numbers $\mathbb{R}$, and the complex numbers $\mathbb{C}$—are all rings. Add to the list the rational numbers, $\mathbb{Q}$. A number is *rational* if it is the ratio of two integers, e.g., $\frac{3}{4}$, or $\frac{355}{113}$, or $-\frac{17}{1}$, and so on. But we are interested in rings because the set of polynomials over any one of the number sets above is a ring. Let $\mathbb{Q}[x]$ denote the set of polynomials in the indeterminate $x$ with rational numbers serving as coefficients, that is, expressions of the form

$$a_n x^n + a_{n-1} a^{n-1} + \cdots + a_1 x + a_0,$$

where $a_n, \ldots, a_0$ belong to $\mathbb{Q}$ with $a_n \neq 0$; $n$ is the *degree* of the polynomial. The $+$ and $*$ operations are the usual polynomial addition and multiplication,

$$\begin{aligned}
&(a_n x^n + \cdots + a_1 x + a_0) + (b_m x^m + \cdots + b_1 x + b_0) \\
&\quad = a_n x^n + \cdots + (a_m + b_m) x^m + \cdots + (a_1 + b_1) x + (a_0 + b_0) \quad \text{if } n \geq m, \\
&(a_n x^n + \cdots + a_1 x + a_0) * (b_m x^m + \cdots + b_1 x + b_0) \\
&\quad = a_n b_m x^{n+m} + (a_n b_{m-1} + a_{n-1} b_m) x^{n+m-1} \\
&\qquad + (a_n b_{m-2} + a_{n-1} b_{m-1} + a_{n-2} b_m) x^{n+m-2} + \cdots .
\end{aligned}$$

The additive identity is the zero polynomial, with all coefficients equal to zero, and the additive inverse of $f(x)$ is the polynomial all of whose coefficients are the negatives of those of $f$. Checking the laws is straightforward but tedious and we omit it.

More generally, the multivariate polynomials form a ring as well. Let $\mathbb{Q}[x_1, x_2, \ldots, x_d]$ denote the set of sums of terms of the form

$$a_{i_1 i_2 \cdots i_d} x_1^{i_1} x_2^{i_2} \cdots x_d^{i_d}, \tag{15.3.1}$$

where the $i_1, i_2, \ldots, i_d$ are nonnegative integers. A polynomial consisting of a single such term, as in (15.3.1), is referred to as a *monomial*. Again $+$ and $*$ are taken as polynomial addition and multiplication. For brevity, we also use the notation $\mathbb{Q}[\mathbf{x}]$ to refer to this space where $\mathbf{x} = (x_1, \ldots, x_d)$.

In the axioms of a ring listed on the previous page, we noted that commutativity for $*$ was not required. But in all of our examples this property is realized. When commutativity for $*$ holds, the ring is *commutative*. In all that follows, we will need only the polynomial rings introduced above, and therefore all our rings are commutative. In what follows, we will refer to $+$ as addition and $*$ as multiplication.

*Ideals play a fundamental role in rings.*

For a commutative ring $S$, an *ideal* $I$ is a subset of $S$ that is *closed* under $+$ and closed over multiplication by elements in $S$. In other words, the following two properties are satisfied

$$a + b \in I \quad \text{if } a, b \in I$$

and

$$a * r = r * a \in I \quad \text{if } a \in I \text{ and } r \in S.$$

We do not consider the empty set an ideal. On the other hand, obviously every ring is an ideal of itself. More importantly, an ideal $I \subset S$ is a ring in its own right. The reason is that the elements of $I$ satisfy the axioms of a ring over $I$, since they do so over $S$. It remains only to see that $I$ is closed under $+$ and $*$ and that $0 \in I$. The first is part of the definition. The second follows immediately, since if $a \in I$ and $b \in I$, then $b \in S$ and so $a * b \in I$ by the second part of the definition. Finally, if $a \in I$, then $0 = a * 0 \in I$, since $0 \in S$.

**Example.** Consider the set $I$ in $\mathbb{Q}[x]$ consisting of all polynomials of the form $f(x) * (x - 1)$, where $f(x) \in \mathbb{Q}[x]$. Obviously, the second property of the definition holds. But also, if $f(x) * (x - 1)$ and $g(x) * (x - 1)$ are two such polynomials, their sum is $(f(x) + g(x)) * (x - 1)$ and so belongs to $I$, too. Hence this set is an ideal; it is the ideal generated by $(x - 1)$.

The example above is typical. Let $\mathcal{F}$ be a collection of polynomials. Specifically, let $\mathcal{F} \subset \mathbb{Q}[\mathbf{x}]$ be a subset of the ring of polynomials in one or more indeterminates. The ideal *generated by* $\mathcal{F}$, denoted by $\langle \mathcal{F} \rangle$, is the set of all polynomial linear combinations of elements in $\mathcal{F}$,

$$\langle \mathcal{F} \rangle = \{h_1 f_1 + \cdots + h_n f_n : f_1, \ldots, f_n \in \mathcal{F},\ h_1, \ldots, h_n \in \mathbb{Q}[x]\}.$$

It is possible for two subsets $\mathcal{F}$ and $\mathcal{F}'$ to generate the same ideal,

$$\langle \mathcal{F} \rangle = \langle \mathcal{F}' \rangle.$$

In fact, by the Hilbert basis theorem, every ideal is finitely generated.

**Theorem 1 (Hilbert basis theorem).** *Every infinite set of polynomials $\mathcal{F}$ in $\mathbb{Q}[\mathbf{x}]$ has a finite subset $\mathcal{F}' \subset \mathcal{F}$ such that $\langle \mathcal{F} \rangle = \langle \mathcal{F}' \rangle$.*

The theorem says more than promised. To see that an ideal $I$ is finitely generated, take $\mathcal{F}$ to be the ideal itself.

*A Gröbner basis makes it easier to work with ideals.*

It will make a difference in which order the terms of a polynomial are written. The monomial written first is its *leading term.* For a single indeterminate, we write the terms in order of higher to lower degree. For multivariate polynomials we use *lexicographic monomial order*, signified by $\succ$. This means that in order to decide the largest monomial in a set, we use the degree of $x_1$ without regard for any other indeterminate, unless more than one term has the same highest degree in $x_1$. In that case, we decide between these according to the highest degree of $x_2$, and so on. It is like the order of words in a dictionary. Nonzero coefficients are ignored. For example, among the monomials $-5x_1^3x_2^5x_3^4$ and $7x_1^2x_2^7x_3^4$, the former is larger in $\succ$-order. But among $-5x_1^3x_2^5x_3^4$ and $4x_1^3x_2^6$, the latter is larger in $\succ$-order.

We say that $\mathcal{G} = \{g_1, g_2, \ldots, g_r\}$ is a Gröbner basis for an ideal $I$ if $\mathcal{G}$ generates $I$ and if every polynomial $f$ in $I$ has its leading term divisible by the leading term of some $g_i$. MAPLE can be used to calculate Gröbner bases.

**Example.** In the polynomial ring $\mathbb{R}[x, y, z]$ (real coefficients allowed although this is not germane to the problem), let $I$ be the ideal generated by $x^2 - y$ and $x^3 - z$. We will use lexicographic monomial order with $x \succ y \succ z$. This is communicated to MAPLE by the order used in listing the indeterminates and by the keyword `plex`:

```
MAPLE
> with(grobner):
> gbasis(x^2-y,x^3-z,[x,y,z],plex);
```

The result is

$$\{-y + x^2, -z + xy, -y^2 + xz, -z^2 + y^3\}$$

(each written in low to high order). Thus every polynomial in $I$ must have its leading term divisible by $x^2$ or $xy$ or $xz$ or $y^3$. Note that the last three of these are in the ideal although they were not among the original generators. For example, we have the following polynomial linear combination:

$$-z^2 + y^3 = (y^2 + yx^2 + x^4)(y - x^2) + (-z - x^3)(z - x^3).$$

*Varieties are the zero sets of polynomials.*

Let $f \in \mathbb{Q}[x_1, \ldots, x_d]$ be a multivariate polynomial with rational coefficients. The *variety* $\mathcal{V}(f)$ is the set of points $(z_1, \ldots, z_d)$ in $d$-dimensional complex space where $f$ is zero,

$$\mathcal{V}(f) = \{(z_1, \ldots, z_d) \in \mathbb{C}^d : f(z_1, \ldots, z_d) = 0\},$$

in other words, the roots of $f$. Complex numbers are used here because polynomials are sure to have roots if complex numbers are allowed, but not if restricted to rational numbers or even real numbers. If $S$ is a subset of $\mathbb{C}^d$, then define $\mathcal{V}_S(f) = \mathcal{V}(f) \cap S$, that is, the roots of $f$ in $S$.

In applications of varieties studied in this chapter, the functions are often probability calculations and the solutions are expected to satisfy the requirement that they be nonnegative numbers summing to 1; for example,

$$p_1 \geq 0, \ldots, p_m \geq 0, \quad p_1 + \cdots + p_m = 1. \tag{15.3.2}$$

The set of points in $m$-dimensional space satisfying (15.3.2) is called an *$m$-dimensional simplex* (or just a *simplex* if the dimension is understood). A simplex in 3-space is a triangular portion of the plane lying in the first octant and passing through the three points (1, 0, 0), (0, 1, 0), and (0, 0, 1); see Figure 15.3.1. The figure was made using the following MAPLE commands:

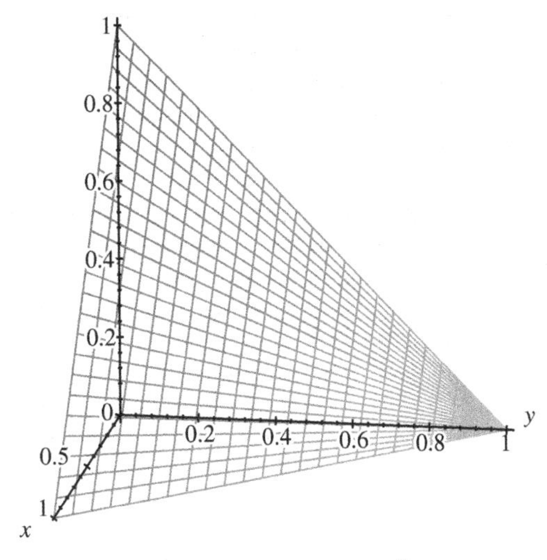


**Fig. 15.3.1.** Simplex in $\mathbb{R}^3$.

MAPLE
```
> with(plots):
> plot3d(1-x-y,0),x=0..1-y,y=0..1,axes=normal,scaling=constrained,orientation=[10,75]);
```

We will denote the $m$-dimensional simplex by the notation $\Delta^m$, or just $\Delta$ if $m$ is understood. Note that the $m$-dimensional simplex is a *hypersurface* in $m$-space, that is, a surface of dimension 1 less than that of the space. Using the notation introduced above, $\mathcal{V}_\Delta(f)$ is the zeros of $f$ that are meaningful as probabilities.

The variety of a polynomial is generally more than a finite set of points. For example, the variety of $f(z_1, z_2, z_3) = 4z_1z_2 - z_3^2$ is the *hypersurface* of $\mathbb{C}^3$ for which $z_3 = \sqrt{4z_1z_2}$. If $S = \mathbb{R}^3$, the subset of real numbers, then $\mathcal{V}_S(f)$ is the set of points in ordinary 3-space such that $x_3 = \sqrt{4x_1x_2}$ and is reminiscent of a saddle-surface defined only for $x_1 \geq 0$ and $x_2 \geq 0$ or $x_1 \leq 0$ and $x_2 \leq 0$.

Now let $\mathcal{F} \subset \mathbb{Q}[\mathbf{x}]$ be an arbitrary set of polynomials; $\mathbf{x}$ could be a vector of indeterminates, $\mathbf{x} = (x_1, \ldots, x_d)$. By the *variety* $\mathcal{V}(\mathcal{F})$ we mean the intersection of all hypersurfaces $\mathcal{V}(F)$ for all $F \in \mathcal{F}$. Put differently, $\mathcal{V}(\mathcal{F})$ is the set of all points $(z_1, \ldots, z_d) \in \mathbb{C}^d$ that are roots of all $f \in \mathcal{F}$,

$$f(z_1, \ldots, z_d) = 0 \quad \text{for all } f \in \mathcal{F}.$$

Let $I$ be an ideal in $\mathbb{Q}[\mathbf{x}]$ and suppose that $I = \langle \mathcal{F} \rangle$. Then $\mathcal{V}(I) = \mathcal{V}(\mathcal{F})$ because every polynomial $g \in I$ can be written as a polynomial linear combination

$$g = h_1 f_1 + \cdots + h_r f_r, \quad f_1, \ldots, f_r \in \mathcal{F}.$$

So any point $\mathbf{z}$, a zero of $f_1, \ldots, f_r$, is also a zero of $g$.

*The image of a polynomial map is also a variety.*

The setting for our next result is that of a vector-valued function $\mathbf{f}(\mathbf{z})$ having $m$ component functions, $f_1, \ldots, f_m$, each of which is defined for a $d$-dimensional variable $\mathbf{z} = (z_1, \ldots, z_d)$. This sort of setting occurs so often that a special notation is universally used for it; we write

$$\mathbf{f} : \mathbb{C}^d \longrightarrow \mathbb{C}^m.$$

The space $\mathbb{C}^d$ is called the *domain space* of $\mathbf{f}$, and the space $\mathbb{C}^m$ the *range space* or just the *range*.

**Theorem 2 (implicitization).** *Let* $\mathbf{f} : \mathbb{C}^d \longrightarrow \mathbb{C}^m$ *be a function whose components are multivariate polynomials in* $\mathbf{z}$. *Then the topological closure of the image of* $\mathbf{f}$ *is a variety in* $\mathbb{C}^m$.

The following example involves only polynomials of degree 1, *affine functions*, to simplify the calculations, but nonetheless captures the substance of the theorem.

**Example.** Define $\mathbf{f} : \mathbb{C}^2 \longrightarrow \mathbb{C}^3$ by

$$p_1 = 2\theta_1 - 3\theta_2 + 1,$$
$$p_2 = -\theta_1 + \theta_2 + 5,$$

$$p_3 = \theta_1 + 2\theta_2.$$

The image is the set of all points

$$\begin{bmatrix} p_1 \\ p_2 \\ p_3 \end{bmatrix} = \theta_1 \begin{bmatrix} 2 \\ -1 \\ 1 \end{bmatrix} + \theta_2 \begin{bmatrix} -3 \\ 1 \\ 2 \end{bmatrix} + \begin{bmatrix} 1 \\ 5 \\ 0 \end{bmatrix}$$

as $\theta_1$ and $\theta_2$ vary. This is a plane defined by two lines. The first has the direction of the vector $[2 \quad -1 \quad 1]^T$, and the second has the direction $[-3 \quad 1 \quad 2]^T$; both lines pass through the point $(1, 5, 0)$. To obtain this image as a variety, we eliminate the parameters $\theta_1$ and $\theta_2$. Add the second and third equations to get

$$3\theta_2 = p_2 + p_3 - 5.$$

Substitute this into the first two equations (multiplying the second by 3 avoids fractions):

$$p_1 = 2\theta_1 - p_2 - p_3 + 6,$$
$$3p_2 = -3\theta_1 + p_2 + p_3 + 10.$$

Now eliminate $\theta_1$, and we have

$$3p_1 + 7p_2 + p_3 - 38 = 0.$$

Hence the image of $\mathbf{f}$ is the variety of $3p_1 + 7p_2 + p_3 - 38$. As already noted, this shows that the image is a hypersurface.

Note that the theorem does not say that the image itself is necessarily a variety, but that its topological closure is. In another example, consider the mapping $\mathbf{f} : \mathbb{C}^2 \to \mathbb{C}^2$ defined by $p_1 = z_1$ and $p_2 = z_1 z_2$. Name any point in the range $(\hat{p}_1, \hat{p}_2)$; the point $z_1 = \hat{p}_1$, $z_2 = \frac{\hat{p}_2}{\hat{p}_1}$ maps to it, that is, except when $\hat{p}_1 = 0$. So the image of $\mathbf{f}$ is the entire plane except for the $p_2$-axis itself, and even there the point $(0, 0)$ is in the image.

The *topological closure* of a set is the set itself together with its *boundary points*. In the example above, points on the $p_2$-axis are boundary points of the image.[2] Therefore, the closure of the image in this example is the entire $(p_1, p_2)$-plane. This is the variety of the zero polynomial (in $(p_1, p_2)$-space).

## 15.4 Algebraic Analysis of Maximum Likelihood

The philosophy of algebraic statistics is that statistical models are algebraic varieties. In this section, we show how the maximum likelihood problem can be cast in these

[2] This is because for any point on the $p_2$-axis, a sequence of points in the image leading to it can be found.

terms. The development serves as a prototype for other statistical problems that occur in biology.

The methodology goes as follows. Each derivative of the log-likelihood equation, e.g., (15.2.15), can be put into the form of a polynomial in the unknown parameters, $\theta$. One is interested in knowing where these polynomials are simultaneously zero. As we have just seen in Section 15.3, the set of such zeros is called a variety. To analyze a variety, one tries to find the simplest system of polynomials that generate it. This is the purpose of a Gröbner basis. The resulting system of polynomials is easier to solve.

Note that the computations of this section are algebraic and often symbolic. So MATLAB will be able to do the calculations only if the symbolic package has been purchased. Although this package is from the people who produce MAPLE, the MAPLE code of this section has not been tested within MATLAB.

*Interior nodes lead to a hidden Markov model.*

The two-claw tree problem from before is too simple to illustrate the ideas and techniques of the algebraic method. Instead, we will analyze the rooted three-leaf tree of Figure 15.4.1.

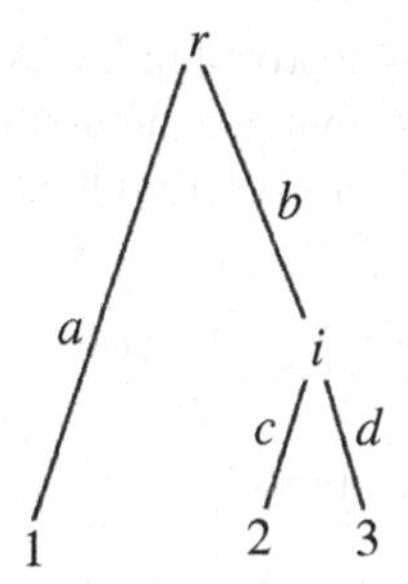


**Fig. 15.4.1.** Rooted three leaf tree with internal node.

It is important to note that this problem is fundamentally different from the two-claw problem because of the interior node $i$. We cannot know its state, only those of the leaf nodes. For this reason, interior nodes are said to be *hidden*, and the model is referred to as a *hidden Markov model*. Outcomes at the leaf nodes depend on the state of the hidden nodes, but the data will not be able to pin down those states directly; their values will have to be inferred. In calculating probabilities at the leaf nodes, allowance will have to be made for all possible states of the hidden nodes.

As before, we will assume Jukes–Cantor rates. Hence matrices $a$ and $b$ are defined in terms of their mutation rates $\alpha$ and $\beta$ by (15.2.9); likewise, $c$ and $d$ are given by similar equations in terms of their rates $\gamma$ and $\delta$, respectively. And, as before, we will calculate probabilities in terms of the parameters $a_0$, $a_1$, $b_0$, $b_1$, and now $c_0$, $c_1$ and $d_0$, $d_1$.

Another difference between this problem and the two-claw problem is that there are now three leaf nodes. Observed outcomes here are triples of the nucleotides A, C, G, and T. Consequently, the number of possible outcomes is $4^3 = 64$, making for a vector of length 64,

$$p = [p_{AAA} \;\; p_{AAC} \;\; \cdots \;\; p_{TTT}].$$

But by symmetries of the Jukes–Cantor model, many of the components are the same. We can see what they are by noting that the probabilities are invariant under any shuffling of the letters A, C, G, and T. That gives us the following equalities:

$$\begin{aligned} &p_{AAA} = p_{CCC} = p_{GGG} = p_{TTT}, && 4 \text{ terms},\\ &p_{AAC} = p_{AAG} = \cdots = p_{TTG}, && 12 \text{ terms},\\ &p_{ACA} = p_{AGA} = \cdots = p_{TGT}, && 12 \text{ terms},\\ &p_{CAA} = p_{GAA} = \cdots = p_{GTT}, && 12 \text{ terms},\\ &p_{ACG} = p_{ACT} = \cdots = p_{CGT}, && 24 \text{ terms}. \end{aligned} \tag{15.4.1}$$

Accounting for symmetries leaves only the five output probabilities shown in (15.4.1). They are $p_{123}$ for all three nucleotides the same, $p_{12}$ for only the first two the same; similarly define $p_{13}$ and $p_{23}$. Finally, $p_{dis}$ denotes the case in which all three are distinct.

We may now calculate the output probabilities. As noted above, $p_{123}$ can occur in four ways; pick one, say AAA, compute it, and multiply by 4. As before, we assume that the root node could be A or C or G or T with equal probability, $\frac{1}{4}$; hence

$$p_{123} = \frac{4}{4}(a_0b_0c_0d_0 + 3a_0b_1c_1d_1 + 3a_1(b_1c_0d_0 + b_0c_1d_1 + 2b_1c_1d_1)). \tag{15.4.2$_1$}$$

In the same way, we calculate the others:

$$\begin{aligned} p_{12} = \frac{12}{4}[&a_0(b_0c_0d_1 + b_1c_1d_0 + 2b_1c_1d_1) + a_1(b_0c_1d_0 + b_1c_0d_1 + 2b_1c_1d_1)\\ &+ 2a_1(b_1c_0d_1 + b_1c_1d_0 + b_0c_1d_1 + b_1c_1d_1)], \end{aligned} \tag{15.4.2$_2$}$$

$$\begin{aligned} p_{13} = \frac{12}{4}[&a_0(b_0c_1d_0 + b_1c_0d_1 + 2b_1c_1d_1) + a_1(b_1c_1d_0 + b_0c_0d_1 + 2b_1c_1d_1)\\ &+ 2a_1(b_1c_1d_0 + b_1c_0d_1 + b_0c_1d_1 + b_1c_1d_1)], \end{aligned} \tag{15.4.2$_3$}$$

$$\begin{aligned} p_{23} = \frac{12}{4}[&a_1(b_0c_0d_0 + 3b_1c_1d_1) + a_0(b_1c_0d_0 + b_0c_1d_1 + 2b_1c_1d_1)\\ &+ 2a_1(b_1c_0d_0 + 2b_1c_1d_1 + b_0c_1d_1)], \end{aligned} \tag{15.4.2$_4$}$$

$$\begin{aligned} p_{dis} = \frac{24}{4}[&a_0(b_0c_1d_1 + b_1c_0d_1 + b_1c_1d_0 + b_1c_1d_1)\\ &+ a_1(2b_1c_1d_1 + b_0c_0d_1 + b_1c_1d_0) + a_1(2b_1c_1d_1 + b_1c_0d_1 + b_0c_1d_0)\\ &+ a_1(b_1c_1d_1 + b_1c_0d_1 + b_1c_1d_0 + b_0c_1d_1)]. \end{aligned} \tag{15.4.2$_5$}$$

Before continuing, we note that these equations can be significantly simplified by invoking the observation made at the end of the previous section, that because the states of the root are equally likely, we can determine only the product matrix $ab$ and not $a$ and $b$ individually. Therefore, we should be able to simplify this equation by using (15.2.16) and defining the matrix $e = ab$; then

$$e_0 = a_0b_0 + 3a_1b_1, \qquad e_1 = a_0b_1 + a_1b_0 + 2a_1b_1. \tag{15.4.3}$$

Of course, the product $ab$ is also stochastic, so its rows sum to 1:

$$e_0 + 3e_1 = 1. \tag{15.4.4}$$

The equivalent (unrooted) tree is shown in Figure 15.4.2.

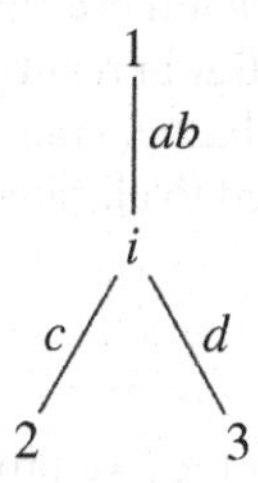


**Fig. 15.4.2.** Equivalent phylogenetic tree.

To incorporate (15.4.3) into (15.4.2), we let MAPLE do the work. Note that the new variables are in uppercase and we use the MAPLE command `algsubs` in place of `subs`.

## Code 15.4.1.

MAPLE

```
> #first define the old variables
> p123:=(a0*b0*c0*d0+3*a0*b1*c1*d1+3*a1*(b1*c0*d0+b0*c1*d1+2*b1*c1*d1));
> p12:=(12/4)*(a0*(b0*c0*d1+b1*c1*d0+2*b1*c1*d1)+a1*(b0*c1*d0+b1*c0*d1+2*b1*c1*d1)
        +2*a1*(b1*c0*d1+b1*c1*d0+b0*c1*d1+b1*c1*d1));
> p13:=(12/4)*(a0*(b0*c1*d0+b1*c0*d1+2*b1*c1*d1)+a1*(b1*c1*d0+b0*c0*d1+2*b1*c1*d1)
        +2*a1*(b1*c1*d0+b1*c0*d1+b0*c1*d1+b1*c1*d1));
> p23:=(12/4)*(a1*(b0*c0*d0+3*b1*c1*d1)+a0*(b1*c0*d0+b0*c1*d1+2*b1*c1*d1)
        +2*a1*(b1*c0*d0+2*b1*c1*d1+b0*c1*d1));
> pdis:=(24/4)*(a0*(b0*c1*d1+b1*c0*d1+b1*c1*d0+b1*c1*d1)+a1*(2*b1*c1*d1+b0*c0*d1+b1*c1*d0)
        +a1*(2*b1*c1*d1+b1*c0*d1+b0*c1*d0)+a1*(b1*c1*d1+b1*c0*d1+b1*c1*d0+b0*c1*d1));

> #now compute the new variables
> P123:=algsubs(a0*b0=e0-3*a1*b1,p123): P123:=expand(P123):
  P123:=algsubs(a0*b1=e1-a1*b0-2*a1*b1,P123);
> P12:=expand(p12): P12:=algsubs(a0*b0=e0-3*a1*b1,P12): P12:=expand(P12):
  P12:=algsubs(a0*b1=e1-a1*b0-2*a1*b1,P12);
> P13:=expand(p13): P13:=algsubs(a0*b0=e0-3*a1*b1,P13): P13:=expand(P13):
  P13:=algsubs(a0*b1=e1-a1*b0-2*a1*b1,P13);
> P23:=expand(p23): P23:=algsubs(a0*b0=e0-3*a1*b1,P23): P23:=expand(P23):
  P23:=algsubs(a0*b1=e1-a1*b0-2*a1*b1,P23);
> Pdis:=expand(pdis): Pdis:=algsubs(a0*b0=e0-3*a1*b1,Pdis): Pdis:=expand(Pdis):
  Pdis:=algsubs(a0*b1=e1-a1*b0-2*a1*b1,Pdis);
```

Alternatively, one may argue directly from Figure 15.4.2. Either way, the simplified probabilities are these:

$$\begin{aligned}
P_{123} &= e_0c_0d_0 + 3e_1c_1d_1, \\
P_{12} &= \frac{12}{4}(e_0c_0d_1 + e_1c_1d_0 + 2e_1c_1d_1), \\
P_{13} &= \frac{12}{4}(e_0c_1d_0 + e_1c_0d_1 + 2e_1c_1d_1), \\
P_{23} &= \frac{12}{4}(e_0c_1d_1 + e_1c_0d_0 + 2e_1c_1d_1), \\
P_{\text{dis}} &= \frac{24}{4}(e_0c_1d_1 + e_1c_0d_1 + e_1c_1d_0 + e_1c_1d_1).
\end{aligned}$$

On the surface it would appear that these five probabilities are functions of six variables, $e_0$, $e_1$, $c_0$, $c_1$, $d_0$, and $d_1$. But in reality, (15.4.4) holds between $e_0$ and $e_1$. Since matrices $c$ and $d$ are also stochastic, similar relationships hold for the $c$s and $d$s. These dependencies could be used to eliminate, say, $e_0$, $c_0$, and $d_0$ throughout,

$$e_0 = 1 - 3e_1, \qquad c_0 = 1 - 3c_1, \qquad d_0 = 1 - 3d_1. \tag{15.4.5}$$

With these substitutions, we see that the five probabilities are a function of a three-dimensional parameter vector $\theta$,

$$\theta_1 = e_1, \qquad \theta_2 = c_1, \qquad \theta_3 = d_1. \tag{15.4.6}$$

But it is preferable to retain the homogeneous coordinates, that is, both $e$s, both $c$s, and both $d$s, as long as possible and invoke (15.4.5) as the last step. When we refer to $\theta$, it will be as if the substitutions (15.4.5) and (15.4.6) had been carried out.

Our development so far can be summarized in terms of a vector-valued function, $\mathbf{f}$, mapping the three-dimensional parameter space of $\theta$ into a five-dimensional outcome space with the five probabilities as its component functions. In the notation of the last section, we have $\mathbf{f} : \mathbb{C}^3 \to \mathbb{C}^5$, with component functions $f_1 = p_{123}$, $f_2 = p_{12}$, and so on,

$$\mathbf{f}(\theta) = (f_1, f_2, f_3, f_4, f_5) = (p_{123},\ p_{12},\ p_{13},\ p_{23},\ p_{\text{dis}}).$$

Notice that each component function $f_i(\theta)$ is linear as a function of the homogeneous coordinates and that these components sum to 1.

Now suppose we have three aligned DNA sequences, each $n$ bases long, that correspond to the three leaves of the tree. Out of the $n$ places, suppose that $u_1$ places match in all three sequences, $u_2$ places match in the first two sequences only, $u_3$ match in the first and third only, $u_4$ match in the second and third only, and in $u_5$ places all three sequences are different. Thus the vector $\mathbf{u} = [u_1\ u_2\ u_3\ u_4\ u_5]$ constitutes the observed data. The likelihood function for this outcome is[3]

$$L(\theta) = \left(\frac{n!}{u_1!u_2!u_3!u_4!u_5!}\right) f_1^{u_1} f_2^{u_2} f_3^{u_3} f_4^{u_4} f_5^{u_5}.$$

To maximize this, take the logarithm of both sides, set its derivative with respect to each parameter to zero, and solve the resulting system,

$$\frac{\partial \log(L(\theta))}{\theta_1} = \frac{\partial \log(L(\theta))}{\theta_2} = \frac{\partial \log(L(\theta))}{\theta_3} = 0,$$

where, for $i = 1, 2, 3$,

$$\frac{\partial \log(L(\theta))}{\partial \theta_i} = \frac{u_1}{f_1}\frac{\partial f_1}{\partial \theta_i} + \frac{u_2}{f_2}\frac{\partial f_2}{\partial \theta_i} + \frac{u_3}{f_3}\frac{\partial f_3}{\partial \theta_i} + \frac{u_4}{f_4}\frac{\partial f_4}{\partial \theta_i} + \frac{u_5}{f_5}\frac{\partial f_5}{\partial \theta_i}. \tag{15.4.7}$$

*The likelihood variety.*

Recall that each function $f_i$ is multilinear as a function of the homogeneous coordinates. It follows that by combining the terms of (15.4.7) with a common denominator, the result is a ratio of polynomials in the homogeneous coordinates or in the $\theta_i$ as well. For example, invoking (15.4.5) and (15.4.6), we have

$$\begin{aligned} f_1 &= (1 - 3e_1)(1 - 3c_1)(1 - 3d_1) + 3e_1c_1d_1 \\ &= (1 - 3\theta_1)(1 - 3\theta_2)(1 - 3\theta_3) + 3\theta_1\theta_2\theta_3 \end{aligned}$$

and

$$\frac{\partial f_1}{\partial \theta_1} = -3(1 - 3\theta_2)(1 - 3\theta_3) + 3\theta_2\theta_3.$$

Similar results hold for the other components.

The *critical points* of the problem are the points $\theta$ in three-dimensional space where the functions (15.4.7) vanish, that is, equal zero, but the denominators of these equations are not zero. Then to be a solution to our problem, a critical point $\hat{\theta}$ must also be a vector of probabilities, that is, each component must lie between 0 and 1, the 3-simplex.

Discounting, temporarily, points where the denominators are zero, the set of critical points is a variety in 3-space called the *likelihood variety*. The maximum likelihood solution we want, the solution in terms of $\theta$, is the computation of this variety.

---

[3] The point of showing this equation is to note the relationship between the components $f_i$ and the data $u_i$. How does the ratio of factorials come about? The argument is the same as our derivation of the combinations factor in Section 2.8. For example, there are $u_1$ places in the three DNA sequences where the bases match, say, all are $A$s. One first imagines that these $A$s are different, say, $A_1, \ldots, A_{u_1}$. This contributes to the $n!$ in the numerator, but too much so because these $A$s are, in fact, not distinct. Since there are $u_1!$ ways to rearrange the $A$s, dividing by it corrects the overcount. Argue similarly for the other factors.

In the following MAPLE code, we enter the component probabilities in terms of `t1`, `t2`, `t3` representing $\theta_1$, $\theta_2$, and $\theta_3$. MAPLE calculates the derivatives and combines terms over a common denominator to form the three numerators `p1`, `p2`, `p3`. The denominator is not needed for the critical points of the likelihood derivatives, but is needed for the *Hessian* (see next) and to check that a root is admissible. We also compute the Hessian, or the second derivative matrix, in order to check the nature of a critical point. If the eigenvalues of the Hessian are all negative, then the critical point is a maximum point of the surface. If the eigenvalues are all positive, then the critical point is a minimum. If the eigenvalues are of mixed sign, then the critical point is a saddle-point and corresponds to a saddle-surface in parameter space there. (Like the flat spot on a saddle, this means that there are nearby points of the surface both higher and lower than at the flat spot.) The numerical root finder, `fsolve`, is used to solve the system but, owing to the multiplicity of roots, is shown to need additional help.

In the following, assume that the observed data are $u_1 = 31$, $u_2 = 5$, $u_3 = 7$, $u_4 = 11$, $u_5 = 13$.

**Code 15.4.2.**

```
MAPLE
> f1:=(1-3*t1)*(1-3*t2)*(1-3*t3)+3*t1*t2*t3;
> f2:=3*(1-3*t1)*(1-3*t2)*t3+3*t1*t2*(1-3*t3)+6*t1*t2*t3;
> f3:=3*(1-3*t1)*t2*(1-3*t3)+3*t1*(1-3*t2)*t3+6*t1*t2*t3;
> f4:=3*(1-3*t1)*t2*t3+3*t1*(1-3*t2)*(1-3*t3)+6*t1*t2*t3;
> f5:=6*(1-3*t1)*t2*t3+6*t1*(1-3*t2)*t3+6*t1*t2*(1-3*t3)+6*t1*t2*t3;
> cden:=f1*f2*f3*f4*f5;
  #fix t3 to reduce computation time, then p3 won't be needed
> t3:=1/10;
> u1:=31: u2:=5: u3:=7: u4:=11: u5:=13:
> p1:=u1*diff(f1,t1)*cden/f1 + u2*diff(f2,t1)*cden/f2 + u3*diff(f3,t1)*cden/f3 + u4*diff(f4,t1)*cden/f4
                              + u5*diff(f5,t1)*cden/f5;
> p2:=u1*diff(f1,t2)*cden/f1 + u2*diff(f2,t2)*cden/f2 + u3*diff(f3,t2)*cden/f3 + u4*diff(f4,t2)*cden/f4
                              + u5*diff(f5,t2)*cden/f5;
  #the next for the Hessian matrix
> h11:=cden*diff(p1,t1)-p1*diff(cden,t1);
> h12:=cden*diff(p1,t2)-p1*diff(cden,t2);
> h21:=cden*diff(p2,t1)-p2*diff(cden,t1);
> h22:=cden*diff(p2,t2)-p2*diff(cden,t2);
> simplify(h12-h21); #check that the mixed partials are equal
  #use fsolve to find a root
> S1:=fsolve({p1,p2},{t1,t2},{t1=0..1,t2=0..1});
  #see if this is a root of cden, check size of cden against p1 and p2
> assign(S1); p1; p2; cden;
  #if this is a root of cden, try another
> t1:='t1'; t2:='t2'; #reset t1 and t2
> S2:=fsolve({p1,p2},{t1,t2},{t1=0..1,t2=0..1},avoid={S1});
  #and avoid={S1,S2} if another round necessary, etc.,
  #also check if the eigenvalues of the Hessian are negative
> assign(S2);
> h:=array([[h11,h12],[h21,h22]]);
> evalf(Eigenvals(h));
```

As the above shows, MAPLE's numerical root finder might find a root that is also a root of the denominator. Then it is necessary to search for another. By contrast, an algebraic root finder proceeds in a very different way. The public domain computer algebra system *SINGULAR* is specialized to deal with these kinds of problems. (SINGULAR is obtainable free of charge from the website www.singular.uni-kl.de.)

By casting the roots as the variety of an ideal, a more suitable basis, a Gröbner basis, may be used in place of the original polynomials. Furthermore, there is a clever way to avoid roots of the common denominator.

Let $z_i$ play the role of $\frac{1}{f_i}$; then $z_i f_i - 1 = 0$. Adjoin the $z_i$ to our ring and work in the space $\mathbb{Q}[\theta_1, \ldots, \theta_d, z_1, \ldots, z_m]$. Let $J_u$ be the ideal generated by the maximum likelihood polynomials and these reciprocal relations for the $z_i$,

$$J_u = \left\langle z_1 f_1 - 1, \ldots, z_m f_m - 1, \sum_{j=1}^{m} u_j z_j \frac{\partial f_j}{\partial \theta_1}, \ldots, \sum_{j=1}^{m} u_j z_j \frac{\partial f_j}{\partial \theta_d} \right\rangle.$$

A point $(\theta, z) \in \mathbb{C}^{d+m}$ belongs to the variety $\mathcal{V}(J_u)$ of $J_u$ if $\theta$ is a root of the maximum likelihood equations and if $f_i(\theta) z_i = 1$, for all $i$; therefore, $f_i(\theta) \neq 0$. Since we are not interested in the $z$s, only the $\theta$s, put

$$I_u = J_u \cap \mathbb{Q}[\theta_1, \ldots, \theta_d]$$

to eliminate the $z$s; $I_u$ is the likelihood ideal. Here is the SINGULAR code for this problem.

**Code 15.4.3.**

```
SINGULAR
> ring bigring = 0, (t1,t2,z1,z2,z3,z4,z5), dp; number t3 = 1/10;
> poly f1 = (1-3*t1)*(1-3*t2)*(1-3*t3)+3*t1*t2*t3;
> poly f2 = 3*(1-3*t1)*(1-3*t2)*t3+3*t1*t2*(1-3*t3)+6*t1*t2*t3;
> poly f3 = 3*(1-3*t1)*t2*(1-3*t3)+3*t1*(1-3*t2)*t3+6*t1*t2*t3;
> poly f4 = 3*(1-3*t1)*t2*t3+3*t1*(1-3*t2)*(1-3*t3)+6*t1*t2*t3;
> poly f5 = 6*(1-3*t1)*t2*t3+6*t1*(1-3*t2)*t3+6*t1*t2*(1-3*t3)+6*t1*t2*t3;

> int u1=31; int u2=5; int u3=7; int u4=11; int u5=13;
> ideal Ju = z1*f1-1, z2*f2-1, z3*f3-1, z4*f4-1, z5*f5-1,
  u1*z1*diff(f1,t1)+u2*z2*diff(f2,t1)+u3*z3*diff(f3,t1) +u4*z4*diff(f4,t1)+u5*z5*diff(f5,t1),
  u1*z1*diff(f1,t2)+u2*z2*diff(f2,t2)+u3*z3*diff(f3,t2) +u4*z4*diff(f4,t2)+u5*z5*diff(f5,t2);
> ideal Iu = eliminate(Ju,z1*z2*z3*z4*z5);
> ring smallring = 0, (t1,t2), dp;
> ideal Iu = fetch(bigring,Iu); Iu;
  // dim(G)=dimension of G, vdim(G)= #roots if dim(G)=0
> ideal G = groebner(Iu); dim(G); vdim(G);
  // 20 digits of precision
> ideal G = groebner(Iu); LIB "solve.lib"; solve(G,20);
```

Of the 16 solutions, only three are in the range $0 < \theta_1, \theta_2 < 1$. And only one of those has a negative definite Hessian (making it a maximizing point), as shown in the MAPLE calculation above.

## 15.5 Characterizing Trees by Their Variety, Phylogenetic Invariants

We now view the problem of the last section in a completely different way. Instead of studying the problem of maximizing the log-likelihood function from the perspective of parameter space $\theta$, one can analyze it from the standpoint of the range space of probabilities. In our three-leaf problem, this is the five-dimensional space of the points

$(p_{123}, p_{12}, p_{13}, p_{23}, p_{\text{dis}})$. In this section, we no longer regard these as probabilities but merely as points $(p_1, p_2, p_3, p_4, p_5)$ in 5-space. In this analysis, we would like to characterize the *image* of $\mathbf{f}$,

$$\text{image}(\mathbf{f}) = \{\mathbf{p} = (p_1, p_2, \ldots, p_5) : \mathbf{p} = \mathbf{f}(\theta) \text{ for some } \theta\}.$$

The rationale is that different tree structures will give rise to images in $\mathbb{C}^5$ having different surface structures no matter what the values of the parameters may be. We seek to characterize these surface structures. Our best hope for this is to regard them as varieties and compute their generating polynomials. Polynomials that are zero on the image of $\mathbf{f}$ are called *phylogenetic invariants*.

*Phylogenetic invariants characterize the tree without having to solve it.*

As we saw in the algebra section, the closure of the image of $\mathbf{f}$ is an algebraic variety. Let $I_{\mathbf{f}}$ denote the ideal of polynomials in $p_1, \ldots, p_m$ that vanish on this variety. If $h$ is one such polynomial, then

$$h(p_1, \ldots, p_m) = 0, \quad \text{where } \mathbf{p} = \mathbf{f}(\theta), \quad \theta \in \mathbb{C}^d.$$

The problem of finding generating polynomials for $I_{\mathbf{f}}$ is the problem of implicitization. As we saw earlier, it amounts to eliminating the $\theta$s from the component functions $f_i$. We illustrate the method for the three-leaf tree of the previous section, (15.4.2).

One begins by transforming the equations to a simpler form as prescribed by *Fourier analysis*. The theory underlying the *Fourier transformation* is well known, but its study is beyond the scope of this text. The transformation for this problem and for all other small trees is given at the *small trees website*,

http://www.math.tamu.edu/~lgp/small-trees.

In this example, the transformed coordinates are simple products of the factors $(e_0 - e_1)$ and $(e_0 + 3e_1)$ and the same in $c$ and $d$.

In fact, it can be verified that

$$q_{111} = p_{123} - \frac{1}{3}p_{12} - \frac{1}{3}p_{13} - \frac{1}{3}p_{23} + \frac{1}{3}p_{\text{dis}} = (e_0 - e_1)(c_0 - c_1)(d_0 - d_1). \tag{15.5.1$_5$}$$

Denote this combination by $q_{111}$. Similarly, it can be verified that

$$q_{000} = p_{123} + p_{12} + p_{13} + p_{23} + p_{\text{dis}} = (e_0 + 3e_1)(c_0 + 3c_1)(d_0 + 3d_1), \tag{15.5.1$_1$}$$

$$q_{011} = p_{123} - \frac{1}{3}p_{12} - \frac{1}{3}p_{13} + p_{23} - \frac{1}{3}p_{\text{dis}} = (e_0 + 3e_1)(c_0 - c_1)(d_0 - d_1), \tag{15.5.1$_2$}$$

$$q_{101} = p_{123} - \frac{1}{3}p_{12} + p_{13} - \frac{1}{3}p_{23} - \frac{1}{3}p_{\text{dis}} = (e_0 - e_1)(c_0 + 3c_1)(d_0 - d_1), \tag{15.5.1$_3$}$$

$$q_{110} = p_{123} + p_{12} - \frac{1}{3}p_{13} - \frac{1}{3}p_{23} - \frac{1}{3}p_{\text{dis}} = (e_0 - e_1)(c_0 - c_1)(d_0 + 3d_1). \tag{15.5.1$_4$}$$

The coordinates $q_{000}, q_{011}, \ldots, q_{111}$ are called the *Fourier coordinates*, and the equations (15.5.1) constitute the *Fourier transform*. They have been indexed by the subgraphs of the tree; $q_{000}$ corresponds to the empty tree, $q_{111}$ to the full tree, $q_{011}$ to the span of leaves 2 and 3, and so on. An excluded edge corresponds to a factor such as $(e_0 + 3e_1)$, an included edge to a factor such as $(e_0 - e_1)$.

With this simplification we work to eliminate the parameters. First, eliminate $e_0 + 3e_1$; from (15.5.1$_1$) and (15.5.1$_2$),

$$\frac{q_{000}}{(c_0 + 3c_1)(d_0 + 3d_1)} = e_0 + 3e_1 = \frac{q_{011}}{(c_0 - c_1)(d_0 - d_1)}.$$

Next eliminate $c_0 + 3c_1$. Solve the first and third members of this for $c_0 + 3c_1$ and use (15.5.1$_3$) to get

$$\frac{q_{000}(c_0 - c_1)(d_0 - d_1)}{q_{011}(d_0 + 3d_1)} = c_0 + 3c_1 = \frac{q_{101}}{(e_0 - e_1)(d_0 - d_1)}.$$

Use the first and third members of this to solve for $d_0+3d_1$ and combine with (15.5.1$_4$),

$$\frac{q_{000}(c_0 - c_1)(d_0 - d_1)^2(e_0 - e_1)}{q_{011}q_{101}} = d_0 + 3d_1 = \frac{q_{110}}{(e_0 - e_1)(c_0 - c_1)}.$$

And finally incorporate (15.5.1$_5$) into this; we get

$$q_{000}q_{111}^2 = q_{011}q_{101}q_{110}.$$

Therefore, the ideal $I_{\mathfrak{t}}$ is generated by the homogeneous third-degree polynomial

$$I_{\mathfrak{t}} = \langle q_{000}q_{111}^2 - q_{011}q_{101}q_{110} \rangle.$$

*The molecular clock assumption yields a different ideal.*

Recall from Section 15.2 the molecular clock assumption: For any subtree and each path from the root of that subtree to any leaf, the products of the transition matrices corresponding to the edges of the path are identical. As applied to the three-taxa problem we have been following, Figure 15.4.1, this means that

$$c = d, \qquad a = bc = bd.$$

Or, in terms of the individual parameters,

$$d_0 = c_0, \qquad d_1 = c_1,$$
$$a_0 = b_0c_0 + 3b_1c_1, \qquad a_1 = b_0c_1 + b_1c_0 + 2b_1c_1.$$

In calculating the five probabilities, the assumption that $a$ and $b$ cannot be distinguished is no longer valid (since their edges are different distances from the root).

There are now only two independent matrices, $b$ and $c$, and four homogeneous parameters or two independent parameters. Also note that $p_{12}$ and $p_{13}$ will be the same, because $c = d$, and the one chosen must be counted twice. Hence the data space is only four-dimensional here. Starting from (15.4.2), we use MAPLE for the calculation. Denote the output probabilities by $p_1$, $p_2$, $p_3$, and $p_4$:

```
MAPLE
#recall that p123,p12,...were defined in Code 15.4.1
> p1:=subs({d0=c0,d1=c1,a0=b0*c0+3*b1*c1,a1=b0*c1+b1*c0+2*b1*c1},p123): p1:=simplify(p1);
> p2:=subs({d0=c0,d1=c1,a0=b0*c0+3*b1*c1,a1=b0*c1+b1*c0+2*b1*c1},p12):
> p2:=2*simplify(p2); #for emphasis
> p3:=subs({d0=c0,d1=c1,a0=b0*c0+3*b1*c1,a1=b0*c1+b1*c0+2*b1*c1},p23): p3:=simplify(p3);
> p4:=subs({d0=c0,d1=c1,a0=b0*c0+3*b1*c1,a1=b0*c1+b1*c0+2*b1*c1},pdis): p4:=simplify(p4);
```

This gives

$$\begin{aligned} p_1 &= b_0^2c_0^3 + 3b_0^2c_1^3 + 6b_0b_1c_0^2c_1 + 6b_0b_1c_0c_1^2 + 12b_0b_1c_1^3 + 3b_1^2c_0^3 \\ &\quad + 6b_1^2c_0^2c_1 + 6b_1^2c_0c_1^2 + 21b_1^2c_1^3, \\ p_2 &= 6b_0^2c_0^2c_1 + 12b_0c_0^2b_1c_1 + 84b_1c_1^2b_0c_0 + 102b_1^2c_0c_1^2 + 84b_1^2c_1^3 \\ &\quad + 6b_0^2c_1^2c_0 + 48b_0c_1^3b_1 + 30b_1^2c_1c_0^2 + 12b_0^2c_1^3, \\ p_3 &= 3b_0^2c_0^2c_1 + 42b_0c_1^3b_1 + 6b_1c_0^3b_0 + 21b_1^2c_0c_1^2 + 12b_0c_0^2b_1c_1 + 60b_1^2c_1^3 \\ &\quad + 3b_0^2c_1^2c_0 + 12b_1c_1^2b_0c_0 + 21b_1^2c_1c_0^2 + 6b_0^2c_1^3 + 6b_1^2c_0^3, \\ p_4 &= 24b_0c_0^2b_1c_1 + 18b_0^2c_1^2c_0 + 60b_1c_1^2b_0c_0 + 114b_1^2c_0c_1^2 + 60b_0c_1^3b_1 \\ &\quad + 78b_1^2c_1^3 + 24b_1^2c_1c_0^2 + 6b_0^2c_1^3. \end{aligned}$$

From the small trees website, we can look up the Fourier transformation for this problem; the Fourier coordinates are

$$\begin{aligned} q_{0000} &= p_1 + p_2 + p_3 + p_4 = (b_0 + 3b_1)^2(c_0 + 3c_1)^3, \\ q_{0011} &= p_1 - \frac{1}{3}p_2 + p_3 - \frac{1}{3}p_4 = (c_0 + 3c_1)(b_0 + 3b_1)^2(c_0 - c_1)^2, \\ q_{0111} &= p_1 + \frac{1}{3}p_2 - \frac{1}{3}p_3 - \frac{1}{3}p_4 = (c_0 + 3c_1)(b_0 - b_1)^2(c_0 - c_1)^2, \\ q_{1111} &= p_1 - \frac{1}{3}p_2 - \frac{1}{3}p_3 + \frac{1}{3}p_4 = (b_0 - b_1)^2(c_0 - c_1)^3. \end{aligned}$$

As before, the Fourier coordinates are indexed according to the portions of the subtree included and excluded. It is easy to check that $q_{0011}q_{0111}^2 = q_{0000}q_{1111}^2$; therefore, the ideal is generated by

$$I_f = \langle q_{0011}q_{0111}^2 - q_{0000}q_{1111}^2 \rangle.$$

## 15.6 Constructing the Phylogenetic Tree

Previously, we have seen how to compute branch lengths for an existing phylogenetic tree using genomic alignments. In this section, we take up the study of how to construct the tree itself. For this we need a matrix of pairwise distances, called a *dissimilarity map*, between the taxa of the tree. The map can be the result of a multiple alignment between genomes. For example, it could be calculated by the public domain software, MAVID, written for this purpose.

*The number of possible trees for n taxa is exponential in n.*

In Figure 15.6.1(a), we show a two-leaf unrooted tree. Also shown is an equivalent form more commonly seen in phylogenetic studies. The tree may be described by the notation (1, 2), indicating the leaves of the tree and their connectivity. This tree has $n = 2$ leaves, no interior nodes, and one edge. There is only one such tree.

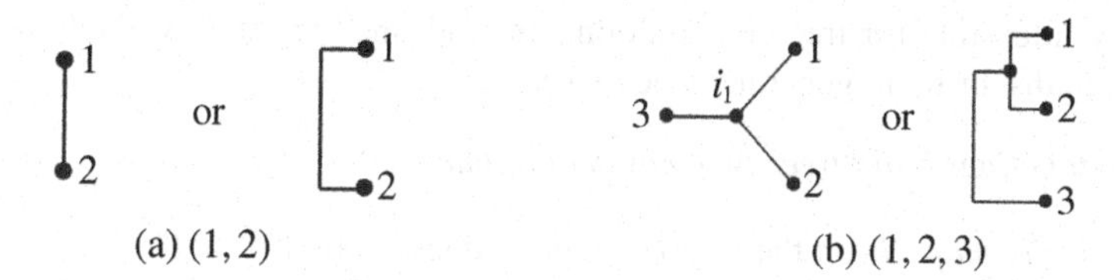


**Fig. 15.6.1.**

By adding a new leaf, attaching it to the single existing edge, we obtain the $n = 3$ leaved tree of Figure 15.6.1(b). The notation (1, 2, 3) indicates that the leaves share a common interior node, $i_1$. By adding the new leaf, we have also added an interior node. In general, one new interior node will be added for each new leaf, and so inductively, the number of interior nodes is given by $n - 2$, equal to 1 here. It is also seen that by dividing an edge for the leaf addition, we have created two new edges. In general, the number of new edges created by the addition of a new leaf is two, and so inductively, the number of edges of a tree is given by $2n - 3$, equal to 3 here. The number of such three-leaved trees is one.

Once again we add a new leaf. This time there are three edges at which to make the attachment, giving rise to three different tree structures. The possibilities are shown in Figure 15.6.2. In the first case, the leaf is added to the edge between the interior node and leaf 3, in the second case to the edge between the interior node and leaf 1, and in the third case to the edge between the interior node and leaf 2. Each has $n - 2 = 2$ interior nodes and $2n - 3 = 5$ edges. As noted, for this $n = 4$ case there are three different trees.

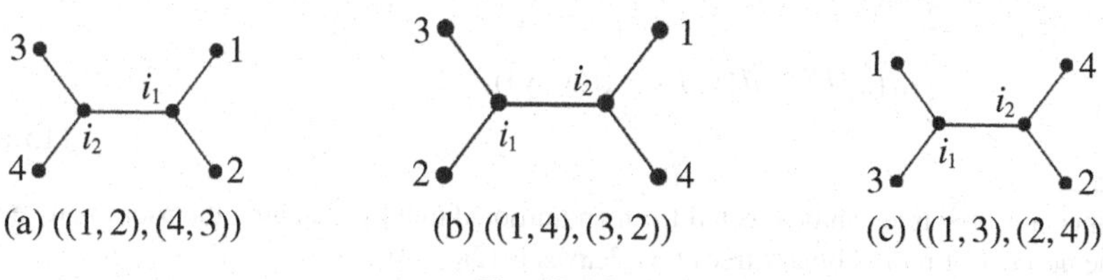


**Fig. 15.6.2.**

In general, moving from the $n$ to the $n+1$ case, the number of different possibilities is equal to the number of existing edges. Hence the number of tree structures increases according to

$$1 \cdot 3 \cdot 5 \cdot 7 \cdots (2n-5) = (2n-5)!!.$$

The right-hand side of this equation defines a notation for the left-hand side. These relationships between the number of leaves, interior nodes, edges, and tree structures hold for unrooted phylogenetic trees having three edges adjacent at each interior node.[4]

On the basis of this analysis, we see that the number of trees that have to be searched in calculating a phylogenetic tree grows very rapidly with $n$; for $n = 6$, it is 105 trees, but for $n = 10$ the number is 2,027,025. As a result, maximum likelihood is usually infeasible for tree construction. Instead, methods are available utilizing branch lengths, or more generally tree metrics.

*Distance functions build trees two leaves at a time.*

A *dissimilarity map*, $d$, on the first $n$ integers, denoted by $[n] = \{1, 2, \ldots, n\}$, is a symmetric nonnegative-valued function satisfying

$$d(i, j) = d(j, i) \geq 0, \qquad d(i, i) = 0.$$

The matrix of a dissimilarity map is the $n \times n$ matrix $D$ whose $(i, j)$th element is $d_{ij}$.

A dissimilarity map is a *metric* on $[n]$ if it satisfies the *triangle inequality*,

$$d(i, j) \leq d(i, k) + d(k, j), \quad i, j, k \in [n].$$

A dissimilarity map $d$ on $[n]$ is a *tree metric* if there is a tree $T$ with $n$ leaves and a nonnegative length for each edge such that for every pair of leaves $i$ and $j$, the length of the unique path from $i$ to $j$ equals $d(i, j)$. An example is given in Figure 15.6.3 with corresponding distances presented in (15.6.4). For a tree metric, fix two leaves $i$ and $j$ and let $k$ be some leaf. It is easy to see that $d(i, j) \leq d(i, k) + d(k, j)$ by considering the subtree spanned by $i$ and $k$. So a tree metric is a metric.

A *cherry* of a tree is a pair of leaves both adjacent to the same node, their common ancestor. Let $(x, y)$ be a cherry of a tree, let $v$ be their common ancestor, and let $k$ be any other leaf. Then[5]

$$\begin{aligned} d(v,k) &= \frac{1}{2}(d(v,k) + d(v,k)) \\ &= \frac{1}{2}(d(k,v) + d(v,x) + d(k,v) + d(v,y) - (d(x,v) + d(v,y))) \\ &= \frac{1}{2}(d(x,k) + d(y,k) - d(x,y)). \end{aligned} \tag{15.6.1}$$

[4] Since a rooted tree can be created from an unrooted tree by attaching the root to any edge, the number of rooted binary trees on $n$ leaves is $(2n-3)!!$.

[5] Use equality here because $v$ is on the unique path from $k$ to either $x$ or $y$.

**Theorem 1 (Saitou and Nei).** *Let $d$ be a tree metric on $[n]$. For every pair $i, j \in [n]$, put*

$$Q_d(i, j) = (n-2)d(i, j) - \sum_{k \neq i} d(i, k) - \sum_{k \neq j} d(j, k). \tag{15.6.2}$$

*The pair $x, y \in [n]$ that minimizes $Q_d(i, j)$ is a cherry of the tree. Note that $Q_d(i, j)$ will be negative if $d$ is a metric, since $d(i, j) \leq d(i, k) + d(k, j)$.*

Given a dissimilarity map on $n$ taxa, one can invoke the theorem to identify the two taxa, $x$ and $y$, most related, i.e., most likely to be a cherry of a phylogenetic tree on these taxa. Let $v$ denote their common node. To continue the construction, we use (15.6.1) to define the distance from $v$ to the other leaves of the tree. The construction is now continued as if $v$ were a leaf of the reduced set. We are led to the *neighbor-joining algorithm.*

**Neighbor-joining algorithm.**

*Step* 1. Compute the $\binom{n}{2}$ values $Q_d(i, j)$ of (15.6.2); let $x$ and $y$ give the minimum. Add $x$, $y$, and their common node $v$ to the tree.

*Step* 2. Remove $x$ and $y$ from the list $[n]$, but add $v$ to the list. Extend the dissimilarity map to $v$ by defining

$$d(v, k) = \frac{1}{2}(d(x, k) + d(y, k) - d(x, y)) \tag{15.6.1}$$

for all remaining leaves $k$ in the list.

*Step* 3. If the reduced list is of length 3 or more, return to Step 1; otherwise, join the last two with a common edge, completing the tree.

A tree metric, $d_T$, can be created for the tree recursively. Add Step 1.5 between Steps 1 and 2 as follows:

*Step* 1.5 Pick an arbitrary element of the list, $r$, different from $x$ and $y$, and define

$$\begin{aligned} d_T(x, v) &= \frac{1}{2}(d(x, y) + d(x, r) - d(y, r)), \\ d_T(y, v) &= d((x, y) - d_T(x, v). \end{aligned} \tag{15.6.3}$$

If $d$ is already a tree metric for some tree $T$, then the algorithm will find it, and the metric $d_T$ constructed in Step 1.5 will be again $d$. Otherwise, we hope that $d_T$ will be close to $d$.

**Example.** To illustrate the algorithm, let a dissimilarity map be given by the matrix

$$D = \begin{array}{c} \\ 1 \\ 2 \\ 3 \\ 4 \end{array} \begin{array}{c} \begin{array}{cccc} 1 & 2 & 3 & 4 \end{array} \\ \begin{pmatrix} - & 10 & 11 & 14 \\ 10 & - & 3 & 12 \\ 11 & 3 & - & 13 \\ 14 & 12 & 13 & - \end{pmatrix} \end{array}. \tag{15.6.4}$$

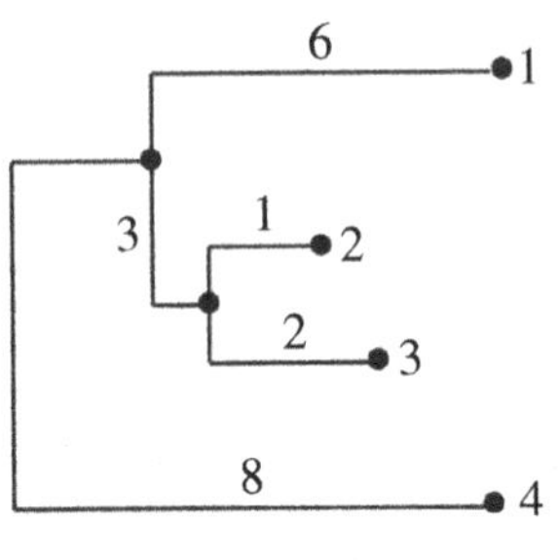


**Fig. 15.6.3.**

Since this derives from a tree metric (see Figure 15.6.3), the algorithm should regenerate the tree.

First, calculate the $Q_d$ matrix using (15.6.2). For example,

$$Q_d(1,2) = (4-2)*10 - (10+11+14) + (10+3+12) = -40.$$

In full, the matrix is

$$Q_d = \begin{bmatrix} - & -40 & -40 & -46 \\ -40 & - & -46 & -40 \\ -40 & -46 & - & -40 \\ -46 & -40 & -40 & - \end{bmatrix}.$$

The minimal value is $-46$ in two places, 1, 4 and 2, 3. This means that we may choose either; the resulting graphs will be equivalent. Selecting 1 and 4, join them by an internal node, $i_2$ say, and remove them from the list, Step 2. In their place, add $i_2$ to the list with distances constructed using (15.6.1),

$$d(i_2,2) = \frac{1}{2}(d(2,1) + d(2,4) - d(1,4)) = 4,$$
$$d(i_2,3) = \frac{1}{2}(d(3,1) + d(3,4) - d(1,4)) = 5.$$

To figure the tree distances $d_T(1,i_2)$ and $d_T(4,i_2)$, select as "root" $r = 2$. Then from (15.6.3),

$$d_T(1,i_2) = \frac{1}{2}(d(1,4) + d(1,2) - d(2,4)) = 6,$$
$$d_T(4,i_2) = d(1,4) - d_T(1,i_2) = 8.$$

At this point the construction is as shown in Figure 15.6.4(a), and the new dissimilarity distances figured above are noted in $D'$,

$$D' = \begin{matrix} & \begin{matrix} 2 & 3 & i_2 \end{matrix} \\ \begin{matrix} 2 \\ 3 \\ i_2 \end{matrix} & \begin{pmatrix} - & 3 & 4 \\ 3 & - & 5 \\ 4 & 5 & - \end{pmatrix} \end{matrix}.$$

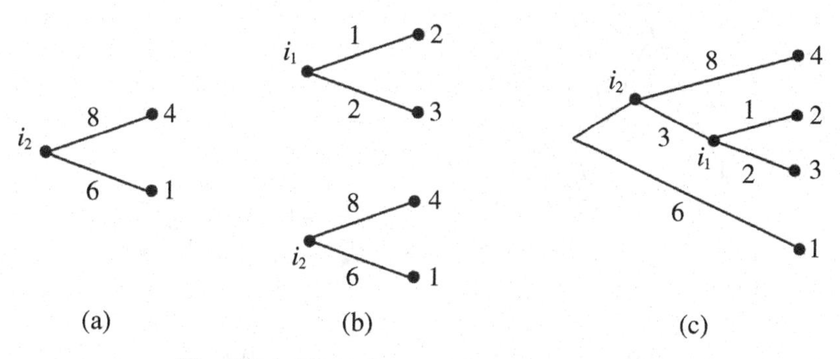


**Fig. 15.6.4.** (a)–(c) Stages of the tree construction.

The next round continues with $\{2, 3, i_2\}$. Invoking (15.6.2), now with $n = 3$, gives $-12$ for all terms; e.g.,

$$Q_d(2, 3) = 1 \cdot 3 - (3 + 4) - (3 + 5) = -12.$$

In fact, at this final stage, each term of $Q_d$ equals the negative of the sum of the distances, $-(3 + 4 + 5) = -12$. Therefore, any pair may be selected, and each choice gives an equivalent graph. So we choose 2 and 3 and join them through a common vertex $i_1$. The tree distances are calculated via (15.6.3) using, say, $r = 1$ as root,

$$d_T(2, i_1) = \frac{1}{2}(d(2, 3) + d(1, 2) - d(3, 1)) = 1,$$
$$d_T(3, i_1) = d(2, 3) - d_T(2, i_1) = 2.$$

This gives the tree of Figure 15.6.4(b).

What remains is $i_1$ and $i_2$; their distance is figured using (15.6.1),

$$d(i_1, i_2) = \frac{1}{2}(d(2, i_2) + d(3, i_2) - d(2, 3)) = 3.$$

These final two elements are simply joined, finishing the tree, Figure 15.6.4(c).

## Exercises/Experiments

1. What is the branch length between the HSPs (high scoring pairs) of the *Latimeria chalumnae* Hoxa-11 gene (AF287139) and the *Polyodon spathula* Hoxa-11 gene (AY661748.1) as calculated by (15.2.8)? Use blastn to get the "identities" (match percent) between these DNA segments.

2. What is the substitution matrix $a = e^{Q_{K80}}$ for the Kimora-80 rate matrix? Notice that there are three distinct terms in $a$. Labeling these $a_0$, $a_1$, and $a_2$, work out the probabilities $p_{\text{same}}$ and $p_{\text{dif}}$ for the two-claw problem.

3. Under the Jukes–Cantor model, work out the probabilities $p_{\text{same}}$ and $p_{\text{dif}}$ for the two-claw problem with an interior node on one edge; see Figure 15.6.5(a).

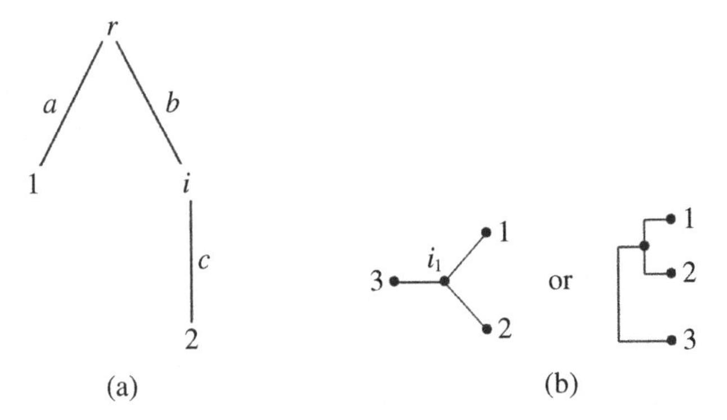


**Fig. 15.6.5.**

**4.** In the text, the maximum likelihood solution for the three-leaved tree with internal node (Figure 15.4.1 or, equivalently, Figure 15.6.5(b)) was worked out. The solution for the observed data, $u = [31\ \ 5\ \ 7\ \ 11\ \ 13]$, can be computed from the MAPLE program on p. 526 or the SINGULAR program on p. 527. Find the solution to obtain $\theta_1 = e_1$, $\theta_2 = c_1$, and $\theta_3 = d_1$. From these find the branch length from the internal node $i$ to 2; recall from (15.2.6) that

$$\text{branchlen}_2 = 3\gamma t, \qquad \text{and} \quad c_1 = \frac{1}{4}(1 - e^{-4\gamma t}),$$

the branch length from $i$ to 3; similarly,

$$\text{branchlen}_3 = 3\delta t, \qquad \text{and} \quad d_1 = \frac{1}{4}(1 - e^{-4\delta t}),$$

and the branch length from 1 to $i$,

$$\text{branchlen}_1 = 3(\alpha + \beta)t, \quad \text{and} \quad e_0 = \frac{1}{4}(1 + 3e^{-4(\alpha+\beta)t}).$$

**5.** Assume that the following matrix defines a dissimilarity function among the proteins: 1 = At1g20880, 2 = Hs20556011, 3 = CE13934, 4 = Hs14192947, and 5 = At5g53680. Construct the phylogenetic tree of these proteins based on this dissimilarity function. Compare with KOG0149 of the Clusters of Orthogolous Groups at NCBI,

$$\begin{bmatrix} - & .96 & .46 & .54 & .38 \\ .96 & - & .64 & .55 & .43 \\ .46 & .64 & - & .33 & .33 \\ .54 & .55 & .33 & - & .39 \\ .38 & .43 & .33 & .39 & \end{bmatrix}.$$

### Questions for Thought and Discussion

1. Discuss the problem of speciation: How do new species arise? How does a new organism find a mate (speciation is defined in terms of mating)? How are new species confirmed (how can you say that a species is new)? What problems are there in verification?

2. What evidence do fossils and the fossil record provide in helping to fix phylogenetic trees?

## References and Suggested Further Reading

[1] H. B. Stenzel, Successional speciation in paleontology: The case of the oysters of the sellaeformis stock, *Evolution*, **3** (1949), 33–50.
[2] J. Felsenstein, *Inferring Phylogenies*, Sinauer, Sunderland, MA, 2004.
[3] N. Eldredge and J. Cracraft, *Phylogenetic Patterns and the Evolutionary Process*, Columbia University Press, New York, 1980.
[4] R. D. M. Page and E. C. Holmes, *Molecular Evolution: A Phylogenetic Approach*, Blackwell Science, Oxford, UK, 1998.
[5] L. Pachter and B. Sturmfels, *Algebraic Statistics for Computational Biology*, Cambridge University Press, Cambridge, UK, 2005.
[6] C. Zimmer, What is a species?, *Sci. Amer.*, **298**-6 (2008), 72–79.

# Code Index

# Index